The United Kingdom's Natural Wonders

This book guides readers through the most iconic and geologically significant scenery in the United Kingdom, points out features of interest, explains what they are seeing, and describes how these features came to be. It illustrates numerous regions, explaining classic locations in the development of geology and paleontology in the United Kingdom, and giving readers a tour through sites of special scientific interest. The author puts the geology of Britain into a plate tectonic context and discusses the history of sedimentary basins and mountain building. Volcanism and glacial features of the region are also explained, and the effects of the Ice Age are discussed.

Features:

- Clearly explains the geology of regions with emphasis on landscape formation.
- Addresses the outsized role of Scottish and English geologists in developing the science of geology.
- Lavishly illustrated with numerous colorful maps and breathtaking geological landscapes and their various features.
- Describes the major geologic features of the United Kingdom through the device of a geologic tour for those without any geology training as well as professionals.
- Written in easy-to-understand language, the author brings his experience to readers who want to explore and understand geologic sites firsthand.

The United Kingdom's Natural Wonders is an inviting text giving individuals with no background in geology the opportunity to understand key geologic aspects of local landscapes. It also serves as a guide to undergraduate- and graduate-level students taking courses in earth science programs, such as geology, geophysics, geochemistry, mining engineering, and petroleum engineering. Teachers of these courses can also use this book to better understand their local geologic environment and geography.

Geologic Tours of the World

Series Editor: Gary Prost

PUBLISHED TITLES

North America's Natural Wonders: Appalachians, Colorado Rockies, Austin-Big Bend Country, Sierra Madre
Gary Prost

North America's Natural Wonders: Canadian Rockies, California, The Southwest, Great Basin, Tetons-Yellowstone Country
Gary Prost

The United Kingdom's Natural Wonders: Scotland and Northern Ireland, Lake District and Yorkshire Dales, Wales and West Midlands, England
Gary Prost

For more information about this series, visit https://www.routledge.com/Geologic-Tours-of-the-World/book-series/GTW

The United Kingdom's Natural Wonders

Scotland and Northern Ireland, Lake District and Yorkshire Dales, Wales and West Midlands, England

Gary L. Prost

CRC Press
Taylor & Francis Group
Boca Raton London New York

CRC Press is an imprint of the
Taylor & Francis Group, an **informa** business

Designed cover image: © Gary Prost

First edition published 2023
by CRC Press
6000 Broken Sound Parkway NW, Suite 300, Boca Raton, FL 33487-2742

and by CRC Press
4 Park Square, Milton Park, Abingdon, Oxon, OX14 4RN

CRC Press is an imprint of Taylor & Francis Group, LLC

© 2023 Taylor & Francis Group, LLC

Library of Congress Cataloging-in-Publication Data
Names: Prost, Gary L., 1951- author.
Title: The United Kingdom's Natural Wonders : Scotland and Northern
Ireland, Lake District and Yorkshire Dales, Wales and West Midlands, England / Gary L. Prost.
Other titles: Scotland and Northern Ireland, Lake District and Yorkshire
Dales, Wales and West Midlands, England
Description: First edition. | Boca Raton : CRC Press, 2023. |
Series: Geologic tours of the world; vol 3 | Includes bibliographical references and index.
Identifiers: LCCN 2022056554 (print) | LCCN 2022056555 (ebook) |
ISBN 9781032495064 (hardcover) | ISBN 9780815349013 (paperback) |
ISBN 9781351165600 (ebook)
Subjects: LCSH: Geology–Great Britain. | Great Britain—Description and travel. |
Natural history–Great Britain.
Classification: LCC QE261 .P96 2023 (print) | LCC QE261 (ebook) |
DDC 554.1–dc23/eng/20230221
LC record available at https://lccn.loc.gov/2022056554
LC ebook record available at https://lccn.loc.gov/2022056555

ISBN: 9781032495064 (hbk)
ISBN: 9780815349013 (pbk)
ISBN: 9781351165600 (ebk)

DOI: 10.1201/9781351165600

Typeset in Times
by codeMantra

Dedication

To my wife Nancy, who lets me entertain my passion for geology; my daughter Elizabeth, who consents to accompany me on my geologic jaunts; and my good friend Guy Peasley, always excellent company on these travels.

Contents

Preface

Who would have thought that such a small island would have such an outsized impact on the science of geology? And not just the people and concepts that came from here. This land, with a mild climate and abundant rainfall, is not known for great outcrops, rugged peaks, and precipitous gorges. Most of it is covered by rich soil, is deeply weathered, and has verdant forests, fields, and moors. However, the coasts, battered by the sea, have spectacular cliffs, and glaciers have scraped bare and uncovered many of the harder rocks. So there is a whole lot to see.

From Durness to The Lizard, from Land's End to Dover, this geological guidebook presents field trips across some of the best-known natural wonders in the United Kingdom. The purpose is to explain to the layman, the interested general public, the rock hound, and the curious what they are seeing when they walk along the Jurassic Coast, see the Giant's Causeway, wonder where the stones came from at Stonehenge, ramble over Salisbury Crags, explore a Roman mine at Great Orme, or holiday in the Lake District. In terms anyone can understand, I explain what you are looking at, how it came to be, and why it is important. And yet there should also be enough information to keep a geologist interested. Along the way historical context is provided, a word or two is said about plants and animals, and you may learn where to find interesting minerals and fossils.

These trips can be run in either direction, but most start where there is easy access (a major town or airport). The transects as laid out can take up to 5 or 6 days, but these trips can easily be broken into smaller bits that can take as little time as a few hours. Times and distances are provided, but times are meant as a guide only and will change depending on traffic and weather conditions. Although described stops are provided, there are many more stops that can be made along these routes: don't feel constrained to use only those in the guidebook. Be aware that some routes may be closed in winter. They should all be accessible with a standard road car; four-wheel or all-wheel drive is not necessary for these trips.

I recommend taking a full day to see a single stop, such as the Triassic red cliffs at Coryton Cove or fossil hunting at Lyme Regis. Or do two or three stops in a day, as in a visit to Land's End and The Lizard, or Birling Gap to Dover. You don't have to do an entire tour in one go.

Acknowledgments

The assistance and good company of Guy Peasley (Chapters 1 and 2) and Elizabeth Prost (Chapters 3 and 4) was instrumental in allowing me to complete the field work on this tour and is gratefully acknowledged. Chapter-specific acknowledgments follow.

Chapter 1: Suggestions for sites to include in this tour came from Samuel Eguiluz (Hutton's Section at Salisbury Crags), Catherine Skilliter (Durness and Sango Bay), and Rida Aslam (Kilt Rock, Mealt Falls, Storr, Brother's Point, Torridon, Ardvreck Castle, and Glen Coe). Steven Finney helped proofread parts of this manuscript. Thanks to "Dominik" for providing access to the Fossil Grove in Glasgow, Michael Dempster for photos of the Portrush ammonites, and Diane Housley for photos of the Leiths Skye Marble Quarry.

Chapter 2: Thanks are directed to Ross Collin of Saint Gobain/British Gypsum for providing photos of the Birkshead Mine. Thanks to Samuel Eguiluz for suggesting the Combs Quarry unconformity stop.

Chapter 3: Suggestions for sites to include in this tour came from Samuel Eguiluz (Peak District National Park, Mam Tor, and Whitesands Bay), Sarah Roberts of the Iron Bridge Gorge Museum, and Nick Bunn (Aust Cliff, Ogmore-by-Sea, and Little Haven to Broad Haven).

Chapter 4: Suggestions for sites to include in this tour were provided by Jo Sovin (Lyme Regis and the Jurassic Coast).

Author

Gary L. Prost, PhD, earned his BSc in Geology from Northern Arizona University in 1973 and an MSc (1975) and PhD (1986) in Geology at Colorado School of Mines. Over the past 45 years, he has worked for Norandex (mineral exploration), Shell U.S.A. (petroleum exploration worldwide), the U.S. Geological Survey (geologic mapping and coal), the Superior Oil Company (mineral and oil exploration), Amoco Production Company (worldwide oil exploration, remote sensing, and structural geology), Gulf Canada (international new ventures), and ConocoPhillips Canada (Canadian Arctic exploration, gas field development, oil sand development, and reservoir characterization). He spent over 20 years working as a satellite image analyst in the search for hydrocarbons and minerals in more than 30 countries. During this time, he applied structural geology and remote sensing to exploration, development, and environmental projects. The second half of his career was spent working on regional studies, new ventures and frontier exploration, and oil and gas field development. Now retired, his most recent work has been in public outreach, leading field trips and educating the public on topics of geological interest. He is the principal geologist for G.L. Prost GeoConsulting of El Cerrito, California, and has been a registered professional geologist in Wyoming (United States) and in Alberta and the Northwest Territories (Canada). He has published five books, including *North America's Natural Wonders Volumes 1 and 2* (Taylor & Francis, 2020), *The Geology Companion: Essentials for Understanding the Earth* (Taylor & Francis, 2018), *Remote Sensing for Geoscientists: Image Analysis and Integration* (third edition, Taylor & Francis, 2013), and *English–Spanish and Spanish–English Glossary of Geoscience Terms* (Taylor & Francis, 1997). He is currently working on his next volume, *Geologic Tours of the World – South America's Natural Wonders.*

Introduction and a Few Words about Driving in the United Kingdom

I must confess to some trepidation in preparing this volume. It is one thing to do this in North America, where I am familiar with the places being described. It is quite another to take on a different continent, and specifically a land with a rich history of the gentleman geologist and citizen scientist, a land that has numerous Geoparks and World Heritage sites and where the public routinely goes rambling through the countryside hunting fossils and minerals. Britain is a "target rich" environment: almost everywhere you look is a geologic point of interest. I accepted this challenge and hope the result is worthwhile.

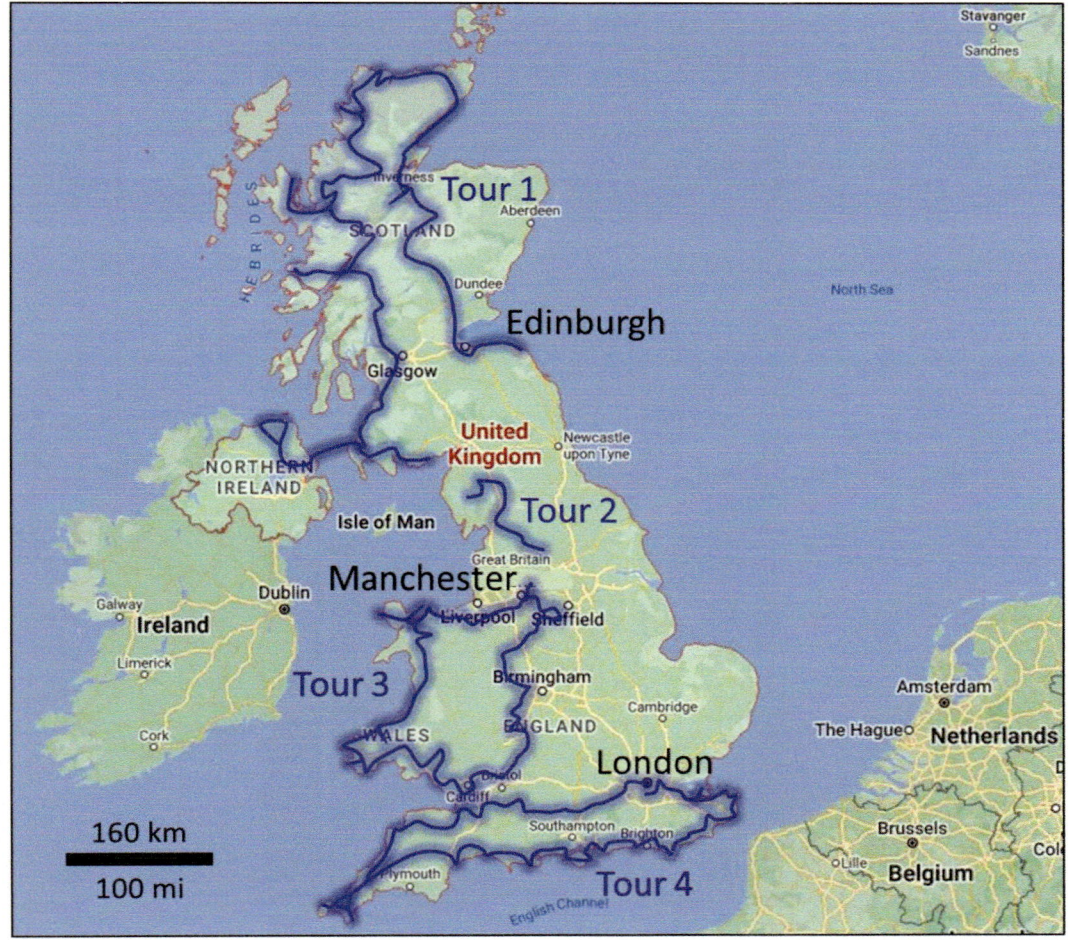

Overview of geologic transects in this volume.

Charts are provided to show what rock strata exist in each area, how old the layers are, and what they are made of. Each region is placed in a plate-tectonic context, so that you know how and when mountains and basins formed. The geologic influences on the economy are discussed so that you know, for instance, the reasons that the industrial revolution began near Ironbridge Gorge, and why the first geologic maps were made by a surveyor of coal canals.

Units are provided in metric first, and then their Imperial equivalent.

Abbreviations are explained, and then used. For example, millions of years is abbreviated to Ma (for "mega-annum"), Ga is used for billions of years, Fm is used for Formation, Gp for Group, mbr for member, and BBOE for billions of barrels of oil equivalent.

Road descriptions, pullouts, travel times, mine and other special tours, contact names, websites, and emails are current as of the time of writing. Contact a park or museum website for current hours and entry fees. All entrance fees and tour prices mentioned in the text are current as of 2022.

A geologic timescale is provided as a useful reference for those not familiar with the different geologic periods referred to in the trip descriptions.

EON	ERA	PERIOD		EPOCH		Ma
Phanerozoic	Cenozoic	Quaternary		Holocene		0.01
				Pleistocene	Late	0.8
					Early	1.8
		Tertiary	Neogene	Pliocene	Late	3.6
					Early	5.3
				Miocene	Late	11.2
					Middle	16.4
					Early	23.7
			Paleogene	Oligocene	Late	28.5
					Early	33.7
				Eocene	Late	41.3
					Middle	49.0
					Early	54.8
				Paleocene	Late	61.0
					Early	65.0
	Mesozoic	Cretaceous		Late		99.0
				Early		144
		Jurassic		Late		159
				Middle		180
				Early		206
		Triassic		Late		227
				Middle		242
				Early		248
	Paleozoic	Permian		Late		256
				Early		290
		Pennsylvanian				323
		Mississippian				354
		Devonian		Late		370
				Middle		391
				Early		417
		Silurian		Late		423
				Early		443
		Ordovician		Late		458
				Middle		470
				Early		490
		Cambrian		D		500
				C		512
				B		520
				A		543
Precambrian	Proterozoic	Late				900
		Middle				1600
		Early				2500
	Archean	Late				3000
		Middle				3400
		Early				3800?

Geologic time scale. (Wiki Commons, diagram courtesy of the U.S. Geological Survey, https://commons.wikimedia.org/wiki/File:Geologic_time_scale.jpg.)

The author examines rocks at the Victoria and Albert Museum, London.

Regarding driving in the United Kingdom, I strongly recommend getting a GPS navigation system such as Tom Tom, Google Maps, or Apple Maps. These were a lifesaver for me. Failing this, get a good navigator.

If you learned to drive on the right side of the road, take some time to get accustomed to driving on the left side. Practice driving a standard transmission shifting with your left hand. Learn about traffic signs that indicate right-of-way. Familiarity with road signs may save your life. Watch for speed limits as they are enforced by traffic cameras.

Perhaps for historical reasons, there are few straight roads. Also, there appear to be two kinds of Britons: those who take the train, and those who think they are race car drivers. One moment you have the road to yourself, and the next moment someone zips up behind you and passes at the slightest opportunity. Beware too the gigantic tractors that lumber out of fields and trundle down narrow country roads.

Roundabouts can be a nightmare when busy. Check the arrows on the pavement when approaching a roundabout to see which lane to be in. Expect to be honked at. Green traffic lights at a roundabout entrance do not guarantee you the right-of-way: always look to your right. Signs at the exits may have 5 or 6 road designations, too many to read as you fly by.

Just when you think you have mastered the roundabout, there is the double roundabout: one roundabout immediately after the other. Go slow – if you exceed 4 miles an hour, you will miss your exit on one or the other.

You will encounter "tree tunnels" and "trench roads" as well as single-track roads with turnouts every hundred yards or so. You will know what these are when you see them.

While certainly doable, driving after dark compounds these challenges. The things I found difficult during the day (judging lane position, distance from oncoming traffic, and reading street signs) are harder at night.

Even in the remotest places, expect to see a dozen cars parked at your destination. The reason for this becomes clear when you realize that this island is half the size of California but has nearly 30 million more inhabitants.

Most parking areas are small and make you pay. They may accept only coins, or phone payment, or card payment.

If you plan to take a car ferry, book in advance. Drive-ups only work if the ferry operates every 10 or 20 minutes.

About driving in London: don't. Find a hotel outside of town and take a train.

Be careful when pulling off the road, especially on narrow shoulders or curves in the road. Watch for traffic at all times. Safety first!

1 Scotland and Northern Ireland

Mealt Falls and Kilt Rock seen from Brother's Point. © User: Colin/Wikimedia Commons/CC BY-SA 4.0, https://commons.wikimedia.org/wiki/File:Panorama_of_Mealt_Waterfall_with_Kilt_Rock,_Isle_of_Skye.jpg

OVERVIEW

Starting at Siccar Point, where Hutton derived the concept of deep time and unconformities, the tour moves to Hutton's Section in Edinburgh, where intrusion of molten rock was first recognized. Glacial features are examined at Dulnain Bridge in Cairngorms National Park, and then we cross the Great Glen Fault and stop at the world-famous Devonian fish quarry at Achanarras. We examine the Caledonian Orogenic zone from Loch Ness to Durness and see the Moine Thrust at Loch Glencoul and Knockan Crag. We encounter some of the oldest rocks in Europe in North West Highlands Global Geopark, and some of the most spectacular sea cliffs and young volcanics related to the opening of the Atlantic on the Isle of Skye. The ancient volcanic centers at Glen Coe and Ardnamurchan are where the concept of caldera collapse was developed. Stunning glacier-carved glens and moors are all around as we pass through Lochaber Global Geopark. We cross the Highland Boundary Fault in Loch Lomond and the Trossachs National Park, examine a petrified forest in Glasgow, and visit beach terraces at Ballantrae that rose as the Ice Age glaciers retreated. Entering the Southern Uplands at the Southern Uplands Fault, we cross the Irish Sea to the Giant's Causeway World Heritage Site. The tour ends at turbidites between Monreith, Wigton Bay, and Southerness that were deformed during the Variscan Orogeny.

ITINERARY

Begin – Siccar Point
Stop 1 Siccar Point, Unconformities, and Deep Time
Stop 2 Edinburgh and Holyrood Park
      2.1 Ballagan Formation
      2.2 Lion's Haunch Agglomerate
      2.3 Samson's Ribs
      2.4 Salisbury Crags and Hutton's Section

DOI: 10.1201/9781351165600-1

Key stops on the Scotland/Northern Ireland geo-tour.

A BRIEF HISTORY OF SCOTLAND

Scotland, like the rest of Great Britain, has a long history. We cover it briefly, focusing on the Scottish Enlightenment, which greatly influenced the development of geology as a science.

People have inhabited Scotland for at least 10,000 years. Hunter-gatherers left flint arrowheads, Iron Age people built forts on peaks, and medieval farmers planted crops in the valleys. The earliest people hunted elk and deer, collected hazelnuts and honey, and found fish and shellfish along the coast. By around 5,000 years ago, people would have been growing crops such as barley and domesticating cattle for milk and meat (Miller, 2012). The earliest prehistoric tools found in Scotland date from 3,000 BC (Scotland.org).

Scotland's recorded history begins with the arrival of Romans in 124 CE. They built two massive fortifications, Hadrian's Wall to defend the northern border of Britannia and the Antonine Wall across Central Scotland. The Romans never conquered the Scoti and Picts in the area they called Caledonia. Facing invasions from the east, the Romans withdrew from Britain around 410 (Scotland.org).

Angles and Saxons settled in southeastern Scotland in the mid-400s to early 600s, but never conquered the interior.

The Gaelic kingdom of Dál Riata was founded on the west coast of Scotland in the 6th century. In 596, Pope Gregory sent English missionaries to the Pictish king Nechtan, who abolished most Celtic practices in favor of the Roman religion, thus avoiding war with Anglian Northumbria (southeastern Scotland). Irish missionaries converted many of the remaining pagan Picts during the 700s.

In the late 700s, Vikings from Norway and Denmark crossed the North Sea and established settlements across southern Scotland. The invasions forced the Picts and Gaels to unite in the 9th century under the House of Alpin, forming the Kingdom of Alba (Scotland). Shakespeare's Macbeth, probably the best-known early Scottish king, reigned from 1040 till his death in battle in 1057.

The last of this line died without an heir in 1286. The English, under Edward I, took advantage of the ensuing chaos to invade Scotland, leading to the Wars of Scottish Independence. In 1297, Edward's army planned to cross the River Forth at Stirling Bridge. The Scots, led by William Wallace, forced the English to retreat. A fictionalized version is portrayed in the movie *Braveheart*.

During the 12th century, the Kingdom of Alba evolved into a feudal society. More land was cultivated, trade with Europe stimulated the economy, and monasteries and abbeys flourished. Scotland passed back and forth between the House of Balliol and the House of Bruce. In 1306, Robert the Bruce was crowned the king of Scotland. Fighting with the English continued until 1314 at the Battle of Bannockburn, where Robert the Bruce finally defeated Edward II.

The Declaration of Arbroath, signed by Scottish Barons and Nobles and sent to Pope John XXII in 1320, declared Scotland's status as an independent state. Though mainly symbolic, the declaration is considered a milestone in Scottish history, and many historians believe it inspired America's Declaration of Independence 456 years later.

In 1371 Robert II, grandson of Robert the Bruce, established the House of Stuart, which would rule Scotland for the next three centuries. James VI, Stuart king of Scotland, also inherited the throne of England in 1603. Stuart kings and queens thus ruled both kingdoms until the Acts of Union in 1707 merged them into the Kingdom of Great Britain. Queen Anne, the last Stuart monarch, died in 1714, after which the house of Hanover and Saxe-Coburg and Gotha (Windsor) ruled Scotland as part of Great Britain (Wikipedia, History of Scotland).

During the Scottish Enlightenment and Industrial Revolution, Scotland became one of the commercial, intellectual, and industrial powerhouses of Europe. The Scottish Enlightenment was a period during the 18th- and early 19th centuries characterized by numerous intellectual and scientific accomplishments. Scotland established a network of parish schools in the Lowlands and four universities (St Andrews, Aberdeen, Glasgow, and Edinburgh) that encouraged intellectual inquiry. Sharing the humanist and rationalist outlook of the European Enlightenment, Scottish intellectuals stressed the importance of reason combined with a rejection of authority. The Scottish Enlightenment was characterized by practicality and benefit for the individual and society.

The modern science of geology was born in Edinburgh at this time. The effects of the Scottish Enlightenment were felt far and wide because of the esteem in which Scottish achievements were held outside Scotland, and because the ideas and attitudes it generated were carried around the world as part of the Scottish diaspora, and by foreign students studying in Scotland (Wikipedia, Scottish Enlightenment).

In recent decades, Scotland has enjoyed a cultural and economic renaissance, fueled in part by the financial services sector and the discovery of oil and gas in the North Sea in 1967. Nationalism became a political movement during the 1960s. When Scotland merged with England, the two parliaments merged into the Parliament of Great Britain. After a referendum in 1997, when the Scots voted for devolution (transfer of power to the Scottish Parliament), the Scottish Parliament was established by the Scotland Act of 1998. The Scottish Parliament met for the first time in 300 years in May 1999 (Wikipedia, History of Scotland). In 2012, the Edinburgh Agreement allowed a referendum on Scottish independence in 2014 by confirming the Scottish Parliament's authority to hold such a vote. In September 2014, the people of Scotland defeated the referendum by 55% to 45% (Scotland.org).

SCOTLAND'S OUTSIZED ROLE IN GEOLOGY

Scottish geologists were instrumental in developing the new science of geology in the 1700–1800s. James Hutton (1726–1797), a farmer and naturalist, is considered the "father of modern geology." He noticed that every year his fields lost some soil to erosion that the local rivers carried to the sea, where it was deposited as layers of sand and mud. He figured it must take a very long time to build the thick layers of sandstone and mudstone he saw on the sea cliffs. At Siccar Point, he noticed gently inclined beds of red sandstone overlying nearly vertical beds of a different sandstone. The contact surface between the two formations, Hutton felt, represented a gap in deposition. Below the gap, the originally horizontal sandstone must have been tilted, eroded, and then re-submerged in the sea. New sands were deposited on the erosion surface. The new sands became the red sandstone that Hutton saw above the gap. Here he developed the concept of unconformity, a gap in deposition.

Erosion and deposition, followed by tilting and uplifting of the land, must be a slow process that took large amounts of time. In 1788, Hutton wrote that the history of the Earth must be so long that "we find no vestige of a beginning, no prospect of an end." The lengths of time involved were beyond our comprehension. Before Hutton, most Europeans assumed the Earth was only 6,000 years old based on the Bible. In fact, in 1650, Bishop Ussher of Trinity College, Dublin, worked backward through the lives of the patriarchs and claimed that Earth was created on the evening of October 22, 4004 BCE. Based on the Bible, most Europeans believed not only that Earth was young, but also that all rocks had formed when the Earth was created. Hutton argued that if the processes we see today have always been at work, then the Earth must be much older. This has come to be known as the concept of "deep time."

Hutton's *Theory of the Earth*, published in 1788, stated, "In examining things present, we have data from which to reason with regard to what has been; and, from what has actually been, we have data for concluding with regard to that which is to happen hereafter." Charles Lyell, another Scottish geologist, later condensed this idea into "the present is the key to the past." In other words, we can explain what happened in the past by observing processes at work today, because those processes have not changed throughout Earth's history. Lyell (1797–1875) published his *Principles of Geology* built on Hutton's ideas. For example, Lyell's theory of uniformitarianism was built on Hutton's concept that present processes have been uniform over time. Lyell interpreted geologic change as the steady accumulation of small changes over enormously long timespans. Robert FitzRoy, captain of *HMS Beagle*, loaned Darwin a copy of the *Principles* before they set out on their voyage in the *Beagle*. The *Principles* had a strong influence on Darwin and many of his observations.

John Playfair (1748–1819) was a mathematician who developed an interest in geology from his friend Hutton. His *Illustrations of the Huttonian Theory of the Earth*, published in 1802, helped spread Hutton's ideas.

Sir Roderick Murchison (1792–1871) investigated and mapped rocks of the Silurian period in south Wales, which he published as *The Silurian System* in 1839. Silurian rocks are also present in the Southern Highlands of Scotland. Murchison worked on rocks of Cambrian and Ordovician age in England, and studied the Permian deposits of Russia, proposing the Permian System in 1841. He published *The Geology of Russia and the Ural Mountains* in 1845. Murchison was knighted in 1846 and in 1855 was appointed director of the Geological Survey and of the Royal School of Mines, London.

Hugh Miller (1802–1856) was a stonemason and self-taught geologist. His *The Old Red Sandstone* (1841) became a bestseller. He described and was fascinated by the fossils in these rocks, although he opposed the concept of evolution because of his religious beliefs.

Sir Archibald Geikie (1835–1924) was the first to recognize that there had been multiple glaciations in Scotland, and his *On the Glacial Drift of Scotland* (1863) was key to acceptance of the recently proposed theory of glaciation. He became Director-General of the Geological Survey of the United Kingdom in 1888.

More recently Arthur Holmes (1890–1965), born in England but Regius Professor of Geology at the University of Edinburgh, wrote *Principles of Physical Geology* (1944). In it, he proposed that convection currents in the Earth's mantle were the driving force behind "continental drift," now called plate tectonics. He also pioneered the discipline of geochronology, the dating of rocks using radioactive decay, isotope ratios, paleomagnetism, and fossils. He is considered one of the most influential geologists of the 20th century.

GEOLOGY OF SCOTLAND

Scotland is divided into three main geographical areas: the Highlands north and west of the Highland Boundary Fault; the Central Lowlands/Midland Basin; and the Southern Uplands, which lie south of the Southern Uplands Fault.

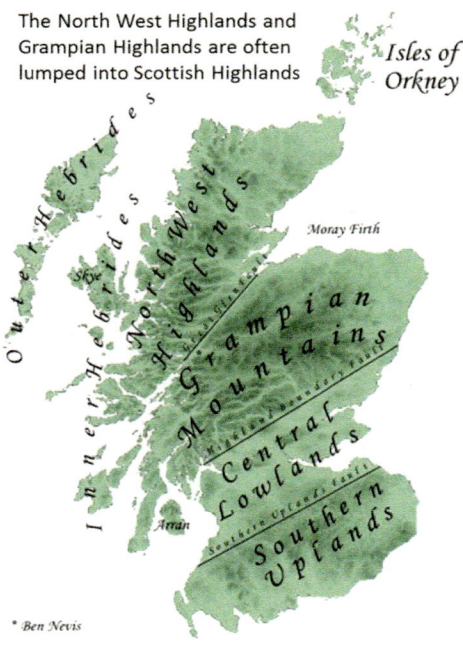

The North West Highlands and Grampian Highlands are often lumped into Scottish Highlands

Isles of Orkney

Outer Hebrides

Inner Hebrides

North West Highlands

Moray Firth

Grampian Mountains

Central Lowlands

Southern Uplands

Arran

* *Ben Nevis*

Physiographic subdivisions of Scotland. Each terrane is separated by a major regional fault: from the north, they include the Great Glen Fault, the Highland Boundary Fault, and the Southern Uplands Fault. (Courtesy of Rab-k, https://en.wikipedia.org/wiki/File:Scotland_(Location)_Named_(HR).png)

Rocks in Scotland range in age from Archean gneiss, metamorphic beds punctuated by granite intrusions created during the Caledonian Orogeny, Cambrian through Carboniferous sedimentary units, to fairly recent Paleogene volcanics (Wikipedia, Geology of Scotland).

During early Paleozoic time, prior to the Ordovician Period (460 Ma), sandstone, mudstone, and limestone were deposited in the shallow Iapetus Sea on the margin of Laurentia (proto-North America-Scotland).

The Scottish Highlands lie north and west of the Highland Boundary Fault, which runs northeast from Arran to Stonehaven. This part of Scotland consists of Precambrian and Cambrian rocks that were uplifted and deformed during the Caledonian Orogeny. The Caledonian Orogeny was a mountain-building event that extended from New York to Newfoundland to Ireland-Scotland to Scandinavia during the Ordovician to Early Devonian (490 to 390 Ma). The Caledonian Mountains were raised up by the collision and suturing of the early continents of Laurentia, Baltica (proto-Scandinavia and Eastern Europe), and Avalonia (proto-New England and the Canadian Maritimes) to form the northern supercontinent Laurussia (also known as Euramerica and the Old Red Sandstone Continent; Wikipedia, Laurasia). The merging of continents coincides with the closing of the Iapetus Ocean: marine sediments eroded from the surrounding landmasses were folded, faulted, and uplifted in the newly-created Caledonian Mountains. The remants of these mountains in Scotland form the Scottish Highlands and Southern Uplands.

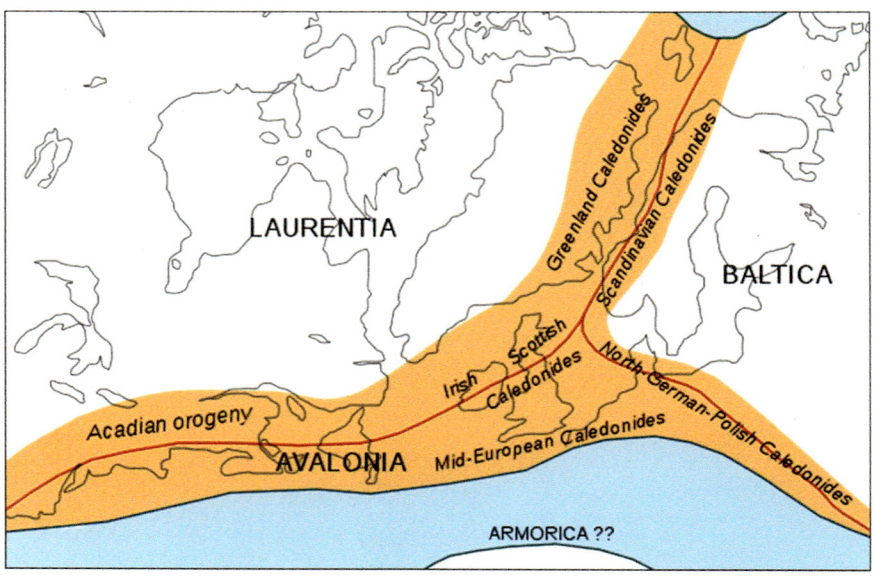

Caledonian/Acadian mountains during the Early Devonian. Present-day coastlines are shown for reference. Red lines are suture zones. Precursor continents that collided during the Caledonian Orogeny are shown. (Courtesy of Woudloper, https://commons.wikimedia.org/wiki/File:Caledonides_EN.svg)

This basement was injected by Silurian to early Devonian intrusions and is overlain by Devonian sedimentary rocks, primarily the Old Red Sandstone along Moray Firth. Igneous intrusions form massifs, a group of mountains, such as at Skye Cuillins and Cairngorms.

The main structures in the Highlands are the Moine Thrust and Great Glen Fault. Thrust faults are nearly horizontal faults that push older rock over younger rock.

The continent-continent collisions generated a series of thrust faults in the North West Highlands, the most prominent of which is the Moine Thrust. The Moine Thrust is a series of related thrusts whose surface traces run north-northeast from the Sleat peninsula on Skye to Loch Eriboll on the north coast. The thrusts formed during the later stages of the Caledonian Orogeny and carried the much older Northern Highlands Terrane (exposed east and south of the thrust zone) over the younger Hebridean Terrane (exposed to the west and north).

The thrust sheet consists primarily of Lewisian Complex metamorphic rocks (mainly granite gneiss 3.0–1.7 Ga, some of the oldest rocks in the world), and the overlying Morar Group of the Neoproterozoic Moine Supergroup (which may be the lateral equivalent of the Torridonian Supergroup exposed along the northwest coast). The thrust sheets were transported anywhere from 40 to as much as 200 km (24–120 mi) to the northwest. Neoproterozoic is the period from 1,000 to 541 Ma.

The stratigraphic section of the Hebridean Terrane consists, from base to top, of Lewisian Complex, unconformably overlain by Neoproterozoic Torridonian Supergroup (1 Ga–750 Ma; up to 8 km of mainly sandstone), and the Cambro-Ordovician Ardvreck Group (mainly quartz sandstone) and Durness Group dolomite with some limestone and chert beds.

Topographically, the exposed thrust zone separates rolling hills developed on metamorphic rocks to the east from rugged glaciated mountains comprising igneous, sedimentary, and metamorphic rocks to the west of the zone (Wikipedia, Geology of Scotland). The Caledonian Mountains eroded fairly quickly and deposited sediments in low-lying areas. These new formations include the Devonian Old Red Sandstone and the coal fields of central Scotland (Scottish Geology Trust, 51 Best Places).

Regional northeast-trending strike-slip faults (faults where the two sides move laterally past one another) also formed during the Caledonian episode. These include the Highland Boundary Fault, separating the Lowlands from the Highlands, the Great Glen Fault that separates the North West Highlands from the Grampians, the Southern Uplands Fault, and the Iapetus Suture, which runs from the Solway Firth to Lindisfarne and which marks the final closing of the Iapetus Sea.

The Great Glen Fault formed toward the end of the Caledonian Orogeny with the collision of the Laurentia and Baltic tectonic plates (430–390 Ma). The movement at that time was left-lateral. In a left-lateral fault the opposite side of the fault moves to the left with respect to the observer. In a right-lateral fault the opposite side of the fault moves to the right. A second movement episode during the Carboniferous caused a right-lateral offset along the fault. The final phase of the movement, from Late Cretaceous to Early Tertiary, is estimated to have had about 104 km of lateral offset (64 mi). Quaternary erosion along the fault zone during the last Ice Age formed Loch Ness. The fault is essentially inactive today, although occasional tremors have been felt (Wikipedia, Geology of Scotland).

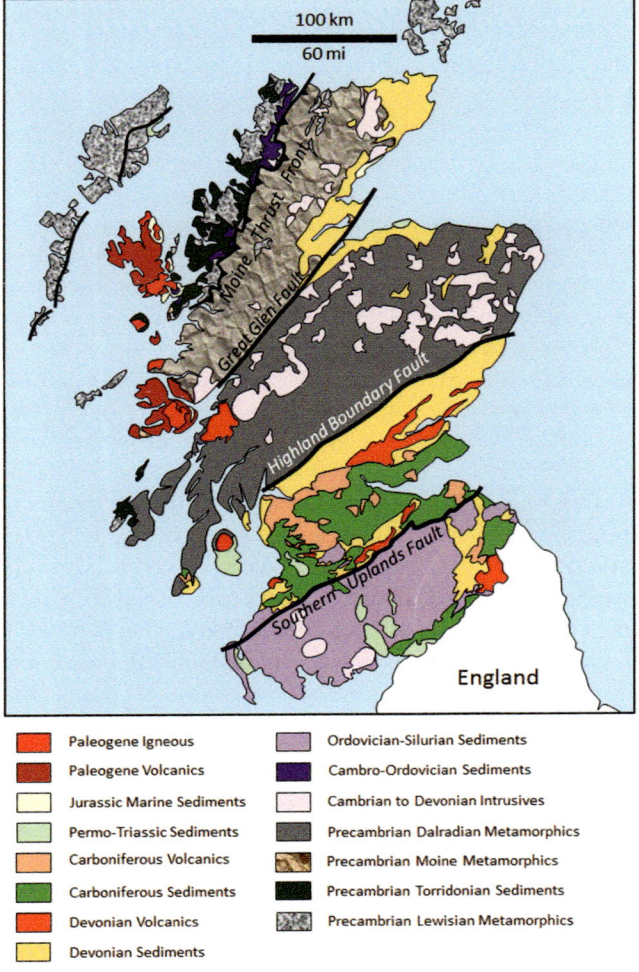

Geologic map of Scotland. (Modified after Scottish Geology Trust, © Image reproduced with kind permission of National Museums Scotland.)

Volcanism accompanied the closing stages of the Caledonian Orogeny across Scotland. These 400 Ma igneous rocks are expressed as Late Silurian-Early Devonian volcanics in southern Scotland, and as large granitic plutons at Glen Coe and the Cairngorms in the Highlands (Wikipedia, Geology of Scotland; Scottish Geology Trust, 51 Best Places). A pluton is an igneous body that crystallized in the subsurface.

As you might guess from the name, the Highlands contain the highest peaks in the United Kingdom, with the highest at Ben Nevis (1,344 m, or 4,409 ft).

Often referred to as the Central Lowlands, the Midland Valley is an early Paleozoic rift containing Paleozoic sedimentary units. This is the region that contains important iron-bearing rocks and Carboniferous coal deposits that fueled Scotland's industrial revolution. The Lowlands also experienced volcanism, as seen in the Carboniferous Campsies, the volcano at Arthur's Seat, and Castle Rock in Edinburgh. Ice Age glaciers smoothed the landscape, forming drumlins (an elongated hill of glacial debris) and other glacial features.

The Southern Uplands in the south of Scotland are a range of hills almost 200 km (120 mi) long, interspersed with broad valleys. They lie south of the Southern Highlands Fault that runs from Ballantrae to Dunbar. The basement rock consists of Cambro-Ordovician volcanic rocks and Silurian sediments deposited 400–500 Ma (Wikipedia, Geology of Scotland). A much younger episode of volcanism occurred on the west coast related to rifting and opening of the Atlantic Ocean. As North America separated from Europe, a thick pile of lava flows and large volcanoes erupted during the early Paleogene (63–52 Ma), centered on the Isle of Skye, Ardnamurchan, the Isle of Mull, the Firth of Clyde, and Northern Ireland (Scottish Geology Trust, 51 Best Places).

During Late Carboniferous time Gondwana, the southern continent, was colliding with Laurussia to create the supercontinent of Pangea. The Laurussia–Gondwana collision is called the Variscan Orogeny, with the northern limit of the fold belt (the Variscan Front) extending from mainland Europe across southern England. Southern Scotland was well north of the main deformation front, but some folding occurred as far north as the Southern Uplands. Northwest-directed shortening reactivated many of the old, basin-bounding normal faults as reverse faults, a process known as basin inversion (Stone et al, 2012a, b). Normal faults drop rocks down; reverse faults push rocks up. This period is also known for volcanism across Scotland, and reactivation along the Great Glen Fault.

Laurussia, or the Old Red Sandstone Continent, became a part of the supercontinent Pangea during the Permian (299–252 Ma). This had little or no impact on the geology of Scotland. At the start of Jurassic time (201–145 Ma), Pangea began to split into two continents, Gondwana and Laurasia. In the Cretaceous (145–66 Ma), Laurasia split into the continents of North America and Eurasia. Scotland, part of the Eurasian tectonic plate, separated from the North American tectonic plate to form the Atlantic. Rifting in the North Sea area led to deposition of algae-rich sediments that, in time, generated large deposits of North Sea oil and natural gas. During the Cretaceous, sea levels rose globally and limestone (chalk) was deposited over much of low-lying Scotland (and Europe).

The whole of Scotland was covered by ice sheets during the Pleistocene Ice Ages (the last 2.6 Ma). Evidence for glaciation can be seen in common glacial landforms, including drumlins, *roche moutonnees*, glacial polish, glacial scour, moraines, U-shaped valleys, boulder clays, and tors (large, free-standing outcrops rising abruptly from the surrounding gentle slopes of a hill or ridge) among others.

Since the last Ice Age ended around 12,000 years ago, the land has been slowly rising due to isostatic rebound, because the weight of the ice sheets has been removed. Coupled with a post-Ice Age rise in sea level, this has led to some shorelines having coastal wetlands, and others having multiple levels of uplifted, wave-cut terraces (Scottish Geology Trust, 51 Best Places).

GEOLOGY OF WHISKEY

No review of the geology of Scotland would be complete without a discussion of what gives whiskey its flavor. Unlike wine, terroir (the characteristic flavor imparted by the environment in which it is produced) is not a major influence. This is because whiskey is a distilled spirit (often double or triple distilled), and the factors that influence flavor are based less on the geology and more on the distillation and aging process.

The earliest historical evidence from Scotland dates the distilling of malted barley to the middle of the 16th century. The distillate was known originally as uisge beatha (water of life), which evolved to "whiskey" [the Scots spell it "whisky"]. The distilling process in Scotland was brought under government regulation in the 1830s. By law, Scotch whiskey must be derived from malted and/or unmalted cereals and be matured in oak barrels in Scotland for at least 3 years.

Whiskey-making involves a brewing phase, followed by distillation. The result of the first stage is the production of a liquid equivalent to a strong beer. The first part of this process is the malting of barley, where the grain is allowed to germinate for several days to initiate the enzyme processes that break starch down into sugars. That process is stopped by heating. Historically, some peat smoke has permeated the malt during heating, with the amount of permeation reflected in the smokiness of the final whiskey. The second stage involves mixing the malted barley with fresh local water to extract the sugars and produce a liquid ready for fermentation. Fermentation uses a combination of beer and wine yeasts to raise the alcohol content to around 8%. The resulting liquid is then heated in copper stills, normally in a two-stage process that raises the alcohol content first to 25%, and then to a final alcohol content between 55% and 65%. Following this, the spirits are aged in oak barrels.

There are three groups of whiskeys: grains, malts, and blends. Grain whiskey is manufactured using malted and unmalted grains. Malt whiskeys are made only from malted barley. Blended whiskeys contain 40%–60% grain whiskey, and the remainder is malt. Malt whiskey is generally considered the finest (Cribb, 2005b). All Scotch whiskey was originally made only from malted barley. Distilleries began introducing whiskey made from grain (wheat, corn, rye) in the late 1700s. Distillers like to keep some character of the source grain, so they don't distill all the way to a true neutral spirit (Selfbuilt, Source of Whisky's Flavor). Thus, the type of grain used and percent of each has an effect on the resulting flavor.

The size of stills and the number of times the liquid is distilled have some bearing on the final flavor, as more contact with the copper of the still creates lighter and fruitier notes. Copper appears to act as a filter, getting rid of some less palatable aromas.

Yeast also imparts flavor, and different strains are added to achieve different tastes.

The size of the barrels affects surface area-to-volume: the smaller the cask, the faster the whiskey interacts with the wood, whereas larger casks prevent the wood's qualities from overwhelming the spirit (Rnjak, 2017).

The purpose of aging is to allow air access to the whiskey to allow oxidation, evaporation, and concentration of the spirit. Aging allows the wood in the barrel to remove some flavors and add others (Selfbuilt, Source of Whisky's Flavor). Whiskeys are aged in wood barrels, usually oak, that frequently contained other spirits (bourbon, wine, or sherry). The oak can be new wood, European oak generally from the sherry industry in Spain and American white oak previously used as bourbon barrels. According to some experts, the bourbon casks impart vanilla, cherry, and spice notes. Wine casks, made of European oak, impart clove, orange, and dried fruit flavors (Rnjak, 2017).

The wood affects the flavor by soaking up impurities, as well as delivering chemical compounds from the wood and previous contents to the whiskey. The wood is porous and permeable and "breathes," allowing gasses to move into and out of the barrel. This includes evaporation of the alcohol, the "Angel's share," that concentrates flavor and makes the final product taste milder over time.

The inside surface of the barrels is frequently charred or "toasted." The purpose of charring is to filter out unpleasant compounds (e.g., sulfur). Charring introduces new flavors into the whiskey: it releases vanillins, imparts toasty and caramelized notes, and gives color to the final product (Rnjak, 2017; Selfbuilt, Source of Whisky's Flavor).

Thus, the crucial factors that combine to produce the wide variety of whiskey flavors include the degree to which peat is used during malting (the smoky flavor), the combination of grains used, the yeast used, the still, the size and types of barrels, the aging process, and the chemistry of the water used (Cribb, 2005b).

Location has some influence on flavor as a result of local soil, weather, and water. Soil may influence the type of grain that goes into a whiskey. Weather/temperature/environment has an effect on how much the casks breathe. As the casks are mostly stored in rooms without temperature control, the liquid expands and contracts with the seasons, seeping in and out of the wood and taking on more of its qualities (Rnjak, 2017). Some say that coastal areas impart a spicy or salty character to the spirit.

However, it is in the water that, some argue, location and geology have the most impact. Water chemistry affects three areas of the production process: the extraction of sugars from the malt, the nature and type of fermentation, and the chemistry of the distillation process (Cribb, 2005b). Also, despite distilling, there is always some water remaining in the distilled product, and some may be added before casking.

Water is taken from wells, rivers, springs, and lakes, and every water source is unique in terms of its chemistry, dissolved solids, and trace elements. For example, soft water (low dissolved solids) tends to flow from granites, quartzites, and schists. Water derived from Proterozoic-Ordovician Dalradian dolomite or the Devonian Old Red Sandstone tends to be hard, containing substantial amounts of calcium and magnesium carbonates, a factor that distilleries claim has a significant effect on the nature of the whiskey (Cribb, 2005a).

There are five whiskey regions in Scotland today, and each region has a characteristic flavor. According to the Whisky Foundation (2016), in addition to the equipment and methods each distillery uses, climate and location add distinct flavors to whiskey.

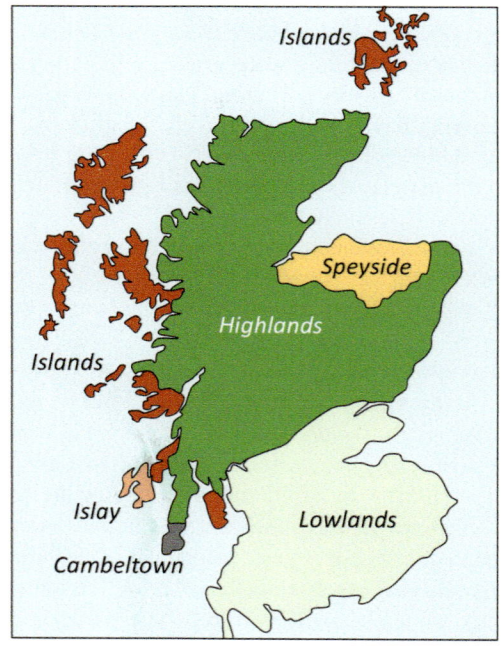

Scottish whiskey regions.

HIGHLANDS

The highlands are the largest whiskey region in Scotland and home to some of the most famous distilleries. In the Western Highlands, the malts are less peaty and sweeter. Whiskeys in the North Highlands, influenced by the coast and soil, tend to be light-bodied with a spicy, sometimes salty character. Central Highland whiskeys are considered to be sweeter and have "fruity" flavors.

The Islands is a sub-region of the Highlands. Whiskeys produced here share the maritime feel of Islay whiskeys, but are sweeter and less peaty.

SPEYSIDE

Speyside is home to about half of Scotland's whiskey distilleries. Speyside whiskeys are typically characterized as the sweetest and most complex of Scotch whiskeys, commonly having fruity flavors and sweet aromas. With age Speyside whiskeys often turn rich and heavy, delivering dried fruit flavors and spicy notes

LOWLANDS

The Lowlands, in the southernmost part of Scotland, still contain some single-malt distilleries which produce whiskeys that are typically dry, fresh, and light, with floral notes and fragrances.

CAMPBELTOWN

Campbeltown, on the west coast of Scotland, only has three remaining distilleries. They produce whiskeys considered to be full-bodied with a distinctive, salty finish. Some of the whiskeys are quite peaty.

ISLAY

Many believe the whiskey produced here is affected by the sea, wind, and rain. A large percentage of the land consists of peat, and use of this in malting gives Islay whiskeys their unique peaty character. Islay whiskeys are considered the strongest flavored of all single malts, regarded as dry and smoky (Whisky Foundation, 2016).

START: SICCAR POINT

Take the A1 to the A1107 and exit to the east. Take A1107 1.4 km (0.8 mi) to turnoff on left to Pease Bay and Old Cambus Quarry; turn left (northeast) and drive 1.1 km (0.7 mi) to a free car park on the left side of the road before the gates into the Drysdales' site. From here, two information boards guide you ~960 m (3,150 ft) northeast along the cliff top path to Stop 1, Siccar Point (55.931433, −2.301720). Climb down the steep slope to the shore for the best views.

STOP 1 SICCAR POINT, UNCONFORMITIES, AND DEEP TIME

Like the Grand Canyon and Thingvellir, Siccar Point is a place of pilgrimage. Geologists come from around the world to see the rocky promontory that is famous for Hutton's Unconformity and its role in developing the geologic concept of Uniformitarianism and Deep Time.

In 1750, James Hutton inherited and began working two farms in the Scottish Borders. After witnessing the processes of erosion and sediment deposition on his farms, he became interested in geology. Hutton returned to Edinburgh in 1767, where he developed and finally published his geological theories. Hutton's *Theory of the Earth* was presented in 1785 to the Royal Society of Edinburgh. Later, during a boat trip in 1788, Hutton observed the angular unconformity (the abrupt change in the angle of the layers) exposed at Siccar Point. He felt that what he saw was conclusive evidence of "uniformitarianism," the theory that the natural laws and processes we see today on the Earth must be the same as have operated in the past.

Here, Silurian graywacke sandstones (dirty sandstones) of the Gala Group (about 435 Ma) were deposited as turbidites (submarine landslide deposits) in the Iapetus Ocean. The layers were later rotated to near-vertical during the Caledonian Orogeny, and then eroded. The Devonian Stratheden Group (about 370 Ma), consisting of red river channel sandstone and conglomerate, lies over the older strata. We know now that about 65 million years of rock is missing between the lower vertical and the upper horizontal layers. At the time, Hutton knew only that it takes a long time to deposit thick layers of sand, one grain at a time, then bury them and convert the layers to stone, then uplift and deform the sandstone, erode it down, and deposit another layer above it. He understood this to mean that an enormous span of time must have passed between deposition of the two beds (Edinburgh Geological Society, Siccar Point; Wikipedia, Siccar Point). This was bitterly disputed by those who believed in the biblical 6,000-year-old Earth, but ultimately Hutton was proved right.

South of Siccar Point the graywacke sandstone was quarried for road metal. The Old Cambus Quarry is now the site of the Drysdale vegetable distribution warehouse complex.

Excellent coastal outcrops east of here, between Fast Castle and St Abb's Head, were mapped by Holdsworth et al. (2002). The exposures are mainly Silurian Gala Group turbidites. These units have been deformed into upright to northwest vergent (inclined) folds that record regional northwest-southeast shortening. The similarity of deformation style here to that seen in southwestern Scotland indicates to these workers that the entire Southern Uplands were subjected to northwest-directed back thrusting and folding, and a transition from compression to left-lateral transpression (combined compression and shear) during the Early Silurian part of the Caledonian Orogeny.

Siccar Point angular unconformity (dotted line). Silurian graywacke is near-vertical below the unconformity; Devonian Stratheden Group sandstone lies above it. (Photo courtesy of Dave Souza, https://commons.wikimedia.org/wiki/File:Siccar_Point_red_capstone_closeup.jpg.)

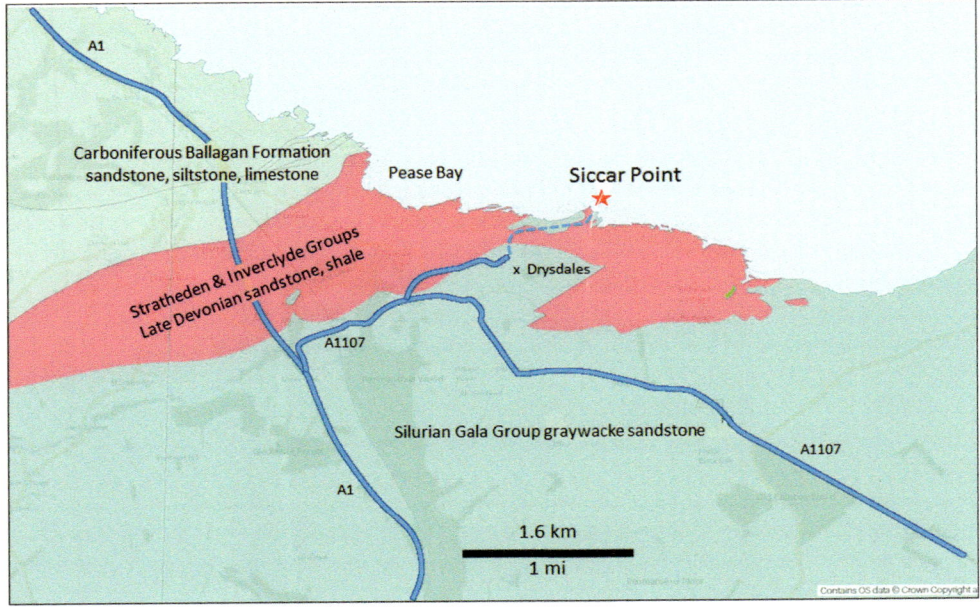

Geologic map of the Siccar Point area. (Modified after BGS, Geology of Britain Viewer. Contains British Geological Survey materials © UKRI 2022. Base mapping provided by ESRI.)

Age	Group	Formation
Devonian / Upper Old Red Sandstone	Inverclyde	Ballagan
		Kinnesswood
	Stratheden	Greenheugh Sandstone
		Redheugh Mudstone
Silurian	Gala	Queensberry
		Garheugh

Stratigraphy at Siccar Point. (Modified after Browne et al., 2002; Stone et al., 2012e.)

STOP 2 EDINBURGH AND HOLYROOD PARK

Edinburgh was established as a settlement on the high ground around the Castle by 500 CE, and by 1128, the local king had established the Abbey of Kelso near Holyrood Palace at the foot of the Royal Mile, opposite Edinburgh Castle. Scottish legends say that it was at the foot of Arthur's Seat, covered by the forest of Drumselch, that Scotland's 12th-century King David I encountered a stag while out hunting. His horse threw him and, about to be gored, a vision of the cross appeared between the animal's antlers. The stag, turned away by divine intervention, left him unharmed. Whereupon King David founded Holyrood Abbey on the spot (Wikipedia, Arthur's Seat). The abbey was the site of a parliament held by Robert the Bruce in 1326 and was likely the royal residence by 1329. Holyrood Palace was built adjacent to the abbey by James IV in 1501. In 1544, the Earl of Hertford, during the War of the Rough Wooing, sacked and burned Edinburgh, including Holyrood. After the Scottish Reformation in the 1500s, when the Scots took up the Calvinist Presbyterian Church, the abbey was neglected and parts were demolished in 1570.

James VI took up residence at Holyrood Palace at the age of 13 in 1579. When he became king of England in 1603 he moved to London. The palace was no longer the permanent royal residence. In 1670, two doctors established a small garden to grow medicinal herbs. The garden eventually became the Edinburgh Botanic Gardens.

After the union of England and Scotland in 1707, the palace was used only to elect Scottish representatives to the House of Lords. The roof of the abbey collapsed in 1768, leaving it as it is today. Holyrood Park became a sanctuary during the 16th century, a place where debtors could escape imprisonment, and a small community grew there. Among the debtors who resided there was the youngest brother of Louis XVI of France, the Comte d'Artois, who resided there after the French Revolution.

Queen Victoria renovated the palace and began staying there in 1871. However, it was George V who upgraded the palace by installing central heating and electrical lighting for his visit in 1911. During the 1920s, the palace became the official residence of the monarch while in Scotland, as well as the site of royal functions and ceremonies.

Edinburgh's landscape is typical of the landscape throughout the Central Scottish Lowlands. The hills and valleys are a function of the different rocks found in the region. Sedimentary rocks erode easily, whereas igneous rock is resistant to erosion. Salisbury Crags, Arthur's Seat, and other hills in the Edinburgh area are all igneous rocks (Miller, 2012).

Edinburgh lies in the Midland Valley, an ancient rift, or pull-apart basin formed during the Devonian and Carboniferous (roughly 410 to 280 Ma). Sediments poured into the basin from bordering highlands. The oldest sediments in the area are thick red sandstones called the Upper Old Red Sandstone (Kinnesswood and Ballagan formations of the Upper Devonian Stratheden Group) deposited about 370 Ma when the area was arid mountains and deserts. Windblown sand, eroded from older rocks, was deposited in valleys along with river deposits and shallow lakes. The mountains were gradually worn down to plains with shallow lagoons and seas during the Carboniferous (McMillan et al, 1999; McAdam, 2003).

The Carboniferous was a time of rapid environmental changes, leading to alternating layers of rock in regular cycles that repeated roughly the same sequence. These were known as cyclothems. During two periods the lush forests led to thick coal deposits. Tropical seas fostered massive coral reefs and abundant sea life that led to thick limestones. Lagoons with organic-rich muds developed repeatedly, giving rise to the oil shales of West Lothian. Some sands accumulated along beaches and in dunes; others were deposited in river channels (McAdam, 2003).

This is as good a place as any to discuss the Old Red Sandstone. The term "Old Red Sandstone" is an informal name that describes a suite of sedimentary rocks, dominantly terrestrial sandstone, deposited in a number of basins across northern Scotland from Late Silurian to earliest Carboniferous (Mississippian) time. Although it is mostly red sandstone, it can be gray, green, and purple as well and includes conglomerates, siltstones, mudstones, and thin limestones deposited in river channels, alluvial fans, sand dunes, and lakes. The sediments in this unit were derived from erosion of the Caledonian Mountains to the north (Wikipedia, Old Red Sandstone). Basins such as the Midland Basin, located along the northern edge of the former Avalonia, deepened in response to crustal flexure south of the rising Caledonian Mountains. An abundant supply of sediment from the new highlands caused the basins to fill rapidly, replacing marine conditions with terrestrial.

Although the unit was described by James Hutton when he visited Siccar Point in 1788, the term Old Red Sandstone was first used in 1821 by Scottish naturalist Robert Jameson to refer to the red rocks below the "Mountain Limestone" (Carboniferous Limestone) that was thought at that time to be the British equivalent of Germany's Rotliegendes Formation (which turned out to be Permian, a bit younger; Wikipedia, Old Red Sandstone).

The majority of this unit does not contain fossils, making it difficult to correlate the layers from one depositional basin to another. Of those fossils that do exist, there is a diverse assemblage of trace fossils, including worm burrows and tracks generated by arthropods (insects, spiders, and crustaceans) (Kendall, 2017). Locally, beds may contain fish fossils, as at the Achanarras Quarry. Many of the layers are correlated using palynomorphs (fossil pollen).

The sandstone-dominated Scottish Old Red Sandstone was originally divided into the Lower, Middle, and Upper Old Red Sandstone by Sir Roderick Murchison in 1859. The Lower Old Red Sandstone (latest Silurian to early Devonian) was mainly fluvial (river deposits) and volcanic; the Middle Old Red Sandstone (mainly mid-Devonian) was dominantly lake deposits; the Upper Old Red Sandstone (mainly Late Devonian to earliest Carboniferous) was chiefly river deposits.

The limited extent of any given bed has led to a proliferation of formations within the sequence. An attempt to coordinate names (detailed in Browne et al., 2002) led to a reduction in the number of formations and assigning them to a limited number of groups. For example, the Lower Old Red Sandstone in the northern Midland Valley was laid down in three sedimentary basins including, from oldest to youngest, the Stonehaven (Group), Crawton (Dunnottar-Crawton Group), and Strathmore (Arbuthnott-Garvock and Strathmore groups) basins. The number of formations was reduced to 19. The Middle Devonian is not represented in the northern Midland Valley as a result of non-deposition or erosion during the Acadian Orogeny. Late Devonian strata all belong to the Stratheden Group (Browne et al., 2002).

Units of the same age and depositional environments have been found in Scandinavia, the United Kingdom, and eastern North America. They were all deposited on the margins of Laurussia and were later separated when the landmass broke apart into today's continents (AMNH, Old Red Sandstone).

About 350 Ma, the area around Edinburgh was a flat tropical coast. Eruptions formed a volcano where Castle Rock now stands. The volcano spread ash as well as lava, building a volcanic cone several hundred meters high. Other eruptions farther east in Holyrood Park poured out black basalt flows. Volcanic vents and the surrounding area filled with irregular blocks of lava and volcanic ash known as agglomerate. You can see this in the roadcuts at Lion's Haunch, Queen's Drive. Lava flows that cooled slowly formed cooling cracks that, when eroded, appear as hexagonal columns. These columns, similar to those at the Giant's Causeway (Ireland), can be seen at Samson's Ribs on Queen's Drive. Some of the magma didn't make it to the surface, pushing between layers of sedimentary rocks and prying them apart. These became dolerite sills (dolerite being a basalt that cooled slowly underground). The best known of these sills are the Salisbury Crags. The crags are a bit more precipitous today than they originally were because of quarrying during the 1800s to pave the streets of London. Most of the hills on the west side of Edinburgh contain erosion-resistant dolerite sills.

After the volcanism ended, the land subsided and was buried under thousands of meters of sediments. Much later, tectonic forces uplifted and tilted the region to the northeast. Most sediments deposited after the Carboniferous were eroded and carried away. Much of the volcanics were eroded, leaving excellent exposures at Arthur's Seat, Salisbury Crags, and Castle Rock. Alternating hard lava flows and soft volcanic ash give the hill its tilted, stepped form. Outcrops along Queen's Drive show the alternating layers of ash and lava with this eastward tilt.

Like all of northern Britain during the last 2 million years, Edinburgh was repeatedly buried under an ice cap hundreds of meters thick. The latest ice cap melted around 12,000 years ago. Material eroded by the ice was left as glacial till (boulder clay), a mixture of clay with pebbles and boulders of volcanic rock, limestone, and sandstone. Glacial till covers much of the low ground in Edinburgh (McAdam, 2003).

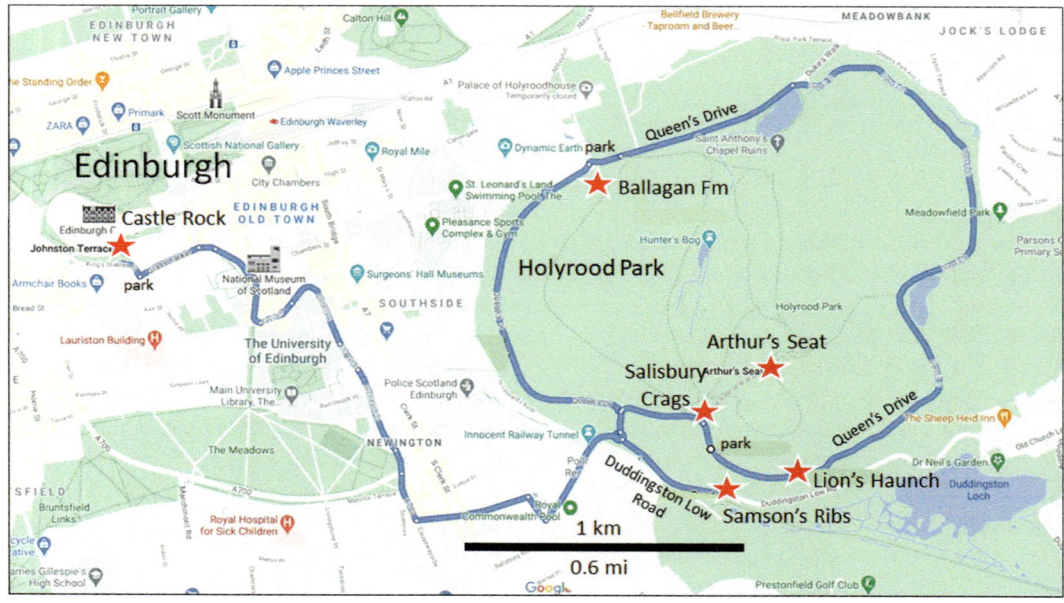

*Siccar Point to Ballagan Formation: Return west to the A1 and turn right (north); drive on the A1 to the eastern outskirts of Edinburgh; at Duddingston Crescent turn left (west) onto A1/ Duddingston Crescent/Milton Road and drive to the A1/Wolseley Terrace; turn left (west) onto Wolseley Terrace; turn left (southwest) onto Meadowbank; at the roundabout, take the 2nd exit (southwest) onto Duke's Walk; continue straight onto Queen's Drive and pull into the car park on the right (north) side of the road just east of the Dynamic Earth venue. This is **Stop 2.1, Ballagan Formation** (55.948631, −3.169589), for a total of 61.7 km (38.3 mi; 50 min). There is ample "pay- and-display" parking here.*

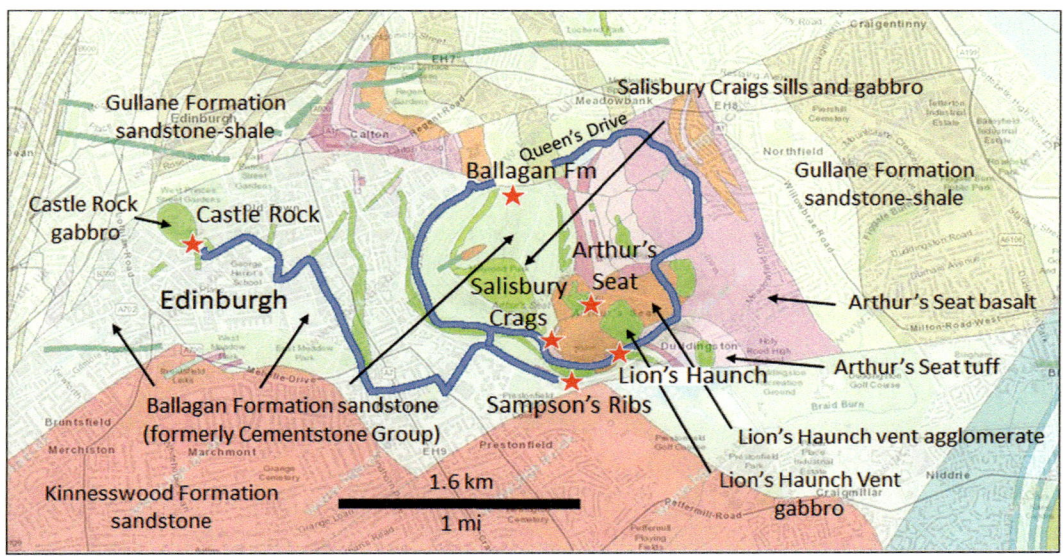

Geologic map of Edinburgh showing stops. All units are Carboniferous except the Kinnesswood Formation, which is Late Devonian. (Modified after BGS, Geology of Britain Viewer. Contains British Geological Survey materials © UKRI 2022. Base mapping provided by ESRI.)

Stop 2.1 Ballagan Formation

There are not many good outcrops of the Carboniferous Ballagan Formation sandstone (formerly known as the Cementstone Group). They are scattered in Holyrood Park, and the layers are inclined about 25° to the east. You may be able to see sedimentary features such as ripple marks, mud cracks, and worm tracks, and if you're lucky, the small crustacean *Estheria peachi* (Ball, 2014; Black, 2015).

The Ballagan Formation is widespread in the low-lying areas around Edinburgh. The forma- tion consists of sandstone, shale, and muddy limestone ("cementstone") and locally thin layers of gypsum. These suggest that they were deposited on a coastal plain or nearshore environment with changing salinities and drying episodes (McMillan et al., 1999).

Ballagan Formation outcrop in Holyrood Park.

Age	Group	Formation
Triassic	Sherwood Sandstone	Sandstone of Spynie
		St Bees Sandstone
Permian	Appleby	Sandstone of Hopeman
		Mauchline Sandstone
		Locharbriggs Sandstone
		Thornhill Sandstone
		Corncockle Sandstone
Carboniferous (Westphalian)	Coal Measures	Upper Coal Measures
		Middle Coal Measures
		Lower Coal Measures
Carboniferous (Namurian)	Clackmannan	Passage
		Upper Limestone
		Limestone Coal
		Lower Limestone
Carboniferous (Viséan)	Strathclyde	West Lothian Oil Shale
		Gullane
		Arthur's Seat Volcanics
	Inverclyde	Ballagan
		Kinnesswood
Devonian	Upper Caithness Flagstone	Spittal
	Arbuthnott	Dundee

Stratigraphy of the Edinburgh area. (Modified after McMillan et al., 1999.)

*Ballagan Formation to Lion's Haunch: On leaving the car park turn left (east) on Queen's Drive; bear right to stay on Queen's Drive south (one way); continue south and west around the park on Queen's Drive to **Stop 2.2, Lion's Haunch Agglomerate** (55.940442, −3.160176), for a total of 3.1 km (1.9 mi; 13 min). Parking on the side of the road may be difficult, so you may have to drive another 400 m (1,312 ft) to a wide spot and walk back.*

STOP 2.2 LION'S HAUNCH AGGLOMERATE

A nice example of Lion's Haunch Vent agglomerate is exposed in a roadcut on Queen's Drive on the south side of Arthur's Seat. Agglomerate is a rock composed of poorly sorted volcanic rock fragments.

Lion's Haunch agglomerate. You can see large clasts (broken rock fragments) in the ash-like volcanic rock.

*Lion's Haunch to Samson's Ribs: Walk down the steps below the road, cross the grassy slope to Duddington Low Road, and walk east on the road to **Stop 2.3, Samson's Ribs** (55.939827, −3.164582), about 580 m (1,900 ft; 10 min).*

STOP 2.3 SAMSON'S RIBS

A spectacular exposure of columnar basalt can be seen in an outcrop at the south end of Arthur's Seat. This exposure, known as Samson's Ribs, is a basalt flow at the margin of the Lion's Haunch Vent agglomerate (Ball, 2014). Columnar basalt is a result of slow cooling and contraction of the lava that caused extensional jointing in the form of hexagonal or pentagonal columns.

Columnar jointing in basalt, Samson's Ribs.

*Samson's Ribs to Salisbury Crags: Continue driving west on Queen's Drive 400 m (0.2 mi; 2 min) and pull over on the left side of the road. This is **Stop 2.4, Salisbury Crags** (55.941293, −3.165219). Take the trail uphill about 60 m (200 ft) to the crags.*

Stop 2.4 Salisbury Crags and Hutton's Section

Salisbury Crags is a small cliff formed by a resistant, altered basalt (variously called dolerite, diabase, teschenite, or microgabbro) sill that was injected as a molten rock between layers of Carboniferous sandstone (the Cementstone Group, now called the Ballagan Formation). Although the sill is also Carboniferous, Hutton studied this rock and determined that it had once been molten and was injected long after the sand layers had been turned to stone (Jenkins, Hutton's Section). The two rock types had been formed at different times and by different processes. This supported his conclusion that the Earth is very old and constantly changing (Miller, 2012). We now know that the Ballagan Formation was deposited as sand and mud on a coastal plain around 360 Ma and that the sill was shallow magma forced between the sandstone layers around 327 Ma (Miller, Holyrood Park).

This site helped resolve a great controversy in geology in the 1700s. At the location called Hutton's Section, you can see sandstone in contact with a volcanic sill. This location shows clear evidence that the sandstone is older than the sill, that the sill had once been molten, and that during injection, it had cut across the older sandstone layers. This was contrary to the accepted wisdom in the late 1700s and early 1800s, a theory called "Neptunism," that stated all rocks, even lavas and granites, were deposited or precipitated in a primeval sea.

There are two locations where the sill cuts across the sandstone. At one the sediment against the sill is crumpled; at the other, a wedge of sill rock intruded beneath a block of sediment, rotating it upward and partly engulfing it. At the west end of the section, the sill immediately above the contact was quickly chilled and turned into glass up to a centimeter thick. Above the glass, the sill is very fine-grained, but the crystals coarsen upward, indicating that they cooled more slowly. In the rock face to the southeast of Hutton's Section, large rafts of sediment can be seen within the sill. The upper contact of the sill is exposed in an embayment in the crags. Here, mineral crystals within the sill become markedly smaller, and the sill becomes vesicular (has gas bubbles) as you approach the contact (Black, 2015).

Much later, but still in Carboniferous time, tectonic movements caused uplift and tilting to the east. Then, 300 million years of slow erosion began to strip away hundreds of meters of overlying rock. In the last 2 million years, ice has sculpted this landscape (Miller, Holyrood Park).

Salisbury Crags looking northwest.

Hutton's Section, Salisbury Crags, Holyrood Park. This spot shows intrusive features at the base of Salisbury Crags sill, including a chilled sill margin and sediments that were broken, twisted, and baked by the intrusion. (Courtesy of Anne Burgess, https://commons.wikimedia.org/wiki/File:Hutton%27s_ Section_-_geograph.org.uk_-_3503433.jpg.)

Salisbury Crags to Arthur's Seat: Find the trail to Arthur's Seat and walk roughly 100 m (300 ft) to the peak. This is *Stop 2.5, Arthur's Seat* (55.943852, −3.161861).

STOP 2.5 ARTHUR'S SEAT

Arthur's Seat in Holyrood Park rises above Edinburgh to a height of 250.5 m (822 ft). It, along with Castle Rock and Calton Hill, has been designated the Arthur's Seat Volcano Site of Special Scientific Interest (SSSI), areas set aside for the preservation of spots of special geologic and/or biological interest (Wikipedia, Arthur's Seat).

Between 335 and 341 Ma, eruptions at Arthur's Seat built up layers of basaltic lava, volcanic ash, and ash with large rock fragments (agglomerate) into a cone that reached perhaps 200 m (650 ft; Miller, 2012; Miller, Holyrood Park). In all, there were at least 13 different layers of flows and agglomerates spreading out from the volcanic crater. After the volcanism ended, the entire region was tilted about 25° to the east. Some of the feeder dikes are exposed where they cut the sandstone, and the flows and agglomerates can be seen in roadcuts on the southern margin of the volcano and as outcrops near the summit (Ball, 2014).

View northeast to Arthur's Seat at the crest of the hill.

*Arthur's Seat to Castle Rock: Return to Queen's Drive and drive west to the first roundabout; take the 1st exit (south) 75 m to the next roundabout; take the 2nd exit (west) onto Holyrood Park Road; drive southwest on Holyrood Park to Dalkeith Road/A7; turn right (northwest) on Dalkeith Road and drive to Preston Street/A7; turn left (west) on Preston Street and drive to A700/Lord Russell Place/Summerhall Place; turn right (north) on A700; continue straight onto Hope Park Crescent; continue straight onto Baccleuch Street; continue straight onto Chapel Street; continue straight onto Potterrow; continue straight onto Teviot Place; turn right (north) onto Forrest Road; bear left onto George IV Bridge; turn left (west) onto Lawnmarket; at the roundabout take the 2nd exit (south) onto Johnston Terrace (one way); drive west on Johnston Terrace to the parking area on the right at the base of Castle Rock. This is **Stop 3, Castle Rock** (55.947812, −3.198680), for a total of 3.8 km (2.2 mi; 15 min).*

STOP 3 CASTLE ROCK

Castle Rock, beneath Edinburgh Castle, is a volcanic plug that rises 80 m (262 ft) above the center of the city. The dolerite plug intruded during the Carboniferous, sometime between 340 and 350 Ma, and would have been the feeder pipe for a volcano on that site. Subsequent erosion and glacial scouring removed all other traces of that particular volcano (Edinburgh Geological Society, Castle Rock; Wikipedia, Castle Rock). Surrounded by cliffs on three sides, the only easy way to the summit was by the gentle ramp of sedimentary rock on the east side (the Royal Mile). This is known as "crag and tail" topography, and it is a result of east-flowing glacial ice over the last 2 million years scouring away the easily-eroded bedrock on three sides and leaving the resistant volcanic rock and its sedimentary tail in the lee of the butte (Miller, 2012). The defensive advantages of this site would have been obvious to early settlers, and this explains why the castle was built there.

The section above Johnston Terrace on the south side of Castle Rock contains both Lower Carboniferous Ballagan Formation sandstone/mudstone and the basaltic volcanic neck. The west side of the section is basalt, forming the bulk of Castle Rock. The contact against the Ballagan Formation is a sharp, subvertical fault. Some 50 m (164 ft) east of the contact, the Ballagan Formation beds dip gently to the east. Within about 50 m of the contact with the Castle Hill intrusion, the Ballagan strata are folded into a gentle anticline and dip slightly west. Within 10–20 m (33–66 ft) of the contact, the strata are rotated to near-vertical, either by faulting or by drag along the intrusive margin. Chain link netting has been installed in this steep section of the hill to prevent rockfalls.

The main contact between the Castle Rock basalt intrusion and the Ballagan Formation is a subvertical fault. The fault scarp is well exposed at the base of the outcrop. The main fault runs toward the west end of Half Moon Battery. One part of the fault surface had subhorizontal slickenlines, indicating that at least some strike-slip movement occurred along the fault. The grooved fault plane is not netted and is well visible from Johnston Terrace (Krabbendam and Callaghan, 2012).

Castle Rock basalt plug (beneath the castle) and east-dipping Ballagan Formation sandstone (foreground) above Johnston Terrace.

Castle Rock to Dulnain Bridge: Return south on King's Stable Road to West Port and turn right (southwest); take West Port to East Fountainbridge/A702; bear left onto East Fountainbridge and continue southwest to Semple Street; turn right (north) on Semple Street; continue straight on

B700/Morrison Street to Torpichen Place; turn right (north) onto A8/Torpichen Place; continue straight onto Palmerston Place; continue straight onto Douglas Gardens; bear left (west) onto Belford Road; continue straight onto Queensferry Terrace; at the roundabout take the 1st exit (west) onto A90/Queensferry Road; continue straight (northwest) on A90/Seaforth Terrace; continue straight on A90/Hillview; continue straight on A90/Hillhouse Road; continue straight (west) on A90/Queensferry Road; continue straight (west) onto the M90; continue straight (north) onto the A9; just north of Silverglades, turn right (east) onto the A95 and drive northeast to the turnoff to Carrbridge and Dulnain Bridge; turn left (north) onto the A938 and drive to the pullout on the right (north) side of the road. This is Stop 4, Dulnain Bridge (57.304396,−3.657878), for a total of 221 km (137 mi; 2 h 40 min). Take the path about 50 m (160 ft) west to the display and rôche moutonnée monument.

STOP 4 DULNAIN BRIDGE *RÔCHE MOUTONNÉES*

This stop, in Cairngorms National Park, is all about how glaciers influenced the landscape. Here you can see classic *rôche moutonnées* formed by moving ice during the last Ice Age. In fact, this is where these features were first described and named. The name means "rock wigs." In 1787, the alpine explorer/geologist de Saussure thought that these groups of whale-back rocks resembled the wavy wigs or "*moutonnées*" that men wore at that time, which were slicked down with mutton tallow to keep them in place.

The ice sheet, several hundred meters thick, moved from southwest to northeast, grinding down the underlying bedrock into asymmetric ridges that are gentle where they faced the oncoming ice and steep in the lee of the moving ice. The alignment of the ridges and grooves scoured into the bedrock by the moving ice indicates the direction of ice movement. At this stop, we are along the side of one such ridge, maybe 6 m high and 30 m long (20 ft high and 100 ft long), and another can be seen to the southwest. These features can be up to 400 m (1,312 ft) high, as at Ord Bàn near Aviemore (Cairngorms National Park, *Rôche Moutonnée*, Dulnain Bridge; Wikipedia, Dulnain Bridge).

The bedrock in this area is Neoproterozoic Grantown Formation gneissic slate and Grampian Group metasandstone (BGS, Geology of Britain Viewer "classic").

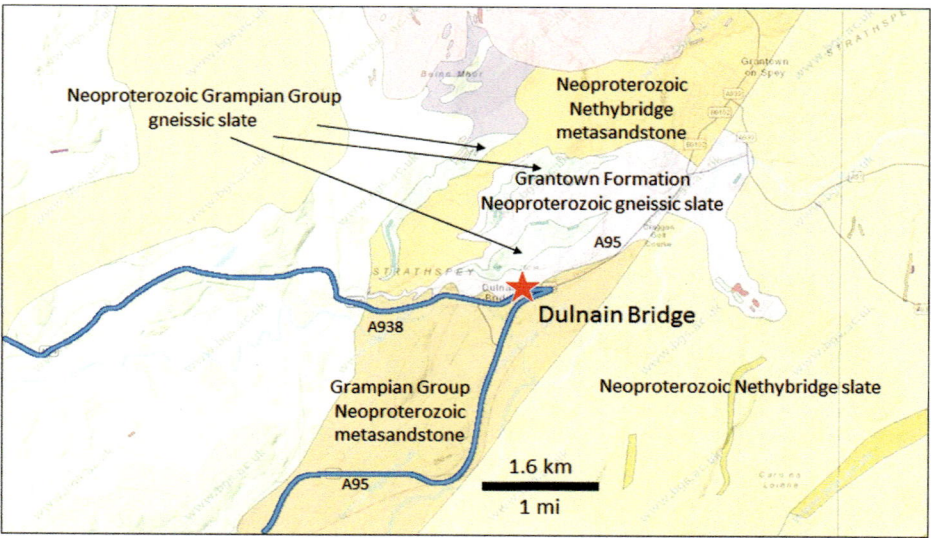

Geologic map of Dulnain Bridge. (Modified after BGS, Geology of Britain Viewer. Contains British Geological Survey materials © UKRI 2022. Base mapping provided by ESRI.)

Rôche Moutonnées, Dulnain Bridge, looking west. Ice movement was from left to right.

*Dulnain Bridge to Loch Ness: Continue west on A938 to the A9; turn right (west) on the A9 and drive to Inverness; turn left (south) on the A82/Longman Road and drive to Tomnahurich Street; turn right (south) on Tomnahurich Street; continue straight on the A82/Glenurquhart Road; cross the River Ness and continue driving southwest on the A82 past Lochend to the parking lot on the left. This is **Stop 5, Loch Ness** (57.39193, −4.36367), for a total of 64.6 km (40.1 mi; 52 min).*

There are a number of turnouts along the road here. Look for the one with the interpretive display panel. It is dangerous to cross the road, so rather than risk it, walk to the lake shore, and examine the pink granite, gneiss, and Old Red Sandstone sandstone and conglomerate boulders there.

STOP 5 LOCH NESS AND THE GREAT GLEN FAULT

The Great Glen is a large linear valley slashing diagonally from southwest to northeast across the Scottish Highlands for a distance of at least 480 km (300 mi). Several lakes lie within the valley, most notably Loch Ness. The valley and lakes are there because of the Great Glen Fault (GGF), a regional strike-slip fault with ancient lineage. The fault divides the Northern Highlands to the west and north from the Grampian Highlands to the south and east. Extending northeast from Inverness, the fault cuts along the northern shore of Moray Firth before curving north to Shetland, where it is called the Walls Boundary Fault. Some workers say there is only left-lateral displacement estimated at 104 km (64 mi; Gazetteer for Scotland, Great Glen Fault), whereas others estimate that there was about 129 km (80 mi) of left-lateral displacement in the Devonian followed by 29 km (18 mi) of right-lateral offset since then, for a net 100 km (62 mi) left-lateral offset (Docherty, 1999). There is a component of normal offset that is down to the southeast. Rocks within the fault zone range from mylonite (recrystallized fault rock formed by ductile deformation) to cataclasite (broken fault rock formed by brittle deformation). They have been extensively altered due to fluid movement along the fault zone (Stewart et al., 1999).

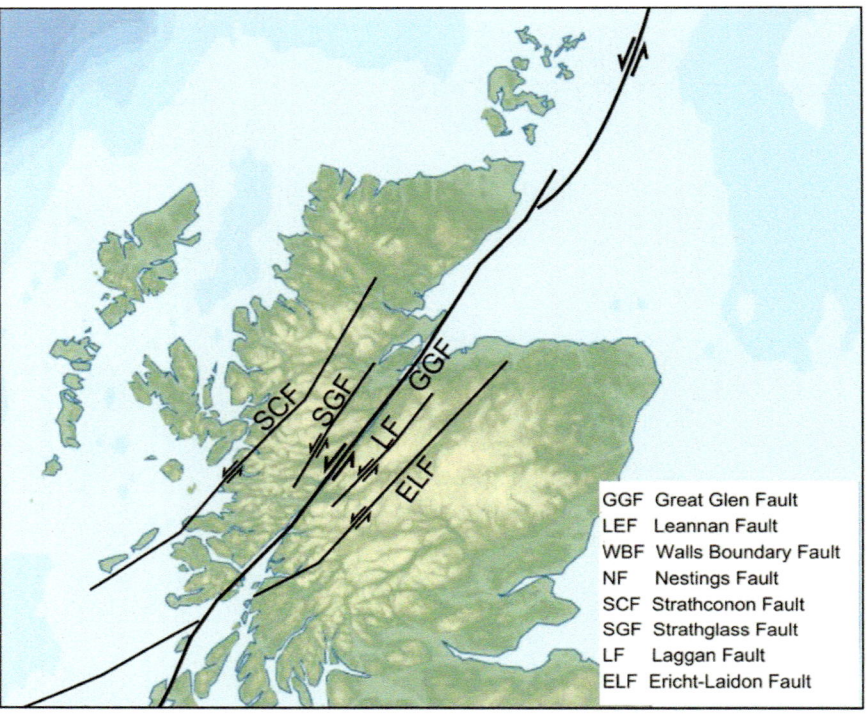

Map of Late Caledonian left-lateral strike-slip faults in Scotland. (Courtesy of Mikenorton, https://commons.wikimedia.org/wiki/File:Great_Glen_Fault_map.png.)

The origin of the Great Glen Fault is shrouded in the mists of time. What we know for sure is that it existed by the end of the Caledonian Orogeny (428–390 Ma) and was active until the Devonian around 370 Ma. The Great Glen Fault Zone resulted from the Middle to Late Silurian oblique collision of the Scottish segment of Laurentia with Baltica during the closing of the Iapetus Ocean. Northwest of the zone is 3.0–1.7 Ga Lewisian basement (originally part of Laurentia); southeast of the zone is 1.8–1.7 Ga Rhinnian/Grenvillian basement (originally part of Avalonia). By Middle Devonian time, the currently exposed levels of the fault zone were at the surface, forming the basement on which the Old Red Sandstone was being deposited (Stewart et al., 1999).

There are indications that the fault was reactivated during the Jurassic, around 170 Ma. Limited movement may still be occurring on the zone, as seen in small-scale seismic activity, truncated glacial scour alignments, and scarplets in moraines, fluvial deposits, and alluvial fans (Stewart et al., 1999; Picardi, 2014; Gazetteer for Scotland, Great Glen Fault).

Fault zone rocks exposed at the surface originally were at depths of about 9–16 km (5-10 mi), indicating that they experienced this amount of uplift and erosion over about 38 Ma of the late Caledonian Orogeny.

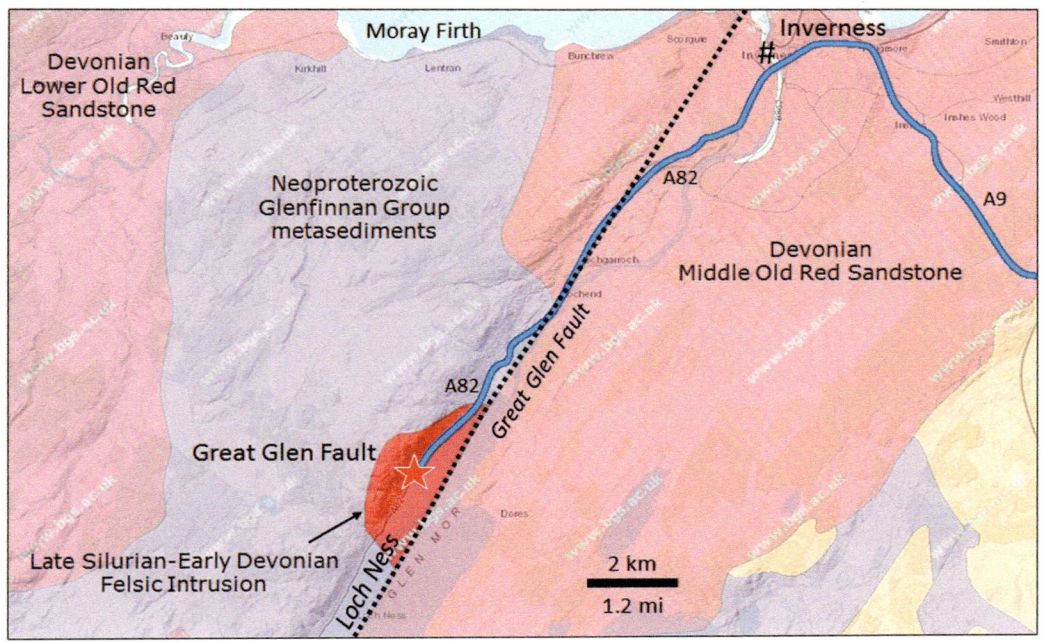

Geologic map of the northeast end of Loch Ness and Inverness area. The felsic intrusion is pink granite. (Modified after BGS, Geology of Britain Viewer. Contains British Geological Survey materials © UKRI 2022. Base mapping provided by ESRI.)

The Great Glen Fault lies along the right (north) shore of long, linear Loch Ness. Boulders of gneiss, granite, and Old Red Sandstone can be seen on the lakeshore. View southwest.

Fractured pink granite along the GGF.

The Great Glen originated during the last Ice Age, between 2 million and 12,000 years ago, as a result of glaciers carving out the softer and shattered rock along the fault zone. Loch Ness is a large lake, 37 km (23 mi) long by 1.6 km (1 mi) wide and averaging 183 m (600 ft) deep. The lake surface is at 15 m (49 ft) elevation; the lake bottom actually extends 183 m (600 ft) below sea level (Docherty, 1999). The valley likely extends below Inverness into Moray Firth, but is filled by glacial sediment that dams the lake. Around the margins of Loch Ness, there are kames (mounds of glacial sand and gravel), eskers (ridges of glacial sand and gravel), and sheets of gravel outwash from glaciers. The River Ness is considered a braided glacial river (Docherty, 1999).

At this location, you cannot touch a fault plane, but you can see a fractured granite and a linear lake that formed along the eroded trace of the fault.

Loch Ness to Loch Na Fiacail*: Return northeast on A82 to Inverness; at the Longman Roundabout, take the 1st exit (northwest) onto A9; just north of Evanton bear left (north) on B9176/Struie Road; continue straight on A836; just north of Lairg turn left (west) onto A838; at Laxford Bridge turn right (north) onto A838 and drive to Stop 6, Lewisian Gneiss, Loch na Fiacail (58.390045, −5.025279) for a total of 153 km (94.8 mi; 1 h 15 min).*

Side Trip 1, Loch Ness to Achanarras Quarry: Return to the A836 and turn left (north); turn right (east) onto A839/The Main Street; drive east on the A839 to the A9; turn left (northeast) on the A9 and drive to Latheron; in Latheron turn left (north) to stay on the A9 and drive to Mybster; turn left (west) on B870 and drive 1.1 km (0.7 mi) to the unpaved Achanarras Quarry Road on the right; turn right (north) and drive to Stop ST1.1, Achanarras Quarry (58.470460, −3.459894) free parking for a total of 110 km (68.5 mi; 1 h 35 min). There is about a 10 min walk north along a well-marked path from the parking area to the quarry.

SIDE TRIP 1 ACHANARRAS QUARRY AND DURNESS

If you have the time, it is well worth the drive to visit this world-famous fossil fish quarry and the Moine Thrust, white-sand beaches, and sea caves at Durness.

ST1.1 ACHANARRAS FISH QUARRY

Scotland's most important fossil sites are protected as Sites of Special Scientific Interest. Achanarras Quarry is one. It is possible to collect fossils from this special site, provided that you adhere to the best practice set out in the Scottish Fossil Code (http://www.scottishgeology. com/where-to-go/fossil-collecting/fossilcode/).

Source of the famous Caithness flags, this quarry is a world-renowned fossil site. The fish fossils are found in the Middle Devonian Achanarras Fish Bed limestone member of the Lybster Flagstone Formation (383–393 Ma), part of the Old Red Sandstone. The units here dip gently to the west and the Fish Bed member extends north–south in a narrow zone.

Opened in 1870 as a quarry for roofing slate, the site has been closed for years. The waste rock, discarded in piles around the quarry, has provided thousands of specimens representing at least 15 species. One of the most common fish fossils is *Dipterus*, ancestor of modern lungfish. Other fish include agnathans, placoderms, acanthodians, and bony fish including ray-finned fish. The fish lived in a large freshwater lake known as Lake Orcadie on the southern margin of Laurussia. When they died or were killed during a natural disaster, they settled to the bottom and were preserved in the accumulating sediment. The bed consists of thin, flaggy limestone and laminated mud deposited when the lake was deepest. Fossils are found on the bedding planes, so the key to finding them is to carefully split the rock along the bedding (Selden and Nudds, 2012; MacFadyen, Achanarras Quarry).

Visit

Within the quarry, there is a stone shelter with interpretive panels. A timeline trail links the car park and quarry. There is no cost to visit the quarry, and no permissions are needed.

Head protection, goggles for eye protection, and gloves are recommended when hammering rock. Please be careful splitting flagstones in case others are nearby. Flying stone chips can be dangerous.

Collectors have permission to remove a maximum of ten specimens per day. A specimen is a piece of rock with fossils. This permission covers a maximum of 7 days.

To help protect the site, collect only from loose material. Spoil heaps on the site are regularly turned over to improve the chances of new finds being made. Appropriate tools include rock hammers and chisels. The use of power tools, sledgehammers, and crowbars is not permitted.

Address: Nature Scotland, The Links, Golspie Business Park, Golspie, Sutherland KW10 6UB
Phone: 0300 067 6841
Web site: https://www.nature.scot/enjoying-outdoors/naturescot-nature-reserves/achanarras-quarry-nature-reserve

Achanarras Quarry.

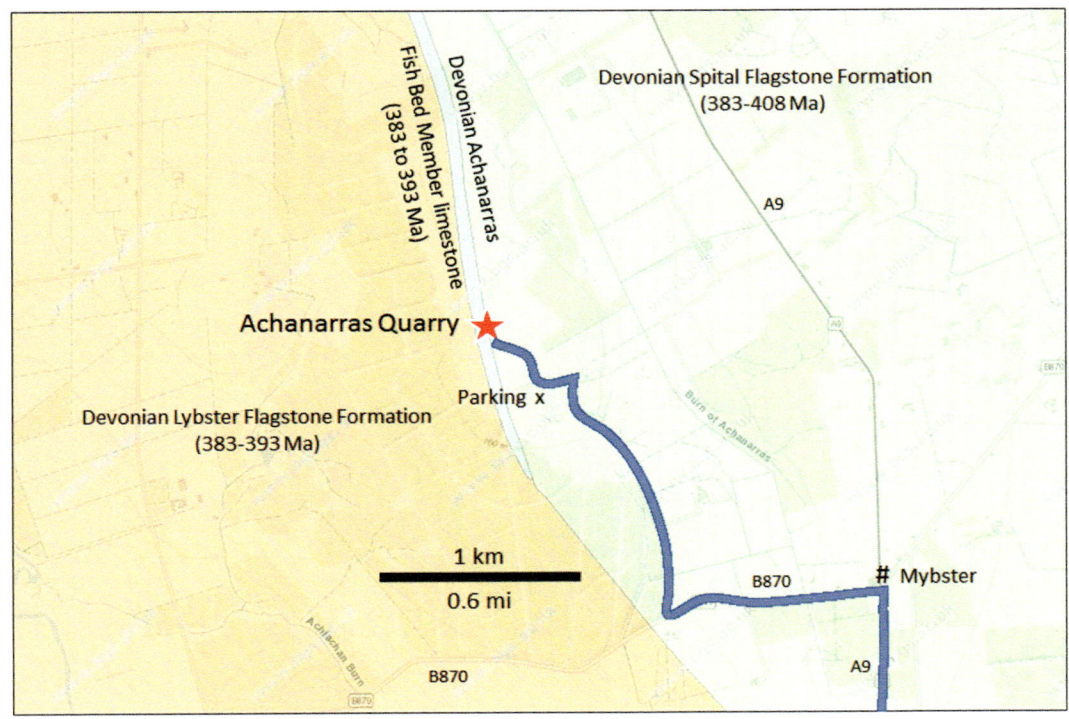

Geologic map of the Achanarras Quarry area. (Modified after BGS, Geology of Britain Viewer. Contains British Geological Survey materials © UKRI 2022. Base mapping provided by ESRI.)

Age		Group/ Subgroup	Formation/ member
Devonian	Givetian	Upper Caithness Flagstone Subgroup	Mey Flagstone
	Eifelian		Spital Flagstone
		Lower Caithness Flagstone Subgroup	Achanarras Member
			Lybster Flagstone
	Emsian	Sarclet Group	

Stratigraphy of the Achanarras Quarry area. (Modified after Newman and Dean, 2005.)

Devonian fish fossil found at the quarry, and an artistic rendering of the original fish. Photo of part of the interpretive display at the Achanarras Quarry.

Side Trip 1, Achanarras Quarry to Smoo Cave: *Return south to B870 and turn right (west); drive west on B870 to the sign for "Westfield 4 mi" and turn left (west); in Westfield turn left (west) and follow the sign to "Reay 8 mi;" at Reay turn left (south) onto A836/Bridge of Isauld; continue* *west on A836 to the A838; turn right (north) on the A838, cross the Kyle of Tongue Bridge, and continue west on A838 to **Stop ST1.2, Smoo Cave** (58.563681, −4.721373) turnout on the right (north) for a total of 126 km (78.4 mi; 2 h 21 min). A short, moderate to steep trail leads into the canyon and to the cave.*

ST1.2 Smoo Cave

Just east of Smoo Cave is one of the most scenic white-sand beaches in Scotland, Ceannabeinne Beach. Take a moment to soak in the out-of-place, almost tropical beauty of this beach.

Ceannabeinne Beach looking northwest. Smoo Cave is just beyond the headland.

Smoo Cave is a combination sea cave and karst feature developed in the Late Cambrian-Early Ordovician Durness Group limestone at the contact between the Sangomore and Sailmhor formations. Sea caves are formed by wave erosion; the karst feature is a cavern dissolved in the dolomite by Alt Smoo Creek after it went underground (Wikipedia, Smoo Cave). Karst landscapes are a result of rainwater dissolution of limestone and dolomite to form sinkholes, caverns, and other features.

The name is thought to derive from the Norse word for a hiding place. Boat tours of the cave are available during the summer.

Visit

Entry to the front chamber is free. A cave tour costs £10 (adult) and £5 (child) and lasts about 20 min. Call 01971 511 492 in the morning (after 9:30 am) to check if tours are operating.

Smoo Cave and the Durness Limestone/dolomite.

Side Trip 1, Smoo Cave to Sango Bay: Continue west on A838 for 780 m (0.5 mi); bear right onto an unnamed road and drive west till encounter A838 once more; turn right on A838 and Stop ST1.3, Sango Bay Visitor Centre parking (58.568103, −4.739648) is immediately to the right for a total of 1.3 km (0.8 mi; 4 min). There is ample "pay-and-display" parking and multiple trails to the beach.

ST1.3 Durness and Sango Bay

Sango Bay combines exposures of the Moine Thrust and north-northeast-trending normal faults that mark the onshore portion of the southern margin of the West Orkney Basin. For this reason, as well as the stunning white-sand beaches, the area is included within the North West Highlands Geopark (Butler, Sango Bay and Smoo Cave).

The southeast end of the bay is defined by a well-exposed normal fault, the Sangobeg Fault, one of several faults dating to the Permian-Triassic development of the West Orkney Basin. The down-to-the-northwest fault, expressed as a breccia zone, drops highly sheared Proterozoic Lewisian gneiss carried here in the hanging wall of the Moine Thrust. The original geologists mapping this rock referred to the crenulated phyllite as "oystershell rock." The upthrown, southeast side of the fault contains Cambro-Ordovician Durness Group dolomite.

A headland in the middle of the bay displays crenulated Lewisian gneiss thrust over Lower Cambrian Quartzite, which in turn lies above Lewisian gneiss; this sequence is separated by the Lochan Riabhach Thrust or the Moine Thrust from the underlying Durness Limestone (Butler, Sango Bay, and Smoo Cave; Holdsworth and Strachan, 2010).

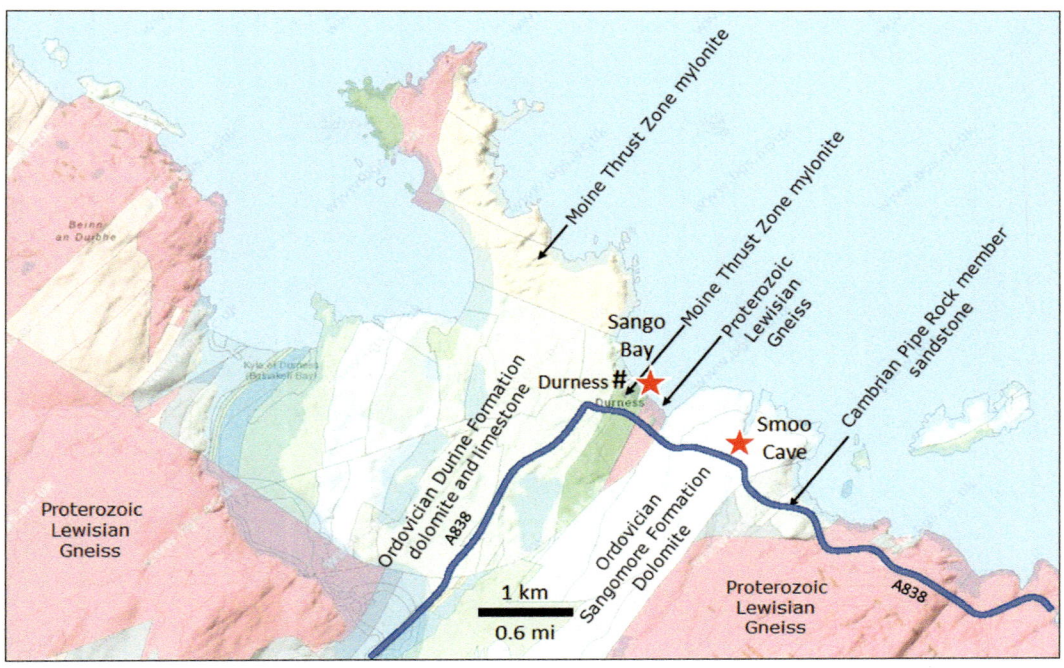

Geologic map of the Durness area. (Modified after BGS, Geology of Britain Viewer. Contains British Geological Survey materials © UKRI 2022. Base mapping provided by ESRI.)

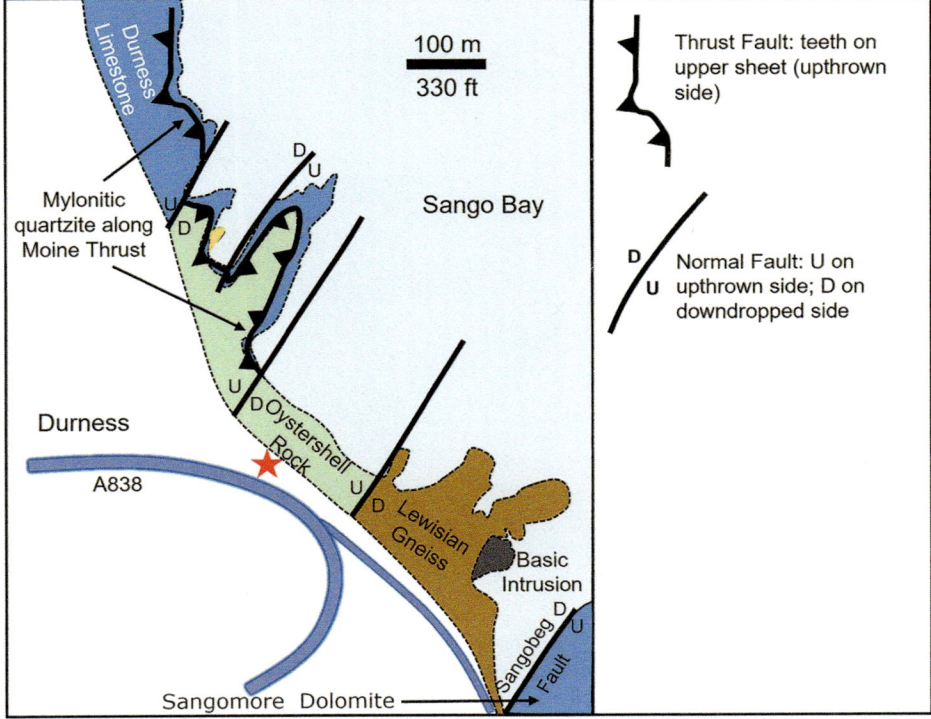

Detailed geology of Sango Bay. (Modified after Holdsworth and Strachan, 2010.)

View northeast to the beach-bounding Sangobeg Fault, down toward the viewer.

View west to the Moine Thrust mylonite zone, Sango Bay.

Continuing north along the beach a small promontory exposes brecciated Durness Limestone over-lain by thrust mylonite, probably derived from Lewisian gneiss. Above this is the Oystershell Rock.

Side Trip 1, Sango Bay to Loch na Fiacail: Continue west and south on A838 for 27.9 km (17.3 mi; 31 min) to Stop 6, Loch na Fiacail (58.390045, −5.025279) parking pullout on the right (west) side of the road.

NORTH WEST HIGHLANDS GEOPARK

You have now entered the North West Highlands Geopark. This is the most sparsely populated corner of Europe and Scotland's first UNESCO European Geopark. Rocks within the park cover almost two-thirds of the Earth's history. The purpose of the park is to promote the local economy, tourism, and geology; to educate the public regarding the geology, history, culture, and economic importance of the area; and to further geologic research in the area.

North West Highlands Geopark is unique as containing the oldest rocks in Europe, the Lewisian gneiss. It contains the Moine Thrust, formed during the collision of ancient continents and the closing of the Iapetus Ocean. This is the first place thrusting was recognized in the United Kingdom, and important research, particularly regarding soil fertility, continues in the area. The Geopark contains the oldest and biggest meteorite impact ejecta in the United Kingdom and rocks that provide evidence of the earliest life on Earth.

Visit

The North West Highlands UNESCO Global Geopark offers one of the best opportunities to explore remote, scenic, and geologically unique landscapes in Europe. The combination of outdoor activities, excellent roads, and educational exhibits is paired with cozy accommodations, roaring fires, and good food. The headquarters at the Rock Stop has interactive displays and a café. The Knockan Crag center has interactive exhibits in a solar-powered, turf-roofed building that fits well into the stunning landscape.

The park staff offer guided geo-tours, geology walks, talks, and Earth Science Week activities, and hosts an annual climbing festival.

Address: North West Highlands Geopark Headquarters, the Rock Stop, Unapool, IV27 2HW
            Also North West Highlands Geopark, Knockan Crag National Nature Reserve, just
            south of Elphin
Phone: 44 (0) 1571 844000
email: info@northwest-highlands-geopark.org.uk
Web site: https://www.nwhgeopark.com/ or www.europeangeoparks.org

STOP 6 LEWISIAN GNEISS, LOCH NA FIACAIL

This is our first stop in the North West Highlands UNESCO Global Geopark. There is an interpretive display in the pullout on the side of the road. This roadcut, known as the "Multi-colored Rock Stop," exposes the Lewisian Complex, ranging in age from 3.0 to 1.7 Ga (BGS, Geology of Britain Viewer "Classic"). As well, you can see cross-cutting relationships and deformation in Lewisian gneiss, the Scourie Dikes, and "younger" (1690–1855 Ma) granite pegmatites (Goodenough and Krabbendam, 2011; Shaw et al., 2016). Pegmatites are very coarse-grained granites. Geologic maps show no other units in the region.

The gray gneiss has well-developed mineral banding dipping gently south. Black gabbro/amphibolite (mainly hornblende) sheets, the Scourie Dikes, cut across the mineral banding at a low angle. On average, the dikes trend about 45° west of north. Both the gneiss and the dikes were again deformed and then intruded by pink granite pegmatites, an igneous rock that cooled slowly, thus forming unusually large mineral crystals. The pegmatites trend about 60° west of north.

The "multi-colored rock stop," Loch na Fiacail. The gray Lewisian gneiss is shot through with dikes (black) and pegmatites (light or pink).

This is a good place to discuss the Lewisian gneiss. The Lewisian is usually divided into Northern, Central, and Southern regions. The Central region comprises the Scourian Complex that formed at 3.0–2.7 Ga. The Northern and Southern regions consist of the Laxfordian Complex that was affected by Laxfordian deformation and metamorphism in three pulses: the 1.86–1.63 Ga main Laxfordian amphibolite-facies metamorphism, with the emplacement of granites and pegmatites at 1.75–1.65 Ga; late Laxfordian metamorphism and deformation at 1.5 Ga; late Laxfordian and/or Grenvillian (1.4–1.1 Ga) brittle deformation and metamorphism (Williams, 2001).

This area contains the older Lewisian Complex. It was originally igneous rocks that were later buried at least 25 km (15 mi) and metamorphosed to orthogneiss, a relatively homogeneous rock with abundant feldspar and quartz, and variable hornblende, biotite, and muscovite. The mineral banding is often extensively folded by later deformation .

Since the gneiss was injected by the black Scourie Dikes, we know that the dikes are younger. The dikes are also slightly metamorphosed, as they contain pink rhodolite garnets, a mineral derived from metamorphism of the mineral biotite. Since the pegmatites cut both the gneiss and dikes, we know it is the youngest of the rocks here. Indeed, the pegmatites were injected during the Paleoproterozoic Laxfordian Orogeny around 1,700 Ma (Hamlet and Harrison, Scourie Bay and Laxford).

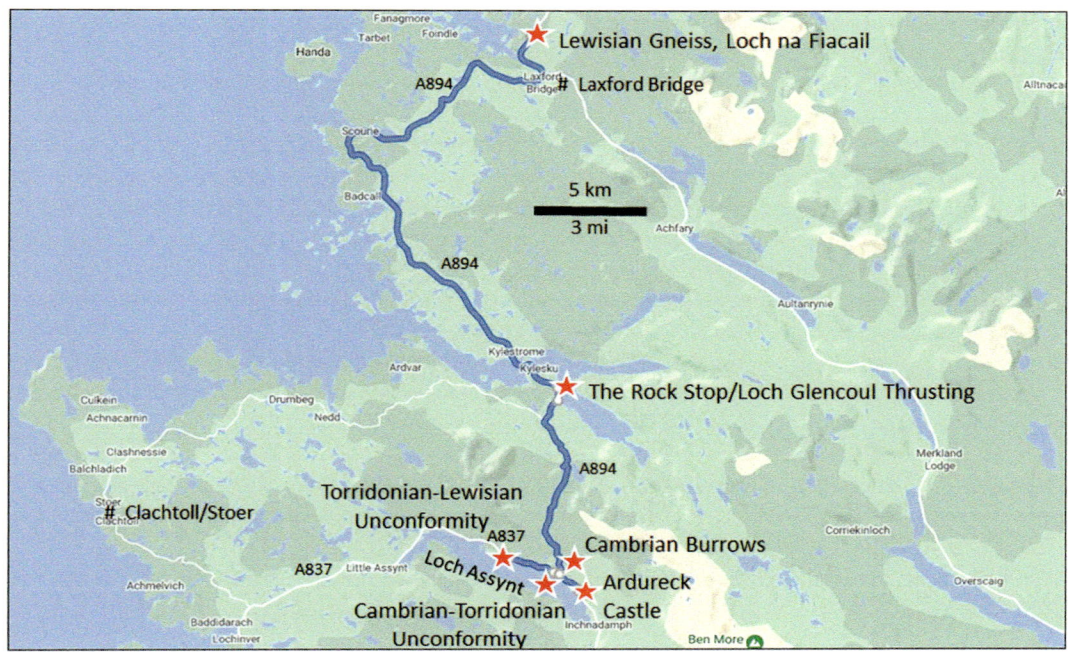

Loch na Fiacail to The Rock Stop: Drive south on A894 for 29.9 km (18.6 mi; 28 min) to **Stop 7, Loch Glencoul** (58.248101, −5.005217).

STOP 7 LOCH GLENCOUL AND THE CONCEPT OF THRUST FAULTING

Beginning in the 1830s, Swiss geologists working in the Alps began to speculate that older rocks above younger rocks were a result of folding and faulting. At about the same time, British geologists started working in the North West Highlands and noticed older Lewisian metamorphic rocks over younger Torridonian and Cambrian sedimentary rocks.

The first geologists to study the sequence here, Roderick Murchison and Archibald Geikie, thought they were a normal sedimentary succession with the oldest rocks to the west and the youngest to the east. They were fooled by the layered nature of the lower part of the Moine rocks, later identified as mylonite (sheared and recrystallized fault zone rock) formed by thrust faulting (Hamlet and Harrison, Knockan Crag NNR). It took a while to resolve this "Highlands Controversy," but in the 1880s, mapping by Charles Lapworth, Ben Peach, and John Horne determined that there was a nearly horizontal fault that put older rocks from the east over the younger rocks in the west (Dryburgh et al., 2014). Horne and Peach documented their findings in the classic *The Geological Structure of the North-West Highlands of Scotland* (1907).

After the British Association for the Advancement of Science met in Dundee in September 1912, geologists from around the world took the opportunity to visit Assynt. After the meeting, 30 delegates set off on a week-long expedition headed by Peach and Horne. During their tour, the party examined the Lewisian, Torridonian, and Cambrian rocks around Loch Assynt and Lochinver, as well as sections in the Moine Thrust Zone at Beinn a Fhuarain, at Knockan Crag, and in the Loch Ailsh/River Oykell area. They also stopped at Loch Glencoul to examine the Glencoul Thrust, and some climbed the base of the Stack of Glencoul to see mylonites associated with the Moine Thrust (Dryburgh et al., 2014). A statue of Horne and Peach was erected at Inchnadamph near the hotel which played a prominent part in the early development of the science of geology (Wikipedia, Geology of Scotland).

The area around Loch Glencoul has a long history of geological investigation and discoveries.

The modest Rock Stop Coffee Shop and Exhibition is the main visitor center for the North West Highlands Geopark and park central headquarters.

Displays explain the geology seen in the area and across the lake. There is a video of the geology of the area, displays of local rocks, and experts to answer questions. The exhibition contains a mountain-building machine, an augmented reality sandbox, and an interpretation of the view across Loch Glencoul. In the interpretation, the Moine Thrust Fault crosses the Stac of Glencoul just below the summit. Below that is an imbrication zone that repeats the quartzite and Lewisian gneiss. A ductile shear zone (mylonite) near the base of the cliff is interpreted as the Moine Thrust itself. The zone below the Glencoul Thrust consists of Fucoid Beds, Salterella Grit, and some Durness Limestone. The Fucoid Beds and limestone weather to neutral soil and good drainage and are covered in grass. This can be seen below the Glencoul Thrust on both sides of the lake. The gneiss above the thrust has heather growing on it (North West Highlands Geopark, Loch Glencoul, Teacher's Sheet).

The café serves tea, coffee, and hot chocolate plus locally made cakes. The gift shop sells books, maps, and a range of locally made arts and crafts.

Address: The Rock Stop, Unapool, IV27 2HW
Phone: +44 1971 488765
Hours: 10:30 a.m.–3:30 p.m. daily
Web site: https://www.nwhgeopark.com/

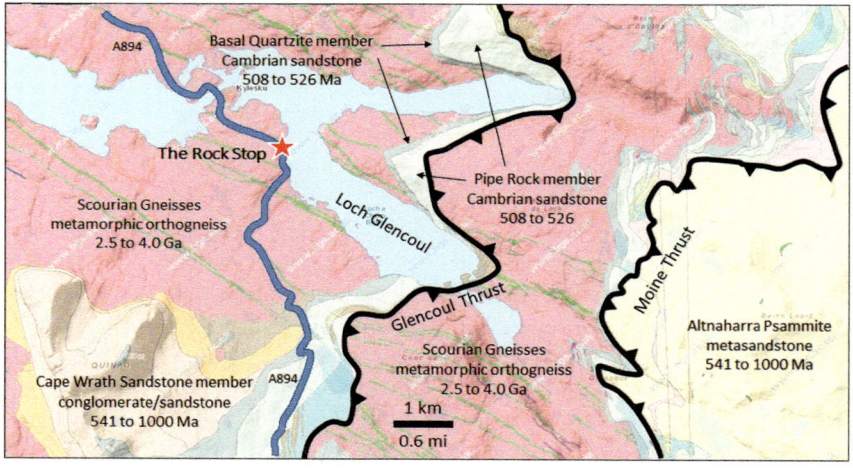

Geologic map of the Loch Glencoul area. (Modified after BGS, Geology of Britain Viewer. Contains British Geological Survey materials © UKRI 2022. Base mapping provided by ESRI.)

On the east side of a glacier-scoured lake is Beinn ard da loch, one of the most important sites in British geology. Geology students come here from around the country, and from around the world, to photograph the Glencoul and Moine thrusts. Beinn ard da loch consists of east-dipping Cambrian Quartzite and conglomerate over a Lewisian gneiss basement. Above the Cambrian, the east-inclined Glencoul Thrust carries Lewisian gneiss westward over the Cambrian units (Hamlet and Harrison, Loch Glencoul). The Glencoul Thrust, which is part of the Ben More Thrust system, lies below the Moine Thrust. It crosses the valley further west, just above the Cambrian Basal Quartzite and Pipe Rock members. The Glencoul Thrust sheet has moved 20–30 km (12–18 mi) west, much less than the Moine Thrust. Underneath the main thrust one often encounters a zone with multiple minor thrusts called an imbricate thrust stack. This stacking up of thrust slices repeats the same rock sequence multiple times. This is what happened with the Ben More Thrust system.

The major Moine Thrust marks the western limit of the Moine Schist. The lowermost part of the Moine Schist has usually been deformed as a ductile shear zone, turning the rock into a mylonite (ground-up and recrystallized rock). This thrusting may have occurred over only 1 or 2 million years during the mountain building 430 million years ago (North West Highlands Geopark, Loch Glencoul, Teacher's Sheet). This thrusting is the result of the Caledonian Orogeny, the mountain-building event associated with closure of the Iapetus Ocean. The Glencoul Thrust was first recognized by Charles Callaway in the early 1880s (Hamlet and Harrison, Loch Glencoul).

Middle Ordovician		Durine Formation dolomites
Early Ordovician	Durness Group	Croisaphuill Formation dolomitic limestones
		Balnakeil Formation limestones
		Sangomore Formation limestones
		Sailmhor Formation dolomites
Cambrian		Eilean Dubh Formation "Durness Limestone"
		Ghrudaidh Formation "Durness Dolomite"
	Ardvreck Group, ~540-500 Ma	Salterella Grit member, An t'Sròn Fm
		Fucoid Beds member, An t'Sròn Fm
		Pipe Rock member, Eriboll Formation sandstones
		Basal Quartzite member, Eriboll Formation
Neoproterozoic	Torridon Group, ~1,000 -541 Ma	Applecross Formation sandstones
		Diabag Formation sandstones
Archean	Stoer Gp ~1,200 Ma	Meall Dearg Formation mudstones
		Bay of Stoer Formation sandstones
		Clachtol Formation sandstones
		Laxfordian gneiss & granite, 1750 Ma
		Scourian gneiss & mafic dikes, 2,700 Ma

Stratigraphic column for the Loch Glencoul-Loch Assynt area. (Modified after Goodenough and Krabbendam, 2011; Waters, 2003.)

Loch Glencoul and Glencoul Thrust. View east to Beinn Aird da Loch.

Loch Glencoul Thrusts to Cambrian Burrows: Continue driving south on A894 for 9.3 km (5.8 mi; 10 min) to Stop 8.1, Cambrian Burrows (58.173194, −5.004017). Turn right (west) onto A837 and pull over on the left side of the road. Walk back to the intersection outcrop (58.173485, −5.002670). Be mindful of traffic!

STOP 8 LOCH ASSYNT

Here, in the foreland of the Moine Thrust, two unconformities can be seen in close proximity on the north shore of Loch Assynt, and the rock section can be seen in its proper stratigraphic, bottom-to-top, old-to-young sequence.

The loch is an elongated lake created by northwestward movement and scouring of bedrock by glaciers during the last Ice Age.

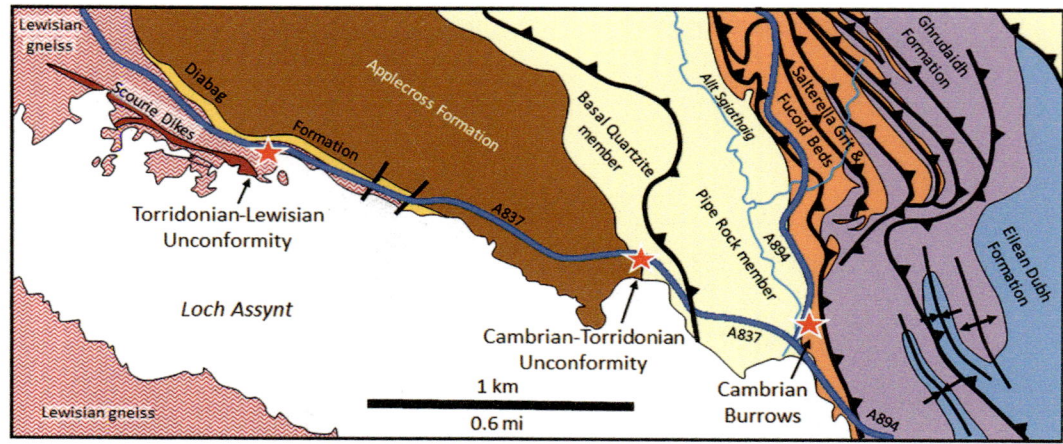

Simplified geologic map, Loch Assynt area. (Modified after Goodenough and Krabbendam, 2011.)

STOP 8.1 CAMBRIAN BURROWS

Just north of the Skiag Bridge (70 m, or 225 ft) is an outcrop containing some of the most photo-graphed trace fossils in the United Kingdom. It is called Pipe Rock because of the vertical tube-like burrows found in it. The normally white Cambrian Pipe Rock member of the Eriboll Formation is here stained pink by iron oxide, but the vertical *Skolithos* burrows stand out as bright white. The *Skolithos* burrows were made by worm-like organisms in a high-energy coastal to shallow marine environment around 517 Ma (BGS, Geology of Britain Viewer "Classic"). The Pipe Rock member is a 130 m (426 ft) thick, coarse to fine-grained quartz-rich sandstone with crossbed sets separated by thin green mudstone layers (Goodenough and Krabbendam, 2011; Harrison, 2017).

Burrowed outcrop of the Cambrian Pipe Rock member. View east.

Close-up of the burrows (vertical light streaks) that characterize the Cambrian Pipe Rock member of the Basal Quartzite. 1£ coin for scale.

Cambrian Burrows to Cambrian-Torridonian Unconformity: *Drive 240 m (0.1 mi; 1 min) west on A837 to **Stop 8.2**, **Cambrian-Torridonian Unconformity** (58.173604, −5.007946) and pull over on the left (south) just before the guardrail. Walk west to view the outcrop and unconformity.*

STOP 8.2 CAMBRIAN-TORRIDONIAN UNCONFORMITY

This outcrop exposes the Cambrian Basal Quartzite member of the Eriboll Formation (508–526 Ma). The youngest beds, at the east end of the roadcut, contain faint *Skolithos* burrows, indicating that the Basal Quartzite–Pipe Rock member contact lies only a few meters above (east of) this outcrop. The west end of the outcrops is just above the base Cambrian unconformity. The unconformity itself is covered by soil and vegetation.

The Basal Quartzite is a 120 m (394 ft)-thick light gray to pink, medium to coarse-grained crossbedded and highly-jointed metasandstone. The rock is 90% quartz with a hematite grain coating that causes the pink color (Harrison, 2017). This quartzite was deposited as a beach, barrier bar, or shallow marine sandstone.

Below the unconformity, about 240 m (780 ft) to the west, lies the Torridonian Applecross Formation sandstone (977–750 Ma). The Applecross Formation was deposited in a fluvial setting that includes channel, floodplain, and levee sandstones (Holmes, 2019; BGS, Geology of Britain Viewer "Classic").

We don't know how thick the Applecross is here. The dominantly quartz sandstone lies conformably above the Diabaig Formation and is in angular unconformity with the overlying Cambrian. It is orange-beige, with the hematite cement providing the color (Harrison, 2017). Applecross outcrops along the road contain glacial striations (scratches) trending 113° that indicate ice moving westnorthwest through the valley. This location also provides a good view northwest to the summit of Spidean Coinich (764 m, or 2,507 ft). Spidean Coinich has white, well-bedded quartz sandstone of the Eriboll Formation dipping ~15° east above the essentially horizontal Applecross Formation. This is the same unconformity as seen here at the road (Goodenough and Krabbendam, 2011).

The basal Cambrian Quartzite lies unconformably over the Applecross Formation. View west.

The unconformity, covered at the road, spans a minimum of 15 million years.

*Cambrian-Torridonian Unconformity to Torridonian-Lewisian Unconformity: Continue driving west on A837 for 2.1 km (1.3 mi; 3 min) to **Stop 8.3, Torridonian-Lewisian Unconformity** (58.179046, −5.041044) and use the pullout on the left.*

Stop 8.3 Torridonian-Lewisian Unconformity

Examine the deeply weathered, pale green to cream-colored Lewisian gneiss in the roadcut at the parking area. The Lewisian is a pyroxene-bearing felsic gneiss. The Diabag, part of the Torridonian Sequence, overlies the gneiss across a major unconformity. The Lewisian here is roughly 3.0 Ga; the Scourie Dikes were intruded between 2.4 and 2.0 Ga; and the Diabaig was deposited around 1.0 Ga. Thus, roughly 2 billion years of rock, about 40% of Earth's history, are missing at the unconformity.

The Torridonian Diabaig Formation sandstone unconformably overlies Lewisian gneiss in this roadcut near the "falling rocks" road sign. Dotted line indicates the unconformity. View east.

From the pullout walk south to the shore of Loch Assynt. Alongside the bay is a 9 m (30 ft) wide, east-southeast-trending ultramafic (dark mineral-bearing) dike. This is one of the Scourie Dikes that were injected into the Lewisian gneiss. The core of the dike is olivine-rich.

Walk east from the pullout and inspect the Diabaig Formation. The Diabaig Formation is a 25–30 m (82–100 ft) thick, dark-brown sandstone, coarse pebbly sandstone, and conglomerate. The conglomerate contains gneiss clasts in a quartz and feldspar matrix, with the color coming from hematite. The sandstone was deposited as fluvial channel, alluvial fan, floodplain, and estuary sediments (Goodenough and Krabbendam, 2011; Harrison, 2017).

Near the "falling rocks" road sign (past the guardrail), you can see Diabaig Formation bedding truncated against a topographic high on the Lewisian gneiss unconformity. The Lewisian gneiss is highly weathered to at least 10 m (33 ft) beneath the unconformity. The variable thickness of the Diabaig is a function of sedimentation on the irregular unconformity surface. You can see the topography developed on top of the Lewisian gneiss south of Loch Assynt as well. Around Quinag, 3.3 km (2 mi) north of Loch Assynt, there is up to 400 m (1,312 ft) of relief on the unconformity.

Continuing east, you come to thick, trough cross-bedded sandstones of the Applecross Formation lying conformably above the thin-bedded Diabaig Formation (Goodenough and Krabbendam, 2011). The Applecross Formation, the main unit in the Torridonian Sequence, is a pebbly sandstone. The Torridonian sediments all dip gently east as a result of tectonic loading by the Moine Thrust (Harrison, 2017).

*Torridonian-Lewisian Unconformity to Ardvreck Castle: Return east on A837 to A894 and turn right (south); drive south on A894 to **Stop 10, Ardvreck Castle** (58.165893, −4.988779) and use the parking area on the right.*

*Side Trip 2, Torridonian-Lewisian Unconformity to Stac Fada: Continue driving west on A837 to the B869 and turn right (northwest); take B869 through Clachtoll to the Stoer Lodge Holiday Cottage on the left (58.200538, −5.338057) for a total of 22.6 km (14.0 mi; 26 min). Park and ask permission to pass, then walk west 700 m (2,300 ft) along the coast to **Stop ST2.1, Stac Fada Impact Breccia** (58.201512, −5.348890).*

SIDE TRIP 2 STAC FADA AND TORRIDONIAN, STOER AND CLACHTOLL

A series of stops in and around Stoer and Clachtoll take us to the Stac Fada impact breccia, Lewisian-Torridonian angular unconformity, and the lower part of the Torridonian succession.

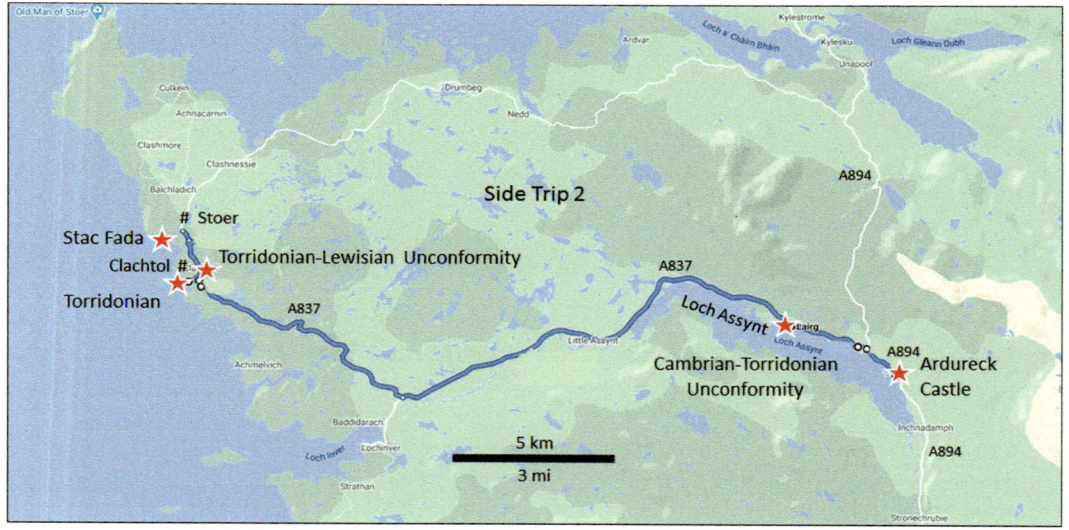

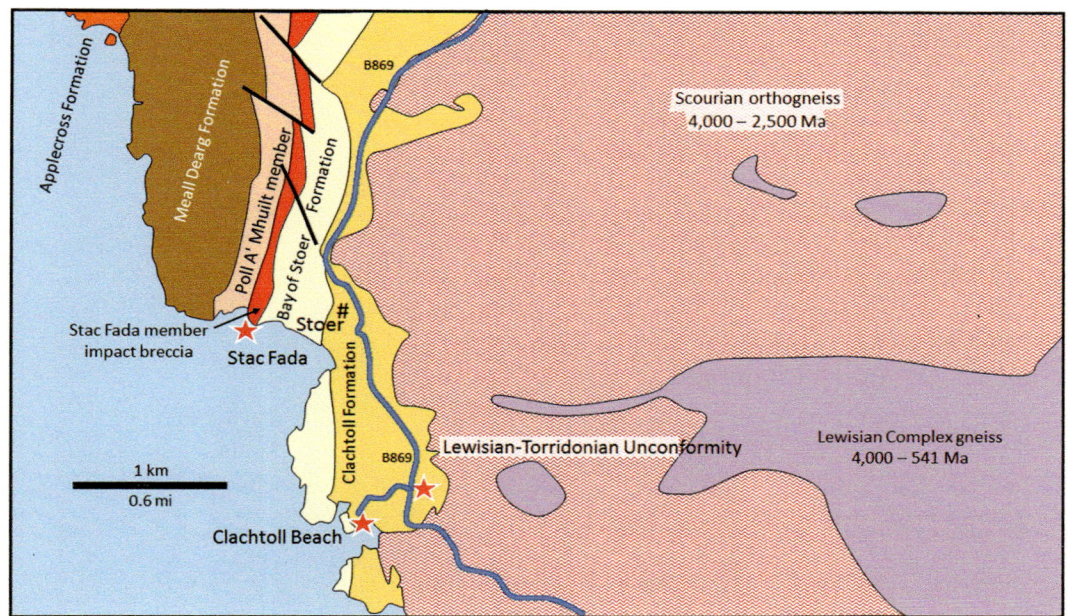

Geologic map of the Stoer area. (Modified after BGS, Geology of Britain Viewer. Contains British Geological Survey materials © UKRI 2022.)

The type section for the Stoer Group occurs at Stoer. The Stoer Group is the lowest division of the Torridonian Supergroup. It consists, from base to top, of the Clachtoll Formation sandstone, the Bay of Stoer Formation sandstone, and the Meall Dearg Formation (mostly mudstone). Below it is the Lewisian gneiss; above it are the Sleat and then Torridon groups (Johnstone and Mykura, 1989).

ST2.1 Stac Fada Impact Breccia

The curious reddish rock with bits of black rock fragments found on the beach here (and in the gentle ridge extending north) were previously thought to be part of a lahar deposit, a volcanic mudflow. More recently, they are described as an impact breccia formed when a meteorite slammed into the Earth around 1.2 billion years ago. One possible site for the crater is beneath Lairg, although this is controversial (BBC, 2016; Simms and Ernstson, 2019; Butler and Alsop, 2019; Amor et al., 2020).

The red material is sandstone rubble; the black fragments are glassy bits of rock molten by the impact; light-colored inclusions are lumps of Lewisian basement.

Google Earth image of the Stac Fada outcrop, Stoer area. (Image © 2022 CNES/Airbus.)

Stac Fada member of the Clachtoll Formation, Stoer Group. This is thought to be an impact breccia. (Photo courtesy of Callan Bentley. From Bentley, 2017a.)

Side Trip 2, Stac Fada to Clachtoll Beach: *Return south on B869 to an unnamed road on the right that leads to Clachtoll Beach; turn right (west) and drive to* **Stop ST2.2, Clachtoll Beach** *(58.190301, −5.337506) parking area for a total of 1.9 km (1.2 mi; 7 min). There is ample parking and facilities.*

ST2.2 Clachtoll Beach Torridonian Section

This stop contains Torridonian sediments dipping ~20° west above the Lewisian gneiss uplift to the east. Local legend has it that the split rock across the bay slipped down the bedding plane one night. What do you think happened?

Should you choose to walk around the bay and examine these rocks, keep an eye out for stromatolites in thin mudstone beds between thicker sandstone layers. These are thin mats caused by cyanobacteria (formerly called blue-green algae), the dominant life form on Earth 1.2 billion years ago. They grew, one layer above another, in clear shallow water along the margins of lakes. There are five or six separate horizons that show mud cracks as if the cyanobacteria were growing as the mud was deposited and then dried out (North West Highlands Geopark, Clachtoll, Teacher's Sheet).

The name Clachtoll is derived from the Gaelic for "hole in the rock" and refers to a previous natural arch that has since collapsed. This is how the "split rock" formed.

Torridonian sediments dipping seaward. View south.

Side Trip 2, Clachtoll Beach to Lewisian-Torridonian Unconformity: *Return east to B869 and turn right (south); drive 60 m (200 ft) and pull over by the gate on the left. This is* **Stop ST2.3, Lewisian-Torridonian Unconformity** *(58.190578, −5.332388) for a total of 400 m (0.3 mi; 2 min).*

ST2.3 Clachtoll Lewisian-Torridonian Unconformity

This stop contains Torridonian sediments dipping west off the Lewisian gneiss highland to the east. The outcrop is on private land, so you will have to view it from the road unless you can get permission from the landowner.

Torridonian sediments sitting unconformably over Lewisian gneiss. View north.

This stop provides a view of the 1 Ga Torridonian sediments dipping gently to the west over an angular unconformity surface above the 1–4 billion-year-old Lewisian gneiss. The unconformity surface is irregular, with relief up to 600 m (2,000 ft).

Non-marine Torridonian sediments were deposited on the Lewisian gneiss along the eastern margin of Laurentia (Williams, 2001; Brasier et al., 2017). Some workers speculate that alluvial fan and braided stream channel deposition occurred in fault-bounded rift basins during a period of extension related to the opening of the Iapetus Ocean in Neoproterozoic time (dates range from 1.2 Ga to 977 Ma; Kinnaird et al., 2007). Other workers interpret the units as having been deposited on a floodplain far distant from the sediment source during a period of widespread subsidence (Williams, 2001).

'Torridonian' refers to the entire Precambrian sedimentary succession exposed along the north-west coast and islands of Scotland. The Torridonian Supergroup is the least deformed and metamorphosed in the North West Highlands, and comprises the Stoer, Sleat, and Torridon groups (Kinnaird et al., 2007). The entire sequence is around 11.5 km (7.1 mi) thick and was deposited between 1 Ga and 750 Ma (Williams, 2001).

Side Trip 2, Lewisian-Torridonian Unconformity to Ardvreck Castle: *Return east on B869; turn left (east) on A837 and drive to **Stop 9, Ardvreck Castle** (58.165893, −4.988779) parking area on the right for a total of 24.7 km (15.5 mi; 30 min).*

STOP 9 ARDVRECK CASTLE AND DEFORMED CAMBRIAN

At this stop, you can observe folding in the Cambrian Durness Group and a number of stacked thrusts. The Ghrudaidh and Eilean Dubh formation dolomites are folded into anticlines and synclines that verge west, indicating a push from the east. These are overthrust by the older Cambrian Pipe Rock member (Eriboll Formation), a fine- to coarse-grained, quartz sandstone with abundant *Skolithos* burrows. A thrust above the Pipe Rock carries the Cambrian Basal Quartzite.

Oh yeah, there are also the ruins of a 16th-century castle. Ardvreck Castle was built around 1590 by the MacLeod clan, a family that had owned Assynt and the surrounding area since the 13th century. The castle walls consist of local stones of Lewisian gneiss, Torridonian sandstone, and Durness Limestone. From the castle walls, you can see west-verging (west-leaning) folds in the Cambrian Durness Limestone. Interestingly, the Ardvreck and Durness groups in this area have the same sequence of Cambrian to Ordovician sandstone-shale-limestone as can be seen in the Appalachians or the Grand Canyon, which tells us that all of these areas were on the margins of Laurentia during a time of rising sea level (Bentley, 2017b).

According to local legend, the MacLeods got Clootie (the Devil) to help them build Ardvreck Castle. In return, Clootie got Eimhir, daughter of the clan chief, in marriage. Upon learning of this arrangement, the girl threw herself into Loch Assynt from the castle tower. She made a new home beneath the water's surface, becoming the elusive Mermaid of Assynt. A variation of the myth has Eimhir hiding from the devil in caverns around the Loch. The legend was used by locals to explain high water years in the loch. The high water level was a result of Eimhir's tears as she mourned her lost life above the water. Some claim to have seen her weeping on the rocks, half woman, half sea creature (Murray, 2012).

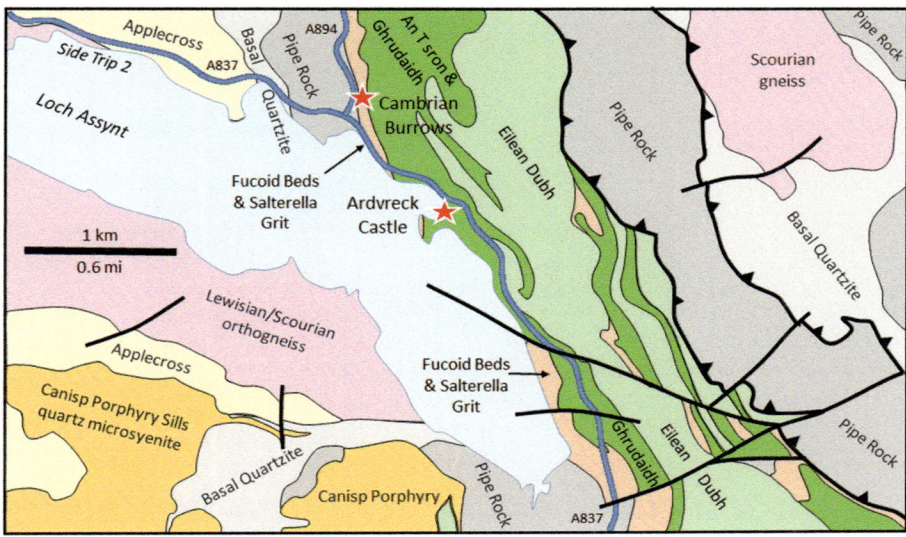

Geologic map of the Ardvreck Castle area. Contains British Geological Survey materials © UKRI 2021.

West-verging folded Cambrian Eilean Dubh Formation (Durness Group) dolomites at Ardvreck Castle. View south across Loch Assynt.

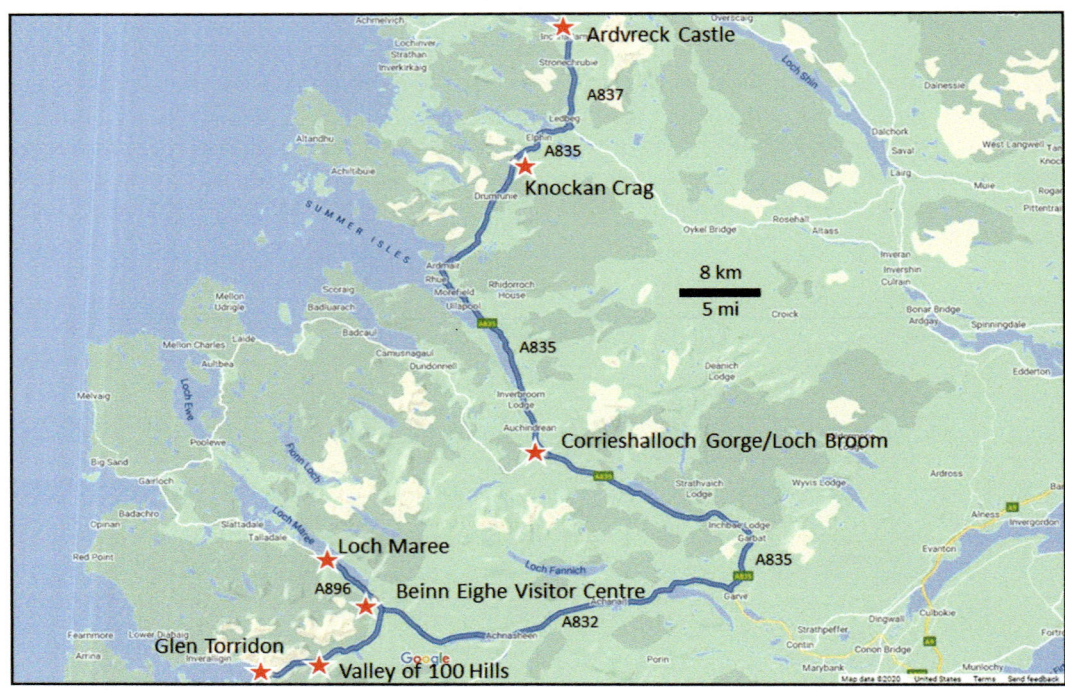

Ardvreck Castle to Knockan Crag: *Continue driving south on A837 to A835; turn right (southwest) onto A835 and drive to **Stop 10, Knockan Crag** (58.033783, −5.070960) parking area on the left (east) for a total of 20.0 km (12.4 mi; 18 min). There is a large car park with restrooms and interpretive displays. Take the north path to the thrust.*

STOP 10 KNOCKAN CRAG AND THE MOINE THRUST

Knockan Crag is a 90 m (300 ft) cliff directly above the A835. The site is owned by Scottish Natural Heritage (SNH) and was declared a National Nature Reserve (NNR) in 2004. Today, it is part of the larger Knockan Cliff SSSI. As the main feature of the Crag is geological, a series of trails take visitors to the main rock exposures. The NNR lies within the North West Highlands Geopark, part of the European Geopark Network.

The first geologists to study the North West Highlands found it difficult to agree on the rock sequence and geological structure of the region. Particularly controversial was the observation that the Moine rocks are found stratigraphically above Cambro-Ordovician sedimentary rocks. Since the geological Law of Superposition states that the oldest rocks are always at the base of the section, it didn't make sense that these metamorphic rocks were at the top of the sequence, apparently the youngest rocks in the North West Highlands. Their metamorphism implied that they had been deeply buried. It was difficult to understand how they could have been metamorphosed, while the underlying, apparently older rocks had not been.

A particularly good place to see this seemingly continuous sequence from unmetamorphosed Durness Group rocks up into metamorphosed Moine rocks is at Knockan Crag. For most of the 1800s geologists struggled to explain this geological enigma. As mentioned earlier, Swiss geologists working in the Alps in the 1830s had encountered this situation and determined that the older rocks had either been folded or thrust over younger rocks. By the 1880s, British geologists had come to understand that the older rocks had been transported from their place of origin to what is now the North West Highlands. It turned out that the Moine rocks had formed many kilometers to the east and had been pushed westwards, while still several kilometers deep, along gently inclined surfaces called thrusts. The main thrust is known as the Moine Thrust, and it can be traced from Loch Eriboll to Skye. Several more large thrusts and many small thrusts were also mapped. This "Zone of Complication," as the early geologists called it, is now referred to as the Moine Thrust Zone (Dryburgh et al., 2014). At Knockan Crag, the Ordovician Durness Group dolomite/limestone is found below the Moine Thrust, whereas the much older Neoproterozoic Morar Group of the Moine Supergroup is found above the thrust.

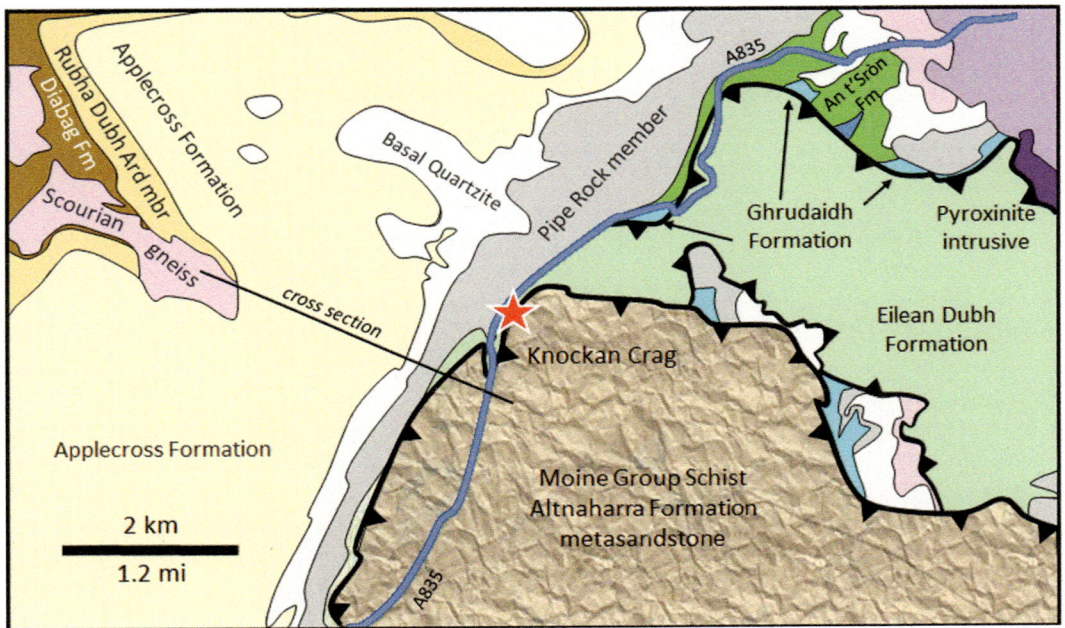

Geologic map of the Knockan Crag area. Contains British Geological Survey materials © UKRI 2021.

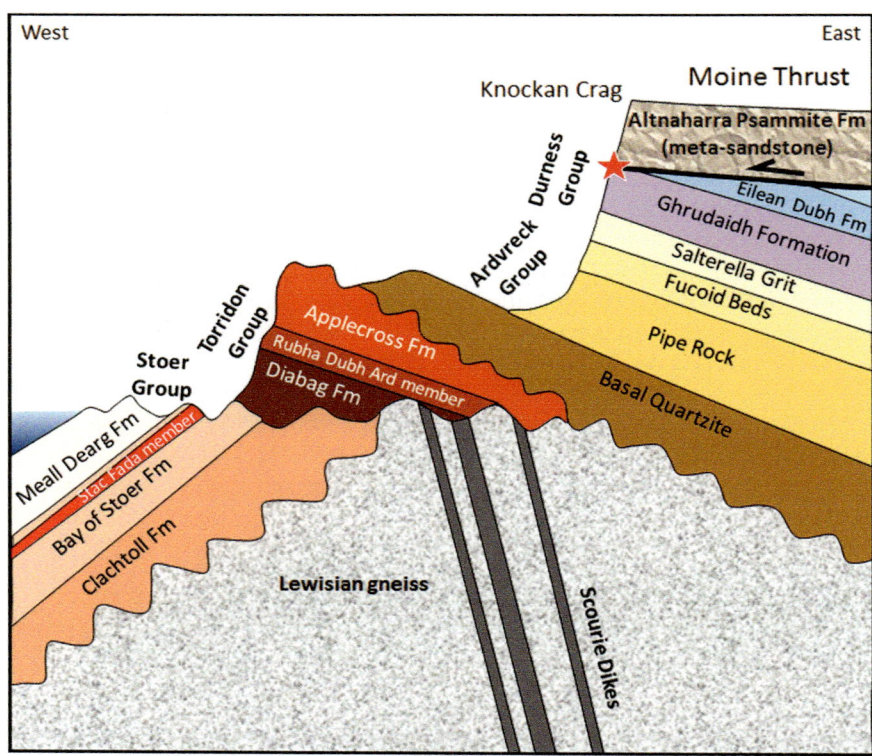

Schematic cross-section through Knockan Crag. (Modified after Bentley (and Cantner), 2017b, American Geosciences Institute.)

The forces required to accomplish this thrusting occurred during the later stages of the Caledonian Orogeny. The driving force was the collision of the Laurentian and Avalonian tectonic plates, thereby closing the Iapetus Ocean during Silurian time. The thrusts suggest tens to hundreds of kilometers of crustal shortening. Dating of crushed and sheared rock in the fault zone indicates that the Moine Thrust Zone was active between about 437 and 408 Ma (Stewart et al., 1999).

At Knockan Crag you can actually touch the thrust surface, a thin zone of crushed rock that places billion-year-old metamorphic rock, formed in the depths of the Caledonian mountain range, over 500 million-year-old limestone.

More than the Moine Thrust is exposed at Knockan Crag: most of the main rock types of the North West Highlands are also represented here. The oldest rock at Knockan Crag is the Cambrian Pipe Rock member (Eribol Formation). It is about 100 m of burrowed white shoreline sandstone. Above it are the Fucoid Beds, 12–27 m (39–89 ft) of mudstone deposited in deeper water. The Victorian geologists who named it thought the unit consisted of compressed Fucus seaweed. In fact, they were seeing the tracks of animals such as trilobites. A 10 m (30 ft) thick layer of Salterella Grit was deposited above the Fucoid Beds. The Grit contains salterella fossils, small shells usually less than 1 cm (0.4 in) long deposited in the intertidal zone. The youngest rock, at the top of the sedimentary section, is a limestone named after the village of Durness. The limestone was deposited on the seabed as thick ooze. The rock at the top of the Knockan Crag section is dark Moine Schist. Sediments that became the Moine Schist were laid down about 1,000 Ma as layers of mudstone and sandstone at the bottom of the Iapetus Ocean. The sediments were then metamorphosed by heat and pressure into schists as the Iapetus Ocean closed around 425 million years ago and the collision of continental plates created a large mountain range. One of the last events associated with the closing of the Iapetus Ocean was the thrust faulting that pushed the Moine Schist up and over the younger rocks at Knockan Crag (Breckenridge, 2007).

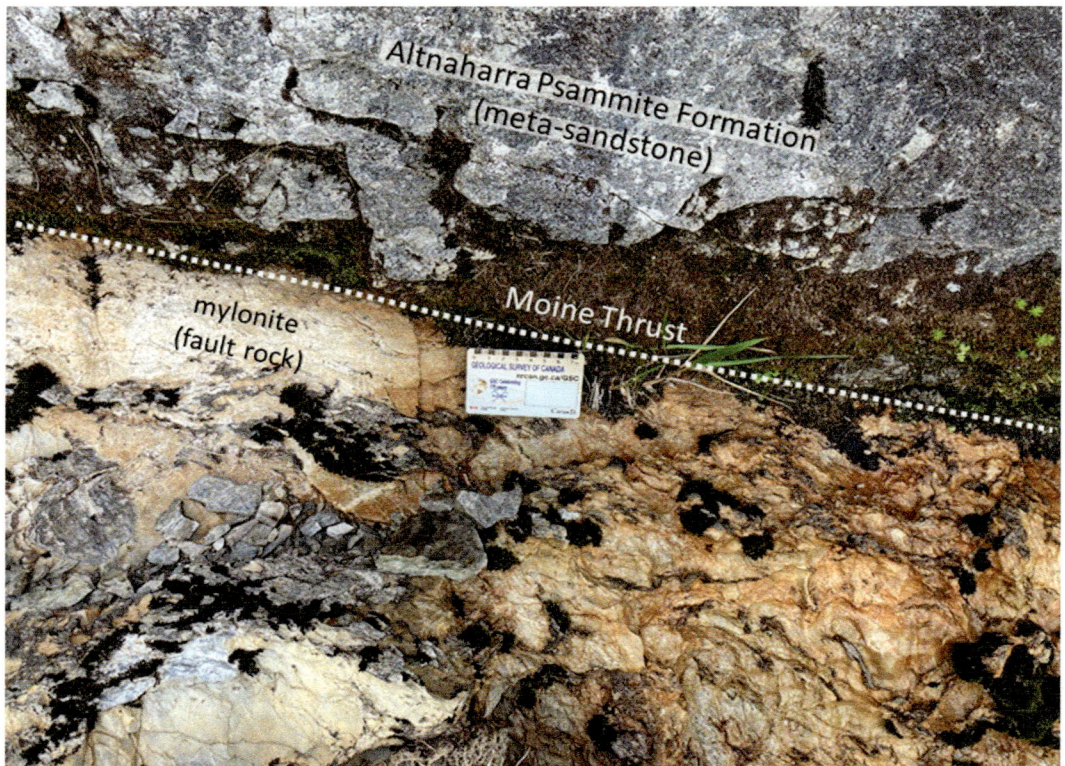

Moine Thrust plane, Knockan Crag. Centimeter scale shown at the fault.

*Knockan Crag to Corrieshalloch Gorge: Continue driving south on A835 to A832/Corrieshalloch Gorge turnoff on the right; drive 900 m (0.6 mi) to **Stop 11.1, Corrieshalloch Gorge** (57.755093, −5.022936) parking on the right for a total of 40.2 km (25.0 mi; 35 min). A short walk on a good path takes you to the Falls of Measach.*

STOP 11 LOCH BROOM

Loch Broom is a northwest-oriented glacial fjord located in a wide valley carved by glaciers. Several narrow, deep gorges feed small streams into it. Corrieshalloch is the main gorge, at the southeast limit of Loch Broom Valley. It was formed at the end of the last Ice Age by a meltwater torrent carrying glacial debris. The current river is much smaller, especially now that there is a hydroelectric project upstream.

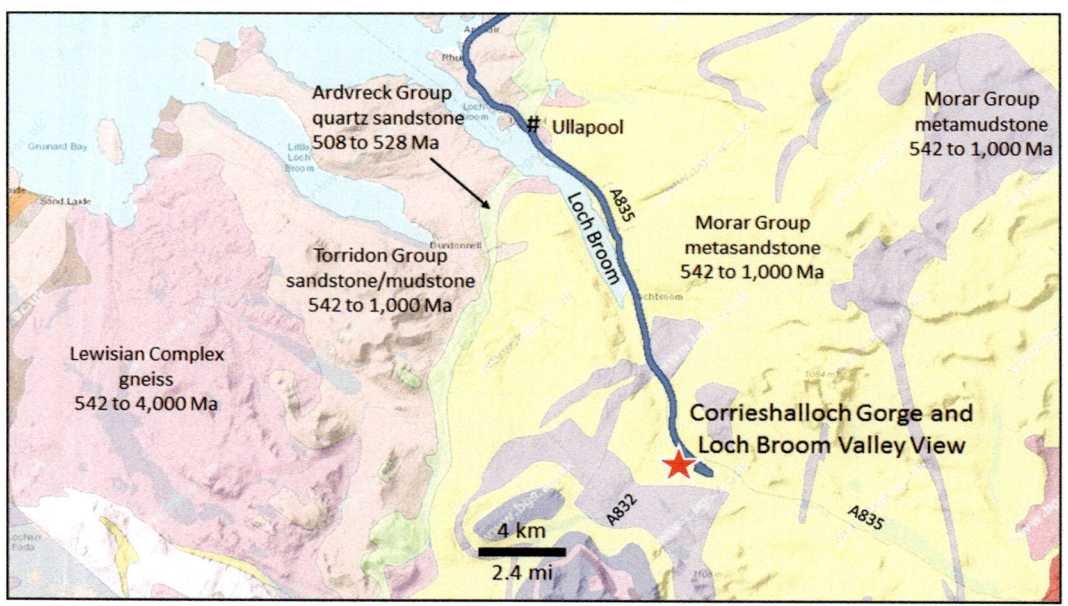

Geologic map of the Loch Broom area. Contains British Geological Survey materials © UKRI 2021.

STOP 11.1 CORRIESHALLOCH GORGE AND THE FALLS OF MEASACH

Spectacular Corrieshalloch Gorge was cut by the River Droma. The name means "ugly hollow" in Gaelic. It is not.

The bedrock at Corrieshalloch is part of the Moine Group carried on the Moine Thrust sheet. At Corrieshalloch, the Moine rocks are mylonite, a type of sheared and recrystallized fault rock that splits easily along flat planes. In mylonite, the grains are flattened and smeared into lens shapes. The shearing and smearing of the constituent grains give the Moine its flaggy appearance.

The bedrock also contains abundant vertical fractures. These break the rock into large blocks, making it easier for meltwater to tear off and remove large slabs of rock to form the gorge. The sharp bends in the gorge are caused by stream erosion along the fractures. Thus, the gorge was cut by grinding and plucking of the bedrock by glacial meltwater as the land was rising due to isostatic rebound (buoying up of the crust after melting and removing the weight of the ice sheets).

A short steep walk takes you to a suspension bridge, from which you can look down at a series of waterfalls. The Falls of Measach, at 45 m (148 ft), can be seen from the suspension bridge or viewed from a platform farther downstream (Harrison, 2020).

Corrieshalloch Gorge is designated a NNR in recognition of the spectacular gorge and the surrounding woodland.

Corrieshalloch Gorge and waterfall.

Corrieshalloch Gorge to Loch Broom View: Continue northwest on A832 for 800 m (0.5 mi) to Stop 11.2, Loch Broom View (57.759048, −5.034147) parking on the right.

STOP 11.2 LOCH BROOM GLACIAL VALLEY

The large amount of glacial debris carried by the meltwater was deposited on the floor of Loch Broom Valley at a time when the sea level was higher (or the land was depressed) and the sea occupied the valley to the base of the gorge. Waves moved the debris around, giving the valley its flat bottom. The valley bottom has been described as a raised beach because the land has rebounded since the ice melted, raising the flat area. You can see the raised beach from the trail at Corrieshalloch Gorge (Harrison, 2020).

Loch Broom U-shaped glacial valley carved into Neoproterozoic Morar Group metasandstone. View north to Loch Broom, an arm of the sea.

Loch Broom View to Loch Maree: *Return southeast on A832 to A835 and turn right (south); drive south on A835 to A832 in Gorstan; turn right (west) onto A832 and drive to* **Stop 12, Loch Maree** *(57.649902, −5.387208) pullout on the right, for a total of 79.4 km (49.3 mi; 59 min).*

STOP 12 LOCH MAREE

Lewisian basement outcrops on the northeast shore of Loch Maree. Above is horizontal Torridonian sandstone, deposited between 1,000 and 541 Ma. These form the Slioch massif. Between the Lewisian and Torridonian is a major unconformity that represents at least 650 Ma of missing time.

The Torridonian sandstones are overlain by the Basal Quartzite and Pipe Rock member sandstones of Cambrian age. The Cambrian sandstones form conspicuous cliffs at the southeast end of Loch Maree. These sandstones weather gray but form bright-white talus slopes. The Torridonian strata form rusty brown outcrops, making them easy to distinguish from the younger Cambrian (Butler, Beinn Eighe NNR and Loch Maree).

Loch Maree itself was eroded along a northwest fault zone and then enlarged by glacial scouring during the last Ice Age. At one time a fjord, it became a lake as the land rebounded and rose as a result of glacial unloading.

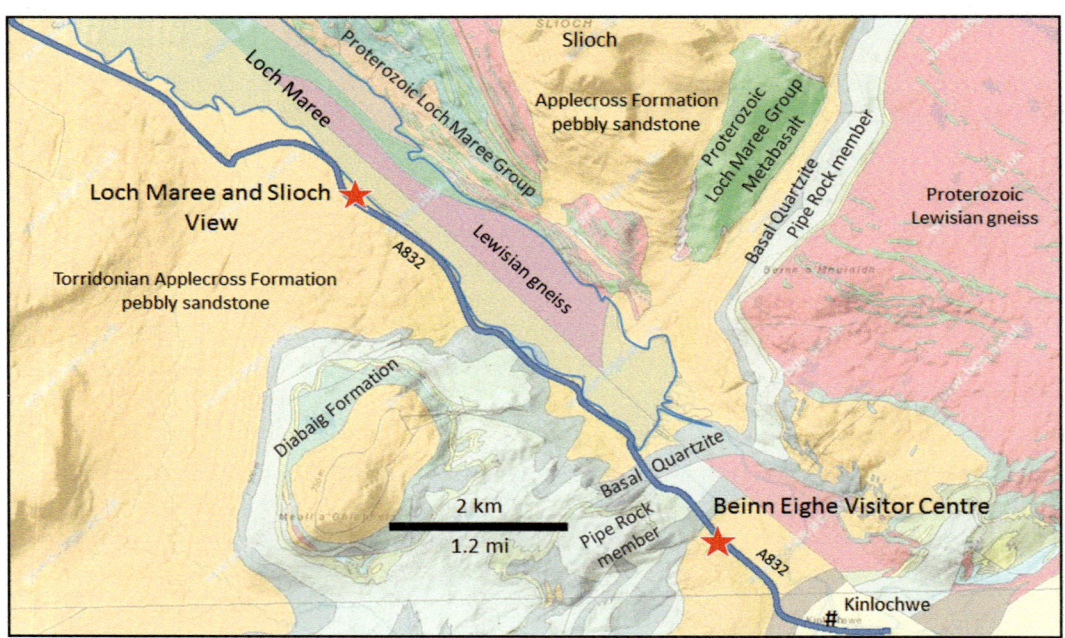

Geologic map, Loch Maree area. Contains British Geological Survey materials © UKRI 2021.

Loch Maree and Slioch. Slioch consists of flat Torridonian Applecross Formation sandstone unconformably over Lewisian gneiss, seen in the foreground around Loch Maree. View northeast.

*Loch Maree to Beinn Eighe Visitor Centre: Return southeast on A832 for 6.1 km (3.8 mi; 6 min) to **Stop 13, Beinn Eighe Visitor Centre** (57.613004, −5.316798) on the right.*

STOP 13 BEINN EIGHE VISITOR CENTRE

Beinn Eighe (pronounced "ben ay"), one of the most spectacular mountain ridges in Scotland, is Britain's oldest NNR. The mountain massif has been sculpted by ice.

The ridge of Beinn Eighe is made up of contrasting Torridonian and Cambrian sandstones in fault contact. The red-brown Torridonian rocks are repeatedly faulted into the gray-white Cambrian sandstones. Today these are interpreted as stacked thrust imbricates that have telescoped the section. The faulting on Beinn Eighe was originally mapped by James Nicol in the 1850s, and he used this as evidence that the North West Highlands are structurally complex. His observations were contested by Sir Roderick Murchison and his assistant, the young Archibald Geikie. In 1860, Murchison and Geikie came here and asserted that the east-inclined strata comprised a normal section with the oldest rocks in the west and the youngest in the east. It wasn't until the early 1880s that Thomas Bonney recognized that the gneisses around Kinlochewe, sitting above the Cambrian sandstones, were actually older Lewisian gneiss. He showed that the Lewisian gneiss had been locally sheared by the faulting that had thrust it up and over the younger Cambrian layers (Butler, Beinn Eighe NNR and Loch Maree).

Beinn Eighe Visitor Centre.

*Beinn Eighe Visitor Centre to Valley of 100 Hills: Continue southeast on A832 for 6.1 km (3.8 mi; 6 min) to Kinlochewe; turn right (southwest) on A896 and drive to **Stop 14, Valley of 100 Hills** (57.555280, −5.414974) car park on the right, for a total of 11.2 km (7.0 mi; 15 min).*

STOP 14 VALLEY OF 100 HILLS (COIRE A' CHEUD-CHNOIC)

This valley contains perhaps the best example of hummocky moraines in Scotland. The moraines were first described by Sir Archibald Geikie in 1863 and further investigated by Peach and Horne in 1913 (Gazetteer for Scotland, Coire a' Cheud-chnoic). Beinn Eighe is part of Britain's oldest and largest NNR (over 4,000 ha; 10,000 ac; Carter, Torridon, and Shieldaig).

The area has been sculpted by ice. Where the recent glaciers deposited the rubble carved from the mountainsides, the resulting hummocky moraines comprise the Coire a' Cheud-chnoic, or Valley of a Hundred Hills. This apparently chaotic arrangement of mounds and ridges actually consists of straight-crested ridges, approximately parallel to each other, extending across the valley. The moraines mark the boundaries of the retreating glaciers (Butler, Beinn Eighe NNR, and Loch Maree).

There has been some debate regarding the significance of the hummocky terrain. Some hummocky moraines, dating from the Younger Dryas (c. 10,000 years), are a result of deposition at glacial margins: the ice melts and the material settles on the ground. Other moraines of this type form by active pushing of the ice, or during glacial surges (Bennett et al., 2020).

At Coire a' Cheud-chnoic, low conical and straight-crested mounds are the dominant landform, covering about 2.5 km^2 (1 mi^2). The moraines consist of unsorted sand to sandy gravel containing rock fragments. Nearly the entire debris zone was pushed by the ice, with the bedrock below acting as the slip surface (Hambrey et al., 1997).

Watch for the local deer that are frequently fed by tourists.

Valley of 100 Hills (Coire a' Cheud-chnoic) looking south. More moraines are seen to the northeast.

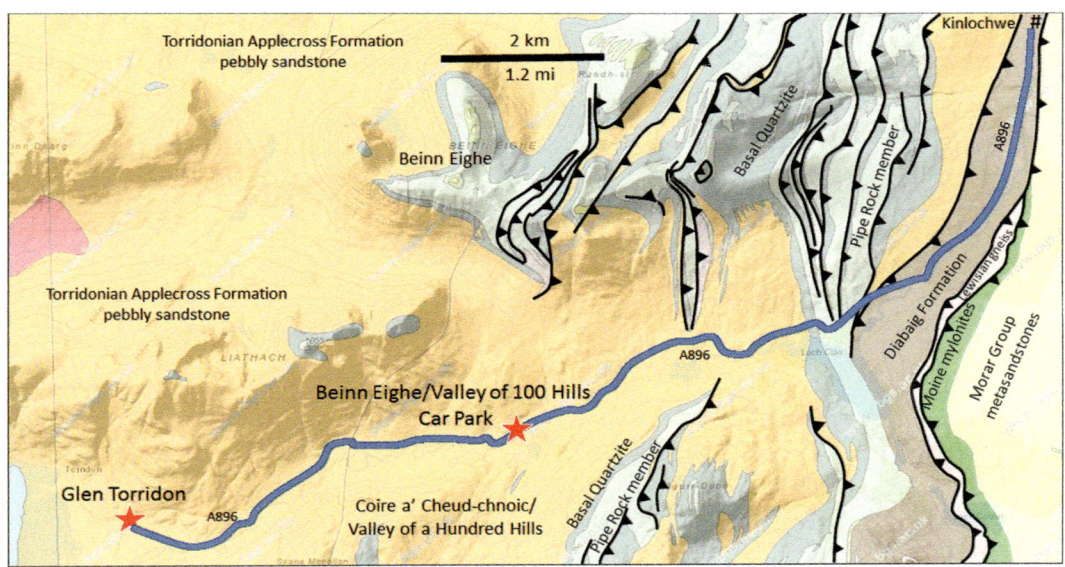

Geologic map of the Glen Torridon-Beinn Eighe area. Contains British Geological Survey materials © UKRI 2021.

Valley of 100 Hills to Glen Torridon: Continue southwest on A896 for 6.1 km (3.8 mi; 6 min) to Stop 16, Glen Torridon (57.54285, −5.5009) and pull over at the bus stop on the right (northeast).

STOP 15 GLEN TORRIDON

This stop near the type section of the Torridonian provides an excellent view of west-dipping Neoproterozoic Applecross Formation sandstone, the major unit in the Torridonian Group.

The name "Torridon Sandstone" was first used in1866, but although sandstone is dominant, there are other rock types as well. This is why Geikie (1892) introduced the name "Torridonian" based on the rocks he saw around Glen Torridon. Some of the most scenic landscapes in Britain, here peaks and crags of flat to gently dipping layers rise 600–900 m (1,970–3,000 ft). Peaks and ridges are commonly capped by resistant Cambrian Quartzite. North of the type area, in Assynt, the Torridonian Sandstone is more eroded, taking the form of narrow ridges and "inselbergs" (island mountains), like castles rising above the rolling gneiss landscape.

The Torridonian was originally eroded from the underlying Lewisian gneiss and deposited as alluvial and fluvial coarse red-brown sandstone in the west, grading to lacustrine (lake) gray shale in the east. These sediments may have been deposited in rift basins that were precursors to opening of the Iapetus Ocean between Laurentia and Baltica. Following deposition, these Precambrian (1,000–750 Ma) units were never subjected to intense deformation or regional metamorphism (Johnstone and Mykura, 1989).

Liathach is the high peak north of Glen Torridon. It has perhaps the best outcrops of Torridonian Applecross Formation sandstone that appear as terraced cliffs. The massive sandstone layers rise 1 km (3,280 ft) above the road. Hard and resistant white Cambrian Quartzite caps the tops of these peaks.

The Cuillin Hills south of Glen Torridon were deformed by Moine-related thrusting. Between the main Moine Thrust and the foreland, they consist of imbricate stacks of Cambrian Quartzite and Torridon Sandstone similar to that seen on Beinn Eighe (Fenton, 2019).

This is a glacially sculpted landscape. The ice-carved U-shaped glens and valleys, caused hanging valleys that are tributary to the main valleys, and left debris piles (moraines) alongside and across the valleys. The entire area is rising due to glacial unloading and has risen here about 70 m (230 ft) in the past 12,000 years.

West-dipping Torridonian Applecross Sandstone, Glen Torridon. View north.

Queen Victoria is said to have loved to travel the road between Torridon and Diabaig, another 14 km (8.6 mi) west down Loch Torridon, in the late 19th century. Along with John Brown and others, she described this area as a fine, wild, uncivilized country, like the end of the world. She noted that "hardly anyone ever comes here" (Carter, Torridon and Shieldaig).

*Glen Torridon to Loch Lochy/Great Glen Fault: Continue west on A896 to A890 near Strathcarron; turn right (southeast) on A890 and drive to A87 at Auchtertyre; turn left (east) onto A87 and drive to the A82 at Invergarry; turn right (south) on A82 and drive to **Stop 16, Loch Lochy/Great Glen Fault** (56.984771, −4.873522) parking area on the right (north) for a total of 146 km (90.5 mi; 2 h 20 min).*

*Side Trip 3 – Glen Torridon to Skye Marble Quarry: Continue west on A896 to A890 at Strathcarron; turn right (south) on A890 and drive to the A87 at Auchtertyre; turn right (west) onto A87 and drive to the B8083 in Broadford; turn left (southwest) onto B8083 and drive to **Stop ST3.1, Skye Marble Quarry** (57.209114, −6.004653) for a total of 94.2 km (58.5 mi; 1 h 38 min).*

SIDE TRIP 3 ISLE OF SKYE

Some of the most spectacular scenery in western Scotland can be found on the Isle of Skye, but it is a little off the beaten track. Also, sometimes it can be foggy or shrouded in low clouds.

Most of Skye consists of the Tertiary plateau lavas that extend from Ireland to western Scotland. At a time when shallow marine, estuarine, and inshore deposits were accumulating in southeastern England, western Scotland could not have been more different. The eruption of thousands of meters of basalt flows was followed by explosive volcanics and intrusions, with the location of these events shifting from area to area. The central mountains of Skye are plutonic masses: intrusive gabbro forms the serrated peaks of Cuillin, whereas the smoother Red Hills are granitic (Richey et al., 1961).

Bits of Jurassic, including some famous fossil locations, outcrop along the east coast and southeast. Precambrian rocks make up the southeastern fringe of the island.

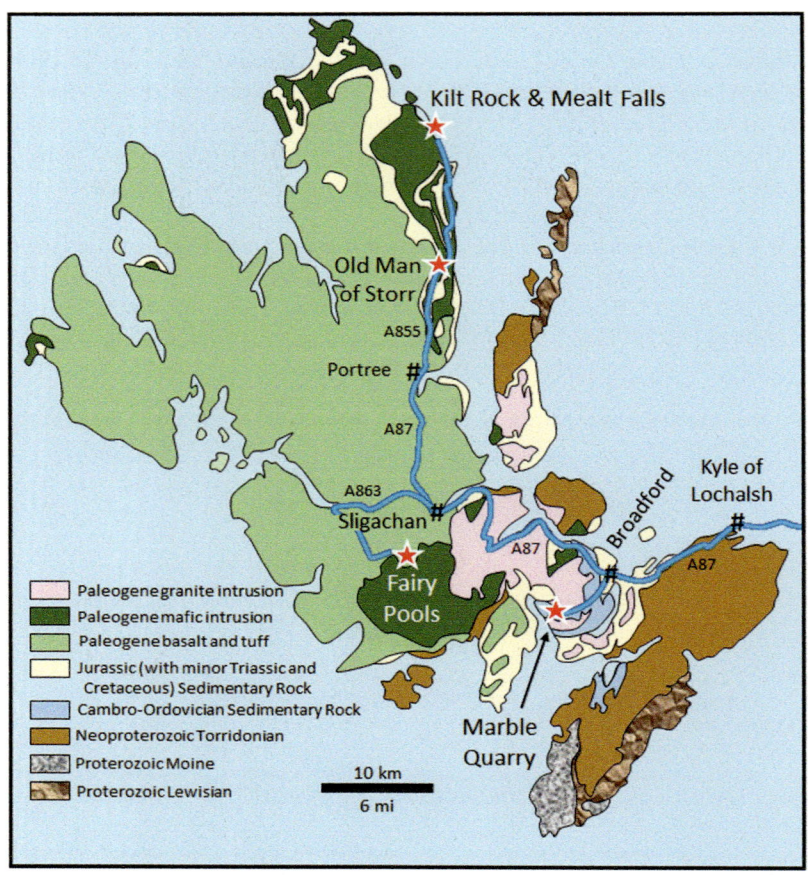

Generalized geologic map, Isle of Skye. (Modified after Stephenson and Merritt, 2006.)

ST3.1 SKYE MARBLE QUARRY

Cambro-Ordovician sedimentary rocks outcrop in a fenster or window in the Kishorn Thrust near Torrin on southern Skye. Durness Group dolomite formed approximately 528 to 464 Ma in shallow marine carbonate banks (BGS, Geology of Britain Viewer "Classic"). The Skye Marble Company quarries the marble that has been metamorphosed through contact with adjacent granitic and gabbroic intrusions (Wikipedia, Geology of the Isle of Skye). In the walls of the quarry, you can see gabbro sills cutting the carbonates.

The rocks were not only affected by contact with the sills. Hot fluids associated with the magma also altered the carbonates. This hydrothermal alteration can be seen in the Skye Marble Quarry. The original dolomite was rich in magnesium and contained chert (silica) nodules, which were altered to produce the green, yellow, and reddish streaks in the marble (Stephenson and Merritt, 2006).

Basalt sills (dark) cut across Durness Group dolomite (light) and altered dolomite (reddish). (Photo courtesy of Diane Housley. Torrin Quarry is operated by Leiths (Scotland) Limited. Skye Marble is extracted and processed at the quarry and is distributed throughout Scotland and into England.)

Visit

This is an active quarry, so you should contact them for a tour and permission if you plan to visit.

Address: Leiths Skye Marble Torrin Quarry, Broadford, Isle of Skye IV49 9BA
Phone: 01471 822265
Email: through website
Web site: http://www.leiths-group.co.uk

*Side Trip 3, Skye Marble Quarry to Old Man of Storr: Return north on B8083 to Broadford and turn left (west) onto A87; drive west on A87 to Portree; turn right (east) onto A855/Bridge Road and drive north to **Stop ST3.2, Old Man of Storr** (57.500572, −6.156112) coach stop on the right, for a total of 60.3 km (37.5 mi; 59 min).*

ST3.2 OLD MAN OF STORR

As you near Brother's Point, look for the large Trotternish escarpment on the left (west) side of the road about 10 km (6 mi) north of Portree. The escarpment marks the edge of a 1.5 km (4,920 ft) thick volcanic plateau deposited about 66–55 Ma. The volcanism was related to opening of the Atlantic

(McGowan et al., Trotternish, Skye). The lavas are basalt flows of the Skye Lava Group (BGS, Geology of Britain Viewer "Classic").

About 1 km before (south of) the coach stop, there is a paved road on the right (east) side that leads to the Scottish Hydro Storr Leathann power station. If you take this road to the end, you are just above Bearreraig Bay Beach, known for its abundant Jurassic fossils. Ammonites, belemnites, and plant remains are common. Be careful while descending the sea cliffs. Collecting is best at low tide. Collect only beach rocks. This is an SSSI, so hammering the bedrock is strictly prohibited.

Along the eastern margin of the plateau massive, rotated lava blocks and pinnacles mark the largest continuous area of landslides in Britain. The "Old Man of Storr" is one such detached block. The landslides are mostly post-glacial (younger than 12,000 years), although the outer zone was modified by the last ice sheet. The detachment is on the underlying Jurassic rocks: the Skudiburgh Formation, Kilmaluag Formation, Duntulm Formation, and Valtos Sandstone Formation, a dominantly mudstone section with subordinate limestone and sandstone, all deposited in the interval from 168 to 166 Ma (McGowan et al., Trotternish, Skye; BGS, Geology of Britain Viewer "Classic").

Old Man of Storr (pinnacle on the right) and Trotternish escarpment. (Courtesy of Paul Lucas, https://commons. wikimedia.org/wiki/File:Old_Man_Of_Storr_(8016352083).jpg.)

Side Trip 3, Old Man of Storr to Mealt Falls: Continue north on A855 for 13.3 km (8.3 mi; 13 min) to Stop ST3.4, Mealt Falls (57.610516, −6.171909) parking area on the right.

ST3.3 Mealt Falls and Kilt Rock

From this stop, you can see Mealt Falls and Kilt Rock. Wispy Mealt Falls, near the outlet of Mealt Loch, plunges 60 m (197 ft) into the Sound of Raasay. When the wind is strong, the water is blown away and it doesn't reach the rocky bottom at all (Allan, Kilt Rock).

The wind also plays the fence around this stop. Like blowing over the top of a bottle, the wind blows over holes in the metal fence posts to create an eerie, yet somehow fitting background music.

Looking north a few kilometers, you can see 90 m (295 ft) high Kilt Rock. The sea cliff consists of columnar basalt (specifically, olivine dolerite sills) resting on flat-lying Middle Jurassic Valtos Formation sandstone, with another sill just above sea level. The intersection of the columnar jointing and sill boundaries gives the appearance of a pleated tartan kilt (Bell and Harris, 1986; Atlas Obscura, Ancient cliffs). The sills are part of the north Skye Sill Complex, which in turn is part of the British Tertiary Volcanic Province (which is part of the Greeland-Scotland Transverse Ridge) formed during a period of intense volcanic activity associated with opening of the Atlantic Ocean (Armando, 2015).

Mealt Falls and Kilt Rock.

*Side Trip 3, Mealt Falls to Fairy Pools: Return south on A855 to Portree; in Portree turn left (south) onto A87/Viewfield Road; drive south on A87 to Sligachan; in Sligachan turn right (south) onto A863 and drive west to B8009; turn left (west) onto B8009 and drive to Merkadale; in Merkadale turn left (southeast) onto an unnamed road and drive to **Stop ST3.5, Fairy Pools** (57.250532, −6.272249) "pay-and-display" car park on the right (north side of the road) for a total of 57.3 km (35.6 mi; 1 h). Take the trail 930 m (3,050 ft) east to the pools.*

ST3.4 Fairy Pools (Glumagan Na Sithichean)

This scenic series of small waterfalls and pools on the River Brittle occurs near the contact between an unnamed Paleogene mafic lava/tuff and an unnamed Paleogene (66–23 Ma) olivine gabbro (BGS, Geology of Britain Viewer "Classic").

A tributary of the River Brittle runs down from the Cuillin mountains into Glenbrittle. The Fairy Pools are beautiful rock pools of icy, crystal-clear spring water fed by a series of waterfalls.

A gravel path from the parking lot takes you to the nearest and largest waterfall. From the parking lot cross the road and take the path labeled "Sligaghan." There are several small river crossings, and the way can be slippery when wet (Atlas Obscura, Fairy Pools). Continue up the path to see some of the smaller pools.

The operation and maintenance of the Fairy Pools car park has been leased to Outdoor Access Trust Scotland, and they set the parking charges. Parking costs £6 per car (2022) and must be paid for by coins or card.

The Fairy Pools looking west over Glen Brittle.

Enjoying the Fairy Pools.

Google Earth oblique view northeast over the Great Glen Fault (GGF). (Image © 2022 Landsat/Copernicus and © 2020 CNES/Airbus; Data SIO, NOAA, U.S. Navy, NGA, GEBCO.)

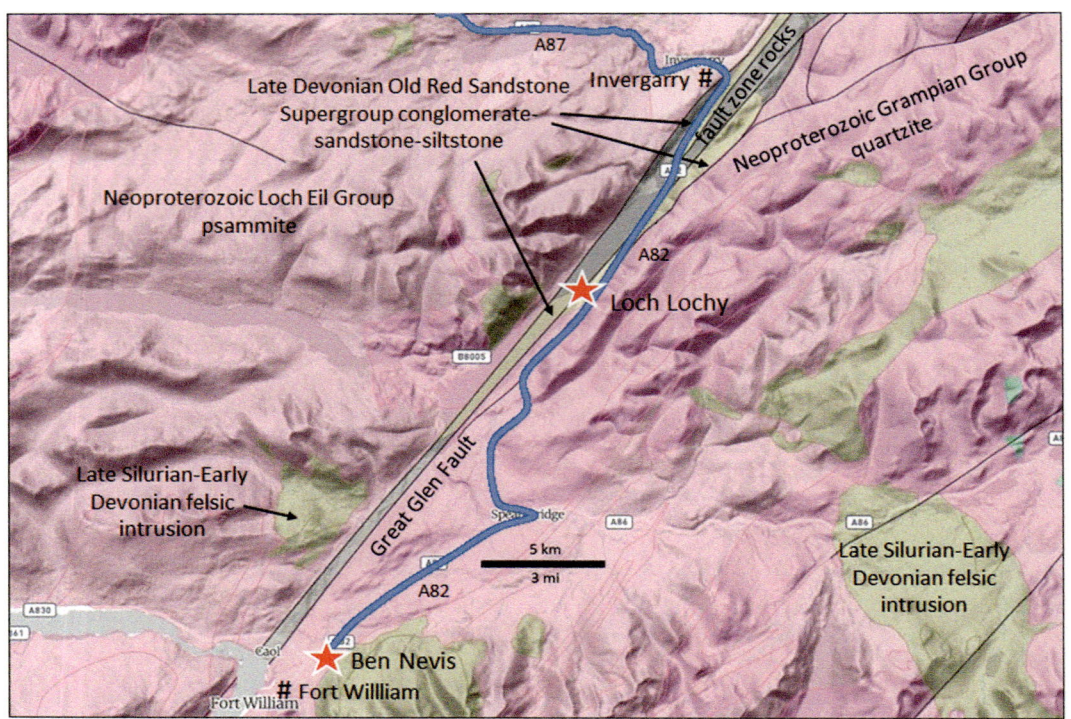

Geologic map of the GGF near Loch Lochy. (MacroStrat map, https://macrostrat.org/map/#x=-4.57&y=57.01&z=9.94.)

Side Trip 3, Fairy Pools to Great Glen Fault: Return north to Merkadale and turn right (east) onto B8009; drive east on B8009 to A863; turn right (east) onto A863 and drive to Sligachan; in Sligachan turn right (east) onto A87 and drive east and south to the A82 at Invergarry; turn right (south) on A82 and drive to **Stop 16, Great Glen Fault, Loch Lochy** *(56.984771, −4.873522) parking area on the right (north) for a total of 149 km (92.6 mi; 2 h 15 min).*

STOP 16 GREAT GLEN FAULT, LOCH LOCHY

Once again we cross the Great Glen Fault (see the discussion at Stop 5). This time we are at the southwest end of this regional terrane-bounding fault. Pull off at the lakeside car park with superb views northeast and southwest along the linear lake that straddles the Great Glen Fault (GGF).

Northwest of the GGF in this area is the West Highland Granite Gneiss Intrusion that solidified approximately 1,000–541 Ma. Southeast of the GGF is Old Red Sandstone Supergroup, a rock formed approximately 424–359 Ma, mainly during the Devonian Period. Originally sedimentary rocks, in this area the rocks were later metamorphosed slightly and then broken up within the fault zone (BGS, Geology of Britain Viewer "Classic").

As mentioned previously (Loch Ness and the Great Glen Fault), the Great Glen was carved during the last Ice Age, between 2 million and 12,000 years ago, as a result of glaciers grinding out and removing the softer, shattered rock along the fault zone.

Loch Lochy and GGF. View northeast.

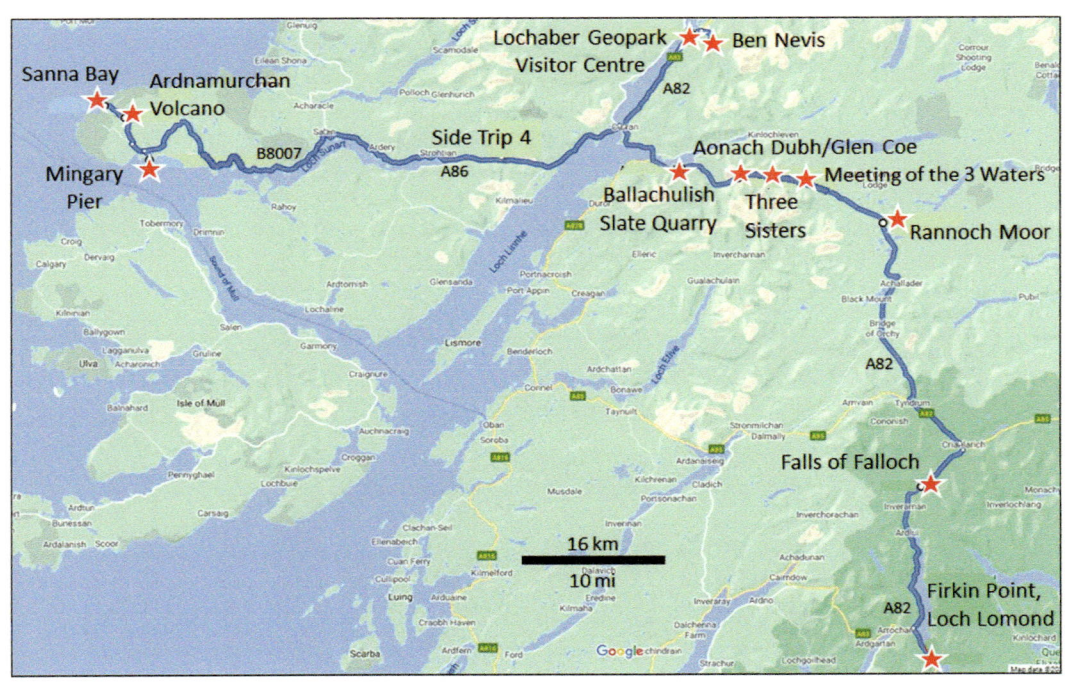

Great Glen Fault, Loch Lochy to Ben Nevis Visitor Centre: *Continue driving southwest on A82 to Inverlochy; in Inverlochy turn left (east) onto C1162/Glen Nevis and drive south to **Stop 17, Ben Nevis Visitor Centre** (56.810818, −5.076833) on the left, for a total of 28.7 km (17.9 mi; 27 min).*

STOP 17 BEN NEVIS VOLCANO AND VISITOR CENTRE

Stop at the Ben Nevis Visitor Centre for a great view of Ben Nevis and to learn more about this ancient volcano. Ben Nevis was an important site in the history of theories explaining how volcanoes develop.

The highest mountain in Britain at 1,344 m (4,410 ft), Ben Nevis, was formed by volcanic activity just over 400 million years ago.

The volcanoes of Ben Nevis and Glen Coe erupted through Dalradian rocks. These basement rocks were deposited as sands, muds, and carbonates in shallow seas and on coastal plains on the margin of Laurentia between 750 and 480 Ma (Stephenson et al., 2012). The resulting sandstones, mudstones, and limestones were later metamorphosed to slates and schists during the Caledonian Orogeny (470–430 Ma).

From Fort William, Ben Nevis has a humped appearance, but the north side is sheer cliffs. The summit, including the cliffs on the north face, consists of volcanic rocks that erupted probably a few million years after the volcanic activity at Glen Coe. The rocks of the lower slopes and surrounding hills are granite, formed in the magma chamber below the volcano.

The lower parts of the north face are made up of fragments of volcanic rock and ash, probably formed in pyroclastic flows and landslides that raced down the slopes of the exploding volcano. The upper cliffs and summit plateau are andesitic lavas (Stephenson and Goodenough, 2007).

The lower slopes of Ben Nevis comprise two different granites. The Outer Granite, which forms the hills of Meall an t-Suidhe and Aonach Mòr, cooled first. This granite porphyry contains pink feldspar crystals up to 2 cm (1 in) long. It is cut by the Etive Dyke Swarm. The Inner Granite, seen on Carn Mòr Dearg, is finer-grained, pink to gray, and is not cut by dikes. Apparently, the Outer Granite had cooled and solidified and had been shot through with dikes, before the Inner Granite solidified.

We know that volcanic rocks erupt at the surface, whereas granites solidify at depths of several kilometers. It is unusual that these lavas and granites are in contact with each other. Resolving this dilemma led to the theory of caldera collapse, the collapse of a surface vent into the magma chamber after the eruption of a great volume of magma. A caldera (a large volcanic crater) probably formed at the surface, but the upper parts of the volcano have eroded away so that only the deep rocks remain (Stephenson and Goodenough, 2007). Subsequently, this area has been eroding for most of the past 400 Ma.

Ben Nevis from the Ben Nevis Visitor Centre. View east.

Ben Nevis Visitor Centre to Lochaber Geopark Visitor Centre: Return north on C1162/Glen Nevis to the roundabout in Inverlochy; continue straight onto A82/Belford Road; take A82 west to Middle Street and turn left (west); take Middle Street west to Stop 18, Lochaber Geopark Visitor Centre (56.817847, −5.111722; 55A High Street) on the left.

STOP 18 LOCHABER GEOPARK VISITOR CENTRE

Lochaber Geopark encompasses 4,468 Km^2 (1,725 mi^2) including the volcanoes at Ardnamurchan, Ben Nevis, and Glen Coe.

Geoparks must have an outstanding or unique geological heritage and a strategy to promote that heritage for the benefit of the local community. Lochaber Geopark is well known for its spectacular scenery. The outstanding qualities of Lochaber were officially recognized in April 2007 when Lochaber was given UNESCO Global Geopark status.

Many geological features of international importance can be seen in Lochaber. The unique features include a record of plate collisions in the Caledonian Orogeny, and later plate rifting to form the Atlantic Ocean. Magmas beneath the Caledonian Mountains caused by subduction gave rise to the famous volcanoes of Ben Nevis and Glen Coe. Much later rifting of Eurasia from North America produced the volcanics at Ardnamurchan, among others. Lastly, the landscape was sculpted by glaciers over the last 2 Ma. The park has many excellent examples of glacial features, including U-shaped valleys, glacial scour, and glacial lake shorelines (LochaberGeopark.org, Lochaber Geopark – Glencoe). Interpretive panels explaining the geology and natural history are located throughout the park.

Ben Nevis Visitor Centre.

VISIT

The Visitor Centre has educational films, good displays, geologic books, brochures, and maps. The expert staff provides geologic tours and day hikes. For information on tours, contact them at https://lochabergeopark.org.uk/product-category/geotours/

Address: Lochaber Geopark Association Visitor Centre, 55A High Street, Fort William, PH33 6DH
Phone: 01397 705314
Email: info@lochabergeopark.org.uk
Website: https://lochabergeopark.org.uk/

Lochaber Geopark Visitor Centre to Ballachulish Slate Quarry: Continue driving southwest on Middle Street; continue straight onto High Street; continue straight onto A82/Achintore Road; continue driving on A82 to Ballachulish; in Ballachulish turn right (south) onto Albert Road and drive 150 m (500 ft) to **Stop 19, Ballachulish Slate Quarry** *(56.678181, −5.129045) on the left for a total of 23.9 km (14.8 mi; 22 min). There is ample parking behind the Visitor Centre across the street from the quarry.*

Side Trip 4, Lochaber Geopark Visitor Centre to Ardnamurchan Volcano: Continue southwest on Middle Street; continue straight onto High Street; continue straight onto A82/Achintore Road; continue driving on A82 to Corran; in Corran turn right (west) onto A861; take the Ardgour-Corran ferry; turn left (south and west) onto A861 and drive to Salen; turn left (southwest) onto B8007; from the B8007 north of Kilchoan, take the minor road on the right toward Achnaha and Sanna Bay; continue along this road for about 4 km (2.5 mi) and park at the old quarry car park approximately 1 km southeast of Achnaha. This is **Stop ST4.1, Ardnamurchan Volcano** *(56.732703, −6.139021) for a total of 88.7 km (55.1 mi; 2 h 14 min). There is room for 4–5 vehicles and free parking but no facilities.*

SIDE TRIP 4 ARDNAMURCHAN VOLCANIC COMPLEX

Head north along the graveled track toward the deserted village of Glendrian. Park at the old quarry and take the trail a hundred meters or so to the top of the rise (there is a gate in the fence) and observe a central low area completely ringed by small hills, the remains of the subterranean magma chamber and its feeder pipe.

The Ardnamurchan volcanic complex is the eroded felsic magma chamber and largely rhyolitic ring dikes of a ~60 Ma Paleogene volcano (Troll et al., 2019). As mentioned earlier, volcanism at Ardnamurchan is related to rifting during opening of the Atlantic. Ardnamurchan volcano is one of five volcanic centers along the west coast of Scotland. The other four were on Mull, Rùm, Skye, and Arran (West Highland Peninsulas, Discover the Geology of the West Highland Peninsulas).

ST4.1 ARDNAMURCHAN VOLCANO

Ardnamurchan is where geologists developed ideas regarding magma emplacement. J.E. Richey in 1920 first described this series of ring intrusions as allowing the surrounding rocks to adjust as magma below a volcano repeatedly punched its way up, erupted, and subsided. Ardnamurchan is said to have the best-developed ring dikes in Britain. There are three sets of large, overlapping rings, each associated with its own volcanic center: an early eruptive center in the east, a later center in the west, and a final center between the earlier two. The gabbroic Great Eucrite is considered by many to be the type example of a ring dike, although this interpretation has been challenged recently. It is now thought that the Great Eucrite may be a lopolith, a funnel, or a dish-shaped intrusion (Butler, Ardnamurchan Ring Complex; Wild About Ardnamurchan, The Ardnamurchan Volcanoes).

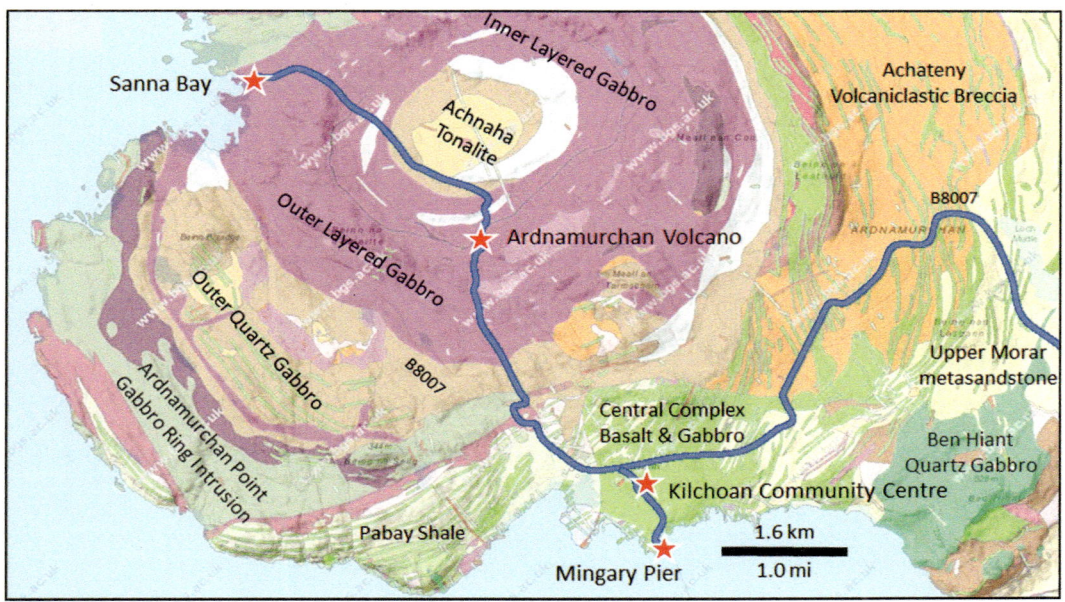

Geologic map of the Ardnamurchan volcanic complex. (Modified after BGS, Geology of Britain Viewer. Contains British Geological Survey materials © UKRI 2022. Base mapping provided by ESRI.)

Ardnamurchan Volcano, central depression, and ring of hills signifying the collapsed magma chamber. The outer ring of hills has historically been known as "the Great Eucrite." The low, inner ring consists of the "Fluxion Biotite Gabbro," and the central knob is quartz monzonite. View north.

*Side Trip 4, Ardnamurchan Volcano to Sanna Bay: Continue northwest on this unnamed road for 2.9 km (1.8 mi; 6 min) to **Stop ST4.2, Sanna Bay** (56.745142, −6.175753). There is ample free parking and facilities at the beach.*

ST4.2 SANNA BAY

Since you are here, you might as well explore the beautiful, remote sandy beaches and turquoise waters of Sanna Bay. Sanna Bay exposes gabbros formed in a second volcanic center; they form the low headlands along the coast (Wild About Ardnamurchan, The Ardnamurchan Volcanoes). You do get a nice panoramic view of Ardnamurchan Volcano from here as well.

Sanna Bay. (Photo courtesy of Wendy Kirkwood, https://commons.wikimedia.org/wiki/File:Sanna_Bay_-_geograph.org.uk_-_890264.jpg.)

*Side Trip 4, Sanna Bay to Mingary Pier: Return southeast on this unnamed road to B8007; turn left (southeast) onto B8007 and drive to **Stop ST4.3, Mingary Pier** (56.688770, −6.094610) for a total of 9.8 km (6.1 mi; 18 min). There is ample free parking and facilities near the ferry terminal.*

ST4.3 MINGARY PIER

Explanations of the local geology can be found on a Lochaber Geopark interpretive panel at Mingary Pier by the ferry terminal. Take the single-track road B8007 toward Kilchoan. At Kilchoan, follow the signs to Mingary Pier. A panel at the pier explains the geology and shows the various intrusions associated with the volcanic centers. At the pier parking area, a sign welcomes you to "West Ardnamurchan, the most westerly point on the UK mainland."

Dikes, where magma forced its way vertically into Jurassic limestone, can be seen along the shore here.

Mingary Castle, a short distance to the east, was built on two sills, horizontal layers of magma that were injected between bedding. A granophyre (quartz-feldspar) sill lies above a dolerite (basaltic) sill: both intrude Jurassic sediments. The Jurassic Blue Lias and Breakish formations consist of interfingering limestone, sandstone, siltstone, and mudstone deposited in shallow carbonate seas. Mingary Castle, dating to the 13th or 14th century, was built by either the MacDougalls or the MacDonalds of Ardnamurchan (Wikipedia, Mingary Castle).

Mingary Castle, built on stacked sills, as seen from Mingary Pier.

*Side Trip 4, **Mingary Pier to Kilchoan Community Centre**: Return north on B8007 for 650 m (0.4 mi) to **Stop ST4.4, Kilchoan Community Centre** (1 Pier Road, Kilchoan; 56.696363, −6.100294) parking on the right. There is ample free parking at the Community Centre.*

ST4.4 KILCHOAN COMMUNITY CENTRE

Stop at the Community Centre at Kilchoan to see a geological map of the Ardnamurchan peninsula and local rock samples (Wild About Ardnamurchan, The Ardnamurchan Volcanoes).

A display in the Kilchoan Community Centre shows rocks associated with the Ardnamurchan Volcano.

*Side Trip 4, Kilchoan Community Centre to Ballachulish Slate Quarry: Continue north on B8007/Pier Road to B8007; turn right (east) on B8007 and drive to A861 in Salen; turn right (south) onto A861 and drive east to the Corran-Ardgour Ferry; take the ferry to the A82; turn right (south) on A82 and drive east to Albert Road in Ballachulish; turn right (south) onto Albert Road and drive 150 m (500 ft) to **Stop 19, Ballachulish Slate Quarry** (56.678181, −5.129045) on the left for a total of 81.1 km (50.4 mi; 2 h). There is ample parking behind the Visitor Centre across the street from the quarry.*

STOP 19 BALLACHULISH SLATE QUARRY

Slate quarrying was, for a time, a major industry in this area. The mines are in the Dalradian Ballachulish Slate Formation, a graphitic and pyritic slate deposited as organic-rich mud in shallow seas off the coast of Laurentia between 750 and 480 Ma. The Dalradian Sequence in this area was metamorphosed to greenschist facies during the Caledonian Orogeny (470–430 Ma; Stephenson et al., 2012).

The slate at Ballachulish was quarried starting in 1693 and lasted a bit over 250 years, till 1955. Production reached a peak in 1875, when the quarry was 151 m (497 ft) deep, employed 587 quarriers, and produced 26 million shingles. The slate was used for roofing shingles and was shipped throughout Scotland, England, Ireland, and even to North America. The shingles are easily recognized by the pyrite cubes, up to a cm (half in) across, that are common in the slate.

The Ballachulish Slate Quarry.

The main quarry at Ballachulish has now been restored, with a visitor's trail and interpretive panels.

Following the last Ice Age, once the weight of ice was removed, the land slowly began to rise. Coastal features such as beaches rose well above sea level. Excellent examples of "raised beaches" can be seen around the margins of Loch Leven and Loch Linnhe, presently about 8 m (26 ft) above sea level. The road through Ballachulish was built on one such raised beach (Stephenson and Goodenough, 2007).

Group	Subgroup	Formation	Lithology
Argyll Group	Islay	Jura Quartzite	
		Bonahaven Dolomite	
		Port Askaig Tillite	
Appin Group	Blair Atholl	Islay Limestone	
		Mullach Dubh Phyllite	
		Lismore Limestone	
		Cuil Bay Slate	
	Ballachulish	Appin Phyllite	
		Appin Limestone / Appin Quartzite	
		Ballachulish Slate	
		Ballachulish Limestone	
	Lochaber	Leven Schist	
		Glencoe Quartzite	
		Binnein Schist	
		Binnein Quartzite	
		Eilde Schist	
		Eilde Quartzite	
Grampian Group	Glen Spean	Brunachan Psammite	
		Beinn Iaruinn Quartzite	
		Glen Fintaig Semipelite	
		Glen Gloy Quartzite	
		Auchievarie Psammite	
		Tarff Banded Psammite	
		Allt Goibhre Semipelite	

Dalradian stratigraphy in the Loch Leven-Glencoe area. (Information drawn from Roberts and Treagus, 1977; modified after Stephenson et al., 2012.)

STOP 20 GLEN COE AND SURROUNDING AREA

Glen Coe, the "glen of weeping," has a terrible history of inter-clan hostility and murder, not least the massacre of 1692 (Blair, Glen Coe, Lochaber). The village that grew up in this valley is known as Glencoe. To confuse things even more, the volcano is referred to as Glen Coe, but the associated caldera is Glencoe.

The earliest historical figure connected with Glencoe was Fingal, one of the greatest Celtic heroes and leader of the Feinn, warriors of Gaelic mythology. The glen was his legendary home and his memory is preserved in a number of place names. Fingal is credited with defeating the Vikings led by King Erragon of Sora, probably during the 800s.

Ownership of the glen passed into the hands of his descendants, the powerful MacDougall clan, in the 11th century. Over the next two centuries, the MacDougalls managed to build up a small empire in western Scotland. This came to an end in 1308, when they sided against Robert the Bruce and lost. Robert the Bruce gifted Glencoe to Angus Og, chief of the MacDonald clan. There followed a period of relative calm until the year 1501 when disputes flared up between the Glencoe MacDonalds and the neighboring Argyll Campbells regarding cattle rustling by the MacDonalds and the attempts by the Campbells to extend their territory (Glencoe Scotland, History of Glencoe).

These troubles continued into the mid-1600s. The MacDonalds were associated with the Royalists and rebels, whereas the Campbells sided with the government in London. This all came to a head in 1691, when King William III of England offered a pardon to all Highland clans who had fought against him on the condition that they take an oath of allegiance by January 1, 1692. The alternative was death.

MacIain of Glencoe, chief of the MacDonalds, agreed to take the oath, but mistakenly went to Inverlochy in Fort William instead of Inveraray near Oban. He reached Inveraray on January 6th, well after the deadline. MacDonald believed his clan was safe, despite the delay. Unknown to him, the king had assembled a force to exterminate the entire clan. The force, led by Robert Campbell, left for Glencoe on February 1. Campbell requested lodging for his soldiers and, unaware of his orders, the MacDonalds hosted and entertained them for 10 days.

On February 12, Campbell received orders to kill all MacDonalds the next morning. In the early hours of the morning, the soldiers massacred their hosts. This act of treachery in response to hospitality makes the massacre so heinous. Although "only" 40 were murdered, many that escaped to the hills later died of hunger and exposure.

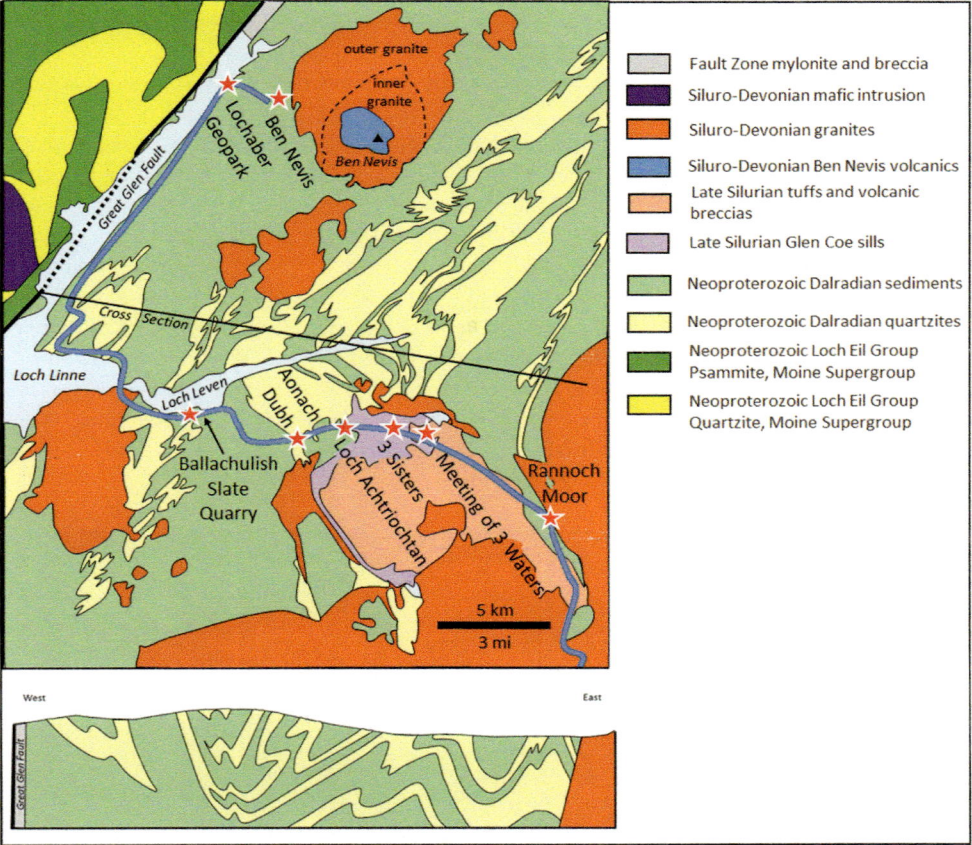

Geologic map and cross-section of the Ben Nevis-Glen Coe area. (Modified after Stephenson and Goodenough, 2007.)

A monument to the fallen MacDonalds is located in Glencoe village. Signal Rock, where the order was given to begin the massacre, is just a few hundred yards west of the Clachaig Inn on the north bank of the River Coe.

Glencoe remained isolated until the first road was built through the glen in 1785. Between 1750 and 1860, large numbers of inhabitants of the Highlands, including the Glen Coe area, were evicted, a period known as The Clearances. By the death of the 17th Chief, Ewen MacDonald in 1837, the influence of the clan was insignificant. In 1935, to prevent commercial exploitation of the valley, the National Trust for Scotland bought 518 ha (12,800 ac) of the glen. With the help of donations, the Trust's holding has been extended until today it covers most of the glen from Rannoch Moor to Loch Leven (Glencoe Scotland, History of Glencoe).

The spectacular highland scenery has made this area a favorite for moviemakers. Among the films shot in the glen are Harry Potter and the Prisoner of Azkaban, Highlander 1 and 3, Braveheart, and Rob Roy (Discover Glencoe, Scottish Hollywood).

Ballachulish Slate Quarry to Aonach Dubh: *Return north to A82 and turn right (east); drive east on A82 to the sign for Clachaig Inn on the left; turn left (west) onto Glencoe Studio Gallery Road and drive to the Clachaig Inn parking area. This is **Stop 20.1, Aonach Dubh** (56.664572, −5.055840), for a total of 7.8 km (4.9 mi; 8 min).*

STOP 20.1 AONACH DUBH

Aonach Dubh and Ardnamurchan volcanoes are where the concept of "cauldron subsidence," now known as "caldera collapse", was developed in the late 1800s (Blair, Glen Coe, Lochaber). The classic view of Aonach Dubh is from near the Clachaig Inn. Looking southeast, you see igneous sills and sedimentary rocks that form the slopes around Loch Achtriochtan as well as the lower half of the west face. The dark lower slopes comprise short, near-vertical steps of ledge-forming andesite sills. These are separated by grassy, slope-forming metamorphic rocks, the Dalradian Leven Schist that hosted the sills. The upper part of Aonach Dubh is distinctly different. The pink upper cliffs, some of Britain's best rock climbing, rise 150 m (492 ft). They are silica-rich tuff formed by pyroclastic flows, the result of violent eruptions of the Glen Coe volcano. Following the early, non-violent emplacement of the sills, a series of massive explosions fueled by gas-charged magma sent pumice, ash, and rock fragments rushing from the vent as flaming clouds of molten particles. The resulting ignimbrites form most of the mountain tops on the south side of Glen Coe, including the Three Sisters (Stephenson and Goodenough, 2007).

Aonach Dubh, view east-southeast.

Glen Coe U-shaped glacial valley from Loch Achtriochtan. View east-southeast.

Aonach Dubh to Loch Achtriochtan Car Park: Return east to A82; turn left (east) onto A82 and drive to Stop 20.2, Loch Achtriochtan Car Park (56.668629, −5.027764) on the left (north) for a total of 1.9 km (1.2 mi; 3 min).

STOP 20.2 LOCH ACHTRIOCHTAN CAR PARK

At the pullout just beyond Loch Achtriochtan, you are well inside the caldera. Andesitic lavas from the volcanic eruptions are clearly visible on the steep western slopes of Aonach Dubh.

Aonach Dubh from Loch Achtriochtan car park, view south.

As mentioned previously, the Glen Coe volcano erupted on a surface of Dalradian metamorphic rocks about 420 million years ago. Volcanism began with the injection of andesitic magma as horizontal sills in the near-surface. Later, the same magma poured out as lava flows (Blair, Glen Coe, Lochaber). At the same time, basaltic magma was injected into northeast-trending vertical fractures, creating a great number of dikes. The dikes are darker than andesite and rhyolite and often softer than the surrounding lava. They erode to form deep gullies on the mountainsides (Discover Glencoe, Lochaber Geopark).

Above the andesites is a great pile of rhyolite lava and pyroclastic tuff. As the magma was exhausted, the roof of the magma chamber collapsed along circular ring faults to form a large crater, or "caldera". Glen Coe was the first place in the world where this "cauldron subsidence" was recognized in ancient rocks (Blair, Glen Coe, Lochaber). The geologic map of this area clearly shows an oval volcanic center. The A82 cuts through this caldera from west to east.

After the subsidence of the Glencoe caldera, granitic magma continued to accumulate, eventually forming a large body, the Clach Leathad Pluton. The pluton cuts across the volcanic rocks of Glen Coe and forms the mountains of lower Glen Etive to the south. It is cut by numerous northeast-trending dikes known as the Etive Dyke Swarm. The dikes cut across all the rocks of Glen Coe, forming deep chasms and gullies (Stephenson and Goodenough, 2007).

By 400 Ma, the volcanic activity here was over and erosion began the long task of wearing away the mountains. The final sculpting of these mountains occurred during the recent Ice Age. For the past 2 million years, the climate here has fluctuated between cold glacial periods and warmer interglacials. A large ice sheet formed, centered at Rannoch Moor, and glaciers flowed west through Glen Coe, Glen Nevis, Loch Leven, and Glen Etive (Stephenson and Goodenough, 2007). The view east up Glen Coe is an excellent example of a glacial U-shaped valley. Glaciers scraped grooves (striae) in the bedrock, carried blocks of rock (erratics) that were left behind as the ice melted and dumped ridges of loose rock along their margins and ends (moraines).

While many of the highest peaks were ice-free, mountain glaciers up to 600 m (1,970 ft) thick slid down the valleys from the high peaks, especially south of Glen Coe, leaving sharp ridges and hanging valleys. Around 18,000 years ago, the glaciers began to retreat. As the glaciers melted, freezing and thawing of water in fractures shattered the bedrock. Loose blocks piled up on steep slopes to form talus and scree. Cliffs collapsed, producing rockfalls. A spectacular rockfall occurred close to Coire Gabhail, where house-size boulders are scattered across the glen. This event is considered the largest single rockfall in all of Great Britain. The fallen boulders dammed the valley so that the stream could no longer flow to the River Coe. Behind this dam, the valley has a completely flat floor consisting of stream sediment deposited behind the slide (Discover Glencoe, Lochaber Geopark).

Loch Achtriochtan Car Park to Three Sisters Car Park: *Continue east on A82 for 2.6 km (1.6 mi; 2 min) to **Stop 20.3, Three Sisters Car Park** (56.667934, −4.986862) on the right (south).*

Stop 20.3 Three Sisters

This stop provides an excellent view south to the Three Sisters peaks. The peaks consist of erosion-resistant ignimbrites (welded rhyolitic tuff). In fact, ignimbrites form most of the mountain tops south of Glen Coe (Stephenson and Goodenough, 2007).

Three Sisters panorama. View southeast (left) to southwest (right).

*Three Sisters Car Park to Meeting of the Three Waters: Continue east on A82 for 700 m (0.4 mi; 1 min) to **Stop 20.4, Meeting of the Three Waters** (56.665571, −4.977094) parking on either side of the road.*

STOP 20.4 MEETING OF THE THREE WATERS

Surging glacial meltwater, carrying large quantities of rock debris, caused rapid erosion along faults and dikes. These eroded to deep, narrow gorges like those at the "Meeting of the Three Waters," where two eroded dikes converge (Stephenson and Goodenough, 2007).

Deep gorges (arrows) eroded in the fractured volcanics at Meeting of the Three Waters. Google Earth image. (Image © 2022 Maxar Technologies.)

Meeting of the Three Waters, panorama from southeast (left) to southwest (right).

***Meeting of the Three Waters to Rannoch Moor**: Continue east on A82 for 15.0 km (9.3 mi; 13 min) to **Stop 21, Rannoch Moor** (56.62919, −4.77550) and pull over on the left (east) side of the road.*

STOP 21 RANNOCH MOOR

Unlike most areas composed of granitic rock, Rannoch Moor is a broad, flat lowland. The Rannoch Moor Pluton is a 444 to 419 Ma (Silurian) granodiorite intruded during the Caledonian Orogeny (BGS, Geology of Britain Viewer "Classic").

Many moraines around Glen Coe were deposited from stationary, slowly melting glaciers. These form hummocky terrane that can cover large areas. Nice examples exist here at Rannoch Moor (Stephenson and Goodenough, 2007).

Rannoch Moor. View east.

21.2 Loch Lomond and The Trossachs National Park

Between Rannoch Moor and the Falls of Falloch, we enter Loch Lomond and The Trossachs National Park. This park, along with Cairngorms National Park, was among the first national parks established by the Scottish Parliament in 2002. Loch Lomond and The Trossachs National Park covers 1,865 km^2 (720 mi^2) of the Southern Highlands. Of particular interest to geologists, the park contains the Highland Boundary Fault, a major regional fault that separates the lowlands to the south from the highlands to the north.

The concept of land worth preserving gained in popularity in Scotland following World War II. A report, published in 1945, designated five areas worth considering as national parks, one of which was Loch Lomond and The Trossachs. By 1981, the areas were designated National Scenic Areas, and in 1990, it was recommended for National Park protection. Despite these recommendations, no action was taken until the Scottish Parliament met in 1999. The park was formally established in 2002. In 2017, there were over 2.9 million visits to the park (Wikipedia, Loch Lomond and The Trossachs National Park).

Rannoch Moor to the Falls of Falloch: Continue south on A82 for 39.2 km (24.4 mi; 30 min) to Stop 22, Falls of Falloch (56.349990, −4.697228) and pull into well-marked parking on the left (south). Walk about 50 m to the falls.

STOP 22 FALLS OF FALLOCH

The Falls of Falloch drop the River Falloch about 9 m (30 ft). The river flows across the Ledi Grit Formation of the Southern Highland Group and Dalradian Supergroup. The formation was deposited between 635 and 508 Ma (Ediacaran and Cambrian Periods) as deep marine sandstone and shale and then subjected to low-grade metamorphism during the Caledonian Orogeny (BGS, Geology of Britain Viewer "Classic"). The falls are the result of eroding back of a resistant layer of the Grit.

The River Falloch flows south down Glen Falloch and into the upper end of Loch Lomond.

Landforms in the surrounding countryside reflect the effects of glaciation on the area during the ice ages (Wikipedia, Falls of Falloch; Wikipedia, Geology of Loch Lomond and The Trossachs National Park).

The Falls of Falloch.

Falls of Falloch to Firkin Point*: Continue south on A82 for 23.7 km (14.7 mi; 23 min) to **Stop 23, Firkin Point** (56.170632, −4.677866) and pull into the picnic area on the left (east) side of the road.*

STOP 23 FIRKIN POINT, LOCH LOMOND

Firkin Point provides a 180° view of Loch Lomond. The lake's name may come from the Gaelic leamhan, meaning "Lake of Elms", or it might come from laom (beacon), referring to nearby mountain Ben Lomond. Historically, the Dukes of Lennox were the main landholders in the area.

Loch Lomond is 36.4 km (22.6 mi) long, between 1 and 8 km (0.62–4.97 mi) wide, and has a maximum depth of about 153 m (502 ft). It is the second largest lake by volume in Great Britain, after Loch Ness. The lake is 8 m (26 ft) above sea level.

The rocks at this spot are mapped as the Beinn Bheula Schist Formation, consisting of metasandstone and mudstone deposited between 1,000 and 541 Ma. Originally, these were deep marine sedimentary rocks, but they were altered by low-grade metamorphism during the Caledonian Orogeny.

Loch Lomond lies in a valley carved out by glaciers during the last Ice Age, mostly between 20,000 and 10,000 years ago. The Highland Boundary Fault cuts across the loch, and the difference between the Highland and Lowland geology is reflected in the shape of the loch: in the north glaciers gouged a deep channel into schist, removing up to 600 m (1,970 ft) of bedrock to create a narrow, fjord-like finger lake. Further south, the glaciers spread across the softer Lowland sandstone, creating a wider and shallower lake. Loch Lomond was connected to the sea several times since the last Ice Age. Melting of the glaciers and removing the weight of the ice has led to rebound and uplift that raised the previous shorelines to 13, 12, and 9 m (43, 39, and 30 ft) above present sea level.

There is another pullout about 760 m (2,500 ft) south of the Firkin Point turnoff. This parking area is at (56.166221, −4.667155). Should you choose to park here in order to examine the excellent roadcut in the Southern Highland Group metasediments, please mind the traffic.

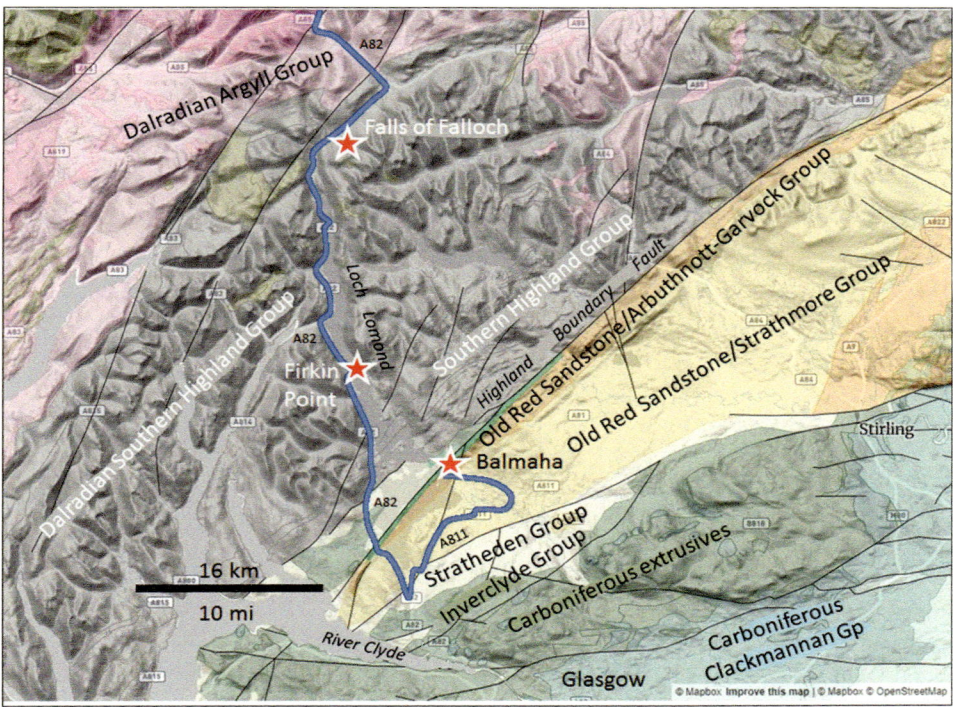

Geologic map of the Loch Lomond area. (Modified after MacroStrat, https://macrostrat.org/map/#x=-4.613&y=56.182&z=9.26.)

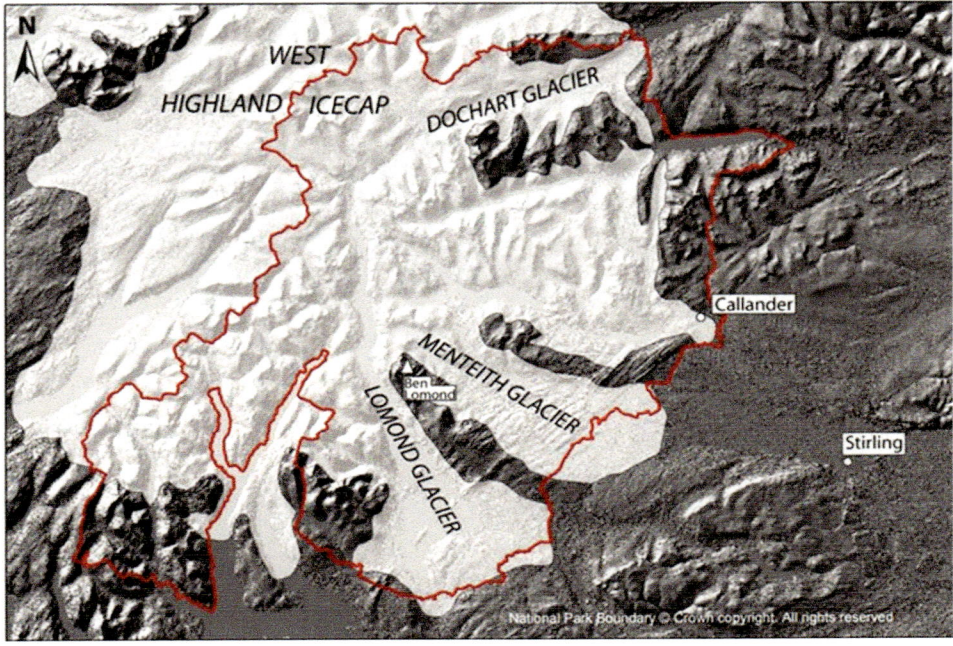

Loch Lomond glaciers during the most recent Ice Age around 25,000 years ago. (From Goodenough et al., 2008. Reproduced with permission of the British Geological Survey.)

View north up Loch Lomond from Firkin Point.

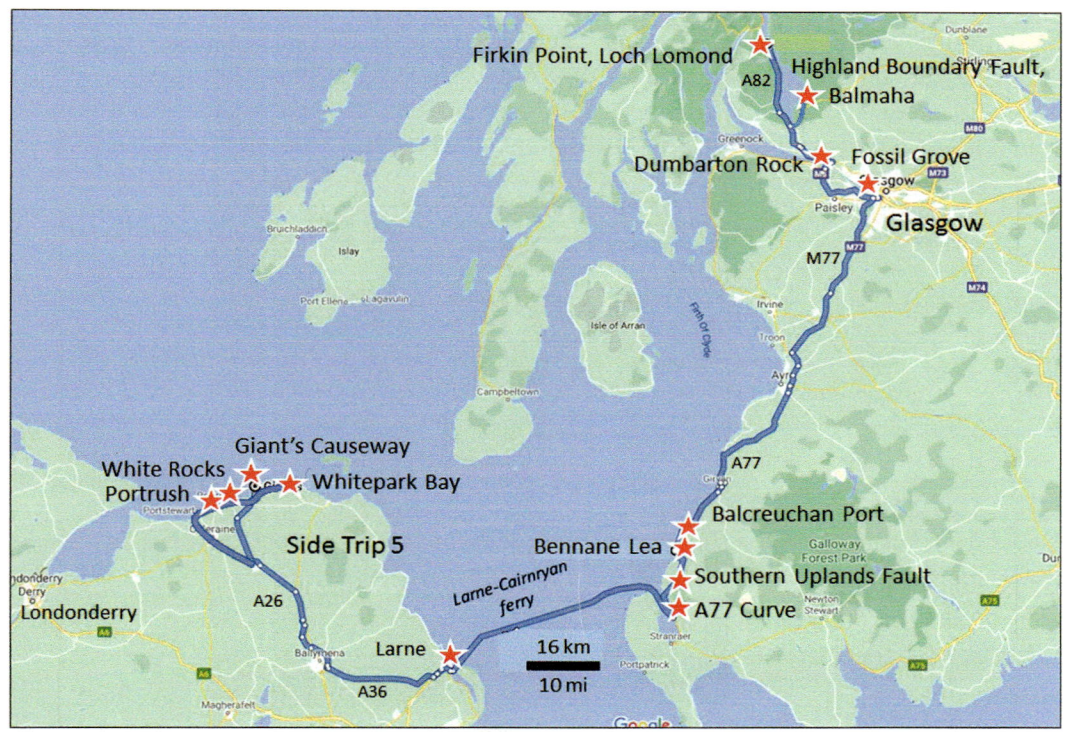

Firkin Point to Balmaha: *Continue south on A82 to the A811; at the roundabout take the 1st exit onto A811; at the next roundabout take the 1st exit onto Old Luss Road; at the next roundabout take the 3rd exit onto Balloch Road; continue straight (east) onto Drymen Road; turn left (north) onto A811/Stirling Road and drive to Main Street in Drymen; turn left (north) on B858/Main Street and drive to Balmaha Road; turn left (west) onto B837/Balmaha Road and drive to* **Stop 24, Balmaha Visitors Center Parking** *(56.085792, −4.538601) on the right (east) for a total of 42.0 km (26.1 mi; 40 min). There is abundant "pay-and-display" parking. Hike up the well-maintained trail till you get above the trees, maybe 1 km (half a mile).*

STOP 24 HIGHLAND BOUNDARY FAULT AT CONIC HILL, BALMAHA

The Highland Boundary Fault Zone divides the Grampian Highlands to the north from the Midland Valley to the south. The zone, extending from Bute in the southwest to Stonehaven in the northeast, separates differing geologic terranes. The Grampian Terrane north and west of the zone is characterized by high mountains and deep valleys in Neoproterozoic-Early Ordovician Dalradian Supergroup metamorphic slates and schists. Late Silurian through Permian sedimentary units occur south and east of the zone and are characterized by low and rounded topography (Cuthbert and Eardley, Balmaha, Loch Lomond).

The fault zone itself contains the exotic Highland Border Complex and Highland Border Ophiolite, Cambrian to Ordovician igneous rock, meta-igneous rock, mudstone, shale, sandstone, marble, and serpentine breccia. These units occur in fault-bounded slivers up to several kilometers in size (Wikipedia, Geology of Loch Lomond and The Trossachs National Park; Stephenson and Gould, 1995; Cuthbert and Eardley, Balmaha, Loch Lomond).

Early work along the fault zone indicated that the Highland Boundary Fault is a reverse fault of Caledonian age (Ramsay, 1962). The Southern Uplands Fault has been interpreted as a graben-bounding normal fault, a major left-lateral strike-slip fault, and a northwest-dipping reverse fault (Wikipedia, Highland Boundary Fault; Tanner, 2008). The 2003 Aberfoyle earthquake solution tells us that, at least today, the zone contains an oblique left-lateral strike-slip fault with reverse off-set. The fault plane was estimated to be dipping at 65° northwest (Wikipedia, Highland Boundary Fault).

The Highland Boundary Fault was active during two phases of the Caledonian Orogeny, the Early Ordovician Grampian episode and the Middle Devonian Acadian phase. Ordovician movement was dominantly vertical (uplift of the Grampian Highlands and subsidence of Midland Valley up to 4,000 m, or 13,000 ft). Evidence for the Acadian displacement is based on garnets in the Lower Old Red Sandstone that were derived from Dalradian sediments. Middle Devonian movement was dominantly left-lateral shear (<30 km, or 18 mi). Devonian offset occurred in conjunction with development of the Strathmore Syncline to the southeast in a transpressive setting (Tanner, 2008).

Post-Acadian displacement is also indicated, based on stratigraphic relationships. The older Old Red Sandstone is unconformably overlain by younger Old Red Sandstone, and the younger Old Red Sandstone is tilted to near-vertical close to the Highland Boundary Fault (Wikipedia, Highland Boundary Fault).

Conic Hill looking west to Loch Lomond.

Balmaha lies on the Highland Boundary Fault Zone. Just north of the village are outcrops of conglomerate, sometimes called "pudding stone", about 420 million years old. The conglomerate is part of the Old Red Sandstone and contains round white cobbles in a red sandstone matrix, like plums in pudding. They lie unconformably on folded and metamorphosed rocks of the Highland Border Complex. The contact between the units is an ancient erosion surface. The Highland Border

Complex had been thrust up from an ocean floor subduction zone to form a mountain range along the margin of Laurentia. In the vastness of time the mountain range was eroded and finally buried under river gravel conglomerate. The whole sequence was then tilted to its current near-vertical attitude. The north side of the Border Complex is truncated against the Gualan Fault, a branch of the Highland Boundary Fault, and this contact can be seen along the northern flank of Conic Hill (Cuthbert and Eardley, Balmaha, Loch Lomond). The Dalradian schists north of the fault zone are mostly covered by soils.

The Devonian Inchmurrin Conglomerate member (~419 to 393 Ma) was deposited by rivers and on alluvial fans. This outcrop is along the Conic Hill trail.

The Balmaha Visitor Centre is located on the Ruchill Flagstone Formation, an interbedded river channel sandstone-siltstone deposited ~419 to 393 Ma (in the Devonian). The Ruchill Flagstone is part of the Arbuthnott-Garvock Group of the Old Red Sandstone. From here, walk the West Highland Way trail 1.9 km (1.2 mi) east to the top of Conic Hill.

Once the trail starts climbing, you are in the Inchmurrin Conglomerate member, deposited in Devonian river channels and alluvial fans.

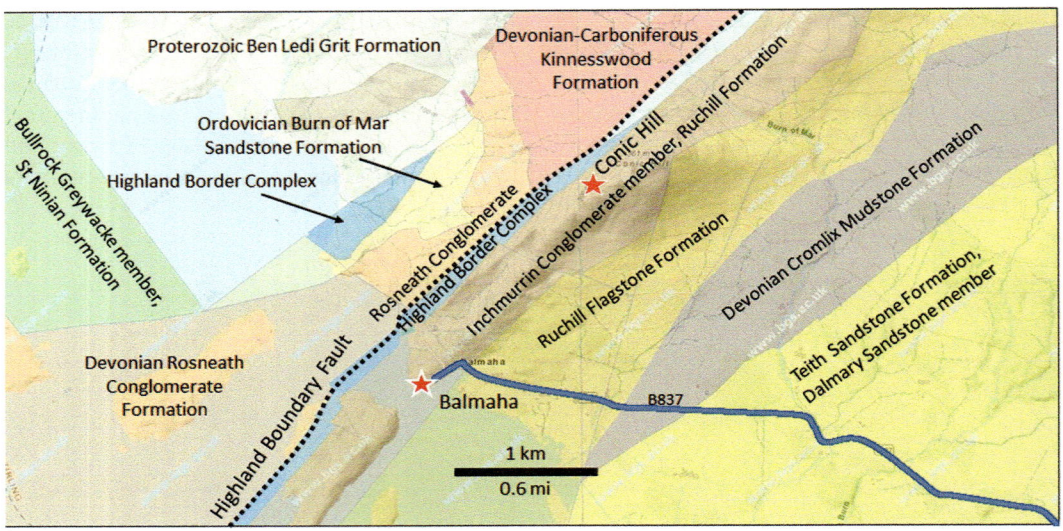

Geologic map of the Balmaha area and Highland Boundary Fault. (Modified after BGS, Geology of Britain Viewer. Contains British Geological Survey materials © UKRI 2022. Base mapping provided by ESRI.)

STOP 25 GLASGOW

At 611,000 inhabitants (2019), Glasgow is the largest city in Scotland and the third most populous in the United Kingdom. The name comes from the Gaelic glas cöü (green hollow). It is located on the River Clyde in the West Central Lowlands of Scotland.

People have occupied the plains around Glasgow for thousands of years, first in hunting and fishing communities, then in farming and trading centers. The Romans built a fort here, as well as the Antonine Wall to keep raiding Celts and Picts from Roman Britain. The actual town is said to have been founded by Saint Mungo, who in the 6th century built a church on the site of the future Glasgow Cathedral. The first Glasgow Fair was held in 1190, as a result of Bishop Jocelin petitioning King William for permission to hold an annual fair free of taxes. This was an occasion for local merchants to sell their wares, everything from livestock to farming implements, from household goods to slaves. The fair, taking place at the cathedral, quickly became an important part of the Glaswegian calendar. The fair continues today as a public holiday held in July in Glasgow Green.

The University of Glasgow was established in 1451, and together with creation of the Archdiocese of Glasgow in 1492, the town became a religious center and center of learning. The economy was based on agriculture, brewing, and fishing, with exports of salmon and herring to Europe and the Mediterranean. Starting in the 17th century, Glasgow became a hub in the international trade of slaves, sugar, tobacco, cotton, and linen. The trade led to Glasgow becoming a center of manufacturing and invention of industrial devices.

As the River Clyde within the city was too shallow for large seagoing ships, the city's Tobacco Lords created a deep water port on the Firth of Clyde. By the late 18th century more than half of the British tobacco trade passed through Glasgow. Opening of the Monkland Canal in 1795 facilitated access to the iron and coal mines of Lanarkshire. After dredging of the River Clyde as far as Glasgow, shipbuilding became a major industry. By the end of the 1800s, it was known as the "Second City of the Empire" and was producing over half of Britain's shipping and a quarter of the locomotives in the world. In addition to being pre-eminent in shipbuilding and engineering, it was now a major center for manufacturing industrial machinery, bridge building, chemicals and explosives, the coal and oil industries, textiles and garments, leather processing, furniture-making, food and drink, cigarette making, and publishing.

Following World War I, the city suffered a post-war recession followed by the Great Depression. The city was bombed during the Clydebank Blitz in World War II, then enjoyed a post-war resurgence that lasted through the 1950s. By the 1960s, competition from Japan and West Germany weakened many of the city's industries, which either shut down or moved elsewhere. As a result, Glasgow entered a period of economic decline. A radical program of rebuilding and urban renewal, involving mass demolition of slums and replacing them with large suburban neighborhoods and tower blocks, extended to the late 1970s. Infrastructure investment, such as roads and the Scottish Exhibition and Conference Centre in 1985, expedited Glasgow's new role as a European center for business and financial services and promoted tourism. In 1990, it acquired the status of European City of Culture (Wikipedia, Glasgow).

The rocks around Glasgow are Carboniferous sedimentary rocks ~360 to 300 Ma. During the Carboniferous, Scotland was part of Laurussia, the continent formed by the collision and amalgamation of several smaller continents during the Caledonian Orogeny. During the Carboniferous and Permian, a new mountain system formed as Laurussia collided with Gondwana during the Variscan Orogeny. Glasgow lay north of the collision zone in an area of shallow seas. The orogeny caused sedimentary basins in this area, then folded and faulted the sediments and injected them with igneous sills, dikes, and volcanoes.

Balmaha to Dumbarton Rock: *Return southeast on B837 to Drymen; turn right (south) on Main Street and drive to A811; turn right (south) on A811 and drive south and west to the A813 in Balloch; at the roundabout take the 1st exit to A813/Carrochan Road and drive south; stay on A813/Stirling Road; at the roundabout take the 2nd exit onto B830/Townend Road; continue straight onto Church Street; at the next roundabout take the 2nd exit onto A814/Glasgow Road and drive southeast to Victoria Street; turn right (south) on Victoria Street; continue straight onto Castle Road and drive to **Stop 25.1, Dumbarton Rock** (55.936007, −4.561593) for a total of 27.6 km (17.1 mi; 32 min). There is limited parking on both sides of the road.*

STOP 25.1 DUMBARTON ROCK

About 334 Ma, during the Early Carboniferous (Mississippian), widespread volcanism occurred throughout the Glasgow area. Dumbarton Rock is one of the feeder vents. The rocks show inward-inclined volcanic agglomerates (broken-up lava), typical of a volcano. After millions of years of erosion, all that is left is this 73 m (240 ft) high basalt plug that had been the root of the volcano. Interestingly, some of the underlying rocks were also broken up and mixed with the lava during the eruptions. Fragments of shale and sandstone of the Devonian Old Red Sandstone are found suspended in the basalt.

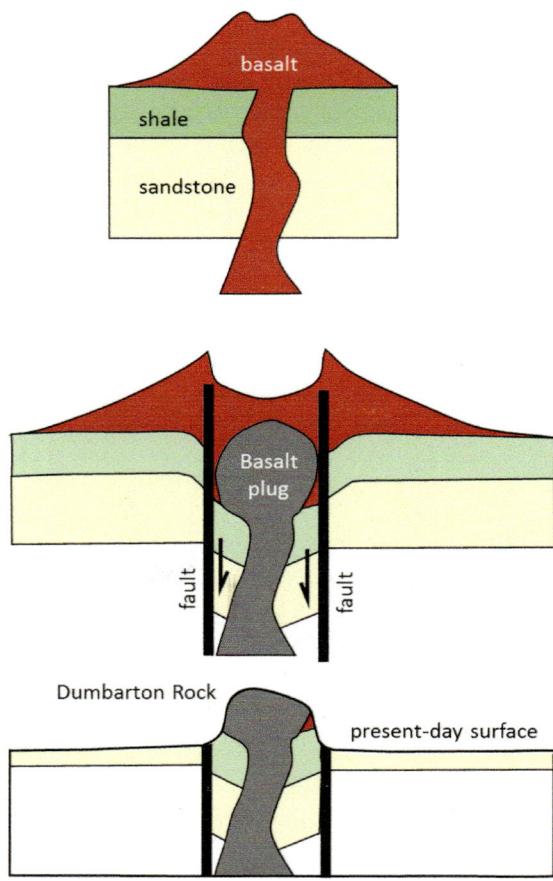

Schematic development of a volcanic plug. Top: Eruption of flows and ash. Middle: magma injection and caldera collapse along ring faults. Bottom: present-day erosion at Dumbarton Rock.

As the lava cooled, it also contracted. The cooling cracks formed the hexagonal columns seen on the flank of Dumbarton Rock. These columnar joints were formed by the same process that formed the Giant's Causeway in Northern Ireland and Samson's Ribs in Edinburgh (Geological Society of Glasgow, Dumbarton Rock; Whyte and Weedon, 1992).

Visit

Dumbarton Castle sits on Dumbarton Rock and can be toured, although at the time of writing (2022) there were no tours due to ongoing renovation work. Dumbarton Castle has the longest recorded history of any stronghold in Scotland. Ptolemy's "*Geography*" (2nd century AD) names some of the places and early peoples of Scotland. At that time, Dumbarton Rock was in the territory of people that Ptolemy calls the Damnonii who lived between the River Clyde and the Firth of Forth and north to Strathearn. The Britons who had a stronghold at Dumbarton were probably descendants of the Damnonii. The name Dumbarton is from the Gaelic Dun Breatann – meaning Fortress of the Britons.

   Bede, in his "*Ecclesiastical History of the English People*" (around 731) mentions Dumbarton. He describes "a very extensive arm of the sea" (the Firth of Clyde), which "runs inland from the west for a very great distance, where there stands Alcluith, a city of the Britons, strongly fortified to this day" and "the city of Alcluith, which in their language means "the rock of Cluith," … stands near a river of that name" (Rig, 2011).

Between the fifth and ninth centuries, the castle was the capital of the independent Celtic Kingdom of Strathclyde. A castle has occupied this strategic high ground since the 1220s. Today all visible traces of the original buildings and defenses are gone. Of the medieval castle, only the 14th-century Portcullis Arch, the foundations of the Wallace Tower, the foundations of the White Tower, and a 16th-century guard house remain.

Dumbarton Rock and castle. View north.

Dumbarton Rock and castle. (Painting by John Stoddart, 1800. https://commons.wikimedia.org/wiki/File:Dumbarton_castle_and_lime_kiln.jpg.)

Most of what exists today was built in the 1700s, including the Governor's House and fortifications. The splendid views from the White Tower Crag and the Beak make it clear why this rock was chosen as a fortress.

Dumbarton Rock is protected by the Scottish Government as a Scheduled Ancient Monument.

Hours: Currently closed for renovation (2022).

Usually open daily from April to October; from 1 November to 31 March, 10:00 a.m. to 4:00 p.m. daily except Thursdays and Fridays

Reservations are required to tour the castle.

Address: Castle Road, Dumbarton
Phone: 01389 732 167
Web Site: https://www.historicenvironment.scot/visit-a-place/places/dumbarton-castle/overview/
Admission:
   Adult (16–59 years)   £6.00
   60 years+ and unemployed   £4.80
   Child (5–15 years)   £3.60
   Family (1 adult, 2 children)   £12.00
   Family (2 adults, 2 children)   £17.00
   Family (2 adults, 3 children)   £20.50

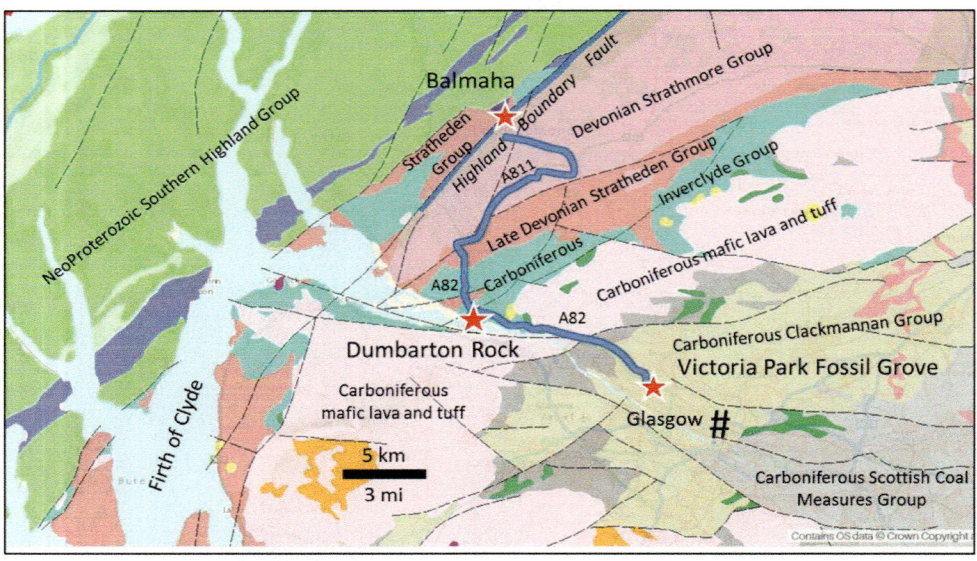

Geologic map of the Loch Lomond to Glasgow region. (Modified after BGS, Geology of Britain Viewer. Contains British Geological Survey materials © UKRI 2022. Base mapping provided by ESRI.)

Dumbarton Rock to Fossil Grove: *Return north on Castle Road; at the roundabout, continue straight on Victoria Street; at Glasgow Road turn right (east) onto A814/Glasgow Road; turn right (east) onto A82/Dumbarton Road; at the roundabout, take the 2nd exit onto A82/Great Western Road; turn right (south) onto Lincoln Ave; continue straight onto Queen Victoria Drive; turn left (southeast) onto Danes Drive; at the roundabout, take the 2nd exit (south) onto Westland Drive and* *drive to **Stop 25.2, Fossil Grove** (55.877295, −4.339213) for a total of 18.6 km (11.6 mi; 23 min). Parking is on the street only. The entrance to the Fossil Grove is at the west end of Victoria Park.*

STOP 25.2 FOSSIL GROVE

The Fossil Grove Museum in Victoria Park displays fossil stumps and root systems of 11 *Lepidodendron*, an extinct genus of tree-like plants related to quillworts and club mosses. *Lepidodendrons* sometimes reached 50 m (160 ft) tall, and the trunks were often over 1 m (3.3 ft) in diameter. They thrived during the Carboniferous Period and died out in the Late Triassic, about 205 Ma. These stumps are in their life positions, having grown in a lowland swamp roughly 325 Ma. They probably died due to flooding of a nearby river: the soft tissues of the trees then decomposed and made the trees and root systems hollow, and subsequent flooding broke off the upper trunks and filled the hollow trees with sand. The sand inside the trunks eventually became solid rock, and the outer bark of the trees became a thin layer of coal. Most of the trunks are 0.3–0.9 m (1–3 ft) in diameter and 0.6–0.9 m (2–3 ft) tall; a single larger stump in the western part of the grove measures 0.9–1.2 m (3–4 ft) in diameter (Scottish Geology Trust, Fossil Grove; Wikipedia, Fossil Grove). These are the same trees that left us the great accumulations of Carboniferous coal, the Carboniferous ("carbon-bearing") Period being named for the coal.

Ripple marks on some of the sandstone beds in the Carboniferous Limestone Coal Formation (Clackmannan Group) indicate that the river flowed to the southwest (Strathclyde Conservation Group, The Fossil Grove).

During the Early Permian, about 290 Ma, dolerite (basaltic) sills intruded the sediments and two of the trunks. Up to 8 m (26 ft) of this sill are exposed in "the defile" next to the museum. The base of one sill is exposed in the lower part of the quarry. It was quarrying of this dolerite that led to the discovery of the fossil trees in 1887 (Scottish Geology Trust, Fossil Grove).

The Fossil Grove.

Visit

The Fossil Grove museum is currently closed (2022) due to renovation and/or lack of funding. It was originally opened in 1890, 3 years after the stumps and upper roots of fossil trees were discovered in a quarry during roadbuilding in the new Victoria Park (Scottish Geology Trust, Fossil Grove; Wikipedia, Fossil Grove). Today, the fossil trees are protected as a SSSI. The site is owned and operated by the Land & Environmental Services Department, Glasgow.

Access: Victoria Park is open 24/7, but the Fossil House has restricted hours and is currently closed. There is no dedicated Fossil Grove or Victoria Park parking area, but you can usually find nearby on-street parking.
Address: 51 Victoria Park Dr. S, Glasgow G14 9QR
Phone: +441412875918
Web Site: http://www.fossilgroveglasgow.org/
Admission is free.

STOP 26 BALANTRAE AREA

The rocks exposed between Balcreuchan Port and Bennane Lea are Ordovician ophiolites of the Balcreuchan Group, roughly 500 million years old. Ophiolites are fragments of the Earth's mantle and oceanic crust thrust onto or welded to continental crust in the suture zone between colliding continents. Remnants of the Iapetus Ocean were caught between Avalonia and Laurentia during the Caledonian Orogeny. They are now exposed here on the sea cliffs (Roberts, Ballantrae Bay, Carrick).

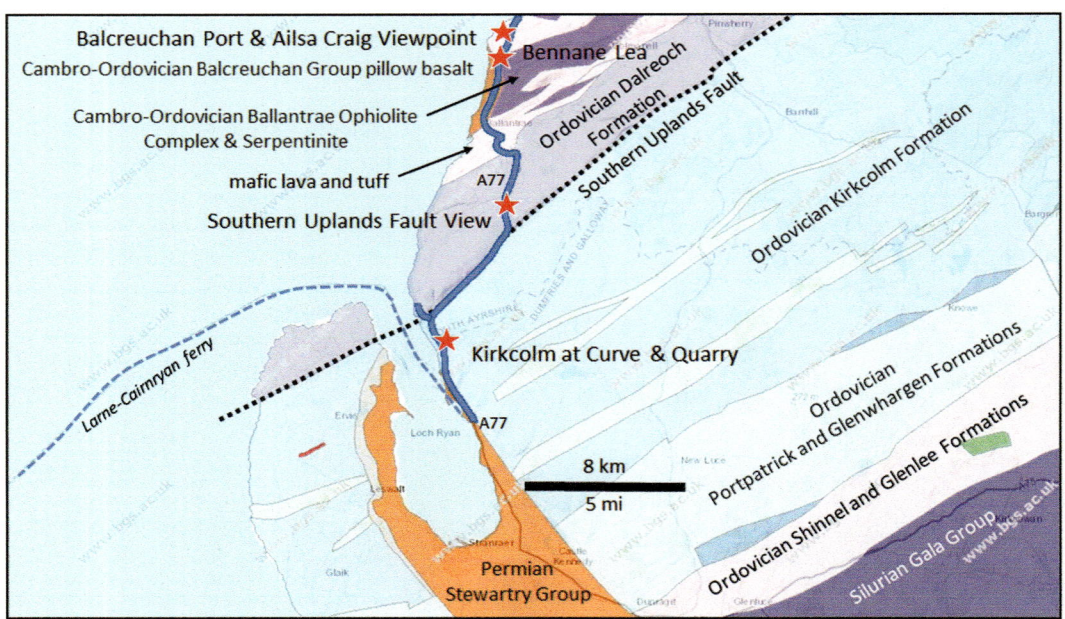

Geologic map of the Ballantrae area and Southern Uplands Fault. (Modified after BGS, Geology of Britain Viewer. Contains British Geological Survey materials © UKRI 2022. Base mapping provided by ESRI.)

Fossil Grove to Balcreuchan Port: return to Victoria Park Drive and turn right (east); turn right (south) onto Balshagray; exit onto A814 to the M8 East/City Centre/Patrick and drive south-east; merge onto the M8; keep left at the fork to merge onto the M77 and follow signs toward Prestwick Airport; keep right to continue on A77; at Glasgow/Prestwick Airport Roundabout take the 1st exit and stay on A77; continue on A77 to Girvan; at the roundabout take the 1st exit (east) to B734; turn right (south) onto Coalpots Road (sign to A77); at Shallochpark Roundabout take the 2nd exit to A77 and drive south to **Stop 26.1, Balcreuchan Port** *(55. 145352, −4.982980) parking area on the right (west), for a total of 109 km (67.5 mi; 1 h 30 min).*

Stop 26.1 Balcreuchan Port Ophiolite and Ailsa Craig Viewpoint

From the car park, there is a prominent island visible to the northwest. This is Ailsa Craig, 16 km (10 mi) west of Girvan. The island is 99 ha (240 ac) and rises to 340 m (1,120 ft) above the sea. The name comes from the Celtic Aillse Creag, meaning "fairy rock."

The island consists of a Paleogene granite, dated at 61.5 Ma, which intruded Permian and Triassic sedimentary units. Multiple joint planes produce spectacular pseudo-columnar structures in the granite that have been confused with cooling joints (Harrison et al., 1987). The granite is in turn intruded by numerous olivine dolerite dikes dated ~58.5 Ma. The island is part of the North Atlantic Igneous Province, formed during the initial opening of the North Atlantic Ocean (Wikipedia, Ailsa Craig).

Today nobody lives there, and the island is a bird sanctuary, mostly for gannets and puffins. What really makes this island unusual is what people do with the rocks found there. Since the early 1800s, the unique composition and texture of the medium-grained granite makes it particularly hard and impact resistant. This has made the rock a favored material for curling stones. Curling is a sport that originated in Scotland, wherein players slide stones on a sheet of ice in an attempt to get the most stones closest to the center of a target. At its peak, the island's quarries produced as many as 1,400 stones/year (Potts and Holbrook, 1987).

Ailsa Craig. View northwest from Balcreuchan Port.

You can find pebbles of the pale speckled granite of Ailsa Craig on the beaches around Ballantrae, probably carried here by glaciers (Roberts, Ballantrae Bay, Carrick).

From the parking area, there is a footpath that leads down to the beach. Be careful as the path can be steep and slippery. Walking along the shore is easy at low tide, but can be hazardous at high tide. The cliffs contain clearly exposed tholeiitic basalt pillow lavas typical of island arcs or oceanic hots spots like Hawaii and Iceland.

At Balcreuchan Port, you can see rocks from the mantle, once dunite (almost entirely the green mineral olivine) and peridotite (olivine and pyroxene), which have been altered to green and red serpentinite. The mantle rocks are interleaved by faulting, probably thrusts, with basaltic pillow lavas, once part of the ocean floor. The basalts have ages ranging from 501 to 476 Ma. Some Ordovician to Silurian sediments are also exposed, from shallow water conglomerate to deep water turbidites as you move up the section. The main tectonic episode, folding and north-directed thrusting of the ophiolite complex, occurred in Late Silurian (Stone et al., 1996b).

A resistant, north-trending basalt dike intruded the softer serpentinites in the Tertiary (Roberts, Ballantrae Bay, Carrick).

A sea cave in the lava sequence was once the hideout of Sawney Bean, a notorious 16th century outlaw and cannibal. South of the cave, on the southwest headland, a fault separates serpentinite from overlying lavas. Continuing southwest, there are sandstones that dip steeply to the west conformably overlain by more pillow lavas. Some pillow lavas are cut by basalt dikes that are considered to be feeders for the next higher pillow lava sequence (Stone et al., 1996b).

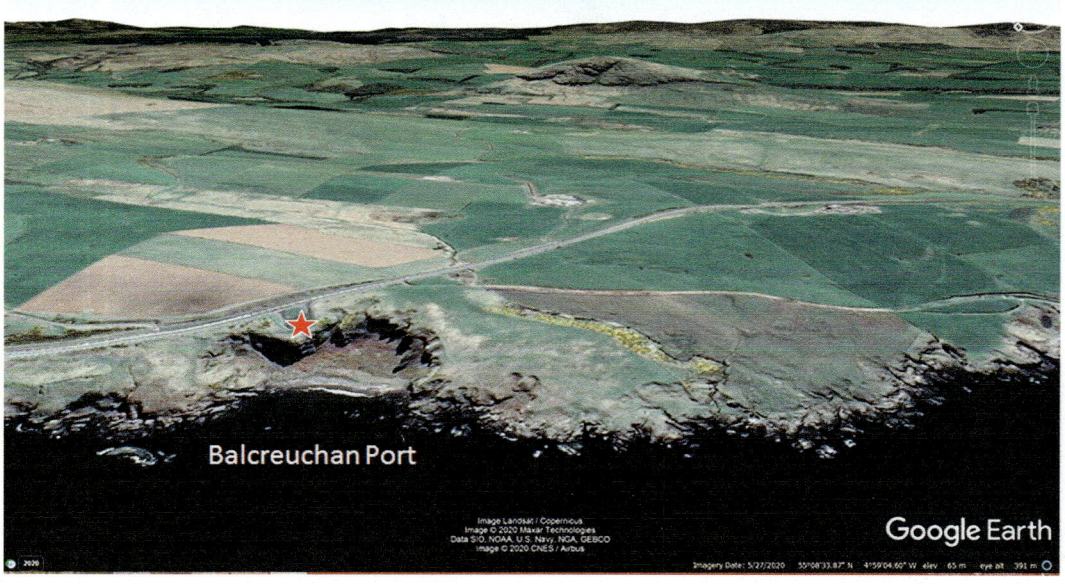

Balcreuchan Port. Google Earth oblique view east. (Image Landsat/Copernicus; Image © 2020 Maxar Technologies; Image © 2020 CNES/Airbus; Data SIO, NOAA, U.S. Navy, NGA, GEBCO.)

*Balcreuchan Port to Bennane Lea: Continue south on A77 for 1.9 km (1.2 mi) and turn right (west) and park at the gate. This is **Stop 26.2, Bennane Lea** (55.132118, −4.994594). Park outside the gate, walk along the old road to the beach, and turn right (north). Old A77 is now a private road. Walk north along the base of the cliffs and beware of falling rocks.*

Stop 26.2 Bennane Lea Pillow Lavas and Raised Beach Terrace

Isostatic rebound due to removal of the weight of glaciers has raised former beaches at Bennane Lea. The uplifted beaches and old sea cliffs can be seen as prominent terraces along the coast from here south to Ballantrae (Roberts, Ballantrae Bay, Carrick). The beach terraces are cut into glacial till (Stone et al., 1996b).

Rocks at this stop belong to the Balcreuchan Group. Bennane Lea is at the faulted contact between resistant, basaltic lava sea cliffs to the north, and less-resistant serpentinite to the south. The fault appears as a thin silicified zone with a massive tuff forming the first outcrop to the north. North of the tuff is a conformable contact with underlying, thin-bedded radiolarian cherts that have been deformed into small-scale disharmonic folds that may be tectonic or the result of soft-sediment deformation. Farther north alternating beds of chert and conglomerate are deformed into large upright anticlines and synclines. Slightly farther north fracture planes appear to be coated with malachite, a green copper carbonate (Stone et al., 1966b).

South of this stop red sandstone beds, dipping gently south, can be seen on the beach at low tide. These belong to the Permian Corseclays Formation. At Bennane Lea, the Permian is unconformably above the Ballantrae Complex, and the basal red sandstone contains fragments of the underlying lava (Stone et al., 1996b).

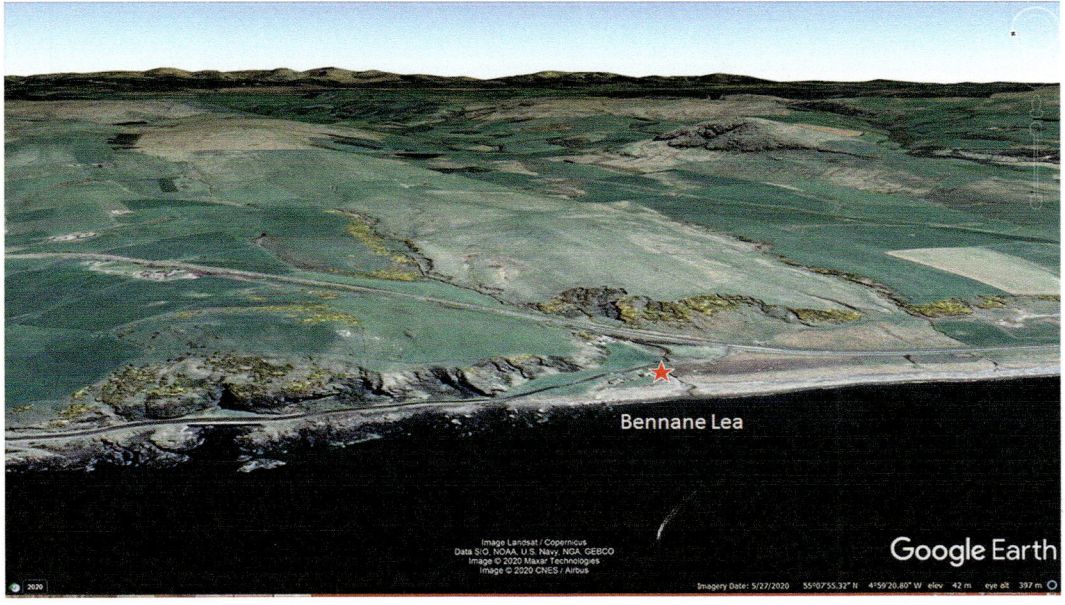

Bennane Lea. Google Earth oblique view east. (Image © 2022 Landsat/Copernicus.)

Bennane Lea sea cliffs and pillow lavas of the Ordovician Balcreuchan Group. View east.

Raised beach terrace, Bennane Lea. View south.

STOP 27 SOUTHERN UPLANDS FAULT AND SOUTHERN UPLANDS

The Southern Uplands Fault extends northeast from Loch Ryan to Dunbar on the northeast coast. It forms the southern boundary of the Central Lowlands and the northern boundary of the Southern Uplands. It is part of the Caledonian collisional suture zone between Avalonia to the south and Laurentia to the north. The Southern Uplands are largely an accumulation of Ordovician-Silurian deep marine sediments scraped off the seafloor and pushed northward onto the new megacontinent (The Geological Society, Southern Uplands; Gazeteer for Scotland, Southern Uplands Fault).

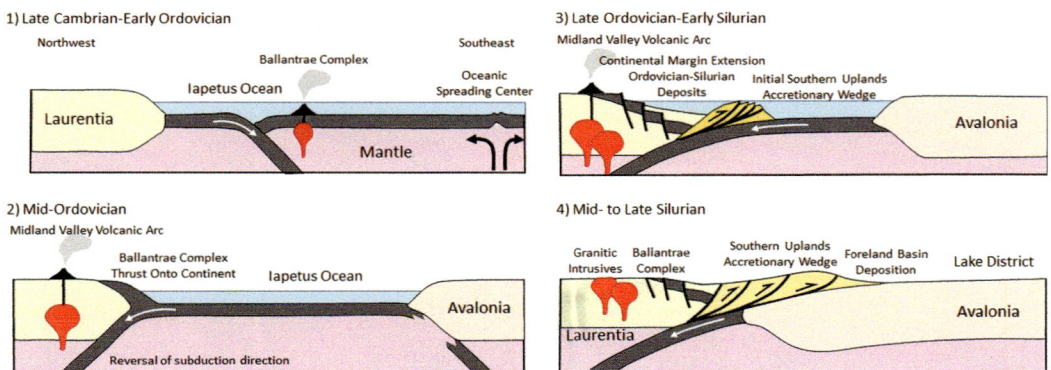

Schematic diagram showing the evolution of the Southern Uplands from Cambrian to Late Silurian time. (Modified after Stone et al., 2012a).

The Midland Valley is a rift, bound on the north by the Highland Boundary Fault and and on the south by the Southern Uplands Fault, that dropped mainly Devonian to Carboniferous rocks as much as 4,000 m (13,123 feet) (Gazeteer for Scotland, Southern Uplands Fault). In fact, both right-lateral and left-lateral offsets as well as normal offsets have been mapped on this complex fault zone (Wikipedia, Southern Uplands Fault). We cross the fault zone at the Glen App Fault, one of several strands that make up the Southern Uplands Fault. Here, we leave behind the Midland Valley and enter the Southern Uplands.

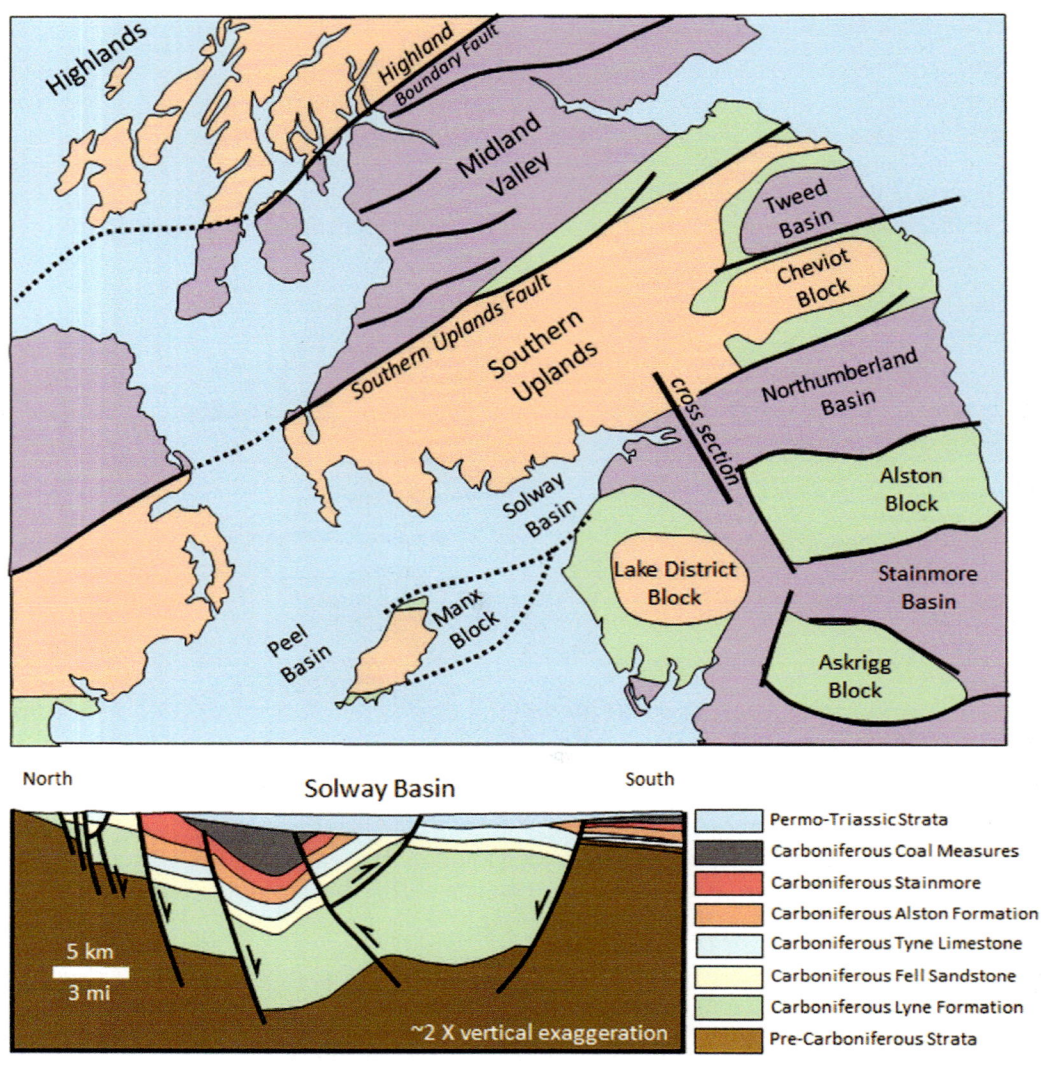

Structure of the Southern Uplands and Solway Basin. (Modified after Stone et al., 2012b.)

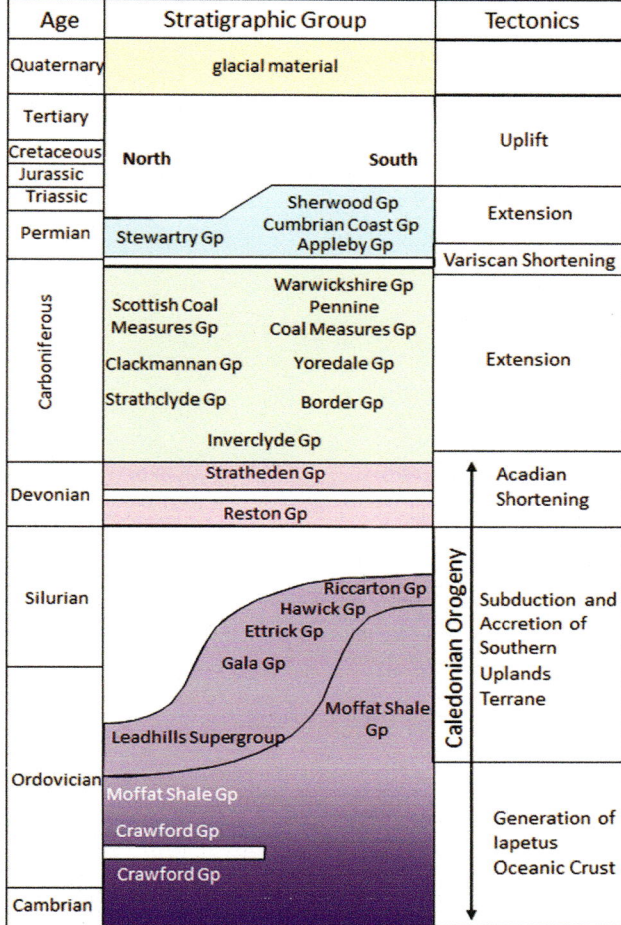

Age	Stratigraphic Group		Tectonics
Quaternary	glacial material		
Tertiary	**North** **South**		Uplift
Cretaceous			
Jurassic			
Triassic	Sherwood Gp		Extension
Permian	Stewartry Gp Cumbrian Coast Gp Appleby Gp		
Carboniferous	Scottish Coal Measures Gp Warwickshire Gp Pennine Coal Measures Gp		Variscan Shortening
	Clackmannan Gp Yoredale Gp		Extension
	Strathclyde Gp Border Gp		
	Inverclyde Gp		
Devonian	Stratheden Gp		Acadian Shortening
	Reston Gp		
Silurian	Riccarton Gp Hawick Gp Ettrick Gp Gala Gp Moffat Shale Gp Leadhills Supergroup	Caledonian Orogeny	Subduction and Accretion of Southern Uplands Terrane
Ordovician	Moffat Shale Gp Crawford Gp Crawford Gp		Generation of Iapetus Oceanic Crust
Cambrian			

Stratigraphic column for the Southern Uplands. (Modified after Stone et al., 2012c.)

Bennane Lea to Southern Uplands Fault View: Return to the A77 and turn right (south); drive south for 10.3 km (6.4 mi; 10 min) to **Stop 27.1, Southern Uplands Fault View** *(55.058339, −4.987271). There is a pullout on the right (west) side of the road.*

STOP 27.1 SOUTHERN UPLANDS FAULT VIEW, DOWNTHROWN SIDE

From this stop, you can look southwest to the escarpment of the Southern Uplands Fault. The pullout is on the downthrown side of the fault, in the Midland Valley/Central Lowlands. The rocks beneath us are Ordovician Dalreoch Formation (Tappins Group) graywacke deposited 458–449 Ma in a deep marine environment.

The Southern Uplands Fault forms this gentle escarpment. View southwest.

*Southern Uplands Fault View to Deformed Kirkcolm Formation: Continue south on A77 for 7.1 km (4.4 mi; 6 min) and Park at the turnoff to Finnart's Bay. Walk 270 m (880 ft) south along A77 to the abandoned and overgrown quarry on the east side of the road. This is **Stop 27.2, Deformed Kirkcolm Formation** (55.007930, −5.047601).*

Park at the turnoff to Finnart's Bay and walk 270 m (880 ft) south along the A77 to the abandoned quarry on the east side of the road. The quarry is getting quite overgrown, but you may be able to see the structures without entering the quarry. Be mindful of the traffic and loose rocks.

Stop 27.2 Deformed Kirkcolm Formation at A77 Curve and Quarry

The quarry contains a north-verging overturned syncline developed in Kirkcolm Formation gray-wackes. At the quarry entrance, the graywacke beds are right-way-up and dip south, whereas to the south along the road the beds dip steeply south but are overturned. Up-section can be determined by sedimentary structures: many of the graywacke beds are clearly graded, and bed bases commonly have flute (scour) and load casts.

Continuing south along the road, the graywacke beds remain uniformly overturned and dip steeply southeast. About 100 m (330 ft) southeast of the quarry, the beds are folded into an anticline–syncline pair.

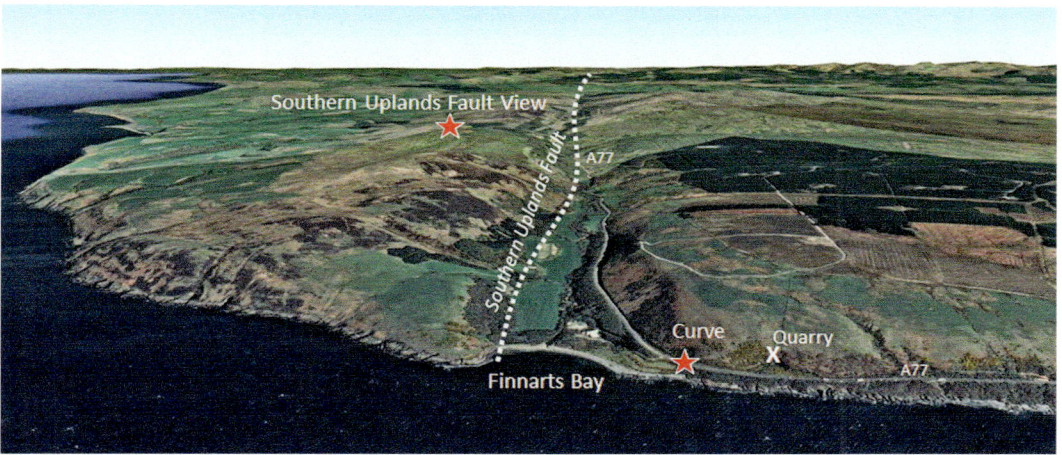

Southern Uplands Fault stops. Google Earth oblique view northeast. (Image Landsat/Copernicus; Image ©️ 2022 CNES/Airbus.)

South-dipping overturned Kirkcolm Formation at the A77 curve, upthrown side of the Southern Uplands Fault. Street View northeast.

Abandoned quarry and folds in Kirkcolm Formation graywackes. Street View east.

Deformed Kirkholm to Monreith: Continue south on A77 to Innermessen; turn left (south) onto A751 and drive to A75; turn left (southeast) onto A75 and drive to A747; turn right (south) onto A747 and drive past Monreith to the sign for St Medan Golf Club on the right; turn right (west) and drive through the golf club to the public car park at the beach. This is **Stop 28, Back Bay Monreith** *(54.723571, −4.540923) for a total of 52.4 km (32.5 mi; 50 min). Walk east along the beach toward the cliff at Back Bay. This area is only accessible at mid to low tide.*

Side Trip 5, Deformed Kirkcolm Formation to Larne: Continue south on A77 to the P&O ferry terminal on the right; take the Larne-Cairnryan ferry to the Larne ferry terminal; exit the terminal, and at the first roundabout, take the 2nd exit; at the next roundabout take the 5th exit (northeast) onto Olderfleet Road; turn left (northwest) onto Fleet Street; bear right onto Coastguard Road; turn right (northeast) onto Tower Road and drive to Larne Leisure Centre parking on the left. This is **Stop ST5.1, Triassic–Jurassic Transition, Larne** *(54.854440, −5.801995), for a total of 68.0 km (42.2 mi; 2 h 16 min). There is abundant free parking at the Leisure Centre. Book the ferry in advance, and be there at least ½ hour before departure or you may not get on.*

SIDE TRIP 5 NORTHERN IRELAND

It should come as no surprise that the geologic history and rocks of Northern Ireland are nearly identical to those across the Irish Sea in Scotland. They have been a part of the same continental and tectonic blocks for the past billion years. Also, as in Scotland, there are substantial areas underlain by Precambrian metasediments, Ordovician through Cretaceous sedimentary rocks, and Paleogene intrusions and volcanics.

Once part of Laurentia, along with much of North America, this part of Ireland shares Ordovician and Silurian sedimentary rocks with their counterparts across the Irish Sea in the Southern Highlands, and in Maritime Canada and New England. They were welded to the rest of Ireland during the Ordovician-Middle Devonian Caledonian Orogeny. Following this mountain-building episode, the area was eroded and subsided such that swamps in the later Carboniferous allowed accumulation of the Coal Measures. The Mercia Mudstone Group tells us the area was a desert during the Triassic, after which the area subsided and nearshore marine mudstones dominated during the Jurassic and, like the Jurassic Coast in England, have abundant marine fossils (for example, at Whitepark Bay). Carbonate banks and reefs were deposited as chalk in warm tropical seas during the Cretaceous, forming white sea cliffs in Northern Ireland's own version of the White Cliffs of Dover at White Rocks.

During the Mesozoic, the Atlantic opened from south to north like a zipper, and the rifting reached Northern Ireland in the Paleogene (Paleocene through Eocene). Not surprisingly, the extension and spreading were accompanied by widespread volcanism. Dolerite sills and basaltic lava flows cover large areas and form some of the iconic geologic features of the country, including the Giant's Causeway and Ramore Head. The continuation of these volcanics can be seen at Fingal's Cave on the Isle of Staffa (Scotland), the counterpart of the Giant's Causeway on the other side of the Irish Sea.

The climate began to cool during the Miocene and cooled significantly during the Pleistocene. Over the past 2 million years glaciers several kilometers thick advanced and retreated several times. They covered much of the land, grinding down the landscape, eroding and smoothing the surface, and depositing glacial debris. When the last glaciers melted roughly 12,000 years ago, the geology looked much as it does today.

ST5.1 Triassic–Jurassic Transition, Larne

The coast at Larne exposes a sequence of sedimentary rocks deposited across the Triassic–Jurassic transition. It is unique in Northern Ireland and is one of the best exposures of this contact in Europe.

The north-dipping rocks get younger as you walk north along the beach. Starting at the Leisure Centre in the south, you see the Mercia Mudstone Group deposited during the arid Triassic Period.

The orange–red delta, tidal flat, and shallow water mudstones were succeeded in Late Triassic time by pale gray–green silty mudstones of the Collin Glen Formation. Above the Collin Glen are gray to black deep marine fossil-bearing mudstones of the Late Triassic Penarth Group. Jurassic rocks (shallow marine mudstone, siltstone, sandstone, and limestone) of the Lias Group are identified by the first appearance of the ammonite *Psiloceras planorbis*. The Penarth Group and Lias Group contain bivalves, crinoids, and ammonites. The most complete example of an ichthyosaur in Northern Ireland was discovered here at Waterloo Bay. In 1999, the "Larne Sea Dragon" was found in the Langport member of the Lilstock Formation, Penarth Group. The backbone, rib cage, bits of the front limbs, the lower jaw, and several teeth were found. These are on display in the Ulster Museum (Wikipedia, Penarth Group).

If you want to see any rocks on the beach, visit during low tide. If the tide is up, walk north along the coastal path to see interpretive panels and outcrops of Jurassic rocks in the sea cliffs.

View north, Waterloo Bay beach. The foreground strata are Triassic, and in the distance, the presence of ammonite fossils marks the transition to the Jurassic. (Photo courtesy of Anne Burgess / Waterloo / CC BY-SA 2.0, https://commons.wikimedia.org/wiki/File:Waterloo_-_geograph.org.uk_-_454845.jpg.)

Gently north-dipping Jurassic sediments, Larne Beach coastal path.

Side Trip 5, Larne to Whitepark Bay: Return southwest on Tower Road to Curran Road; turn right (west) on Curran Road; continue straight on Main Street; at the roundabout take the 2nd exit (north) onto Agnew Street; turn left (west) onto Victoria Road; continue straight on Pound Street; turn right (west) onto A8/Ballymena Road/The Harbour Way; bear right to stay on A8/Ballymena Road; at the roundabout take the 3rd exit (northwest) onto A36/Shanes Hill Road; continue straight on A36/Church Road; continue straight on A36/Moorfields Road; at the Lame Road roundabout take the 5th exit north and merge onto the M2; merge onto the A26 and continue driving north; at roundabout, take the 2nd exit (north) onto A44/Drones Road; at Armoy turn left (west) onto Main Street; then turn right (northwest) onto Carrowreagh Road; bear right (north) onto B147/Ballinlea Road; turn left (west) onto Kilmahamogue Road; turn left (northwest) onto Straid Road; turn right (north) onto Craigalappan Road; turn left (west) onto A2/Whitepark Road, then sharp right to Whitepark Bay Youth Hostel and park. Walk 630 m (2,100 ft) to the beach. This is **Stop ST5.2, Whitepark Bay** *(55.228926, −6.406591), for a total of 82.0 km (50.9 mi; 1 h 8 min).*

ST5.2 WHITEPARK BAY

Rocks of the Lower Jurassic Lias are present on the shores of Whitepark Bay. On the east side of the bay, during low tide, you can see gray mudstones and limestones of the Triassic Waterloo Mudstone Formation. Marine fossils including ammonites, belemnites, and bivalves can be found on the beach. Upper Cretaceous chalk is exposed in the sea cliffs and contains belemnites, sea urchins, and brachiopods.

A time interval of more than 100 Ma is missing between the Liassic mudstones and the overlying Cretaceous rocks, an impressive unconformity.

This attractive beach is dangerous for swimmers due to rip currents.

View northeast across White Park Bay toward Ballintoy headland.

On the east side of Whitepark Bay is Ballintoy Harbor and headland. Between Ballintoy Harbor and Whitepark Bay, the shore contains a metamorphosed chalk. The Cretaceous chalk was covered with and metamorphosed by Paleogene lava from the Bendoo volcanic plug that forms the dark headland.

Ballintoy is known for its raised shoreline. During the last Ice Age, thick sheets of ice depressed the land. As the ice melted, the land began to rebound. The former quarry at Ballintoy Harbor car park contains an uplifted sea cave.

Side Trip 5, Whitepark Bay to Giant's Causeway*: Return to A2/B147/Whitepark Road and turn right (west); turn right (north, then west) onto Causeway Road; turn right (north) onto Giant's Causeway and drive to* **Stop ST5.3, Giant's Causeway** *(55.240579, −6.511423), for a total of 9.0 km (5.6 mi; 16 min).*

ST5.3 Giant's Causeway World Heritage Site, Northern Ireland

The Giant's Causeway is known for the spectacular hexagonal columns of basalt that occur there. It is one of only two World Heritage Sites in Great Britain owing to its geology. The tops of the columns form stepping stones that lead from the sea cliff foot to the shore and beyond. Most of the columns are hexagonal, and the lava cliffs reach 28 m (92 ft) in places (Wikipedia, Giant's Causeway).

It lies about 4.8 km (3 mi) northeast of the town of Bushmills, a village of 1,295 (2011) that is famous for its Irish Whiskey. The Old Bushmills Distillery is worth a tour in its own right (2 Distillery Road, Bushmills, BT5H 8XH). The area has also been featured in films, notably Game of Thrones.

Between 63 and 62 Ma, during the Paleocene, basalt poured out of fissures in the Cretaceous Ulster White Limestone and formed a volcanic pile known as the Thulean Plateau. The fissures were a result of extension and rifting of Laurasia during initial opening of the Atlantic. The early eruptions produced at least six flows. After a period of no activity, a second episode began. The first of these eruptions formed a lava lake up to 90 m (295 ft) deep. These are the flows seen here at the Giant's Causeway. Another 6–8 eruptions completed the Causeway Tholeiite member. The final Upper Basalt Formation erupted following another long quiet period. Feeder pipes for the last eruptions can be seen cutting across the earlier lavas (Lemon, 2012).

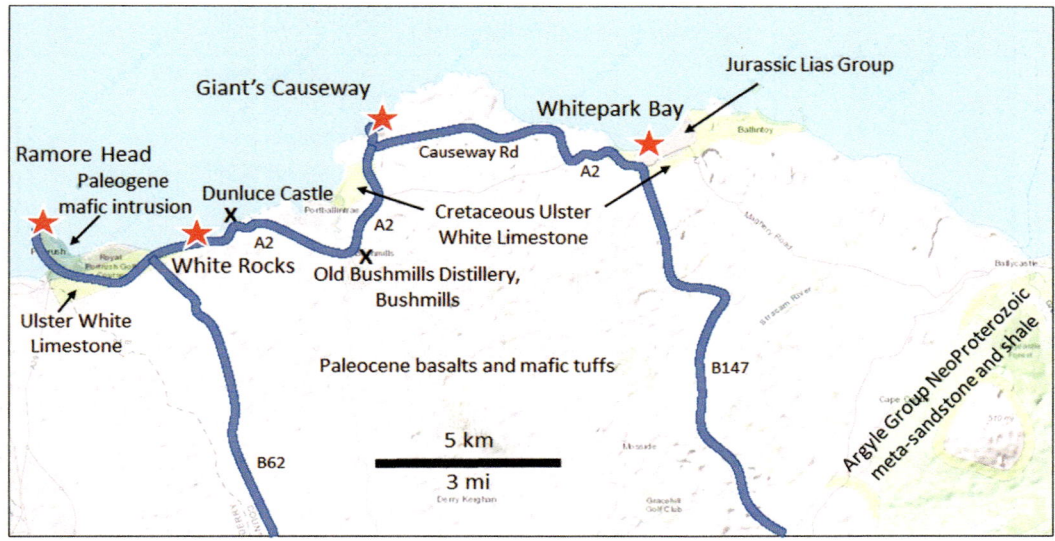

Geologic map of the Giant's Causeway area. (Modified after BGS, Geology of Britain Viewer. Contains British Geological Survey materials © UKRI 2022. Base mapping provided by ESRI.)

So, what caused the columns? The thick pool of lava cooled slowly, and as it did so, the lava contracted and fractured, much like pumpkin pie filling after it is taken out of the oven. Starting at the upper surface of the lava, the vertical fractures propagated downward and inward. It turns out that the least amount of work needed to create fractures occurs when they form at 120° to one another. That just happens to create these astonishingly regular hexagonal (and some pentagonal) cooling fractures.

Of course, before we knew anything about basalt flows and cooling cracks, there were other explanations for this wonder of nature. Pre-Christian myths attributed these to the "stepping stones of the Fomhóraigh," an Irish race of giants. According to more recent legend, the columns are the ruins of a causeway built between here and Scotland by a giant, Fionn mac Cumhaill (Finn MacCool). The same lava flow and jointing occur at Fingal's Cave on the Isle of Staffa across the channel. In one version Fionn was challenged to a fight by the Scottish giant Benandonner; in another version, he wanted to bring over his Scottish girlfriend and Benandonner was his rival for her affections. In any case, Fionn accepted the challenge and built the causeway across the North Channel by throwing boulders into the sea. In one story, Fionn defeats Benandonner. In another, Fionn hides from Benandonner when he realizes that his rival is much larger than he is. Then, Fionn cleverly disguises himself as a baby tucked in a cradle. When Benandonner sees the size of the baby, he reckons its father must be a true monster, and he flees back to Scotland, destroying the causeway behind him (Wikipedia, Giant's Causeway; Sykora, 2019).

Lithology	Age	Group/Formation	Tectonics
	Holocene	Surficial Deposits (alluvium, till)	
			Alpine Orogeny
	Oligocene	Lough Neagh Group	
	Paleocene	Antrim Lava Group; 'Clay with flints'	
			North Atlantic Rifting
	Cretaceous	Ulster White Limestone Formation	
		Hibernian Greensands Formation	
			Cimmerian Uplift
	Jurassic	Waterloo Mudstone Formation	Breakup of Pangaea
	Triassic	Penarth Group	Opening of the Atlantic
		Mercia Mudstone Group	
		Sherwood Sandstone Group	
	Permian	Belfast Group	
		Enler Group	
	Carbonif.	Slievebane/Coal Measures Group	Variscan Orogeny
		Millstone Grit Group	Closing of Iapetus Ocean
		Leitrim/Kilskeery Groups	Caledonian Orogeny
		Tyrone/Omagh Sandstone Groups	
	Silurian	Hawick Group	
		Gala Group	
	Ordovician	Leadhills Supergroup	
	Neoproterozoic	Southern Highlands Group	Opening of Iapetus Ocean
		Argyll Group	

Stratigraphic column for northeastern Ireland. (Modified after Mitchell, 2004.)

The Giant's Causeway first received international attention when Dublin artist Susanna Drury painted it in 1739. The paintings won Drury an award from the Royal Dublin Society. The French Encyclopédie featured an engraving of the area in 1768. In the caption to the plate, the French geologist Nicolas Desmarest was the first to suggest that such features were volcanic (Wikipedia, Giant's Causeway). The site first became popular with tourists after the opening of the Giant's Causeway Tramway, an electric railway opened in 1883. It was declared a UNESCO World Heritage Site in 1986 and became an Irish NNR in 1987.

Visit

One of the most popular attractions in Northern Ireland, visitors can walk over the basalt columns about 1 km (0.5 mi) from the entrance. A new visitor center was officially opened in 2012. Over 998,000 tourists visited the site in 2019.

Parking costs £10, and is pay-by-phone only, which is pretty annoying if you aren't already set up.

Access to the Giant's Causeway is free of charge. It is not necessary to go through the visitor center, which does charge an entrance fee (Wikipedia, Giant's Causeway).

In addition to a gift shop, café, and toilets, the visitor center has interactive exhibits that explore the Giant's Causeway, and outdoor audio guides providing information in nine languages.

A view of the Giant's Causeway, East Prospect. Engraving by Susanna Drury, 1768. (Linda Hall Library of Science, Engineering, and Technology, https://commons.wikimedia.org/wiki/File:Drury_-_View_of_the_Giant%27s_Causeway.jpg.)

Giant's Causeway hexagonal columnar basalt.

Hours: 10:00 a.m. to 4:00 p.m. daily (closed on holidays)
Address: B147 Causeway Road, 2 mi from Bushmills village (55.233242, −6.517344)
Phone: +44 28 2073 3419
Email: northcoastbookings@nationaltrust.org.uk
Web site: https://giantscausewaytickets.com/the-visitor-centre
Visitor Center Admission (online or walk-up)

Adult	£13.00
Child	£6.50
Family*	£32.50
Under 5s	FREE
National Trust members	FREE
Cliff Top Experience Walk	£39.00
(must book in advance)	

*2 adults + up to 3 children (aged 5–17 years)

Side Trip 5, Giant's Causeway to White Rocks: *Return south on Causeway Road; turn right (southwest) on Whitepark Road/A2 and drive to Bushmills; at the roundabout take the 2nd exit (west) to stay on Dunluce Road/A2; continue straight on Bushmills Road/A2; bear right onto road to Whiterocks Beach and park in lot. This is **Stop ST5.4, White Rocks** (55.205716, −6.612965), for a total of 10.9 km (6.8 mi; 19 min).*

ST5.4 WHITE ROCKS

As you approach White Rocks, you can see the ruins of picturesque Dunluce Castle (55.210306, −6.578344) perched on the edge of a basalt cliff 3.2 km (2 mi) west of Bushmills. This site served as "Castle Pyke" of the Iron Islands in Game of Thrones (Sykora, 2019). There is a parking area where you can pull off the road.

Dunluce Castle and Paleogene basalt sea cliff. (Photo thought to be pre-1899. https://picryl. com/media/dunluce-castle-giants-causeway-co-antrim-8e8311.)

White Rocks refers to the white chalk cliffs composed of Cretaceous Ulster White Limestone. A black Paleogene basalt flow caps the chalk. A number of volcanic vents intrude the chalk cliffs at White Rocks. The vents contain agglomerate consisting of rounded and angular blocks of basalt, evidence of the explosive nature of the volcanics. The vents are thought to be related to the Portrush sill, which intruded Jurassic sediments below the chalk.

Fossils in the chalk include belemnites, sea urchins, and brachiopods.

White Rocks, view west.

Side Trip 5, White Rocks to Ramore Head: Return to A2/Dunluce Road and turn right (west); continue straight on A2/Bushmills Road; to Portrush; at the roundabout, take the 1st exit onto A2/Crocknamack Road; at the roundabout, take the 2nd exit (north) onto Eglinton St; drive north on Eglinton and bear slight left onto Kerr Street; continue straight onto Ramore Ave; turn right (north) to continue on Ramore Ave; drive to the parking area on the northernmost point. This is **Stop ST5.5, Ramore Head** *(55.211345, −6.657593), for a total of 5.3 km (3.3 mi; 8 min).*

ST5.5 RAMORE HEAD, PORTRUSH

This is the site of one of the early controversies in the science of geology. During the 1700s, a group known as the "Neptunists" believed that all rocks were deposited or precipitated in the sea. This included igneous rocks. Another group, the "Plutonists" (or Vulcanists), believed that all rocks were the result of igneous processes, that is, began as a melt. At Ramore Head, there was a rock first described in 1799, "Portrush Rock," as a fossil-bearing basalt. The arguments that followed lasted for decades.

It turns out that both were right and both were wrong. The rock looked like basalt but also had Jurassic ammonite fossils. Today we know that the rock is a fossil-bearing mudstone that was subject to metamorphism by heating adjacent to a Paleocene dolerite sill (a shallow basalt intrusion). Technically, the altered rock is a hornfels, a non-foliated fine-grained rock that was recrystallized by metamorphism.

The exposures are part of the Ramore Head and the Skerries Area of Special Scientific Interest (ASSI), protected specifically for its geological features. It is also a NNR (Lemon, 2012). In 2014, several fossils were stolen from this site. According to BBC News (2014), Environment Minister Durkan said, "the reserve was a place of pilgrimage for geologists and an important educational site for geology students because of its historical importance. The damage caused by the thieves will mean that visitors to the reserve will be hard-pressed to locate the fossils which made it famous."

Ramore Head hornfels. 20 pence coin for scale.

Ammonites in Portrush hornfels. (Photo courtesy of Michael Dempster, Northern Ireland Environment Agency. These unique fossils are located in the Portrush NNR and are part of the Ramore Head and the Skerries Area of Special Scientific Interest, protected specifically for its geological features.)

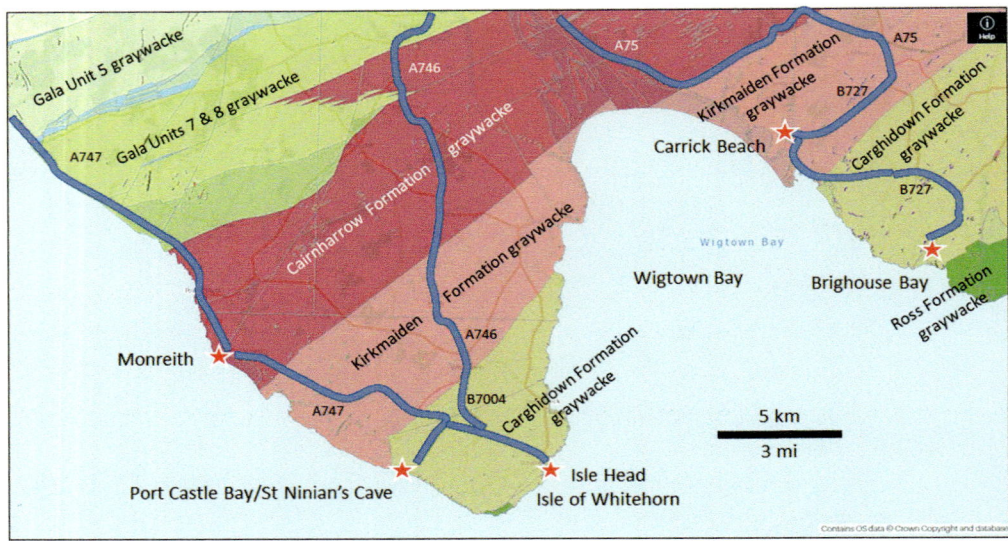

Geologic map of the Monreith Bay-Wigtown Bay area showing Silurian units of the Southern Uplands in southwest Scotland. (Modified after BGS, Geology of Britain Viewer. Contains British Geological Survey materials © UKRI 2022. Base mapping provided by ESRI.)

*Ramore Head to Back Bay, Monreith: Return on the A26 to the A36 to the Larne ferry terminal and take the ferry to Cairnryan; turn right (south) on A77 and drive to the A751 at Innermessen; turn left (south) onto the A751 and drive to the A75; turn left (southeast) onto the A75 and drive to the A747 at Glenluce; turn right (south) onto the A747 and drive about 1.3 km (0.8 mi) past Monreith to the sign on the right for St Medan golf club; turn right (west) and drive through the golf course to the free public car park at the beach. This is **Stop 28, Back Bay Monreith** (54.723571, −4.540923) for a total of 192 km (119 mi; 3 h 55 min). Walk east along the beach toward the cliff at Back Bay. This area is only accessible at mid to low tide.*

STOP 28 MONREITH FOLDED TURBIDITES

We are back in the Southern Uplands. This stop displays complex deformation in the Silurian Hawick Group (444–433 Ma), deep water turbidites deposited on the floor of the Iapetus Ocean. The resulting graywackes were deformed during the Caledonian Orogeny when Avalonia and Laurentia collided and squeezed the intervening ocean floor and sediments between them. In this part of the Southern Uplands, we see southeast-directed thrusting and dominantly southeast-verging folds. It is speculated that there are multiple southeast-directed thrust slivers that thicken the section and confuse age relationships (Barnes, 1996).

The Hawick Group is a sandstone-dominated turbidite sequence. All of the formations in the group consist of alternating turbidite graywackes. The succession is dominated by sandstone with bed thicknesses commonly in the 20–60 cm (8–24 in) range and contains subsidiary silty mudstone up to about 30 cm (12 in). Siltstones commonly contain graptolites, a colonial animal that lived in interconnected tubes. The base of sandstone beds frequently has flute casts (scour marks) that indicate sediment transport to the southwest. Ripple marks are common on top surfaces.

The Hawick Group in the southwest Southern Uplands is divided into four formations based on subtle variations in lithology. From oldest to youngest they are the Cairnharrow, Kirkmaiden, Carghidown, and Ross formations. The Cairnharrow Formation (up to 1,000 m; 3,280 ft) is characterized by thick-bedded graywackes and rare thin, muddy interbeds. The Kirkmaiden Formation

(1,000–1,500 m; 3,280–4,922 ft) is mostly medium-bedded (≤0.5 m, or ≤1.5 ft) graywackes with thin silty mud partings and a few massive sandstones. The Carghidown Formation (1,000–1,500 m; 3,280–4,922 ft) has slightly more abundant mudstone interbeds that locally form up to 40% of the unit. The Carghidown Formation is characterized by the presence of red mudstone along with the normal gray–green mudstone, and by red micas in the sandstone. Soft-sediment deformation is common in the southernmost Carghidown Formation. The youngest unit, the Ross Formation (up to 2,000 m; 6,562 ft) is a turbidite sandstone with occasional mudstone interbeds. It contains a distinctive dark gray, finely laminated carbonaceous siltstone in beds ranging from a few millimeters up to 1.5 m (0.1 in to 5 ft) in thickness (Stone et al., 2012d).

Kirkmaiden Formation (~435 Ma) turbidites are exposed in the sea cliffs here at Monreith. The strata are deformed into southeast-verging folds. Older upright folds appear to be refolded by a younger set with gently dipping axial planes. Recumbent younger folds (folds lying on their sides) are associated with small displacement south-directed thrusts. The headland to the southeast contains a lamprophyre (porphyritic, mafic) dike. This is but one of many such dikes that cut early folds but are folded by later deformation (Barnes, 1996; McMillan, Back Bay, Monreith).

You can also see raised wave-cut terraces here that are a result of isostatic rebound after removal of the weight of glaciers after the last Ice Age. In fact, the beach is covered with colorful cobbles including graywacke from the beach cliffs and granite erratics brought by glaciers from the Galloway Hills in the Southern Uplands (McMillan, Back Bay, Monreith).

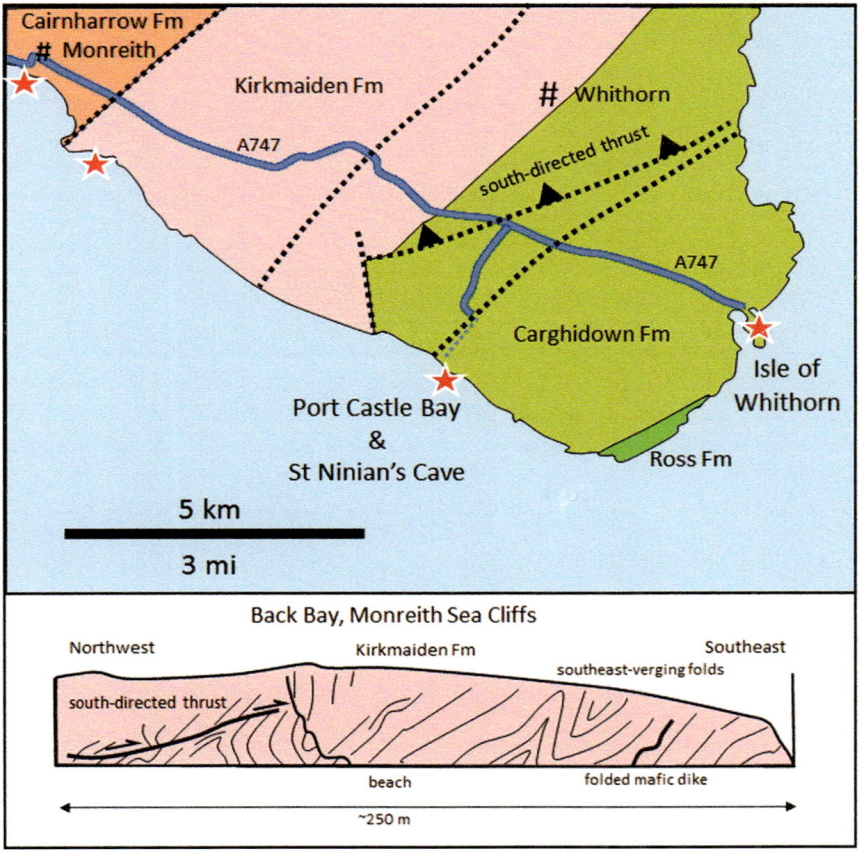

Geologic sketch map, Monreith to Port Castle Bay to Isle of Whithorn. Below is a cross-section sketch of the deformed Kirkmaiden Formation seen in the Back Bay sea cliffs. (Modified after Barnes, 1996.)

Folded Kirkmaiden Formation turbidites, Back Bay, Monreith. See the sketch in the previous figure for a structural interpretation.

Back Bay Monreith to Port Castle Bay: Return to A747 and turn right (south); continue south on A747 for about 9 km (5.6 mi) to an intersection with signs for St Ninian's Cave and Physgill; turn right and after a further 2 km (1.2 mi) park just before Kidsdale farm. This is **Stop 29, Port Castle Bay** *(54.693738, −4.450011) for a total of 11.6 km (7.2 mi; 14 min).*

A public path follows the track through the farmyard then turns right and follows the wooded valley 1 km (0.6 mi) to the beach at Port Castle Bay. St Ninian's Cave is north along the shore, but the interesting geology lies to the south.

STOP 29 PORT CASTLE BAY AND SAINT NINIAN'S CAVE

As you head south along the beach, the outcrop contains thin-bedded graywacke of the Silurian Carghidown Formation, Hawick Group. The Carghidown is characterized by red mudstone inter-beds, sparse here but increasingly common to the south. The turbidite beds are deformed into south-verging folds with crenulation cleavage developed in mudstones in fold hinges. Some of the bedding contains horizontal slickensides indicating bedding-parallel slip that is common in thrusted terrain.

Intervals of disrupted bedding separating packages of well-bedded strata indicate faulting. Two sets of faults have been mapped in this area, one trending northwest; the other north-northeast. The northwest set appears to control the straight section of the coast here. Northwest faults clearly show right-lateral offset of bedding and sills.

Strata at the south end of the beach are cut by several lamprophyre dikes that cross-cut east-northeast-striking, southwest-plunging folds. Other dikes were injected along faults and were deformed by kink bands. A thick dike that forms the southernmost cliff of this bay was injected after folding (Barnes, 1996).

Carghidown turbidite beds are deformed into south-verging folds, Port Castle Bay.

Detail of folding in the Carghidown turbidites, Port Castle Bay.

At the north end of the beach is Saint Ninian's Cave. The cave is a sea cave, carved by wave erosion into steeply dipping Carghidown Formation graywacke. It is 7 m long by 3 m wide by 3 m high

(23 ft×10 ft×10 ft). This is smaller than the original cave, a result of multiple roof falls (Historic Environment Scotland, Saint Ninian's Cave).

Saint Ninian was a missionary to the Picts in what is now southern Scotland. For this reason, he is known as the Apostle to the Southern Picts. It is thought he died in 432, but nothing is known about his teachings, and his identity is uncertain: he has been identified with three other historical figures: Saint Finnian of Moville, Saint Finnian of Clonard, and Saint Finbarr of Cork (Wikipedia, Saint Ninian).

According to local tradition, Ninian was active in the area in the late 300s and used this cave as a retreat. Based on excavations, it may also have been a chapel or stone carver's workshop. Artifacts found here include 18 early medieval carved stones; these are now in the Whithorn Priory Museum. The cave had internal walls and pavements and contained the graves of at least three people. Ten crosses were carved into the cave walls. The cave remains a place of pilgrimage for Catholics (Historic Environment Scotland, Saint Ninian's Cave).

Carghidown Formation at St Ninian's Cave. Layers are overturned and dip 75°–80° southeast. https://commons. wikimedia.org/wiki/File:St.Ninian_Cave_entrance_2007.JPG

Port Castle Bay to Isle Head: Return to B7004 and turn right (south); drive to the Isle of Whithorn; turn right onto Main Street and drive around the harbor, parking in the small car park at the south end of the harbor. This is Stop 30, Isle Head (54.697792, −4.362327), for a total of 7.0 km (4.3 mi; 11 min). Ample free public parking is available.

STOP 30 ISLE HEAD, ISLE OF WHITHORN

The area around Isle Head and to the north is all Silurian Carghidown Formation turbidite gray-wackes. Along the shore below the car park, you can see overturned turbidite beds striking north-east and dipping steeply southeast. Folds here verge to the northwest and are cut by lamprophyre dikes injected along faults, and contain sills injected along bedding. Deformation occurred during the Caledonian Orogeny.

As you walk south you can see a 20 m (66 ft) wide zone of slumping cut by lamprophyre dikes, indicating that the slumping is probably soft-sediment deformation. Continuing south you are in a right-side-up sequence of southeast-inclined beds, indicating we have crossed an anticlinal fold axis. Continuing south we enter another overturned sequence, indicating either we crossed a thrust or passed through a synclinal axis. Look for axial plane cleavage developed in the folded mudstone.

A short distance further south is Chapel Port West Bay, a zone eroded along a northeast-trending fault that crosses the peninsula. South of the bay is a large anticlinal fold. Flute casts (scour marks) can be seen on the base of some beds. Continuing to the point, you cross a zone of strike-parallel faults. You are now in the Ross Formation of the Hawick Group, another Silurian deep marine gray-wacke turbidite sequence. The exact structural and stratigraphic relationships are unclear (Barnes, 1996). Overall, the units on the Machars Peninsula get younger as you move south.

Folded Carghidown Formation turbidites, Isle of Whithorn.

Flute casts or sediment loading marks (arrows) on the base of the Carghidown Formation, Isle of Whithorn.

*Isle Head to Carrick Beach: Return north on Main Street/B7004 to the A746 at Whitehorn; turn right onto the A746/Glasserton Street and drive north to the B7005 at Bladnoch; at the round-about, take the 2nd exit onto B7005 and continue north to the A714 at Culquirk; turn left (north) onto the A714 and drive north to the A75 at Newton-Stewart; at the roundabout take the 3rd exit onto A75 and drive east and south to the A755 and sign for Brighouse Bay; turn right (south) and drive to the B727; turn right (south) onto B727 and drive 0.5 km (0.3 mi) to an unnamed road on the right; turn right and drive 4.2 km (2.6 mi) to an unnamed road on the right; turn right (west) and drive 1.3 km (0.8 mi) to **Stop 31, Carrick Beach** (54.824822, −4.217592) for a total of 69.9 km (43.4 mi; 1 h 6 min). There is ample free parking at the beach.*

STOP 31 CARRICK BEACH TURBIDITES, WIGTON BAY

This stop provides another opportunity to view the Kirkmaiden Formation graywacke turbidites in the Silurian Hawick Group. Sandstone beds are generally less than 0.5 m (1.5 ft) thick, with mud-stone interbeds. Sandstones were deformed into tight folds during the Caledonian Orogeny.

Folded Kirkmaiden Formation, Carrick Beach. View southwest.

*Carrick Beach to Brighouse Bay: Return south and east 1.3 km (0.8 mi) to the intersection with the road to Borgue; turn right (south) and follow the road to Borgue; in Borgue turn right (south) onto B727 and drive 2 km (1.3 mi) to a road on the right; turn right (south) to "Brighouse Bay 1 ¼ mi;" in 2.1 km (1.3 mi) turn right (west) and follow the signs to **Stop 32, Brighouse Bay Beach** (54.788537, −4.124763) car park, for a total of 8 km (5 mi; 13 min). There is ample free parking and facilities. Take the path through the trees most of the way to the point, as walking is difficult among the rocks.*

STOP 32 BRIGHOUSE BAY TURBIDITES

The west side of the bay has excellent beach outcrops illustrating the stratigraphy and structure of the Carghidown Formation turbidite sequence. This area is best seen at low tide. Walk southwest from the car park along the coast to Point of Green. Along the way are numerous folds and shear zones, not all of them obvious to the uninitiated.

Walking southwest, you cross numerous tight folds and shear zones. The thicker sandstones often show flute casts on the base of the bedding. Slaty cleavage is seen in the mudstones and is axial planar to some of the folds. As the folds trend northeast-southwest, the overall shortening direction here is southeast-northwest. There may be a component of left-lateral shear imposed on the shortening (Stone et al., 1996a).

At Dunrod Point (54.776684, −4.137421), steeply plunging folds occur in an anastomosing shear zone.

As you approach Point of Green (54.776622, −4.142388), you pass through a zone of folds separated by shear zones (thrusts). At Point of Green there is a particularly well-exposed fold that appears to be anticlinal but in fact is an overturned syncline as shown by the flute casts on the upper surface of the beds. There is evidence that the thick sandstone was deposited as a channel cutting into the units below (Stone et al., 1996a).

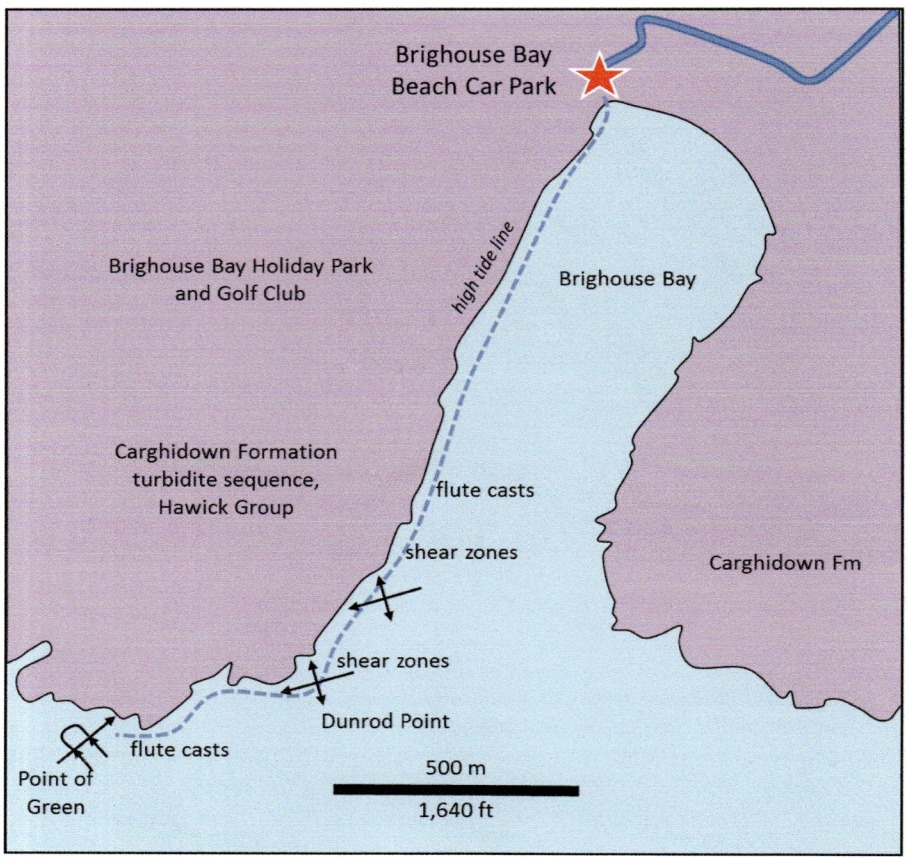

Brighouse Bay sketch map. (Modified after Stone et al., 1996a.)

Steeply dipping turbidites of the Carghidown Formation, Hawick Group, exposed at Brighouse Bay. View northeast.

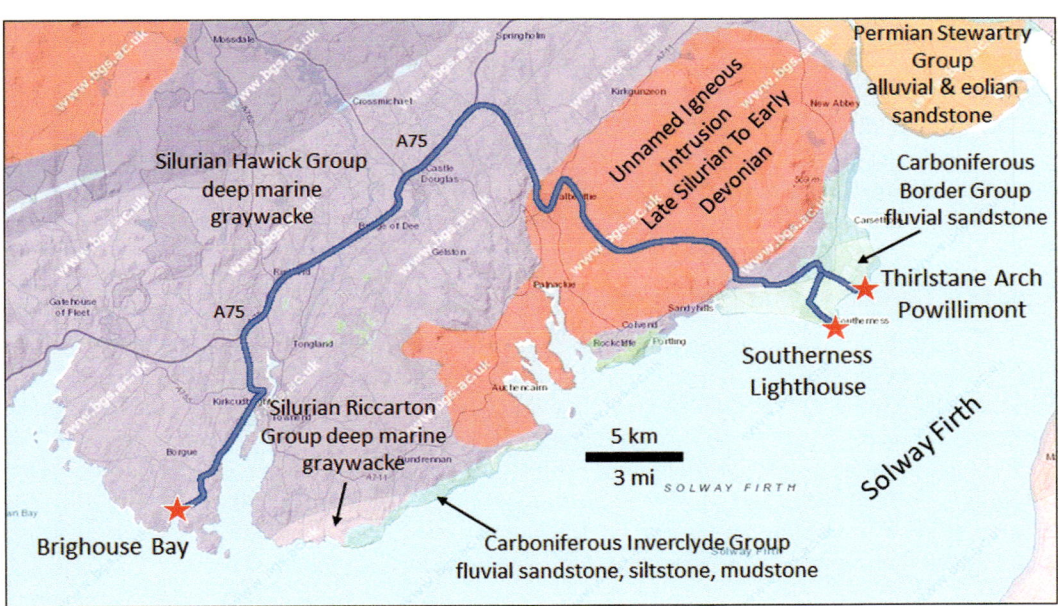

Geologic map of Brighouse Bay to Southerness. (Modified after BGS, Geology of Britain Viewer. Contains British Geological Survey materials © UKRI 2022. Base mapping provided by ESRI.)

STOP 33 SOUTHERNESS AREA

From Brighouse Bay to Southerness, you are driving mostly through the Carghidown Formation (Hawick Group). As you approach Dalbeattie, you are entering the Late Silurian-Early Devonian (~397 Ma) Criffel-Dalbeattie granodiorite pluton. This is described as an intrusion with an inner porphyritic granite and an outer zone of foliated granodiorite, both intruded by aplite (very fine-grained granite) dikes (BGS, Geology of Britain Viewer "Classic"). At Caulkerbush, you enter a narrow zone of Ross Formation before arriving at the Early Carboniferous (340 Ma) Powillimount Sandstone Formation of the Border Group.

The sedimentary rocks (mainly sandstones, mudstones, and limestones) are well exposed on the shore around Southerness. The Early Carboniferous rocks exposed here are on the northern margin of the Solway-Northumberland Basin. This was a subsiding rift basin, bounded by normal faults, that was filled with river and shallow marine sediments in a way very similar to the sedimentation in the Midland Valley to the north of the Southern Uplands.

Recall that during the Late Carboniferous, Gondwana was colliding with Laurussia to create the supercontinent Pangea. The collision caused the Variscan Orogeny. Southern Scotland was well north of the main deformation front, but some folding occurred as far north as the Southern Uplands and involved the Border Group. North-directed shortening and inversion reactivated many of the basin-bounding normal faults as reverse faults (Stone et al., 2012b).

These outcrops are best seen at low tide.

Brighouse Bay to Powillimount: *Return northeast to the B727 and turn right (east); take the B727 northeast to the A755; turn left (west) onto the A755 and drive 650 m (0.4 mi); turn right (north) onto the road to Twynholm and drive to the A75; turn right (northeast) onto the A75 and drive to the B794; turn right (south) onto B794 toward Dalbeattie and drive to Maxwell Street in Dalbeattie; turn left (east) on Maxwell Street and drive to the A711; turn left (east) on A711 and drive north to the B793; turn right (east) and drive south on B793 to the A710 in Caulkerbush; turn left (east) on the A710 and drive to the sign to turn right to Southerness; turn right (south) and follow the signs to the Powillimount car park (just past Paul Jones Cottage). This is **Stop 33.1, Powillimount and Thirlstane Arch** (54.892571, −3.576442), for a total of 59.4 km (36.9 mi; 59 min). There is ample free parking, but no facilities. Stroll north 1.9 km (1.2 mi) toward Arbigland.*

STOP 33.1 DEFORMATION AT POWILLIMOUNT AND THIRLSTANE ARCH

At the Powillimount car park, sandstones and limestones are folded into anticlines and synclines that plunge gently northeast. These are particularly easy to see on Google Earth imagery. The Variscan-age folding, faulting, and soft-sediment deformation are a bit more difficult to make out in the beach outcrops.

Walking north you first encounter the Thirlstane Sandstone member of the Carboniferous Powillimount Sandstone. The Thirlstane Sandstone is the top 25 m (82 ft) of the Powillimount Formation. The sandstone was probably deposited in nearshore, offshore sandbar, beach, and tidal lagoon environments. The sandstone layers become increasingly contorted as you approach Thirlstane Arch (a natural arch in the sandstone caused by wave erosion). These sands show signs of soft-sediment deformation, probably caused by an earthquake and slumping shortly after deposition.

Continuing north along the shore you cross a fault and come to shallow marine limestones, mudstones, and sandstones. Totally out of place on the shore is a huge granite boulder, an "erratic" carried on a glacier and dropped on the beach. It is known locally as the "Devil Stone." Local legend says that the devil bit off this chunk of the hill and, it not being to his taste, spat it out.

Along the point east of Arbigland are limestone and mudstone beds chock full of brachiopod shells, gastropod shells, and corals (including *Lithostrotion clavaticum*). Some of the sandstones contain the burrows of sediment feeders (McMillan, 1996; McMillan, Southerness).

Google Earth oblique view northeast showing folding in the Powillimount Sandstone Formation. Dashed lines are the trace of bedding. (Image © 2022 CNES/Airbus.)

Thirlstane Arch, a natural arch carved by waves in the Powillimount Sandstone.

Cross-bedded Thirlstane Sandstone, Powillimount Beach.

*Thirlstane Arch to Southerness Lighthouse: Return northwest to Arbigland and turn left (west); drive 160 m (0.1 mi) and turn right (northwest) to continue on Arbigland; drive to Southerness and turn left (south); drive to the abandoned lighthouse and park at the beach. This is **Stop 33.2, Southerness Lighthouse** (54.873218, −3.595985), for a total of 5.9 km (3.7 mi; 9 min). Abundant free parking exists in town.*

STOP 33.2 POWILLIMOUNT FORMATION, SOUTHERNESS LIGHTHOUSE

Built in 1749, the Southerness Lighthouse is the second oldest in Scotland.

Walk west from the lighthouse ~500 m (1,600 ft) to see gently dipping and folded sandstone on the beach. The sandstones in Southerness contain plant fossils; the limestones and mudstones were deposited in a shallow water environment. Some of the mudstones have shell and coral fossils (McMillan, Southerness).

Southwest-plunging syncline in the Lower Carboniferous Thirlstane Sandstone, Southerness Beach. View east.

Here we end our geo-tour of Scotland and Northern Ireland. We visited the North West Highlands and Lochaber Geoparks, Giants Causeway World Heritage Site, and five Sites of Special Scientific Interest (Knockan Crag, Arthur's Seat, Fossil Grove, the Old Man of Storr, and Portrush). We have been to sites where several of geology's great concepts were born, including Siccar Point (the concepts of unconformities and deep time), Loch Glencoul and Knockan Crag (the concept of thrust faulting), concepts of magma emplacement (Ardnamurchan Volcano) and caldera collapse (Aonach Dubh and Glencoe), and the concept of magmatic intrusion of sills (Hutton's Section). The theory of ice ages was supported by the recognition of *rôche moutonnées* (Dulnain Bridge), moraines (Valley of 100 Hills), and U-shaped valleys (Glen Coe, Loch Broom). We passed through Loch Lomond and The Trossachs and Cairngorms National Parks and saw breathtaking landscapes and superb geology. We examined ancient continental suture zones (Great Glen Fault) and have seen deformation related to two great episodes, the Ordovician-Devonian Caledonian Orogeny and the Late Carboniferous-Permian Variscan Orogeny. This is well and truly a remarkable corner of the world.

REFERENCES

Allan, J. Kilt Rock. The Skye Guide. Accessed 1 January 2021, https://www.theskyeguide.com/see-and-do-mainmenu-35/27-natural-wonders/177-kilt-rock

Amor, K., S.P. Hesselbo, D. Porcelli, A. Price, N. Saunders, M. Sykes, J. Stevanović, and C. MacNiocaill. 2020. The Mesoproterozoic Stac Fada proximal ejecta blanket, NW Scotland: Constraints on crater location from field observations, anisotropy of magnetic susceptibility, petrography and geochemistry. *Journal of the Geological Society*, v. 176, pp. 830–846.

Armando, P. 2015. Kilt Rock. Accessed 1 January 2021, https://epod.usra.edu/blog/2015/07/kilt-rock.html

Atlas Obscura. Ancient Cliffs Resembling a Kilt on Scotland's Second Largest Island Emit an Entrancing Sound. Accessed 1 January 2021, https://www.atlasobscura.com/places/kilt-rock-and-mealt-falls

Atlas Obscura. Fairy Pools. Accessed 1 January 2021, https://www.atlasobscura.com/places/fairy-pools

Ball, J. 2014. Edinburgh: Arthur's Seat, Salisbury Crags and Hutton's Section. Accessed 30 December 2020, https://blogs.agu.org/magmacumlaude/2014/04/06/edinburgh-arthurs-seat-and-salisbury-crags/

Barnes, R.P. 1996. Whithorn - an excursion. Excursion 13. In Stone, P. (ed), Geology in South-West Scotland: An Excursion Guide. British Geological Survey, Keyworth, Nottingham, 9 p.

BBC. 2014. Rocks Stolen from Portrush National Nature Reserve. Accessed 28 September 2022, https://www.bbc.com/news/uk-northern-ireland-28290468

BBC. 2016. Ancient Meteorite Impact Crater Lies under Scotland. Accessed 31 December 2020, https://www.bbc.com/news/uk-scotland-highlands-islands-37360696

Bell, B.R. and J.W. Harris. 1986. The Kilt Rock, Skye - an excursion. In Bell, B.R. and Harris, J.W. (eds), *An Excursion Guide to the Geology of the Isle of Skye*. Geological Society of Glasgow, Glasgow, 2 p.

Bennett, M.R., M.I. Hambrey, D. Huddart, and N.F. Glasser. 2020. Glacial thrusting & morainemound formation in Svalbard & Britain: The example of Coire a 1 Cheud-chnoic (Valley of Hundred Hills), Torridon Scotland. *Journal of Quaternary Science*, v. 13, suppl. 1, pp. 17–34, ISSN: 0267-8179.

Bentley, C. 2017a. Stac Fada. AGU Blogosphere. Accessed 1 January 2021, https://blogs.agu.org/mountainbeltway/2017/04/06/stac-fada/

Bentley, C. 2017b. Travels in Geology: Geo-diversity and geologic history in the North West Highlands of Scotland. In Earth Magazine. Accessed 30 December 2020, https://www.earthmagazine.org/article/travels-geology-geo-diversity-and-geologic-history-north-west-highlands-scotland

BGS, Geology of Britain Viewer (Classic). British Geological Survey. Accessed 31 December 2020, https://mapapps.bgs.ac.uk/geologyofbritain/home.html

Black, G.P. 2015. Arthur's Seat, Salisbury Crags, Edinburgh – an excursion. Accessed 21 February 2023, https://earthwise.bgs.ac.uk/index.php/Arthur%27s_Seat,_Salisbury_Crags,_Edinburgh_-_an_excursion

Blair, J. Glen Coe, Lochaber: Scene of an Ancient Caldera Collapse. Scottish Geology Trust. Accessed 2 January 2021, https://www.scottishgeology.com/best-places/glen-coe/

Brasier, A.T., T. Culwick, L. Battison, R.H.T. Callow, and M.D. Brasier. 2017. Evaluating evidence from the Torridonian Supergroup (Scotland, UK) for Eukaryotic life on land in the Proterozoic. *Geological Society London Special Publications*, v. 448, no. 1, pp. 121–144.

Breckenridge, J. 2007. The Story of Knockan Crag National Nature Reserve. Scottish Natural Heritage, Ullapool Ross-shire, 16 p.

Browne, A.E., R.A. Smith, and A.M. Aitken. 2002. Stratigraphical Framework for the Devonian (Old Red Sandstone) Rocks of Scotland South of a Line from Fort William to Aberdeen. British Geological Survey Research Report RR/01/04, Keyworth, Nottingham, 60 p.

Butler, R., and G.W.H. Alsop. 2019. Discussion on 'A reassessment of the proposed 'Lairg impact structure' and its potential implications for the deep structure of northern Scotland'. *Journal of the Geological Society*, London, v. 176, pp. 817–829.

Butler, R. Ardnamurchan Ring Complex. The Geological Society. Accessed 1 January 2021, https://www.geolsoc.org.uk/GeositesArdnamurchan

Butler, R. Beinn Eighe NNR and Loch Maree, Torridon: Thrust Chaos and Glaciation. Scottish Geology Trust. Accessed 1 January 2021, https://www.scottishgeology.com/best-places/beinn-eighe/

Butler, R. Sango Bay and Smoo Cave, an Excursion. The Geological Society. Accessed 30 December 2020, https://www.geolsoc.org.uk/GeositesSangoSmoo

Cairngorms National Park, Roche Moutonnée, Dulnain Bridge. Accessed 30 December 2020, https://cairngorms.co.uk/photo-posts/photoposts/13/moreinfo/

Carter, S. Torridon and Shieldaig in the Western Highlands of Scotland. Accessed 1 January 2021, http://www.stevecarter.com/ansh/ansh2.htm

Cribb, S.J. 2005a. Geology of Whisky. In Selley, R.C., Cocks, L.R.M., and Plimer, I.R. (eds), *Encyclopedia of Geology*. Elsevier Academic Press, Amsterdam, pp. 82–85.

Cribb, S.J. 2005b. Geology of Whisky. In Cribb, S.J. (ed), *Encyclopedia of Geology*. Carraig Associates, Elsevier Ltd., Inverness, UK, pp. 82–85.

Cuthbert, S., and G. Eardley. Balmaha, Loch Lomond: Walk the Highland Boundary Fault. Scottish Geology Trust. Accessed 2 January 2021, https://www.scottishgeology.com/best-places/balmaha-loch-lomond/

Discover Glencoe, Lochaber Geopark. Accessed 2 January 2021, https://discoverglencoe.scot/lochaber-geopark-glencoe/

Discover Glencoe, Scottish Hollywood. Accessed 2 January 2021, https://discoverglencoe.scot/key-information/filming-locations/

Docherty, F.M. 1999. A Geological View of Loch Ness and Area. Accessed 31 December 2020, http://www.thefrasers.com/nessie/nessgeo.html

Dryburgh, P.M., S.M. Ross, and C.L. Thompson. 2014. Assynt - The Geologists' Mecca. Edinburgh Geological Society, Edinburgh, ISBN: 978-0-904440-13-3, 39 p.

Edinburgh Geological Society, Castle Rock. Accessed 31 December 2020, https://www.edinburghgeolsoc.org/geological-site/castle-rock/

Fenton, J. 2019. Wester Ross Rocks: The Geology and Scenery of Gairloch and District. Accessed 1 January 2021, https://jeremyfenton.scot/Booklet%20Geology%20lo.pdf

Gazeteer for Scotland, Great Glen Fault. Accessed 31 December 2020, https://www.scottish-places.info/features/featurefirst10625.html

Gazeteer for Scotland, Southern Uplands Fault. Accessed 2 January 2021, https://www.scottish-places.info/features/featurefirst7744.html

Geological Society of Glasgow. Dumbarton Rock. Accessed 2 January 2021, https://geologyglasgow.org.uk/docs/017__074__publications__Final_Printers_Dumbarton_Rock_Leaflet__1329125115.pdf

Glencoe, Scotland. History of Glencoe. Accessed 2 January 2021, https://www.glencoescotland.com/about-glencoe/history/#:~:text=History%20of%20Glencoe, (rock%20of%20the%20Feinn).

Goodenough, K.M., A. Finlayson, and H.F. Barron. 2008. Geodiversity of the Loch Lomond and The Trossachs National Park: Statement of Significance and Identification of Opportunities. British Geological Survey, Geology and Landscape (Northern Britain) Programme Open File Report OR/07/036, 22 p.

Goodenough, K.M., and M. Krabbendam (eds.). 2011. Foreland succession of the Loch Assynt roadside, North-west Highlands - an excursion. In *A Geological Excursion Guide to the North-West Highlands of Scotland*. British Geological Survey, Edinburgh, 205 p.

Hambrey, M.J., D. Huddart, M.R. Bennett, and N.F. Glasser. 1997. Genesis of 'hummocky moraines' by thrusting in glacier ice: evidence from Svalbard and Britain. *Journal of the Geological Society*, v. 154. pp. 623–632.

Hamlet, L., and N. Harrison. Knockan Crag NNR. Scottish Geology Trust. Accessed 31 December 2020, https://www.scottishgeology.com/best-places/knockan-crag/

Hamlet, L., and N. Harrison. Loch Glencoul: Continental Collisions Laid Bare. Scottish Geology Trust. Accessed 31 December 2020, https://www.scottishgeology.com/best-places/loch-glencoul/

Hamlet, L., and N. Harrison. Scourie Bay and Laxford: Oldest Rocks in Western Europe. Scottish Geology Trust. Accessed 31 December 2020, https://www.scottishgeology.com/best-places/scourie-laxford/

Harrison, R.K., P. Stone, I.B. Cameron, R.W. Elliot, and R.R. Harding. 1987. Geology, Petrology and Geochemistry of Ailsa Craig, Ayrshire. London, British Geological Survey Report, v. 16, no. 9, 29 p.

Harrison, N. 2017. Report on the Geology of the Loch Assynt Area. Accessed 23 November 2020, https://www.researchgate.net/publication/331375039_Report_on_the_Geology_of_the_Loch_Assynt_Area

Harrison, P. 2020. Corrieshalloch Gorge NNR and Falls of Measach: The Carving Force of Glacial Meltwater. Scottish Geology Trust. Accessed 1January 2021, https://www.scottishgeology.com/best-places/corrieshalloch-gorge/

Historic Environment Scotland. St Ninian's Cave. Accessed 2 January 2021, https://www.historicenvironment.scot/visit-a-place/places/st-ninians-cave/

Holdsworth, R.E., E. Tavarnelli, and P. Clegg. 2002. The nature and regional significance of structures in the Gala Group of the Southern Uplands terrane, Berwickshire coast, southeastern Scotland. *Geology Magazine*, v. 139, no. 6, pp. 707–717. DOI: 10.1017/S0016756802006854.

Holdsworth, B., and R. Strachan. 2010. Moine geology of Durness and Faraid Head - an excursion. In Strachan, R., Friend, C., Alsop, I., and Miller, S. (eds), *A Geological Excursion Guide to the Moine Geology of the Northern Highlands of Scotland*. Edinburgh Geological Society, Glasgow Geological Society in Association with NMS Enterprises, Edinburgh, 5 p.

Holmes, S.J. 2019. The Pre-Depositional History of the Applecross Formation. MSc(R) thesis, University of Glasgow, Northwest Highlands, Scotland, 148 p.

Johnstone, G.S., and W. Mykura. 1989. Torridonian, Northern Highlands of Scotland. In *British Regional Geology: Northern Highlands of Scotland* (4th edition). British Geological Survey, Keyworth, Nottingham. Accessed 30 December 2020, http://earthwise.bgs.ac.uk/index.php/Torridonian,_Northern_Highlands_of_Scotland

Kendall, R.S. 2017. The Old Red Sandstone of Britain and Ireland – a review. Accessed 21 February 2023, chrome-extension://efaidnbmnnnibpcajpcglclefindmkaj/https://nora.nerc.ac.uk/id/eprint/517284/1/The%20Old%20Red%20Sandstone%20of%20the%20British%20Isles.pdf

Kinnaird, T.C., A.R. Prave, C.L. Kirkland, M. Horstwood, R. Parrish, and R.A. Batchelor. 2007. The late Mesoproterozoic – early Neoproterozoic tectonostratigraphic evolution of northwest Scotland: the Torridonian revisited. *Journal of the Geological Society*, v. 164, no. 3, pp. 541–551.

Krabbendam, M., and E.A. Callaghan. 2012. Geology-CastleRock, Johnstone Terrace, Edinburgh. British Geological Survey, Geology and Landscape Scotland Programme, Commissioned Report CR/13/031, 19 p.

Lemon, K. 2012. Portrush. The Geological Society. Accessed 28 September 2022, file:///C:/WORD/ Publishing/GeoTours/5_References/3_Europe/2_UK/1_UK_Northern/5_Glasgow-Giant'sCauseway/ Lemon_2012_TheGeologicalSociety_Portrush.pdf

LochaberGeopark.org, Lochaber Geopark – Glencoe. Accessed 1 January 2021, https://discoverglencoe. scot/lochaber-geopark-glencoe/

MacFadyen, C. Achanarras Quarry, Caithness: Where Ancient Fish Shoals Lie. Accessed 31 December 2020, https://www.scottishgeology.com/best-places/achanarras-quarry-caithness/

McAdam, D. 2003. Edinburgh & West Lothian – a landscape fashioned by geology. Accessed 21 February 2023, https:// ougs.org/scotland/local-geology/150/a-landscape-fashioned-by-geology-scottish-natural-heritage/

McGowan, A., J. Gordon, S. Brusatte, and N. Clark. Trotternish, Skye: Summit to Sea since the Jurassic. Accessed 1 January 2021, https://www.scottishgeology.com/best-places/trotternish-skye/

McMillan, A.A. 1996. Southerness to Borron Point – an excursion. In Stone, P. (ed.), *Geology in South-west Scotland: An Excursion Guide*. British Geological Survey, Keyworth, Nottingham, 8 p. Accessed 21 February 2023, https://earthwise.bgs.ac.uk/index.php?title=Southerness_To_Borron_Point_-_an_ excursion&oldid=51555

McMillan, A.A., R.J. Gillanders, and J.A. Fairhurst. 1999. Building Stones of Edinburgh. Edinburgh Geological Society, Edinburgh, 14 p.

McMillan, A.A. Back Bay, Monreith: Twisted Rocks of a Long-Lost Ocean Floor. Scottish Geology Trust. Accessed 2 January 2021, https://www.scottishgeology.com/best-places/back-bay-monreith/

McMillan, A.A. Southerness: Explore the Solway's Ancient and Modern Shorelines. Scottish Geology Trust. Accessed 2 January 2021, https://www.scottishgeology.com/best-places/southerness/

Miller, 2012. Historic Scotland. Accessed 15 November 2020, file:///C:/Users/Gary/Downloads/investigating- holyrood-park%20(2).pdf

Mitchell, W.I. 2004. Introduction to the Geology of Northern Ireland. In Mitchell, W.I. (ed), *The Geology of Northern Ireland-Our Natural Foundation*. Geological Survey of Northern Ireland, Belfast, 318 p.

Murray, C. 2012. The Mermaid of Loch Assynt at Ardvreck Castle. Accessed 31 December 2020, https:// archaicwonder.tumblr.com/post/36003372777/the-mermaid-of-loch-assynt-at-ardvreck-castle

Newman, M.J., and M.T. Dean. 2005. A Biostratigraphical Framework for Geological Correlation of the Middle Devonian Strata in the Moray-Ness Basin Project Area. British Geological Survey Geology and Landscape Northern Britain Programme Internal Report IR/05/160, 22 p.

North West Highlands Geopark, Clachtoll, Teacher's Sheet. Accessed 1 January 2021, https://www.nwh- geopark.com/wp-content/uploads/18.-North-West-Highlands-Clachtoll-Teacher-Sheets.pdf

North West Highlands Geopark, Loch Glencoul, Teacher's Sheet. Accessed 1 January 2021, https://www.nwh- geopark.com/wp-content/uploads/19.-North-West-Highlands-Loch-Glencoul-Teacher-Sheets.pdf

Picardi, L. 2014. Post-Glacial Activity and Earthquakes of the Great Glen Fault (Scotland). *Mem. Descr. Carta Geol. d'It.* v XCVI. pp. 431–446.

Potts, P.J., and J.R. Holbrook. 1987. Ailsa Craig - The history of a reference material. *Geostandards Newsletter*, v. 11, no. 2, pp. 257–260.

Rab-k, Relief Map of Scotland with Labels. Accessed 29 December 2020, https://en.wikipedia.org/wiki/ File:Scotland_(Location)_Named_(HR).png

Ramsay, D.M. 1962. The highland boundary fault: reverse or wrench fault? *Nature*, v. 195, pp. 1190–1191.

Richey, J.E., A.G. Macgregor, and F.W. Anderson. 1961. Scotland: The Tertiary Volcanic Districts. Department of Scientific and Industrial Research, Geological Survey and Museum, Edinburgh, 120 p.

Rig, L. 2011. Dumbarton Rock and Castle. Geograph Britain and Ireland. Accessed 22 January 2021, https:// www.geograph.org.uk/article/Dumbarton-Rock-and-Castle

Rnjak, I. 2017. What Gives Whiskey Its Flavor? Essential Facts for Beginners. Accessed 30 December 2020, https://blog.typsy.com/what-gives-whiskey-flavor

Roberts, J. Ballantrae Bay, Carrick: A Rare Relic of an Ancient Ocean. Scottish Geology Trust. Accessed 2 January 2021, https://www.scottishgeology.com/best-places/ballantrae/

Roberts, J.L., and J.E. Treagus. 1977. Dalradian rocks of the South-West Highlands. *Scottish Journal of Geology*, v. 13, Part 2, pp. 165–184.

Scotland.org, History. Accessed 30 December 2020, https://www.scotland.org/about-scotland/history-timeline

Scottish Geology Trust, Fossil Grove, Glasgow: An Ancient Forest Tale. Accessed 2 January 2021, https:// www.scottishgeology.com/best-places/fossil-grove-glasgow/

Scottish Geology Trust, Geologic Map of Scotland. © National Museums Scotland. Accessed 25 May 2022, https://www.scottishgeology.com/geology-of-scotland-map/

Selden, P.A., and J.R. Nudds. 2012. *The Hunsrück Slate, in Evolution of Fossil Ecosystems* (2nd Edition). Accessed 5 December 2020, https://www.sciencedirect.com/topics/earth-and-planetary-sciences/old-red-sandstone

Selfbuilt, Source of Whisky's Flavor. Accessed 30 December 2020, https://whiskyanalysis.com/index.php/background/source-of-whisky-flavours/#:~:text=So%2C%20to%20put%20it%20simply,reason%20%E2%80%93%20to%20flavour%20the%20whisky

Shaw, R.A., K.M. Goodenough, N.M.W. Roberts, M.S.A. Horstwood, S.R. Chenery, and A.G. Gunn. 2016. Petrogenesis of rare-metal pegmatites in high-grade metamorphic terranes: a case study from the Lewisian Gneiss Complex of north-west Scotland. *Precambrian Research*, v. 281, pp. 338–362.

Simms, M.J., and K. Ernstson. 2019. A reevaluation of the proposed 'Lairg Impact Structure' and its potential implications for the deep structure of northern Scotland. Pre-peer review manuscript, *Journal of the Geological Society*, London, v. 176, 28 p.

Stephenson, D., and D. Gould. 1995. Highland Border Complex, Grampian Highlands. In Stephenson, D. and Gould D. (eds), *British Regional Geology: The Grampian Highlands* (4th edition), Reprint 2007. British Geological Survey, Keyworth, Nottingham, 5 p.

Stephenson, D., and J. Merritt. 2006. *Skye, a Landscape Fashioned by Geology*. British Geological Survey and Scottish Natural Heritage, Battleby, Redgorton, Perth, 22 p.

Stephenson, D., and K.M. Goodenough. 2007. *Ben Nevis and Glencoe, a Landscape Fashioned by Geology*. British Geological Survey and Scottish Natural Heritage, Battleby, Redgorton, Perth, 33 p.

Stephenson, D., J. R. Mendum, D.J. Fettes, and A.G. Leslie. 2012. The Dalradian Rocks of Scotland: An Introduction. Accessed 19 January 2021, https://core.ac.uk/download/pdf/16747943.pdf

Stewart, M., R.A. Strachan, and R.E. Holdsworth. 1999. Structure and early kinematic history of the Great Glen Fault Zone, Scotland. *Tectonics* v. 18, no. 2, pp. 326–342.

Stone, P., R.F. Cheeney, and D.E. White. 1996a. Kirkudbright, Excursion 5. In Stone, P. (ed), *Geology in South-West Scotland: An Excursion Guide*. British Geological Survey, Keyworth, Nottingham, 6 p.

Stone, P., R.F. Cheeney, and D.E. White. 1996b. Girvan and Ballantrae, Excursion 8. In Stone, P. (ed), *Geology in South-West Scotland: An Excursion Guide*. British Geological Survey, Keyworth, Nottingham, 6 p.

Stone, P., A.A. McMillan, J.D. Floyd, R.P. Barnes, and E.R. Phillips. 2012a. Southern uplands accretionary complex. In Stone, P., McMillan, A.A., Floyd, J.D., Barnes, R.P., and Phillips, E.R. (eds), *British Regional Geology: South of Scotland* (4th edition). British Geological Survey, Keyworth, Nottingham, 10 p.

Stone, P., A.A. McMillan, J.D. Floyd, R.P. Barnes, and E.R. Phillips. 2012b. Regional structure of the Carboniferous, Southern Uplands. In Stone, P., McMillan, A.A., Floyd, J.D., Barnes, R.P., and Phillips, E.R. (eds), *British Regional Geology: South of Scotland* (4th edition). British Geological Survey, Keyworth, Nottingham, 5 p.

Stone, P., A.A. McMillan, J.D. Floyd, R.P. Barnes, and E.R. Phillips. 2012c. South of Scotland – introduction to geology. In Stone, P., McMillan, A.A., Floyd, J.D., Barnes, R.P., and Phillips, E.R. (eds), *British Regional Geology: South of Scotland* (4th edition). British Geological Survey, Keyworth, Nottingham.

Stone, P., A.A. McMillan, J.D. Floyd, R.P. Barnes, and E.R. Phillips. 2012d. Hawick Group, Silurian, Southern Uplands. In Stone, P., McMillan, A.A., Floyd, J.D., Barnes, R.P., and Phillips, E.R. (eds), *British Regional Geology: South of Scotland* (4th edition). British Geological Survey, Keyworth, Nottingham.

Stone, P., A.A. McMillan, J.D. Floyd, R.P. Barnes, and E.R. Phillips. 2012e. Gala Group, Silurian, Southern Uplands. In Stone, P., McMillan, A.A., Floyd, J.D., Barnes, R.P., and Phillips, E.R. (eds), *British Regional Geology: South of Scotland* (4th edition). British Geological Survey, Keyworth, Nottingham.

Strathclyde Conservation Group, The Fossil Grove. Accessed 7 December 2020, https://geologyglasgow.org.uk/docs/017_070__fossilgroveleaflet_1466951130.pdf

Sykora, S. 2019. The Giant's Causeway (Northern Ireland) and How It Formed? Exploring the Earth. Accessed 2 January 2021, http://exploringtheearth.com/2019/10/29/the-giants-causeway-northern-ireland-and-how-it-formed/

Tanner, G. 2008. Tectonic signicance of the Highland Boundary Fault, Scotland. *Journal of the Geological Society*, v. 165, no. 5, pp. 915–921. https://doi.org/10.1144/0016-76492008-012

The Geological Society. 2012. Plate Tectonic Stories: Southern Uplands, Scotland. Accessed 30 December 2020, https://www.geolsoc.org.uk/Policy-and-Media/Outreach/Plate-Tectonic-Stories/Southern-Uplands-Accretionary-Prism

Troll, V.R., C.H. Emeleus, G.R. Nicoll, T. Mattsson, R.M. Ellam, C.H. Donaldson, and C. Harris. 2019. A large explosive silicic eruption in the British Palaeogene Igneous Province. *Scientific Reports*, v. 9, no. 494, 15 p. DOI: 10.1038/s41598-018-35855-w.

Waters, D.J. 2003. Rocks of NW Scotland - Stratigraphic Section. Accessed 30 December 2020, https://www.earth.ox.ac.uk/~oesis/nws/nws-strat.html

West Highland Peninsulas, Discover the Geology of the West Highland Peninsulas. Accessed 1 January 2021, https://www.westhighlandpeninsulas.com/explore/our-geology

Whisky Foundation. 2016. How Regions Impact Whisky Maturation & Flavour. Accessed 21 February 2023, https://www.whiskyfoundation.com/2016/10/09/regions-impact-whisky-maturation-flavour/

Whyte, F., and D.S. Weedon. 1992. Dumbarton Rock - an excursion. In Lawson, J.D., and D.S. Weedon (eds), *Geological Excursions around Glasgow & Girvan*. Geological Society of Glasgow, Glasgow, 3 p.

Wikipedia, Ailsa Craig. Accessed 21 February 2023, https://en.wikipedia.org/wiki/Ailsa_Craig

Wikipedia, Arthur's Seat. Accessed 3 January 2021, https://en.wikipedia.org/wiki/Arthur%27s_Seat

Wikipedia, Dulnain Bridge. Accessed 3 January 2021, https://en.wikipedia.org/wiki/Dulnain_Bridge

Wikipedia, Falls of Falloch. Accessed 3 January 2021, https://en.wikipedia.org/wiki/Falls_of_Falloch

Wikipedia, Fossil Grove, Victoria Park. Accessed 3 January 2021, https://en.wikipedia.org/wiki/Fossil_Grove

Wikipedia, Geology of the Isle of Skye. Accessed 3 January 2021, https://en.wikipedia.org/wiki/Geology_of_the_Isle_of_Skye

Wikipedia, Geology of Loch Lomond and The Trossachs National Park. Accessed 3 January 2021, https://en.wikipedia.org/wiki/Geology_of_Loch_Lomond_and_The_Trossachs_National_Park

Wikipedia, Geology of Scotland. Accessed 3 January 2021, https://en.wikipedia.org/wiki/Geology_of_Scotland

Wikipedia, Giant's Causeway. Accessed 3 January 2021, https://en.wikipedia.org/wiki/Giant%27s_Causeway

Wikipedia, Glasgow. Accessed 3 January 2021, https://en.wikipedia.org/wiki/Glasgow

Wikipedia, Highland Boundary Fault. Accessed 3 January 2021, https://en.wikipedia.org/wiki/Highland_Boundary_Fault

Wikipedia, History of Scotland. Accessed 3 January 2021, https://en.wikipedia.org/wiki/History_of_Scotland

Wikipedia, Loch Lomond. Accessed 3 January 2021, https://en.wikipedia.org/wiki/Loch_Lomond

Wikipedia, Mingary Castle. Accessed 19 February 2023, https://en.wikipedia.org/wiki/Mingary_Castle

Wikipedia, Old Red Sandstone. Accessed 3 January 2021, https://en.wikipedia.org/wiki/Old_Red_Sandstone

Wikipedia, Penarth Group. Accessed 21 February 21, 2023, https://en.wikipedia.org/wiki/Penarth_Group

Wikipedia, Scottish Enlightenment. Accessed 3 January 2021, https://en.wikipedia.org/wiki/Scottish_Enlightenment

Wikipedia, Siccar Point. Accessed 3 January 2021, https://en.wikipedia.org/wiki/Siccar_Point

Wikipedia, Smoo Cave. Accessed 3 January 2021, https://en.wikipedia.org/wiki/Smoo_Cave

Wikipedia, Southern Uplands Fault. Accessed 3 January 2021, https://en.wikipedia.org/wiki/Southern_Uplands_Fault

Wild about Ardnamurchan, The Ardnamurchan Volcanoes. Accessed 1 January 2021, https://www.wildard-namurchan.com/ardnamurchan-rocks/the-ardnamurchan-volcanoes

Williams, G.E. 2001. Neoproterozoic (Torridonian) alluvial fan succession, northwest Scotland, and its tectonic setting and provenance. *Geological Magazine*, v. 138, no. 2, pp. 161–184.

Woudloper, Location of the Caledonian/Acadian Mountain Chains in the Early Devonian Epoch. Accessed 29 December 2020, https://commons.wikimedia.org/wiki/File:Caledonides_EN.svg

2 Lake District and Yorkshire Dales

Panorama of Keswick, between Derwent Water and the fells of Skiddaw in the Lake District. (Courtesy of Diliff, https://commons.wikimedia.org/wiki/File:Keswick_Panorama_-_Oct_2009.jpg.)

OVERVIEW

This geo-tour crosses the Lake District and Yorkshire Dales. We visit a mining museum and slate quarry in Lake District National Park and World Heritage Site while traversing some of the most scenic areas in Britain. In Yorkshire Dales National Park, we become acquainted with limestone terrain: we see limestone pavements, sinkholes, caves, and unconformities near Ingleton and Malham. The tour ends at Bolton Abbey, where early Carboniferous sediments were deformed during the late Carboniferous Variscan Orogeny.

ITINERARY

Begin – Penrith
Stop 1 Threlkeld Mining Museum and Quarry
Stop 2 Honister Slate Mine
Stop 3 British Gypsum mine, Long Marton
Stop 4 Ingleton Area Sinkholes, Caves, and Unconformities
      4.1 Waterfalls Trail
      4.2 White Scar Cave
      4.3 Gaping Gill
      4.4 Ingleborough Cave
Stop 5 Combs Quarry Unconformity, Horton in Ribblesdale
Stop 6 Malham Area
      6.1 Malham Cove and Pavement
      6.2 Janet's Foss
      6.3 Gordale Scar

DOI: 10.1201/9781351165600-2

Stop 7 Bolton Abbey Folded Carboniferous
    7.1 Folded Bowland Shale, South Flank Skipton Anticline
    7.2 Subsidiary Folds, Bowland Shale
    7.3 Folded Bowland Shale, North Flank Skipton Anticline
    End – Bolton Abbey

Key stops on the Lake District-Yorkshire Dales geo-tour.

PENRITH AND LAKE DISTRICT, LAKE DISTRICT NATIONAL PARK AND WORLD HERITAGE SITE

We begin this trek in the Lake District at Penrith. The mountains and lakes of the Lake District are justly famous for their unique landscape and scenery, which has inspired authors and poets such as Wordsworth, Coleridge, Southey, Potter, Ruskin, and others. The landscape, of course, is directly related to the underlying geology. All the land in England over 915 m (3,000 ft) elevation lies in the Lake District, including Scafell Pike (978 m; 3,210 ft), the highest peak. The district also contains Wast Water, the deepest natural lake, and Windermere, the largest natural lake in England. The region has been subjected to multiple glaciations over the past 2 Ma and displays classic glacial features such as U-shaped valleys, cirques, tarns, and glacial till. Much of the higher landscape is rocky, whereas the lower areas contain peat bogs and moors. Together, these features led to creation of the Lake District National Park in 1951 and the designation of a Lake District UNESCO World Heritage Site in 2017.

PENRITH HISTORY

The origin of the name "Penrith" is a matter of debate, but one suggestion is that *pen* derives from the Welsh "head" and *rhudd* meaning "crimson". The name "red hill" may refer to Beacon Hill, northeast of town. There is also a Redhills southwest of Penrith, near the M6 motorway, and a place called Penruddock, about 9.7 km (6 mi) west of Penrith. These names all reflect the local geology, as red sandstones are abundant in the area.

The origins of Penrith go back to the Neolithic (c. 4500–2350 BCE) or early-Bronze Age (c. 2500–1000 BCE). Stone axes, hammers, knives, and carvings have been found in the Penrith area. Iron Age (c. 800 BCE–100 CE) settlements of the Carvetti tribe have been uncovered in the Eden Valley. The Romans recognized the strategic importance of the location and built a road crossing the Pennines (the present A66) here. They built a fort at Brougham (Brocavum) on another road (the present A6) going north over Beacon Hill. The Roman fort of Voreda was 8 km (5 mi) north of town. Both the forts would have had a settlement nearby for farmers and traders that supported the forts.

Early on Penrith was a Crown possession, though often granted to favored noble families. The town depended largely on agriculture, especially cattle, but also on farming in the fertile Eden Valley.

After the Romans left in about 450, the north became a patchwork of warring Celtic tribes. During the 7th century, the area was invaded by Angles moving west from Northumbria. From about 870, the area around Penrith was settled by Vikings from Dublin and Danes from Yorkshire. The Angles had been farmers; the Vikings mainly raised sheep and cattle.

On 12 July 927, Eamont Bridge (or a place nearby) was the site of a gathering of kings from throughout Britain, among them Athelstan, King of the Anglo-Saxons, Constantín mac Áeda, King of Scots, Owain of Strathclyde, King of the Cumbrians, Hywel Dda, King of Wales, and Ealdred son of Eadulf, Lord of Bamburgh. Several of these kings submitted to Athelstan in order to create an alliance against the Vikings. This is considered the beginning of the Kingdom of England. Penrith was held by the Scottish king until the Norman takeover in 1092. The Norman conquest, under William Rufus, took Carlisle, Penrith, and some other manors near Penrith. Norman and Plantagenet rulers held Penrith as a crown estate ("Penred Regis").

From 1242 to 1295, Penrith was held by the King of Scots in return for renouncing his claims to Northumberland, Cumberland, and Westmorland. Edward I took Penrith and the other manors back for the English Crown. During the Wars of Scottish Independence, Penrith suffered destruction by Scottish forces in 1296 (William Wallace), 1314, 1315–1316, and 1322 (Robert the Bruce); the raids continued until 1388.

Penrith was involved in a rebellion of 1536–1537 known as the Pilgrimage of Grace. Eight town residents were executed as a result. The motives were partly religious, partly to get more protection against Scottish raids.

Penrith was spared from fighting during the English Civil War, the Commonwealth, and the Glorious Revolution. The Union of the Crowns gave Penrith respite from Scottish raids, and peace boosted Penrith's commercial prosperity.

The economy of Penrith was based primarily on cattle rearing, slaughtering, and the processing of leather goods, tanning, and shoemaking (Wikipedia, Penrith).

GEOLOGY

The highlands of the Lake District are a result of a thick Ordovician–Silurian granite intrusion that is relatively buoyant and floats high on the Earth's mantle. The granite was itself a result of the subduction and melting of the Iapetus oceanic crust and sediments beneath the Southern Uplands during the Caledonian Orogeny. The granite massif is exposed in a few places such as near Skiddaw and Carrock Fell. In the northwest Lake District, the granite intruded Ordovician sedimentary rock, chiefly mudstone and siltstone, which was subsequently metamorphosed by heating to become the Skiddaw Slates. These slates are easily eroded and tend to form the smooth slopes of Skiddaw, for example.

The central Lake District contains Ordovician volcanics related to the granitic batholith below. A batholith is a regional magma body; volcanoes are the surface manifestation of this deep magma. Early lavas of the Borrowdale Volcanic Group were followed by pyroclastic (ash and gas) eruptions that deposited tuff (volcanic ash) over a large area. One of the calderas (eruptive centers) is in the area of Scafell Pike. The tuffs are responsible for the rugged landscape typical of the central Lake District.

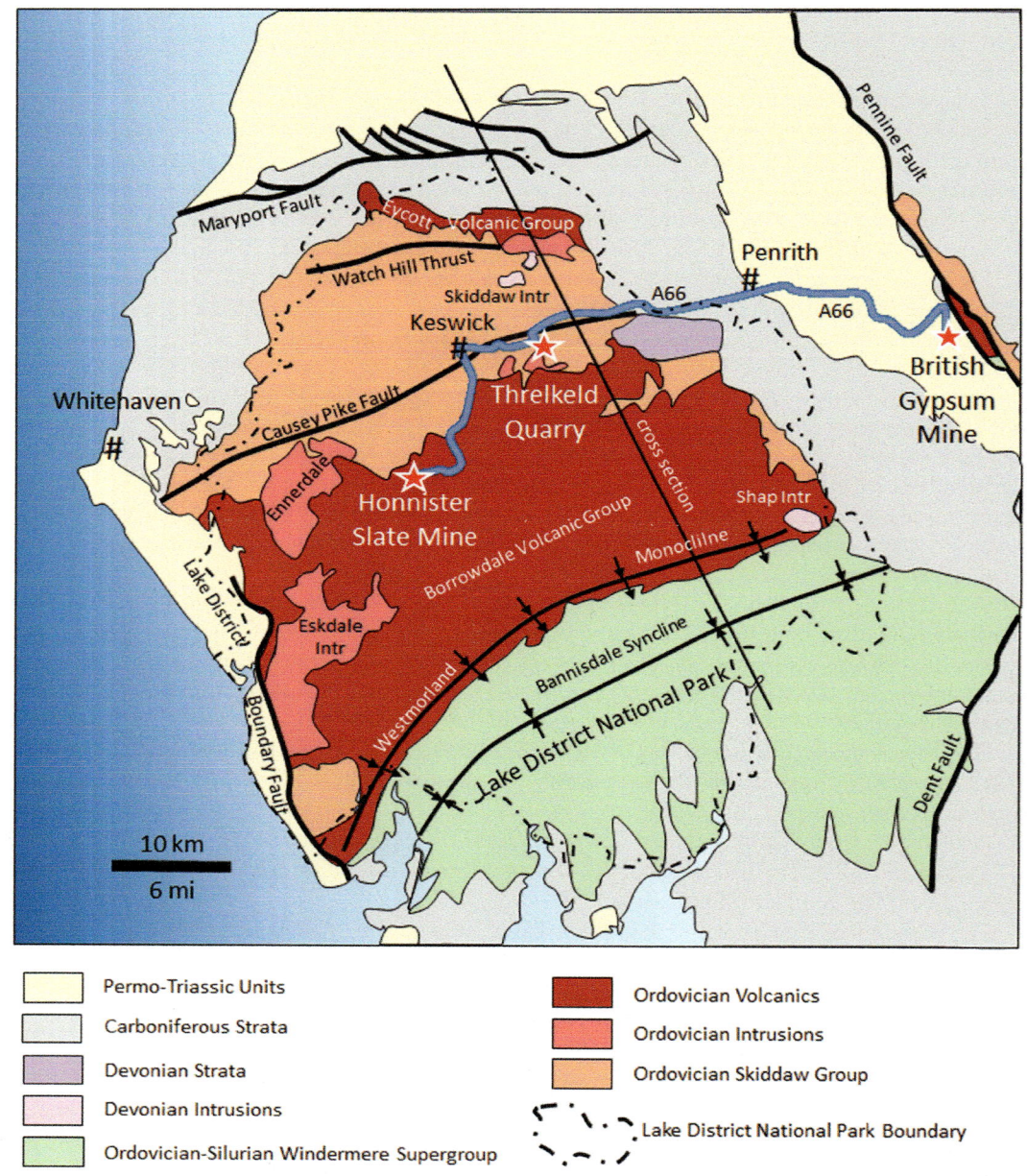

Geological map of the Lake District showing main structures. (Modified after Mikenorton, https://commons. wikimedia.org/wiki/File:Lake_District_Geology_Map.svg.)

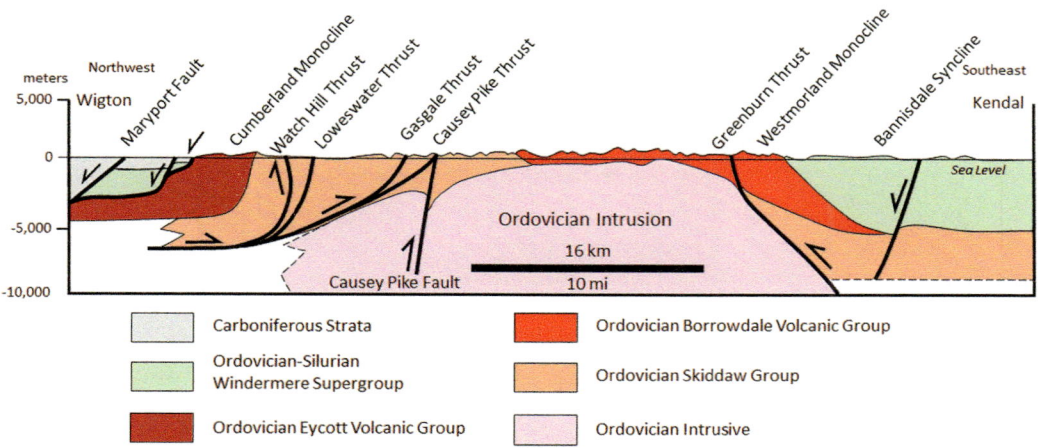

Cross section through the Lake District. The location is shown on the previous figure. (Modified after Stone et al., 2010).

View from Yewbarrow of Scafell massif, showing Scafell Pike in the center, Sca Fell on the right, and Lingmell on the left. (Courtesy of Doug Sim, https://commons.wikimedia.org/wiki/File:Scafell_massif.jpg.)

The southern and eastern Lake District contains graywacke sandstones and shales of the Ordovician–Silurian Windermere Supergroup (the Dent, Stockdale, Tranearth, Coniston, and Kendal groups). These easier-eroded units result in the lower terrain around Coniston and Windermere. Outlying areas to the north, east, and south of the Lake District contain Carboniferous limestones, usually deposited on an unconformity (erosion) surface.

Age	Group	Formation
Carboniferous	Coal Measures	
Carboniferous	Great Scar Limestone	
Carboniferous	Ravenstonedale	
Devonian	Upper Old Red Ss	
Ordovician-Silurian / Windermere Supergroup	Kendal	Kirkby Moor
Ordovician-Silurian / Windermere Supergroup	Kendal	Underbarrow
Ordovician-Silurian / Windermere Supergroup	Kendal	Bannisdale
Ordovician-Silurian / Windermere Supergroup	Coniston	Yewbank
Ordovician-Silurian / Windermere Supergroup	Coniston	Moorhowe
Ordovician-Silurian / Windermere Supergroup	Coniston	Poolscar
Ordovician-Silurian / Windermere Supergroup	Coniston	Latrigg
Ordovician-Silurian / Windermere Supergroup	Coniston	Gawthwaite
Ordovician-Silurian / Windermere Supergroup	Tranearth	Wray Castle
Ordovician-Silurian / Windermere Supergroup	Tranearth	Coldwell
Ordovician-Silurian / Windermere Supergroup	Tranearth	Birk Riggs
Ordovician-Silurian / Windermere Supergroup	Tranearth	Brathay
Ordovician-Silurian / Windermere Supergroup	Stockdale	Browgill
Ordovician-Silurian / Windermere Supergroup	Stockdale	Skelgill
Ordovician-Silurian / Windermere Supergroup	Dent	Ashgill
Ordovician-Silurian / Windermere Supergroup	Dent	Appletreeworth
Ordovician-Silurian / Windermere Supergroup	Dent	Broughton Moor
Ordovician-Silurian / Windermere Supergroup	Dent	Kirkley Bank
Ordovician-Silurian / Windermere Supergroup	Dent	Yarlside Volcanics
Ordovician-Silurian / Windermere Supergroup	Dent	Stile End
Ordovician	Skiddaw Gp	Tarn Moor
Ordovician	Skiddaw Gp	Buttermere
Ordovician	Skiddaw Gp	Kirk Stile
Ordovician	Skiddaw Gp	Loweswater
Ordovician	Skiddaw Gp	Hope Beck
Ordovician	Skiddaw Gp	Watch Hill
Ordovician	Skiddaw Gp	Bitter Beck

Stratigraphy of the Lakes District. (Modified after Stone et al., 2010.)

The collision of the continents Avalonia and Laurentia culminated during the Early Devonian, ending the Acadian phase of the Caledonian Orogeny with the final closure of the Iapetus Ocean and the injection of granites during the last stages of mountain building.

The Lake District has both metallic and non-metallic resources. Early inhabitants used the volcanic tuff found on Langdale Pike as a source for stone axes found across Britain. In the 16th–19th centuries, copper, lead, and silver were mined at sites such as Threlkeld and Goldscope. Graphite found in Ordovician rocks near Borrowdale was used in the pencil industry based in Keswick. Even today slate is mined for roofing material at the Honister Mines.

Today the principal industry of the Lake District is tourism. Tourism to this area began in the late 1700s when Thomas West wrote *A Guide to the Lakes* (1778), and really took off after William Wordsworth's *Guide to the Lakes* was printed in 1810. Arriving first by rail, then by car, tourists flocked to the area in such numbers that the national park was created to protect the environment. As many as 16 million tourists come to the Lake District each year from around the world (Wikipedia, Lake District).

The English Lakes. Selected views from a sketchbook. Mixed media, including watercolor and pencil by John Parker, 1825. (The image is available from the National Library of Wales, https://commons.wikimedia.org/wiki/File:DV342_Ullswater_from_above_Pattersdale.png.)

Visit

Lake District National Park is the most visited national park in the United Kingdom and the second largest in Britain after Cairngorms National Park. The purpose of the park is to protect the landscape by restricting change by industry or commerce. Most of the land in the park is private, with about 55% of that used in agriculture. The National Trust for Places of Historic Interest or Natural Beauty owns around 25% of the area. Access is usually restricted to public footpaths. Much of the uncultivated land has open access (Wikipedia, Lake District National Park).

Park Headquarters Hours: Monday to Friday: 9.30 a.m.–4 p.m.
Address: Lake District National Park Authority, Murley Moss, Oxenholme Road, Kendal LA9 7RL
Phone: 01539 724555

Email: hq@lakedistrict.gov.uk
Web site: https://www.lakedistrict.gov.uk/

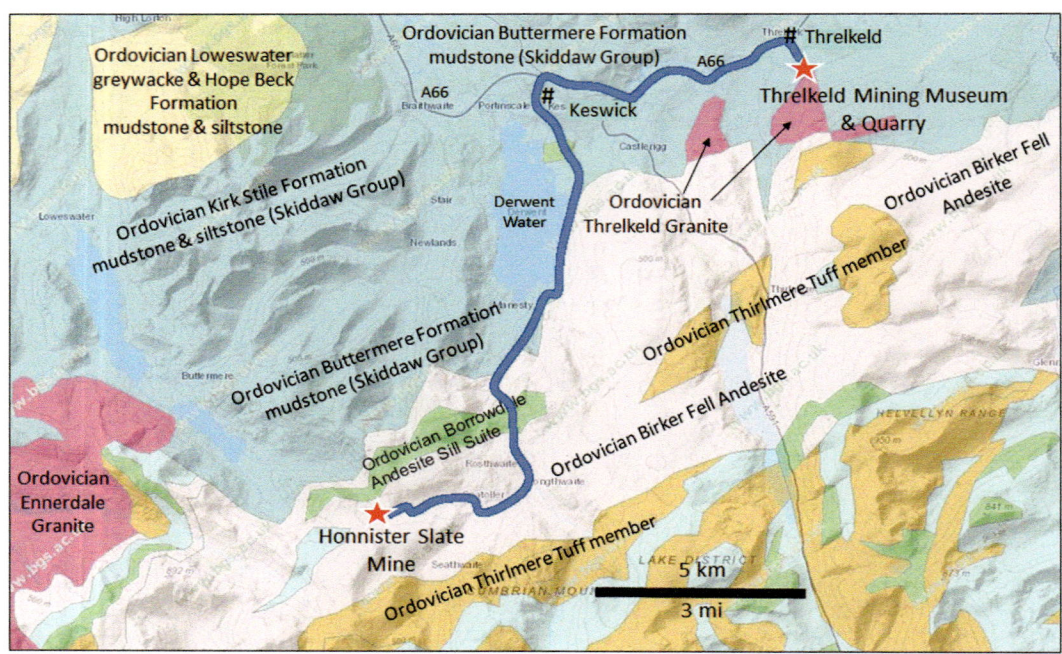

Geologic map of the Threlkeld and Keswick area, northern Lake District. (Modified after BGS, Geology of Britain Viewer. Contains British Geological Survey materials © UKRI 2022. Base mapping provided by ESRI.)

Start – At the junction of the M6 and A66 take the A66 west to B5322/Station Road in Threlkeld; turn left (south) onto B5322 and drive 650 m (0.4 mi) and turn left (east) to the Mining Museum and Steam Train; follow the signs to **Stop 1 Threlkeld Mining Museum and Quarry** *(54.610683, −3.044289) for a total of 131 km (81.5 mi; 1 h 41 min).*

STOP 1 THRELKELD MINING MUSEUM AND QUARRY

The quarry is at the junction of the Skiddaw Slates and Borrowdale Volcanics. At this junction, there was a granite intrusion that was quarried for railway ballast and then road stone. The quarry originally opened in 1870 to supply railway ballast to the Penrith–Keswick line. Later, the stone was used for roadstone and for building facings. The quarry closed in 1982 and now displays vintage excavators and quarry machinery. A narrow gauge railway provides a ride into the inner quarry for visitors to view the quarry faces.

Another feature of note is the mineralized, north-trending Coniston Fault. Lead, zinc, barite, pyrite, and copper were mined east of the fault.

The museum now has an artificial underground mine. A guided tour through the reconstructed lead and copper mine is one of the highlights of the museum. Mineral panning is available.

The mining room exhibits artifacts, plans, and photographs of many of the former mines in the Lake District. The mines produced copper, iron, lead, zinc, graphite, fluorite, and other minerals.

The quarry room explains the relationship between the geology of the Lake District and the quarrying of local slate, limestone, sandstone, and granite. Samples of rocks from all the important local formations are fixed to a large table map (Wikipedia, Threlkeld Quarry, and Mining Museum).

The museum provides free parking and has a gift shop.

Some of the equipment, Threlkeld Quarry and mining museum.

Not much outcrop remains in this part of the overgrown quarry.

VISIT

Hours: The museum is open 7 days a week, 10:00 a.m.–5:00 p.m. from Easter to the end of October. There is no need to prebook.

Steam trains run on the hour from 11:00 a.m.–4:00 p.m. Friday and Saturday. Trains at 11.00, 11.45, 13.00, and 13.45. You must arrive 5 min before your train time. Cost £30.00 per family of 4 or £8.00 per person.

Address: Threlkeld Quarry and Mining Museum, Threlkeld Quarry, Threlkeld CA12 4TT
Phone: 01768 779747
Email: threlkeldquarrymuseum@btconnect.com
Web site: https://www.threlkeldquarryandminingmuseum.co.uk/
Admission
Annual Family Pass: £35 (includes admission and train ride for two adults and two children; valid for 1 year)
Admission to the site (includes the museum):

    Adults £3.00; children under 16 £1.50

Add on **Railway:** Adults £3.00 Children £1.50
Add on **Underground Tour:** (generally 11.30, 1.30, and 3.30)

    Adults £5.00; children £3.00 (minimum age 7)

Add on **Mineral Panning:** £3.00 per pan

Mining Museum to Honister Mine: Return west to B5322/Station Road and turn right (north); drive to the A66 and turn left (west); continue straight onto the A66/Keswick Bypass; at the roundabout north of Keswick, take the 1st exit (south) onto A5271/A591/Crossthwaite Road; at the T intersection, turn left (southeast) onto A5271/A591/High Hill; at the roundabout take the 3rd exit (southwest) onto B5289/Main Street; at the roundabout take the 1st exit (southeast) onto B5289/Borrowdale Road; continue south and then west on B5289 to **Stop 2, Honister Slate Mine** *(54.511375, −3.199618) for a total of 23.4 km (14.5 mi; 35 min). There is ample parking in the "pay-and-display" lot. Parking costs £5.*

STOP 2 HONISTER SLATE MINE

The Honister Slate Mine, a quarry with underground workings centered on Honister Pass in Borrowdale, was opened in 1728 and is still producing. The cleaved slate occurs at several stratigraphic levels within the Late Ordovician (458–449 Ma) Eagle Crag member of the Birker Fell Andesite and volcaniclastics of the Borrowdale Volcanic Group (Stone et al., 2010; BGS, Geology of Britain Viewer "Classic"). Honister slate is a distinctive green color; Lakeland slate comes in a variety of shades and colors. Both are durable and are used for roofs, paving stones, and walls.

Honister's underground workings extend under Honister Crag, south of the pass, with other workings on the opposite side of the valley at Yew Crags. There are also small underground workings at Dubbs Moor.

By 1891 production had reached 2,722 tonnes (3,000 tons) a year, and more than 100 men worked the mines. In 1932, the Dubbs Quarry ceased production largely due to the slowness of transporting finished slate. Production continued through the 1950s and 1960s, although the Yew Crag mine closed in 1966 due to unstable roof conditions. In 1997, the mine was rejuvenated by Mark Weir, a local businessman, who produced small quantities of roofing slate, but also turned the site into a tourist attraction that included a visitor center, England's first via ferrata, and underground tours (Wikipedia, Honister Slate Mine).

Honister Slate Mine (top left). View northwest to Honister Crag.

VISIT

The tour shows the current workings of the mine, at the surface and underground, and displays a mixture of modern and traditional mining methods. You will learn about the 450-million-year-old volcanic green slate, as well as the 350 years of mining history. Special effects enhance the experience. Mine tours take approximately 90 min and involve descending approximately 50 steps. It is suitable for all ages. Bring warm clothing and sensible footwear – this is a walking tour. Safety helmets and lamps are provided (Honister, Mine Tours).

The visitor center has a series of information panels showing the history of the mine. There is a café, and you can browse the gift shop. The center is open year-round, 7 days a week (Visit Cumbria, Honister Slate Mine).

Address: Honister Pass, Borrowdale, Keswick, CA12 5XN
Phone: 017687 77230.
Email: bookings@honister.com
Web site: https://www.visitcumbria.com/kes/honister-slate-mine/
Tour Prices: £9.50 Child, £17.50 Adult, Family Ticket £49.00 (2 Adults and 2 Children)
Times: multiple times daily & runs 7 days a week.

Mined slate, Honister Mine.

Underground at Honister Slate Mine. (From Visit Cumbria, https://www.visitcumbria.com/kes/honister-slate-mine/.)

*Honister Mine to British Gypsum: Return east on B5289 to Seatoller and Rosthwaite; continue straight (north) on B5289 to Keswick; at the roundabout, take the 1st exit onto A5271/A591/Main Street; continue north on Main Street to A5271/A591/Crosthwaite Road; turn right (east) onto Crosthwaite Road and drive to the Crosthwaite Roundabout; take the 3rd exit onto A66/Keswick Bypass and drive east to Penrith; continue east on A66 to Kirkby Thore; turn left (northeast) onto Main Street; turn right (east) to Long Marton and drive 1.7 km (1.0 mi) to **Stop 3, British Gypsum Mine** (54.624742, −2.519467) on the left, for a total of 61.2 km (38.1 mi; 1 h). Note: touring the mine is by permission only.*

STOP 3 BRITISH GYPSUM MINE, LONG MARTON

We are now in Eden Valley.

The Permo-Triassic rocks of the Eden Valley lie in a north-south-elongated basin, roughly 50 km (30 mi) long and 5–15 km (3–9 mi) wide. The valley is bounded to the west by the Lake District and to the northeast by the North Pennines. Strata in the basin are inclined gently to the northeast. The down-to-the-west Pennine Fault and associated North Pennine escarpment form the eastern boundary of the basin. East of the fault are Carboniferous rocks. The Permo-Triassic sequence of the basin wedges out onto Carboniferous strata to the west (Lafare et al., 2014).

The Early Permian Penrith Sandstone Formation was deposited in a basin more-or-less coincident with the present Eden Valley. The Penrith Formation comprises windblown dune sandstones up to 900 m (1,000 ft) thick in the center of the basin.

The late Permian Eden Shale conformably overlies the Penrith Sandstone and consists mainly of mudstone and siltstone deposited in playa (ephemeral) lakes and lagoons subject to seasonal drying. Gypsum and anhydrite, which precipitate as the shallow lake evaporates (hence "evaporites"), are present as beds, nodules, cement, and veins (Stone et al., 2010). The thicker evaporites are being mined here.

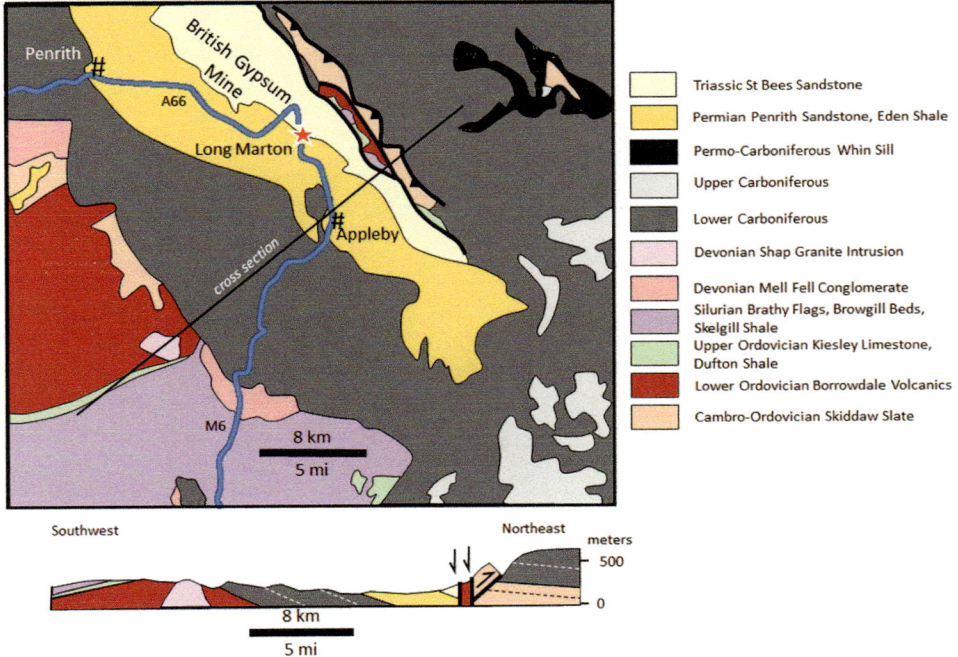

Geologic map and cross section through the Eden Valley at Appleby and Long Marton. Based on mapping by Versey (1942).

The Triassic St Bees Sandstone Formation conformably overlies the Eden Shale Formation. The formation consists mainly of fine-grained sandstone. Mudstone beds are generally subordinate, though increase near the contact with the underlying Eden Shale formation (Lafare et al., 2014).

Gypsum has been quarried or mined around Kirkby Thore for over 200 years. Gypsum and anhydrite mining today is limited to the area near Kirkby Thore. Gypsum in the "A" bed is worked at the Birkshead Mine, and anhydrite of the "B" bed is produced at the Newbiggin Mine. Gypsum is used for plaster, plasterboard, and Portland cement (Visit Cumbria, Gypsum mining in Cumbria).

"Daisy Bed" gypsum formation, British Gypsum Birkshead Mine, Cumbria. (Photo courtesy of Ross Collin, British Gypsum.)

Localized folding in gypsum and shale, Birkshead Mine. (Photo courtesy of Ross Collin, British Gypsum.)

Faulting in the pillar side, gypsum beds in the Birkshead Mine. (Photo courtesy of Ross Collin, British Gypsum.)

The plant at Kirkby Thore has been operating since 1910. The mined gypsum is ground to a powder and heated to evaporate water. Heating to approximately 160°C (320°F) drives off most water bound within the mineral to produce Plaster of Paris. The primary use of this gypsum is in the production of plaster and sheetrock. Heating over 200°C (392°F) drives off all the water to produce anhydrite, which is used in Portland cement.

The British Gypsum plant at Kirkby Thore is a major source of plaster building materials. As well as mined gypsum, the plant processes synthetic gypsum (desulfogypsum), a byproduct of the desulfurization of flue gases generated by coal-fired power stations (Fraser, 2017; Visit Cumbria, Gypsum mining in Cumbria).

Visit

You can visit British Gypsum Headquarters' information center at Kirby Thore. Arrange a visit to the mines by contacting British Gypsum directly.

Address: British Gypsum, Kirkby Thore Nr Penrith, Cumbria CA10 1XU
Phone: 0176 836 6247
Web site: https://www.british-gypsum.com/about-us/contact-us#press_and_media_enquiries

*British Gypsum to Waterfalls Trail: Drive south on the road to Long Marton; continue south on Long Marton Road to Appleby; turn left onto B6542/Battlebarrow and continue south; turn right (west) onto B6260/Bridge Street in Appleby; drive south on R30/Scattergate; continue south on B6260/Parkin Hill past Burrells and Orton; at Tebay Roundabout take the 3rd exit to the M6 going south; at Junction 37 take the A684 exit southeast to Kendall and Sedbergh; at the B6256 turn right (south); continue straight onto the A683; continue south to Kirkby Lonsdale and turn left (east) on the A65/Bentinck Drive; drive southeast on A65 to New Road; turn left (east) at the sign to Ingleton Waterfalls Walk and drive on New Road to Ingleton; bear right onto Holm Head Farm Road and drive to **Stop 4.1, Waterfalls Trail** (54.156180, −2.470939), for a total of 71.7 km (44.5 mi; 1 h 4 min).*

YORKSHIRE DALES NATIONAL PARK

Yorkshire Dales includes much of the uplands of the Pennines. The peaks can be up to 700 m (2,300 ft), but the river valleys (the dales) and hills give it much of its character. Most, but not all of the Yorkshire Dales lie within Yorkshire Dales National Park, an area of 2,178 km² (841 mi²) in the central part of northern England. The National Park was created in 1954, and today it is visited by over 8 million people each year (Wikipedia, Yorkshire Dales National Park).

ROCKS AND STRUCTURE

Structurally, the Yorkshire Dales are divided into the Stainmore Basin in the north, the Askrigg Block in the central area, and the Craven Basin in the south. The Craven Fault system separates the Craven Basin from the Askrigg Block. The Askrigg Block is structurally high because it is more buoyant as it is cored by the Wensleydale Granite. The Askrigg Block subsided as well, but slowly, and tilted northwards. It was covered by thin, shallow water limestones that thicken north into the Stainmore Basin. On the other hand, the Craven Basin subsided quickly and accumulated deep water turbidite mudstones and sandstones. Turbidites are a sediment slurry that flows from shallow to deep water. Whereas the Askrigg Block has on the order of 500 m (1,640 ft) of limestones at the southern margin, the adjacent Craven Basin has over 2,000 m (6,560 ft) of sediments. We will visit the area straddling the North and Middle Craven faults.

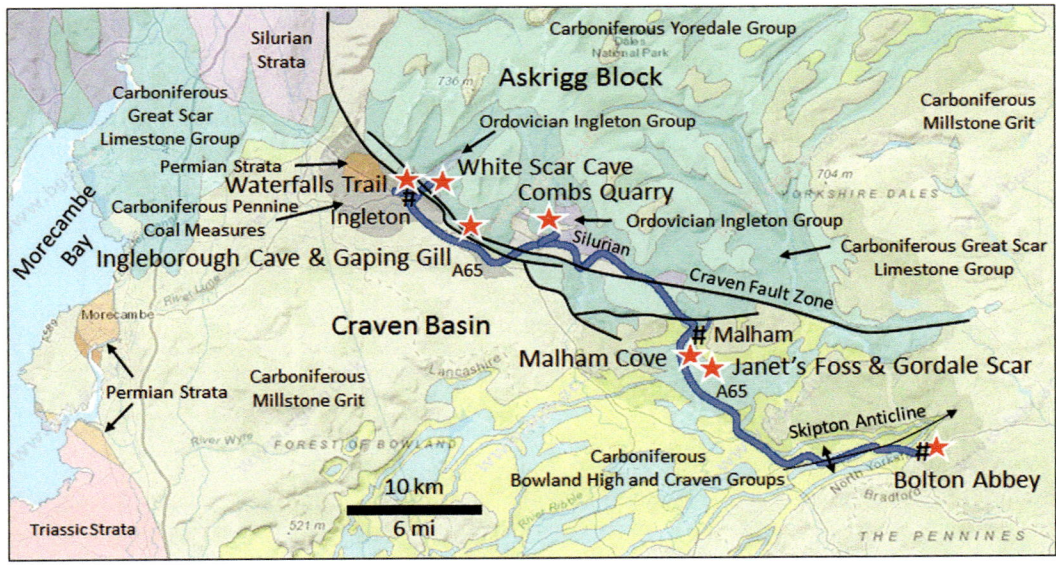

Geologic map of Yorkshire Dales National Park area. (Modified after BGS, Geology of Britain Viewer. Contains British Geological Survey materials © UKRI 2022. Base mapping provided by ESRI.)

The oldest rocks in the area are exposed in a couple of windows through the limestone that covers most of The Dales. The Ordovician Ingleton Group (485–475 Ma), exposed in quarries and waterfalls along the River Doe northeast of Ingleton and at Horton in Ribblesdale, is a deep marine mudstone-siltstone-sandstone turbidite sequence that was folded and eroded before it was overlain in turn by Silurian (446-423 Ma) deep marine turbidites. These rocks were deformed again during a Late Silurian to Middle Devonian mountain-building event, the Acadian Orogeny. Thermal and burial metamorphism caused recrystallization and development of cleavage, turning mudstones and siltstones into slates. By the end of the Devonian time, much of the uplifted rock had been eroded to a point where rising tropical seas flooded the area.

Age	Unit – Askrigg Block	Unit – Craven Basin
Permian	Zechstein Group	Eden Shale
		Penrith Sandstone
Carboniferous		Pennine Coal Measures Group
	Pennine Coal Measures Group	Millstone Grit Group
	Millstone Grit Group	
	Yoredale Group	Craven/Edale Shale Group
	Great Scar Limestone Group	
	Ravenstone Dale Group	Bowland High Group
Silurian	Windermere Supergroup: Horton & Neals Ing Fms	Windermere Supergroup: Kendal Group
	Arcow Fm	Coniston Group
	Austwick Fm	Tranearth Group
	Crummack Fm	Stockdale Group
	Sowerthwaite Fm	Dent Group
	Norber Fm	
Ordovician	Ingleton Group	

Stratigraphy of Yorkshire Dales National Park. (Information drawn from Aitkenhead et al., 2002; Kidd et al., 2006.)

By the start of the Carboniferous Period (~354 Ma), the area lay in tropical latitudes. Rising seas moved over the Silurian and Ordovician rocks and deposited limestones with fringing reefs.

The Great Scar Limestone was deposited as sea shells and corals, nearly pure calcium carbonate almost 200 m (660 ft) thick, in a warm tropical sea about 350 million years ago during the Lower Carboniferous (Mississippian).

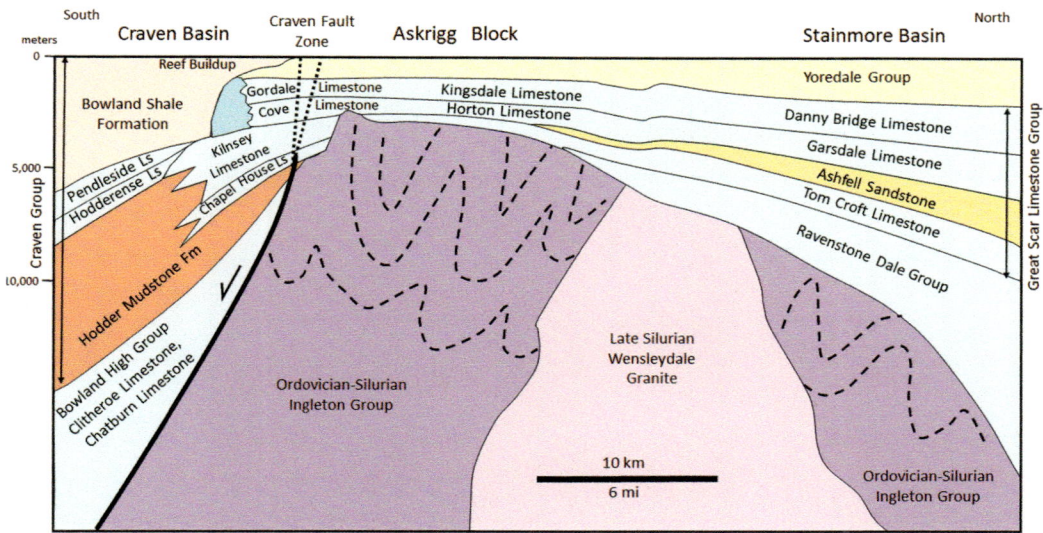

Cross section through the Craven Basin and Askrigg Block. (Information derived from Aitkenhead et al., 2002; Kidd et al., 2006.)

The Yoredale Series overlies the Great Scar Limestone. Rivers to the north brought mud and occasional sand to the sea, resulting in a rhythmic succession of limestone, shale, and sandstone, sometimes with thin coal above, followed by limestone, shale, and sandstone. The units correspond to shallow marine, estuary, delta, and swamp environments, respectively. The Yoredale Series is inclined slightly northeast as a result of later tilting of the region. Recent erosion has created distinctive step-like terraces in the Yoredale Series (Yorkshire Dales National Park Authority, Geology).

The Yoredale Series is capped on the higher fells by Carboniferous (Pennsylvanian; 326–313 Ma) Millstone Grit, a dark, river-deposited coarse sandstone.

About 326 Ma marks the start of a change to dominantly sandstone and mudstone. The Carboniferous sea was being filled in by river, estuary, and delta sediments until the area eventually became a coal swamp.

Any rocks deposited since that time have pretty much been totally removed by erosion.

And yet, the Yorkshire Dales is famous as limestone country. This unique landscape was shaped by the underlying rocks and their history. The limestones were deposited during the Carboniferous (350–327 Ma) in shallow tropical seas over vast expanses of time, eventually forming the massive layers that comprise this elevated plateau (Kidd et al., 2006). Limestone is a hard rock, but it dissolves ever so slowly over time as a result of slightly acidic rainwater. This dissolution causes a distinct topography known as karst, characterized not only by the limestone pavements etched with fractures, but also by scars (cliffs of light-colored limestone), swallow holes (sinkholes), underground rivers, springs, deep gorges, and extensive cave systems. Over 2,500 caves have been discovered; the extensive Three Counties Cave System has been mapped for 87 km (54 mi) (Wikipedia, Yorkshire Dales). Consequently, The Dales is a major caving region. Another limestone feature is reef knolls, conical hills that formed as coral buildups in the shallow sea. Whereas most of the Great Scar Limestone is poor hunting for fossils, fine examples of fossil-rich reef limestone can be found along the south side of the Craven Fault in the southern part of Malhamdale. The Craven Fault scarp along the structurally high southern margin of the Askrigg Block formed high areas on the ancient seafloor that were ideal sites for coral reefs (Kidd et al., 2006).

A typical north Yorkshire landscape. The bedrock is primarily limestone. View north from near White Scar Cave.

GLACIATION

The other major influence on the landscape is the glaciation over the past 2 Ma.

Major ice sheets advanced and retreated across Britain at least three times in the last half million years. Between 478,000 and 423,000 years ago, the whole area was buried under glacial ice. Grinding at the base of glaciers eroded much of the landscape. This continued during the most recent Ice Age, the Devensian, lasting from around 80,000 to 12,000 years ago.

All of the Yorkshire Dales, with the exception of some of the higher peaks, were covered with ice during the Devensian. Existing river valleys allowed the massive ice sheets to spread from the north and west. With each advance of the ice, glaciers scoured and deepened the valleys into a characteristic U-shaped profile. Moving ice shaped the valley bottom gravels into rounded hills called drumlins, best seen in Upper Ribblesdale where they are aligned southeast, the direction of ice flow.

The glaciers carried large amounts of rocks, sand, and clay. Terminal moraines, piles of boulders and gravel at the front of a glacier, mark the farthest advance of the ice sheets. As the ice retreated, the gravel and sand carried by them were left behind as blankets of glacial till, and isolated boulders as glacial erratics. Some of these erratics protected the underlying limestone from solution by rain, leaving them perched as much as 30 cm (1 ft) above the surrounding surface (Yorkshire Dales National Park Authority, Geology; Kidd et al., 2006). As the ice retreated, meltwater joined with rain to carve deep gorges in the uplands. The Ice Age also affected the subsurface: because of the intense cold, there was a lack of flowing water that arrested cave development (Yorkshire Dales National Park Authority, Geology).

RESOURCES

The Yorkshire Dales are rich in resources. Quarries mine slate for roofing stone and limestone for facing stone and aggregate for roadbeds, foundations, and concrete. In the past, coal was mined, as was lead. Lead veins formed as far-field effects of the Variscan Orogeny at the end of the Carboniferous Period (Pennsylvanian). Hydrothermal fluids moved up through faults and fissures in the Carboniferous rocks to form Mississippi Valley-type veins. Lead veins were mined primarily on Grassington Moor, Greenhow Hill, in Arkengarthdale, and Swaledale. The lead and coal mines are all abandoned now, but there are several large-scale, mainly limestone quarries operating within the boundaries of the national park (Kidd et al., 2006).

We start in the Ingleton area, where some of the finest examples of karst features are found.

STOP 4 INGLETON AREA SINKHOLES, CAVES, AND UNCONFORMITIES

From Ingleton to Malham to Gordale Scar, we are in Yorkshire Dales National Park. Our first stops are in the Craven Inliers.

The windows into the Ordovician and Silurian basement rocks are known as the Craven Inliers. They occur on the south margin of the Askrigg Block just north of the North Craven Fault. We see them on the Waterfalls Trail, at White Scar Cave, and near Horton in Ribblesdale. These layers were deposited offshore of the northern margin of Avalonia. The units were uplifted and deformed as a result of the closing of the Iapetus Ocean during the Caledonian Orogeny. The Ordovician here is a mixed sequence of clastics (sandstone, siltstone, and shale) and carbonates (limestone). Silurian units are deep marine turbidites. Shortening (folding and thrusting) during the early Devonian Acadian Orogeny imposed a pervasive cleavage on the rocks. Early Carboniferous (Mississippian) extension formed rifted basins and uplifted blocks, including the high Askrigg Block and the Craven Basin. The basin was filled with turbidites. Continued gradual Carboniferous subsidence led to shallow seas moving across the region and the consequent accumulation of limestones (Scrutton and Powell, 2006).

STOP 4.1 WATERFALLS TRAIL

Ingleton Waterfalls Trail is an 8 km (5 mi) loop trail that begins at the car park west of the River Twiss in north Ingleton. It is a well-maintained footpath with a vertical rise of 169 m (554 ft) that takes you past some of the most magnificent waterfalls in northern England. The trail and bridges were built and opened to the public in 1885. English Nature (a government conservation agency) designated the River Twiss and River Doe along the Waterfalls Trail as a Site of Special Scientific Interest (SSSI) due to the interesting plants, animals, and geological structures there (Wikipedia, Ingleton Waterfalls Trail).

The trail takes you along the River Twiss through Swilla Glen to Pecca Falls, Pecca Twin Falls, Holly Bush Spout, and Thornton Force. On the descent, you walk along the River Doe. The Doe emerges from springs alongside the B6255 about 3.2 mi (5.1 km) north of Ingleton (54.182958, −2.410926) and flows southwest to Beezley Falls Triple Spout, Rival Falls, and Snow Falls.

The rocks along the walk are all in the Ordovician Ingleton Group, the oldest rocks in Yorkshire, and Carboniferous Malham, Garsdale, and Danny Bridge Limestone formations (together known as the Great Scar Limestone) above an angular unconformity. The Ingleton Group was tightly folded, uplifted, and eroded before deposition of the Late Ordovician marine succession, which is not exposed here (Scrutton and Powell, 2006). The Craven Fault Zone is crossed during the walk along the River Twiss (at 54.168830, −2.468625) and along the River Doe (at 54.162283, −2.458880).

Visit

The trail is on private land and takes an average of 2.5 h to complete. You will need to complete the trail before dark.

Address: Ingleton Waterfalls Trail, Broadwood Entrance, Ingleton, Carnforth LA6 3ET
Phone: +44 (0)15242 41930
Email: info@ingletonwaterfallstrail.co.uk
Web site: https://www.ingletonwaterfallstrail.co.uk/information
Entrance fee: £7 adult; £3 child (under 16)

Hours:

March 1st–March 31st	9.00 a.m.–4.00 p.m.
April 1st–August 31st	9.00 a.m.–7.00 p.m.
September 1st–October 31st	9.00 a.m.–4.00 p.m.
November 1st–February 28th/29th	9.00 a.m.–2.30 p.m.

Thornton Force, Waterfalls Trail. The Ordovician-Carboniferous unconformity is just below the lip of the falls. (Courtesy of Val Vannet, https://commons.wikimedia.org/wiki/File:Thornton_Force.jpg.)

Waterfalls Trail to White Scar Cave: Return south to Bell Horse Gate and turn left (east); drive east on Bell Horse Gate to Seed Hill; continue straight on Seed Hill/High Street and drive to B6255; turn left (east) on B6255 and drive northeast to **Stop 4.2, White Scar Cave** *(54.165573, −2.442316) on the right for a total of 3.1 km (1.9 mi; 6 min).*

STOP 4.2 WHITE SCAR CAVE

White Scar Cave is the longest commercial cave in Britain, at 6 km (3.7 mi) long (Wikipedia, White Scar Cave). It was discovered in 1923 by Chris Long, a Cambridge undergraduate student, who was looking for an exit cave for water that sank into the Great Scar Limestone higher up the Chapel-le-Dale valley. He found a dry opening adjacent to the spring ("resurgence") emerging from the limestone and began exploring what is today the White Scar Cave. He found an underground river flowing through large chambers and over waterfalls. The cave was fitted with lights and made safe to enter in April 1925 (Hassan, 2014). The Happy Wanderers Caving Club explored White Scar for years, but in 1971 they actually cleared a passageway and sent Hilda Guthrie (because she was small enough to fit through the hole) into a massive cavern, 91 m (300 ft) long and 18 m (59 ft) wide, which she named "The Battlefield." The Battlefield lies about 70 m (230 ft) below the bottom of a shakehole (depression) on the valley surface, but no connection has yet been found (Yorkshire Dales National Park, White Scar Caves).

White Scar Cave is thought to have formed during warm interglacial periods over the past 600,000 years. These were times when abundant water was available to run through and dissolve the limestone (Hassan, 2014).

Visit

White Scar Cave tours are run year-round. Facilities include a shop and café.

Address: White Scar Cave, Ingleton, North Yorkshire LA6 3AW
Phone: 015242 41244
Email: info@whitescarcave.co.uk
Web site: https://whitescarcave.co.uk/
Hours: 5 December 2020–31 January 2021, Saturdays & Sundays - guided tours @ 11:00 a.m., 1:00 p.m., and 3:00 p.m. Check the website for the latest information.
Admission:
  £12.00 Adult (age 16 or older)
  £8.00 Child (free under 3)
  £34.00 Family (2 adults+2 children)

Entrance to White Scar Cave. The ceiling is at the contact between steep-dipping Ordovician–Silurian Ingleton Group metasediments and flat-lying Lower Carboniferous Garsdale Limestone Formation.

Interior, White Scar Cave. (Courtesy of Russell Greig, https://commons.wikimedia.org/wiki/File:Inside_White_Scar_Caves_-_geograph.org.uk_-_1140760.jpg)

White Scar Cave to Gaping Gill: *Return southwest on B6255 to Ingleton and turn left (southeast) onto Laundry Lane; turn left (southeast) toward Clapham on Old Road; at Clapham bear left (east) onto Eggshell Lane and drive to the Ingleborough Estate Nature Trail trailhead parking (54.120785, −2.390664). There is limited parking, so you may have to park along the street. The trail is 2.1 km (1.3 mi) one way to Ingleborough Cave, and a further 2.1 km (1.3 mi) one way to Stop 4.3 Gaping Gill. There is a £2.50 fee to use the trail.*

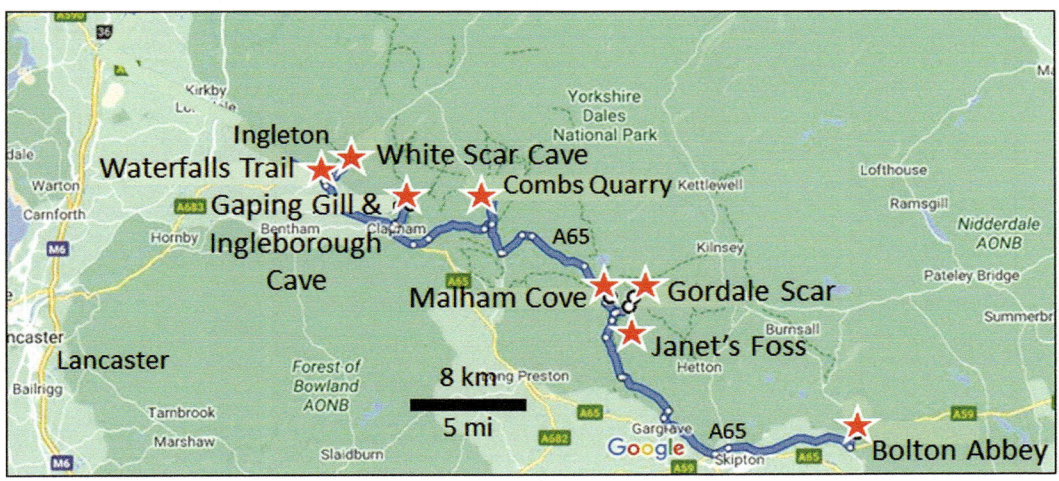

STOP 4.3 GAPING GILL

This stop is worth a gape since you are not likely to be entering the cave. You do not want to stumble into this in the dark. Gaping Gill (or Gaping Ghyll) is where Fell Beck stream drops down a 98 m (322 ft) shaft into a massive underground chamber connected to 15 km (9 mi) of caves and passages (Yorkshire Dales National Park Authority, Geology). The stream eventually resurfaces at Ingleborough Cave. Gaping Gill is the highest unbroken waterfall in England and has the largest underground chamber with a natural connection to the surface.

This is just one of several such cavern systems developed in the Carboniferous limestones along the southern margin of the Askrigg Block near the Craven Fault Zone. At the end of each glacial period, the retreating ice generated large volumes of meltwater. Meltwater downcutting in the valleys created lower water tables after each glaciation, allowing new, lower cave levels to develop. This in turn led to underground waterfalls (Dunlop, Gaping Gill). Caves progressively drained as valley floors were lowered. An estimate of the mean rate of valley incision in the Yorkshire Dales is about 0.15 m/Ka (5.9 in/Ka; Waltham and Long, 2011).

The first complete descent into Gaping Gill was achieved by Édouard-Alfred Martel, who in 1895 lowered himself by a rope ladder using only candles for lighting. Today access is limited: the Bradford Pothole Club around Whitsun May Bank Holiday, and the Craven Pothole Club around August Bank Holiday, set up a winch above the shaft to give rides to the bottom and back for members of the public. A fee is charged (Wikipedia, Gaping Gill; Ingleborough Cave, 450 Million Years of History).

Gaping Gill sinkhole. The stream Fell Beck flows into the subsurface here. (Courtesy of Abcdef123456, https://commons.wikimedia.org/wiki/File:GapingGillSurface.jpg.)

*Gaping Gill to Ingleborough Cave: From Gaping Gill return 2.1 km (1.3 mi) to the cave entrance. From the Eggshell Lane trailhead car park (54.120785, −2.390664) take the trail north along Clapham Beck for 2.1 km (1.3 mi one way) to **Stop 4.4, Ingleborough Cave** (54.134603, −2.377653).*

Stop 4.4 Ingleborough Cave

The 17 km (10.6 mi) of the Gaping Gill cavern system is developed in the Carboniferous Danny Bridge, Garsdale, and Kilnsey limestones (Ingleborough Cave, 450 Million Years of History; BGS, Geology of Britain Viewer "Classic"). Ingleborough Cave (formerly Clapham Cave), where Gaping Gill resurges, is the part of the caverns that is open to the public for tours (Wikipedia, Ingleborough Cave). Fell Beck, the stream that formed the cave, has found a new, lower route and now emerges at Clapham Beck Head.

The entrance to the cave has been known for quite a while, but it wasn't until 1837 when the local landowner, James Farrer, broke through a stalagmite barrier, releasing the water trapped behind it, that the true extent of the caverns began to be explored. The passages revealed stalactites, stalagmites, calcite flowstone drapes, and underground streams and pools. The link between Gaping Gill and Ingleborough Cave was first demonstrated in 1902 when The Yorkshire Geological Society dumped a ton of ammonium salt in the sinkhole. It emerged 5 days later at Clapham Beck Head (Yorkshire Dales National Park, Ingleborough Cave). The actual connection between Gaping Gill and Ingleborough Cave was found by cavers of the Bradford Pothole Club and Cave Diving Group in 1983. Discoveries are still being made. As recently as 2001, the remains of a woolly rhinoceros were discovered (Ingleborough Cave, 450 Million Years of History).

Entrance to Ingleborough Cave. The cave is in the flat-lying Lower Carboniferous Garsdale Limestone Formation.

Visit

The guided cave tour takes guests a total of 1 km (0.6 mi). The passages are well lit, and there is a smooth concrete path. The Cave is cool at 8°C (46°F) throughout the year. Photography is permitted; video is not permitted without specific permission in writing.

A small shop on site provides postcards, fossils, drinks, confections, and ice cream.

Address: Ingleborough Cave Clapham LA2 8EE
Phone: +44 (0) 1524 251 242
Email: info@ingleboroughcave.co.uk
Web site: https://ingleboroughcave.co.uk
Hours: Open Saturdays and Sundays throughout December and January, from 10:00 a.m. until the last entry at 3:00 p.m. The trail is open daily 24/7.
Admission
   Adults – £13.50
   Children (3–15) – £6.50
   Family (2A + 2C) – £37.00
   Senior Citizen (65+)/Student – £11.50

*Ingleborough Cave to Combs Quarry Unconformity: Return south to Clapham and turn left (south) onto the A65; drive southeast and turn left (northeast) onto Clapham Road; drive northeast on Clapham Road/Main Street through Austwick and continue east to the turnoff to the Dry Rig Quarry on the left (north). Park on the side of the road at the junction (54.117572, −2.301421), and walk 1.8 km (1.1 mi) one way along the road and Stories In Stone Geology Trail to **Stop 5, Combs Quarry Unconformity** (54.126105, −2.307869), for a total of 9.9 km (6.1 mi; 20 min). There is free parking on the side of the road. A walking trail begins here and leads about 1.6 km (1 mi) north to the quarry. As a side benefit, a short trail (420 m, or 1,280 ft) leads northwest to the Dry Rigg Quarry overlook, which is still active.*

STOP 5 COMBS QUARRY UNCONFORMITY, HORTON IN RIBBLESDALE

The abandoned Combs Quarry has a face with a spectacular view of the Silurian-Carboniferous angular unconformity. In this location, the horizontal Early Carboniferous (Mississippian) Kilnsey Formation, part of the Great Scar Limestone Group, consists of lagoon and reef slope limestone that rests unconformably on steeply dipping Late Silurian Horton Formation siltstone (Aitkenhead et al., 2002; Campa and Eguiluz, 2010: BGS, Geology of Britain Viewer "Classic"). The unconformity represents about 80–100 million years of missing rock, including all of the Devonian section. The south-dipping Horton beds are on the north flank of the roughly east-west Studrigg-Studfold Syncline. The Horton Formation contains volcanic ash beds that provide evidence for Caledonian volcanism around 400 Ma. Late Caledonian compression folded the Horton Formation, and the entire area was then tilted slightly to the north by movement on the Craven Fault.

Combs Quarry. Horizontal Kilnsey Formation limestone lies unconformably over steeply dipping Horton Formation siltstone. Dotted line indicates the unconformity. View north.

Records indicate that this quarry has been active since at least 1692 (Kidd, 2017).

STOP 6 MALHAM AREA LIMESTONE TERRAIN

The renowned pavements, sinks, and caves of the landscape around Malham are a function of the depositional history of the Carboniferous limestones. The limestones accumulated as reef build-ups along the southern margin of the Askrigg Block, and as ramp carbonates (limestone muds and sands) on the north-tilted block. The southern boundary of the block, the Craven Fault Zone, separated the limestone-dominated section to the north from the mudstone-dominated deep basin deposits in the Craven Basin to the south (Mundy and Arthurton, 2006).

Limestone pavement above Malham Cove. (Courtesy of Lupin, https://commons.wikimedia.org/wiki/File:Limestone_pavement_above_Malham_Cove.jpg.)

*Combs Quarry Unconformity to Malham Cove: Continue east on Austwick Road to intersection with B6479; continue straight (south) on B6479/Sherwood Brow to Church Lane in Stainforth; turn left (northeast) on Church Lane, then turn left (northeast) onto Goat Lane and drive to Henside Road; turn right (southeast) on Henside Road (sign to Malham 6 mi) and drive to Cove Road; turn right (southeast) on Cove Road (sign to Malham 2 ¾ mi) and drive to Malham Riverside Camping on the left for a total of 13.8 km (8.6 mi; 19 min). Park and walk 1,000 m (3,300 ft) on the Pennine Way trail to **Stop 6.1, Malham Cove** (54.072689, −2.158136).*

STOP 6.1 MALHAM COVE AND PAVEMENT

Malham Cove is a spectacular curving limestone cliff with exceptional pavements on top. The scenery is great from the base, but if you have the time, it is worthwhile taking the path to the top to see the pavements.

The majority of this cliff is the Cove Limestone, which is 72 m (236 ft) thick here. Above the Cove is another 10 m (28 ft) of the Gordale Limestone (Mundy and Arthurton, 2006). Both are part of the Great Scar Limestone that dips about 5° northeast and contain fossil corals, crinoids, and brachiopods (bivalve shells), among others. As the Ice Age glaciers melted, runoff flowed over the cliff. It has been mostly dry for the last 12,000 years, although the falls were renewed for the first time in centuries on 6 December 2015, after heavy rainfall from Storm Desmond (Wikipedia, Malham Cove).

Malham Cove. The Cove Limestone forms the lower cliff, whereas the Gordale Limestone forms the upper cliffs. View north.

Malham Beck issues from the base of this cliff. Despite what you might think, only a little of this stream originates at Malham Tarn Water Sinks. Most of the flow derives from the Smelt Mill Sink 1.2 km (0.75 mi) west of the Water Sinks (Mundy and Arthurton, 2006).

The pavement, sinkholes, caves, and resurgences are characteristic of karst weathering. The pavements were scoured bare by moving glaciers (Leeds. The walk from Malham village to Malham Tarn Field Centre). Slightly acidic rain has slowly been etching and dissolving the limestone since the end of the Ice Age. The solution-widened fractures in the pavement are known as grykes; the limestone blocks delineated by grykes are called clints (Dunlop, Malham Cove; Waltham and Long, 2011; Newson, 2015). Some grykes are over 1 m (3 ft) deep (Whittaker, 2020).

Malham Cove is a popular destination for climbers, as it provides a range of difficulties from easy to hard.

Malham Cove to Janet's Foss: Continue south on Cove Road to Pennine Way in Malham; turn left (east) on Pennine Way/Finkle Street; continue straight on Gordale Lane to Stop 6.2, Janet's Foss (54.066090, −2.136396) parking area on the right for a total of 1.9 km (1.2 mi; 4 min). There is limited parking on the roadside. Walk about 100 m (300 ft; 5 min) south to the waterfall and pool.

Stop 6.2 Janet's Foss

Janet's Foss is a small, scenic waterfall and crystal clear pool near Malham. Janet, or Jennet, refers to the fairy said to inhabit a cave behind the falls. "Foss" is the old Norse word for a waterfall, still used in places like Iceland (Wikipedia, Janet's Foss). Gordale Beck flows over a limestone ledge into a moss-covered tufa buildup and a small pond. Tufa is a porous buildup of calcium carbonate (limestone) that precipitates from water. Janet's cave was occupied in the 1600s by laborers working the copper smelters at Pikedaw 1.3 km (0.8 mi) to the west.

Janet's Foss.

The bedrock here is the Gordale Limestone member of the Malham Formation. Beds are inclined about 25° to the northeast. The mottled appearance of the limestone is due to burrows (Mundy and Arthurton, 2006).

*Janet's Foss to Gordale Scar: From the same parking area at Janet's Foss, follow the path north up Goredale Beck for 730 m (2,400 ft; 15 min) to **Stop 6.3, Gordale Scar** (54.072315, −2.131124).*

STOP 6.3 GORDALE SCAR

Gordale Scar is a slot canyon cut into the Carboniferous Malham Formation on the south edge of the Askrigg Block. The cliffs tower 100 m (330 ft) over Gordale Beck, a resurgence that leaves the canyon and flows on to Janet's Foss (Aitkenhead et al., 2002; Wikipedia, Gordale Scar). The gorge exposes about 43 m (140 ft) of the Cove Limestone member and about 94 m (308 ft) of the overlying Gordale Limestone member. The Cove Limestone is massive; the Gordale is bedded and highly fractured. This is the type section of the Gordale member (Mundy and Arthurton, 2006).

Opinions differ as to the origin of the canyon. It may have formed as a meltwater channel beneath glaciers during the last Ice Age (60,000 to 12,000 years ago). Or it may have been a cavern in the limestone where the roof collapsed. Either way, the location of the gorge is controlled by the Middle Craven Fault, which broke the rocks and made them more susceptible to weathering and erosion (Newson, 2015).

You can view the canyon from Gordale Bridge, or walk into the Scar. There is a double waterfall in the gorge, and you can climb it. Upstream from the first waterfall, the gorge turns abruptly northwest, again controlled by minor normal faults developed between the North and Middle Craven faults. These faults are exposed in the canyon (Mundy and Arthurton, 2006). There is a second waterfall farther up the gorge.

The Scar has some of the best examples of tufa in England (Newson, 2015). Calcium carbonate, dissolved from the limestone, is carried in solution in the stream. When it flows over the waterfall, the dissolved minerals come out of solution and precipitate as tufa.

There is a path to the top of the Scar, where you can see impressive limestone pavements.

Gordale Scar, a slot canyon cut into the Carboniferous Malham Formation.

Gordale Scar to Bolton Abbey: *Return southwest on Gordale Lane to Malham; continue straight on Finkle Street to Pennine Way; bear left (southwest) on Pennine Way to Cove Road; turn left (south) on Chapel Gate; continue straight (south) on Kirkby Brow; turn left onto Main Street in Kirkby Malham (sign to Gargrave 6 mi); turn left (east) at Carseylands Hill; continue straight on Pennine Way; turn right (southwest) on Eshton Road; turn left (south) onto Ray Bridge Lane in Gargrave; turn left (east) onto the A65 and drive to the roundabout in Thorlby Stirton and take the 1st exit to A65/A59; drive east on A59 to the roundabout in Bolton Bridge; take the 1st exit (north) to B6160 and drive to* **Stop 7, Bolton Abbey** *(53.980855, −1.892979) car park on the left for a total of 28.7 km (17.8 mi; 34 min). There is ample pay parking for £15.*

From the car park walk through the gateway (the "hole-in-the-wall") into Bolton Abbey, follow the path to the footbridge over the River Wharfe, and walk maybe 100 m upstream to a sharp bend in the river. This cutbank is **Stop 7.1, Folded Bowland Shale** *(53.982245, −1.887056).*

STOP 7 BOLTON ABBEY FOLDED CARBONIFEROUS

At this stop, we come upon the ruins of the Augustinian monastery known as Bolton Abbey. A monastery was founded here in 1120. Bolton Abbey, despite its name, was a priory (monastery). In 1154, the land was given to the Augustinian order by Lady Alice de Romille of Skipton Castle. The priory was closed in 1539 by the "Dissolution of the Monasteries" ordered by King Henry VIII (Wikipedia, Bolton Abbey).

The estate is open to visitors and has several walking paths.

Historic Bolton Abbey sits above folded Bowland Shale on the River Wharfe. Subsidiary folds are exposed in the cut bank to the right of this photo. View west.

Carboniferous units of the Craven Basin are thicker than their equivalents on the Askrigg Block north of the Craven Fault Zone. Whereas those to the north tend to be undeformed and gently-dipping, rocks in the Craven Basin were folded and faulted during the Late Carboniferous Variscan Orogeny. The Craven Basin was inverted, that is, squeezed and uplifted by north–south compression that reactivated old normal faults as high-angle reverse faults and generated northeast-trending folds, the Ribblesdale Fold Belt. There may have been up to 4,000 m (13,000 ft) of local uplift

(Aitkenhead et al., 2002). Skipton Anticline is in the southeastern part of the fold belt at Bolton Abbey. Skipton Anticline, a northeast-plunging fold, indicates north to northwest-directed Variscan shortening and compression. The stops near Bolton Abbey reveal disharmonic accessory folding in the Bowland Shale on the north and south flanks of Skipton Anticline (Varker, 2006).

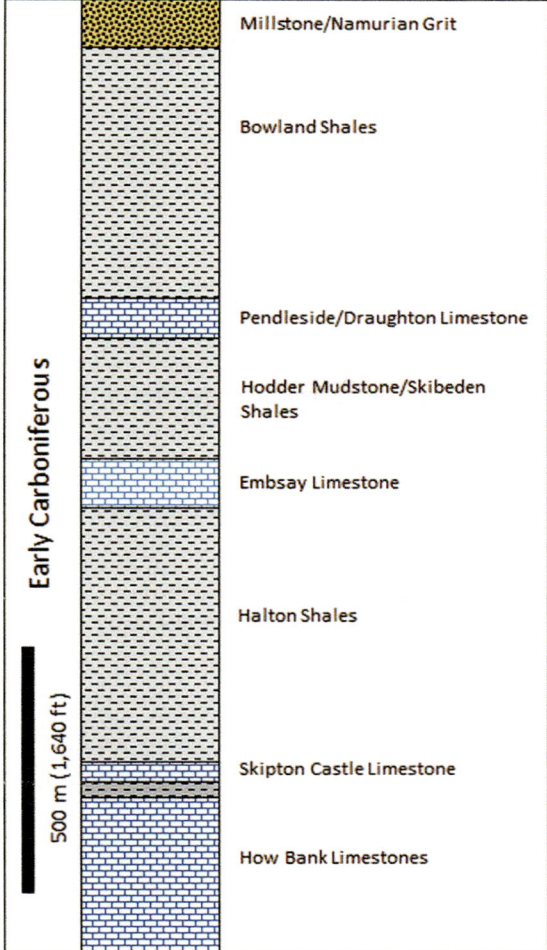

Stratigraphy of the Skipton Anticline–Bolton Abbey area. (Information derived from Metcalfe, 1981; BGS, Geology of Britain Viewer (Classic).)

This walk examines Early Carboniferous (Early Mississippian) mudstones containing thin limestones in the northeastern Craven Basin. We start on the southern flank of the Skipton Anticline and cross the fold axis to the north flank.

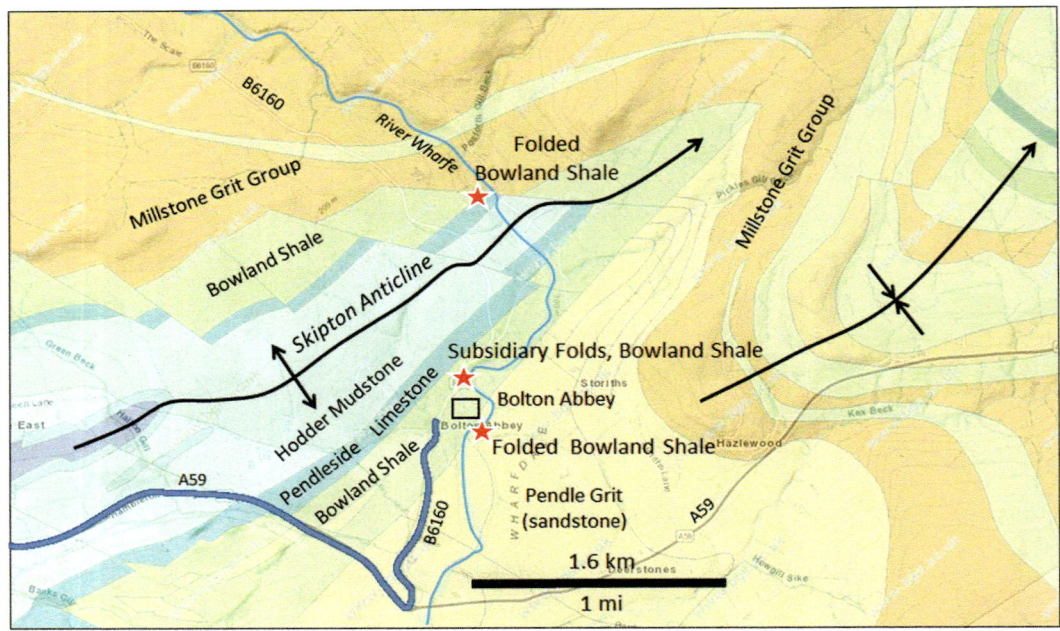

Geologic map of the Bolton Abbey area. (Modified after BGS, Geology of Britain Viewer. Contains British Geological Survey materials © UKRI 2022. Base mapping provided by ESRI.)

STOP 7.1 FOLDED BOWLAND SHALE, SOUTH FLANK SKIPTON ANTICLINE

Downstream from the footbridge, on the west side of the river, is an excellent outcrop of Upper Bowland Shale in a parasitic fold to the Skipton Anticline. This cutbank, approximately 20 m (66 ft) high, is maintained by river erosion, but weathering and vegetation obscure details of the structures. Several folds are present. Farther downstream is massive, well-jointed Skipton Moor Grit. The contact with the shale is a fault; viewed from the footbridge, the fault is near-vertical.

*Walk ~150 m (500 ft) upstream to **Stop 7.2, Subsidiary Folds** (53.985358, −1.888360) at a point across the river from a small anticline on the west bank.*

STOP 7.2 SUBSIDIARY FOLDS, BOWLAND SHALE

This partly overgrown stop contains subsidiary folds to the main Skipton Anticline. Bedding plane slip and flexural flow in the Bowland Shale allowed these small-scale folds to develop between the more competent Pendleside Limestone below and the Pendle Grit above. The fold is highlighted by the thin limey layers (Varker, 2006).

Small-scale subsidiary folding in the Bowland Shale along the River Wharfe at Bolton Abbey.

*Walk 1250 m (4,100 ft) upstream to **Stop 7.3, Folded Bowland Shale** (53.998035, −1.889446). Along the way observe the well-exposed Draughton Limestone on the opposite side of the river. Notice that there are three erosional river terraces, at 1, 2.5, and 5 m above the river, indicating the amount of downcutting since the last glaciation 12,000 years ago.*

STOP 7.3 FOLDED BOWLAND SHALE, NORTH FLANK SKIPTON ANTICLINE

At this stop, you see thinly bedded limestones and shales of the Bowland Shale folded into plunging anticlines and synclines. Dips on the limbs of these folds are on the order of 60°–70°. The folds may not be visible during high water (Varker, 2006).

This completes our geo-tour of the Lake District and Yorkshire Dales. This tour has taken us through the rugged beauty and fascinating geology of the Lake District World Heritage Site and the limestone country of Yorkshire Dales National Park. We have visited museums, mines, waterfalls, caves, sinkholes, pavements, and slot canyons. Whereas the Lake District and Ingleton area (Combs Quarry unconformity) was affected by the Caledonian Orogeny, much of the surface geology of the Yorkshire Dales (Askrigg Block) is essentially undeformed. The geology around Bolton Abbey shows us that the Variscan Orogeny left an imprint on the Carboniferous units of the Craven Basin.

REFERENCES

Aitkenhead, N., W. Barclay, A. Brandon, R.A. Chadwick, J.I. Chisholm, A.H. Cooper, and E.W. Johnson. 2002. *British Regional Geology: the Pennines and adjacent Areas* (4th edition). British Geological Survey, Nottingham, 206 p.

BGS, Geology of Britain Viewer (Classic). British Geological Survey. Accessed 31 December 2020, https://mapapps.bgs.ac.uk/geologyofbritain/home.html

Campa, M.F., and S. Eguiluz. 2010. Excursión Geológica al Reino Unido, Siguiendo las Huellas de William Smith. Universidad Autónoma de la Ciudad de México, Mexico City, 4 p.

Dunlop, L. Gaping Gill. The Geological Society. Accessed 3 January 2021, https://www.geolsoc.org.uk/GeositesGapingGill

Dunlop, L. Malham Cove. The Geological Society. Accessed 3 January 2021, https://www.geolsoc.org.uk/GeositesMalhamCove

Fraser, W. 2017. Geology around Kirkby Stephen. Mid-Week Geology Group. Accessed 22 February 2023, chrome-extension://efaidnbmnnnibpcajpcglclefindmkaj/http://mwggyorkshire.org.uk/pdf/Kirkbystephen%20geology.pdf

Hassan, I. 2014. 100 Great Geosites Nomination: White Scar Cave. Geological Society of London. Accessed 3 January 2021, https://blog.geolsoc.org.uk/2014/07/16/100-great-geosites-nomination-white-scar-cave/

Honister. Mine Tours. Accessed 3 January 2021, https://honister.com/product/mine-tours/

Honister Slate Mine. Accessed 3 January 2021, https://www.visitcumbria.com/kes/honister-slate-mine/

Ingleborough Cave. 450 Million Years of History. Accessed 3 January 2021, https://ingleboroughcave.co.uk/history/

Kidd, A. 2017. Ingleborough Stories in Stone Geo Trails, Helwith Bridge & Horton in Ribblesdale. Brochure by the Ingleborough Dales Landscape Partnership, Led by Yorkshire Dales Millennium Trust, and Supported by the National Lottery Heritage Fund. Accessed 15 May 2022, https://www.storiesinstone.org.uk/Portals/0/adam/Image%20Link/mC_4TSR9bUW4gvnz1AKJyQ/Link/Helwith-Horton%20081019.pdf

Kidd, A., A. Cooper, and J. Brayson. 2006. *The 'Your Dales Rocks Project' – A Draft Local Geodiversity Action Plan (2006–2011) for the Yorkshire Dales and the Craven Lowlands*. North Yorkshire Geodiversity Partnership, Northallerton, North Yorkshire, 39 p.

Lafare, A.E.A., A.G. Hughes, and D.W. Peach. 2014. Geology and hydrogeology of the Eden Valley. In *Eden Valley Observation Boreholes: Hydrogeological Framework and Groundwater Level Time Series Analysis*. British Geological Survey Internal Report, OR/14/041, 5 p.

Leeds. The Walk from Malham Village to Malham Tarn Field Centre. Accessed 3 January 2021, https://www.fbs.leeds.ac.uk/student/biology/docs/Malham%20Walk%20Guide.pdf

Metcalfe, I. 1981. *Conodont Zonation and Correlation of the Dinantian and Early Namurian Strata of the Craven Lowlands of Northern England*. Institute of Geological Sciences, Natural Enyironment Research Council, London, 70 p.

Mundy, D., and R. Arthurton. 2006. Craven Fault Zone — Malham to Settle - an excursion. In Scrutton, C., and Powell, J. (eds), *Yorkshire Rocks and Landscape: A Field Guide* (3rd edition). Yorkshire Geological Society, 8 p.

Newson, R.A. 2015. Marvellous Malham and Gordale Scar! Accessed 3 January 2021, https://geologytrails.wordpress.com/2015/07/14/marvellous-malham-and-gordale-scar/

Scrutton, C., and J. Powell (eds). 2006. *Yorkshire Rocks and Landscape: A Field Guide* (3rd edition). Yorkshire Geological Society, London, 8 p.

Stone, P., D. Millward, B. Young, J.W. Merritt, S.M. Clarke, M. McCormac, and D.J.D. Lawrence. 2010. *British Regional Geology: Northern England* (5th edition). British Geological Survey, Keyworth, Nottingham. Accessed 27 January 2021, http://earthwise.bgs.ac.uk/index.php/British_regional_geology:_Northern_England

Varker, W.J. 2006. Dinantian and Namurian rocks of Bolton Abbey and Trollers Gill - an excursion. In Scrutton, C., and Powell, J. (eds), *Yorkshire Rocks and Landscape: A Field Guide* (3rd edition). Yorkshire Geological Society, London, 7 p.

Versey, H.C. 1942. Geology of the Appleby District. J. Whitehead and Son, Appleby, 39 p. Permission from Geological Magazine, Cambridge University Press.

Visit Cumbria. Gypsum Mining in Cumbria. Accessed 3 January 2021, https://www.visitcumbria.com/gypsum/

Visit Cumbria. Honister Slate Mine. Accessed 3 January 2021, https://www.visitcumbria.com/kes/honister-slate-mine/

Waltham, T., and H. Long. 2011. Limestone plateaus of the Yorkshire Dales glaciokarst. *Cave and Karst Science*, v. 38, no. 2, pp. 65–70.

Whittaker, A. 2020. Limestone Pavement at Malham Cove. Accessed 3 January 2021, https://epod.usra.edu/blog/2020/03/limestone-pavement-at-malham-cove.html

Wikipedia, Bolton Abbey. Accessed 29 September 2022, https://en.wikipedia.org/wiki/Bolton_Abbey

Wikipedia, Gaping Gill. Accessed 3 January 2021, https://en.wikipedia.org/wiki/Gaping_Gill

Wikipedia, Geology of The Lake District. Accessed 3 January 2021, https://en.wikipedia.org/wiki/Geology_ of_the_Lake_District

Wikipedia, Gordale Scar. Accessed 3 January 2021, https://en.wikipedia.org/wiki/Gordale_Scar

Wikipedia, Honister Slate Mine. Accessed 3 January 2021, https://en.wikipedia.org/wiki/Honister_Slate_Mine

Wikipedia, Ingleborough Cave. Accessed 3 January 2021, https://en.wikipedia.org/wiki/Ingleborough_Cave

Wikipedia, Ingleton Waterfalls Trail. Accessed 3 January 2021, https://en.wikipedia.org/wiki/Ingleton_ Waterfalls_Trail

Wikipedia, Janet's Foss. Accessed 3 January 2021, https://en.wikipedia.org/wiki/Janet%27s_Foss

Wikipedia, Lake District. Accessed 3 January 2021, https://en.wikipedia.org/wiki/Lake_District

Wikipedia, Lake District National Park. Accessed 3 January 2021, https://en.wikipedia.org/wiki/Lake_ District_National_Park

Wikipedia, Malham Cove. Accessed 3 January 2021, https://en.wikipedia.org/wiki/Malham_Cove

Wikipedia, Penrith. Accessed 29 September 2022, https://en.wikipedia.org/wiki/Penrith,_Cumbria

Wikipedia, Threlkeld Quarry and Mining Museum. Accessed 3 January 2021, https://en.wikipedia.org/ wiki/Threlkeld_Quarry_and_Mining_Museum

Wikipedia, White Scar Caves. Accessed 3 January 2021, https://en.wikipedia.org/wiki/White_Scar_Caves

Wikipedia, Yorkshire Dales. Accessed 3 January 2021, https://en.wikipedia.org/wiki/Yorkshire_Dales

Wikipedia, Yorkshire Dales National Park. Accessed 3 January 2021, https://en.wikipedia.org/wiki/Yorkshire_ Dales_National_Park

Yorkshire Dales National Park. Ingleborough Cave. Accessed 3 January 2021, https://dalesrocks.org.uk/ geological-processes/cave-formation/ingleborough-cave/

Yorkshire Dales National Park. White Scar Caves. Accessed 3 January 2021, https://www.yorkshiredales.org. uk/places/white_scar_cave/

Yorkshire Dales National Park Authority. Geology. Accessed 3 January 2021, http://www.outofoblivion.org. uk/geology.asp

3 Wales and West Midlands

Cliffs of Penyrafr, looking north from Ceibwr Bay, North Pembrokeshire. These spectacular folds in the Ordovician Dinas Island Formation formed during the Caledonian Orogeny.

OVERVIEW

Spectacular sea cliffs bear witness to the annihilation of an ocean. From deep marine landslide deposits to shallow reef banks and coastal swamps, the rocks of Wales and the West Midlands tell a tale of multiple continental collisions and dramatically changing landscapes.

We begin at Great Orme, a limestone headland with Bronze Age copper mines on the north coast of Wales. Traveling west to Anglesey and Geomôn Global Geopark, there are fossils to be found in the Carboniferous limestones at Caim Beach and spectacular folded rocks at South Stack. Snowdonia National Park has Ordovician volcanic rocks sculpted by glaciers at Cwm Idwal, and a Cambrian slate quarry at Dinorwic. Moving south along the Welsh coast from Borth, we see Silurian and Devonian marine turbidite layers contorted by Caledonian and Variscan mountain-building episodes, and a fossil forest exposed at low tide. Deformed Ordovician layers are exposed in sea cliffs between Cemaes Head and Ceibwr Bay. A side trip takes us to the Neolithic quarry at Carn Goedog, source of the bluestones at Stonehenge. Continuing south in Pembrokeshire Coast National Park we examine Precambrian to Devonian formations revealed on sheer sea cliffs and along pristine white-sand beaches from Whitesands Bay to Marloes Sands. On the Castlemartin Peninsula we again encounter Carboniferous limestone, now deformed, at Stackpole Quay, and visit one of the top 12 beaches in the world at Barafundle Bay. A second side trip takes us to the caverns of Dan yr Ogof in Brecon Beacons National Park. Developed in Carboniferous limestone, this extensive cave system is considered one of the greatest natural wonders in Britain. Jurassic marine fossils and dinosaur tracks are found along the Glamorgan Heritage Coast and Severn shore from Ogmore-by-Sea to Aust Cliff.

DOI: 10.1201/9781351165600-3

Turning north, we traverse the Abberly and Malvern Hills UNESCO Global Geopark, centered on a ridge of resistant Precambrian rocks poking out of the surrounding Silurian and Triassic sedimentary layers. Continuing north, we pass through Black Country Global Geopark, a designation that recognizes the Silurian to Triassic strata and coal resources as well as its heritage as the "birthplace of the Industrial Revolution." The Iron Bridge World Heritage Site spans a gorge along the River Severn whose origins are steeped in controversy. We transit to the Peak District National Park, explore the cave systems at Blue John Caverns, and examine landslides at Mam Tor.

ITINERARY

Begin – Manchester Airport
Stop 1 Rhualt, Llandudno, and Great Orme
    1.1 Rhualt Roadcut
    1.2 Great Orme from Llandudno Pier
    1.3 Great Orme Mines
Side Trip 1 Anglesey and Geomôn Global Geopark
    ST 1.1 Caim Beach
    ST 1.2 South Stack
Stop 2 Snowdonia National Park
    2.1 Cwm Idwal
    2.2 Dinorwic Slate Quarry
    2.3 National Slate Museum
Stop 3 Ceredigion
    3.1 Borth Beach and Submerged Forest
    3.2 Aberystwyth
Stop 4 Pembrokeshire Coast National Park
    4.1 Cemaes Head
    4.2 Ceibwr Bay
Side Trip 2 Carn Goedog Bluestones
Stop 5 South Pembrokeshire Coast
    5.1 Blue Lagoon and Abereiddy Bay
    5.2 Whitesands Bay
    5.3 Caerfai Bay
Stop 6 Broad Haven
Stop 7 Marloes Sands
Stop 8 Castlemartin Peninsula
    8.1 Stackpole Quay
    8.2 Barafundle Bay
Side Trip 3 Dan-yr-Ogof Cave and Brecon Beacons National Park
Stop 9 Glamorgan Heritage Coast
    9.1 Ogmore-by-Sea
    9.2 Dunraven Bay and Heritage Coast Centre
    9.3 Llantwit Major Beach
Stop 10 Severn Coast South of Cardiff
    10.1 Bendricks Rock Dinosaur Tracks
    10.2 Lavernock Point
    10.3 Penarth Dinosaur Tracks
    10.4 Penarth Pier to Cardiff Barrage
Stop 11 Aust Cliff
Stop 12 Abberly and Malvern Hills UNESCO Geopark
    12.1 Gardiners Quarry

Key stops on the Midlands and Wales geo-tour.

THE WEST MIDLANDS AND WALES

GEOLOGY OF WALES

Wales is dominated by Lower Paleozoic rocks – the Cambrian and Ordovician in Llŷn, Snowdonia National Park and north Pembrokeshire, and the Silurian over much of central Wales. The sequence is mainly of marine sedimentary rocks, although Anglesey and the Llŷn Peninsula contain the largest outcrop of Precambrian metamorphic rocks in southern Britain. Smaller Precambrian outcrops occur in Pembrokeshire, Carmarthenshire, and Radnorshire.

Cambrian sedimentary rocks range from silty mudstones, such as those in the north Wales slate belt, to coarse sandstone encountered in the Rhinog Mountains. These were deposited on the northwest coast of Avalonia, a continent on the edge of the Iapetus Ocean (see Geology of Scotland, Chapter 1). During Ordovician times, igneous activity caused thick accumulations of volcanic rocks. The impressive landscape seen in Snowdonia National Park is due largely to the erosion-resistant volcanics. In Pembrokeshire, the coast around St Bride's Bay is a result of erosional differences between tough volcanic rocks and less resistant sedimentary rocks (Howells, 2007). Island arc volcanism resulted from subduction and melting of the Iapetus seafloor beneath Avalonia.

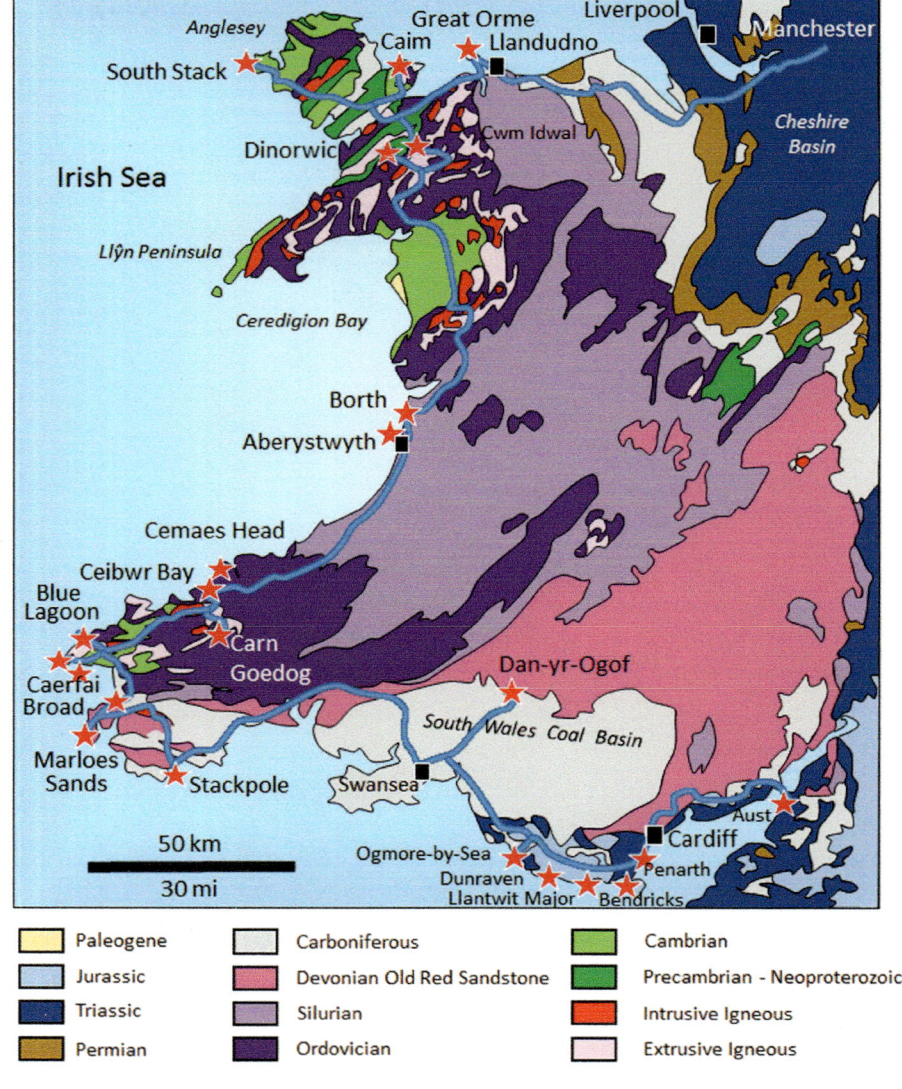

Simplified geologic map of Wales. (Modified after figure 1 in Howells, 2007. Contains British Geological Survey materials © UKRI 2022.)

Age		Unit	Lithology
Tertiary		Eocene-Miocene Undifferentiated	Conglomerate, sandstone, siltstone, claystone
		Paleogene Undifferentiated	Conglomerate, sandstone, siltstone, claystone
Jurassic	Upper	Upper Jurassic Undifferentiated (including Kimmeridge Clay Fm)	Mudstone with subordinate limestone, sandstone
		Corallian Group	Limestone and sandstone with subordinate siltstone, mudstone
	Middle	Mid-Jurassic Undifferentiated (including Kellaways/Oxford Clay Fms)	Mudstone and sandstone
		Great & Inferior Oolite Groups	Limestone with subordinate siltstone, mudstone, sandstone
	Lower	Lias Group (including Blue Lias Fm)	Mudstone with subordinate limestone, siltstone
Triassic	Upper	Penarth Group	Mudstone with subordinate limestone, siltstone
		Mercia Mudstone Group & laterally equivalent Dolomitic Conglomerate (including Sidmouth Mudstone Fm)	Mudstone with subordinate siltstone, sandstone, gypsum, salt
		Sherwood Sandstone Group	Sandstone with subordinate siltstone, mudstone, conglomerate
Permian		Cumbrian Coast and Appleby Groups	Mudstone and siltstone with subordinate sandstone, conglomerate, evaporite
Carboniferous		Warwickshire Group (including Pennant Sandstone Fm)	Mudstone and sandstone with subordinate limestone and coal
		Pennine Coal Measures/South Wales Coal Measures Group	Sandstone, siltstone, mudstone, coal
		Millstone Grit/Marros Group (including Mam Tor Sandstone, Bowland Shale, Bee Low Limestone, Telpyn Point Sandstone Fm Bishopston Mudstone Fm Twrch Sandstone Fm)	Mudstone and sandstone with subordinate siltstone, conglomerate
		Carboniferous Limestone Supergroup (including Pembroke Limestone Group & Clwyd Limestone Group Avon Group mudstones)	Limestone with subordinate mudstone, sandstone
Devonian		Upper Old Red Sandstone Group (including Shrinkle Sandstone Fm)	Sandstone and conglomerate with subordinate siltstone, mudstone
		Lower Old Red Sandstone Group (including Traeth Bach Fm) and laterally equivalent Cosheston Group over Milford Haven Group	Weakly metamorphosed sandstone, conglomerate, siltstone
Silurian		Pridoli Group	Siltstone with subordinate conglomerate, sandstone, limestone
		Ludlow Group	Slate, slaty mudstone, minor sandstone
		Gray Sandstone Group and Wenlock Group (including Wenlock Limestone)	Quartzitic sandstone and subordinate mudstone; limestone and shale
		Coralliferous Group	Mudstones
		Skomer Volcanic Group and laterally equivalent Aberystwyth Grits Group over Borth Mudstone Fm over Gaerglwyd Fm	Basalt and rhyolite flows with volcanic breccias, tuffs, conglomerates Thin, alternating sandstone and mudstone beds
Ordovician		Undifferentiated metasediments (including Yr Allt & Nantmel Mudstone Fms, Ashgill, Caradoc, Llanvirn, & Arenig Groups)	Slate, slaty mudstone, sandstone
		Ordovician metavolcanics undifferentiated	Weakly metamorphosed lavas and tuffs
Cambrian		Cambro-Ordovician metasediments undifferentiated	Weakly metamorphosed mudstone, sandstone, siltstone
		Monian Supergroup (including Lingula Flags Fm and Caerfai Group)	Metasediments and metavolcanics
Precambrian		Neoproterozoic metavolcanics (including Malverns Complex, Uriconian Volcanics)	Metamorphosed lavas, tuffs, breccias
		Precambrian basement (including metasediments and igneous rocks)	Metamorphosed sediments, igneous rocks

Stratigraphy of Wales. (Information drawn from Waters, 2008; Radioactive Waste Management, 2016, and University of Derby, 2020.)

In Late Ordovician and Silurian times, after most volcanic activity ended, mostly silt and mud were deposited throughout central Wales.

In Late Silurian and Early Devonian times, the rocks were folded and uplifted during the Acadian phase of the Caledonian Orogeny or mountain-building event. Wales was part of a continent that extended from the Bristol Channel into northern Britain with the sea to the south. Folding during this orogeny in Wales formed the Harlech and Berwyn domes and the intervening Central Wales Syncline. Fluvial (river), alluvial, and lacustrine (lake) sedimentation produced the Old Red Sandstone of Late Silurian and Devonian age. The Old Red Sandstone is exposed from Pembrokeshire in the west to Herefordshire in the east. Across much of this region, the Old Red Sandstone lies unconformably above an eroded surface of folded Lower Paleozoic basement rocks (Howells, 2007).

Late Caledonian (Devonian) uplift and erosion removed Middle Devonian rocks across much of Wales.

The appearance of marine fossils toward the top of the Old Red Sandstone indicates a marine transgression (sea level rise) moving across the land from the south. By Carboniferous times, the Pennine Basin extended from the Scottish border south to the Wales–Brabant Massif. The Wales–Brabant landmass occupied most of the area of central and northern Wales and extended to around Brussels, Belgium. Extensive Early Carboniferous (Mississippian) limestones were deposited in south Pembrokeshire and South Wales, at Anglesey, and at Great Orme. They form an almost continuous outcrop around the regional east–west basin containing the South Wales Coalfield.

Uplift during Late Carboniferous time caused temporary emergence and much coarse sediment to be deposited in a series of deltas that were periodically flooded by the sea. Tropical rainforests developed on delta plains along the coast. These forests, lagoons, and coastal swamps accumulated plants that became Upper Carboniferous (Pennsylvanian) coals. The many coal seams are a result of repeated uplift and submergence of these environments.

The middle Carboniferous Variscan/Hercynian mountain chain formed as Gondwana, the Southern Continent, collided and merged with Laurussia to create Pangea, a supercontinent. The Variscan Orogeny influenced sedimentation: coarse sandstones, conglomerates, and breccias indicate a nearby sediment source. Deformation reached its peak in the Late Carboniferous when strata were folded and uplifted, forming the Variscide Mountains that extended from southern Ireland across Wales and into northwest Europe. In South Wales, intense folding and faulting can be seen along the coast in Pembrokeshire and the Gower peninsula. Following the Variscan Orogeny, the red sandstones and siltstones of the Permian and Triassic were derived mainly from erosion of Carboniferous rocks. Permo-Triassic outcrops are found in the Vale of Glamorgan in the south, on the west flank of the Cheshire Basin, and in the Vale of Clwyd in the north. Eventually, the Triassic landmass subsided and marine mudstones and thin limestones of the Lias Group, so well-displayed in the coastal cliffs at Dunraven Bay and Llantwit Major, were deposited in the Early Jurassic.

The Cretaceous to Pliocene Alpine Orogeny was a result of the African tectonic plate moving north into the European plate. Largely in southern Europe, the mountain-building episode appears to have affected areas as far west as Wales. Deformation involved uplift and erosion. Marine conditions existed along the Irish Sea through most of the Cenozoic.

Glaciers covered most of Wales in the Pleistocene, as seen in erosional features of upland areas, particularly in Snowdonia. A major ice sheet moved from the Irish Sea south and east across Anglesey and Llŷn in the west and into Cheshire and Shropshire in the east (Howells, 2007).

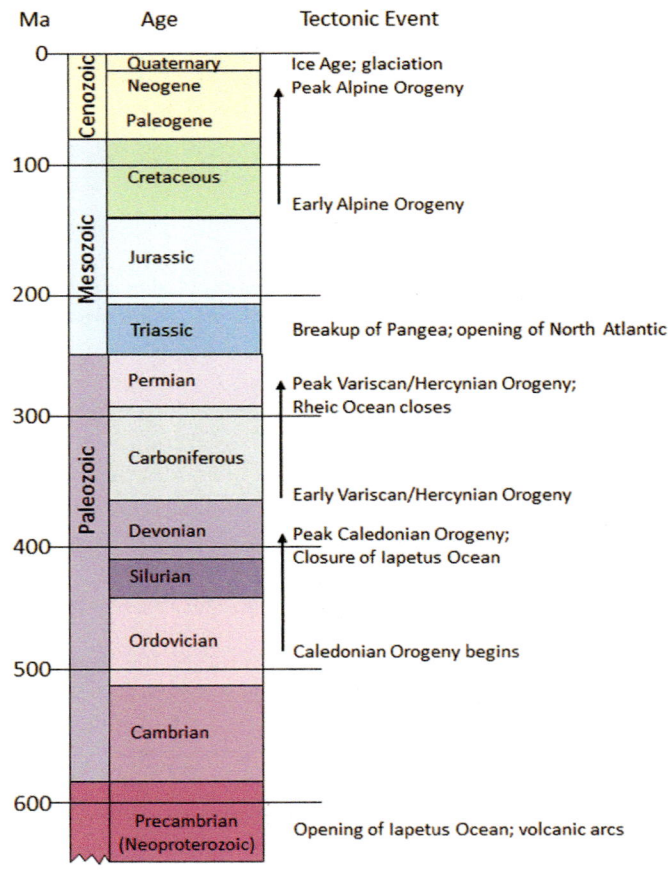

Tectonic event charts of Wales. (Information drawn from Howells, 2007; Talbot and Cosgrove, 2011.)

GEOLOGY OF THE WEST MIDLANDS

The West Midlands contain sedimentary rocks that range in age from Silurian to Triassic and are chiefly sandstone, siltstone, mudstone, and conglomerate, with some limestone and localized coals (Powell et al., 1992). Other than some local variation, the geologic history is much the same as in Wales. The human history of the West Midlands was shaped by the South Staffordshire Coalfield. Coal mining made possible the industrial development of the area. Upper Carboniferous Coal Measures underlie the towns of Wolverhampton, Walsall, West Bromwich, and Dudley (Open University, 2011).

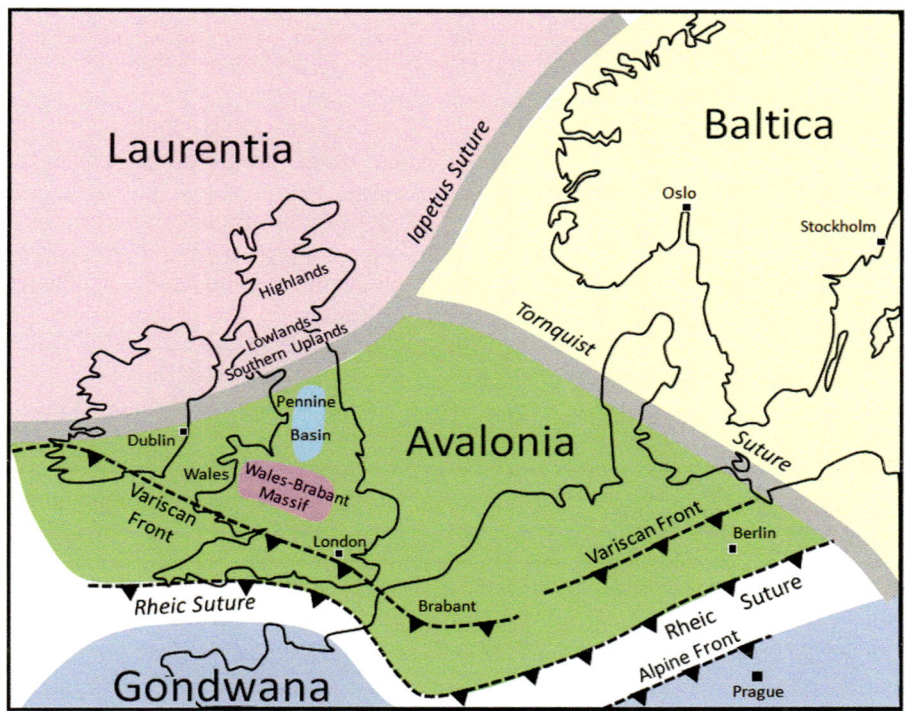

Sutures developed along plate boundaries: the Iapetus Suture, for example, developed during the Caledonian Orogeny as the Avalonian and Baltic plates collided with Laurentia. The middle Carboniferous Variscan/Hercynian mountain chain formed south of the Variscan Front as Gondwana, collided with Laurussia to create Pangea. The largely Cenozoic Alps are south of the Alpine Front. The Alpine Orogeny is a result of the African plate moving north into the European plate. (Modified after Herbosch and Verniers, 2015.)

The oldest rocks in the West Midlands are Silurian strata that were deposited in a warm, shallow sea between about 430 and 410 million years ago. As the sea moved east over the land, it deposited fossil-rich sandstones and shales. These beds are overlain by shallow-water mudstones and silt-stones. Limestones containing abundant shelly fossils were deposited during periods of low sediment input; small reefs containing corals, sponges, and algae developed in the shallow sea. Retreat of the sea in late Silurian to early Devonian time, a result of uplift associated with Caledonian mountain building, replaced marine sediments with continental delta, floodplain, and river deposits. The area was gently folded, uplifted, and eroded during the Caledonian Orogeny, around 380 Ma.

The Pennine Basin, extending from the Southern Uplands of Scotland to the Wales–Brabant Massif in the English Midlands, subsided during the Carboniferous. Sediments deposited within the basin now extend from the Scottish border throughout the Pennines into northern England and northeast Wales (Wikipedia, Pennine Basin). The basin consisted of a number of sub-basins separated by uplifted blocks. Sedimentation was continuous across the blocks but sediment type and thickness varied by sub-basin. The southern margin of the Pennine Basin is defined by the Wales–Brabant Massif (Powell et al, 1992).

Following the Caledonian event, the Coal Measures were deposited in delta and swamp environments along the southern margin of the Pennine Basin. They were deposited over the nearly flat Silurian–Devonian erosion surface. Changes in subsidence, sedimentation, and sea level caused repeated cycles of shale-sandstone-coal sedimentation in the Coal Measures. Typically, a cycle began with marine or brackish water shales (estuary or coastal swamp environment) that grades upward through delta and lake mudstones, siltstones, and sandstones. Delta plains and coastal lagoons were the sites of dense swamp vegetation that, following burial metamorphism, became coals of the Pennine Coal Measures.

As in Wales, during Late Carboniferous (Pennsylvanian) time Gondwana, the Southern Continent, collided with Laurussia to create the supercontinent Pangea. This collision, the Variscan Orogeny, extended across mainland Europe into southern England. As a result, the West Midlands became a river-dominated environment. The Etruria Formation, consisting mainly of red mudstones, was deposited on floodplains cut by river channel deposits of coarse sandstone and conglomerate.

Dolerite magma intruded the Carboniferous strata at shallow depths. These intrusions include dikes (vertical injections) and sills and lopoliths (injections that follow bedding). Some of the magma made it to the surface, and these volcanic rocks are preserved in the area.

Following an early phase of the Variscan Orogeny, alluvial and lacustrine sediments of the Halesowen Formation were deposited. Renewed uplift south of the Pennine Basin between Late Carboniferous and Early Permian times caused north-flowing rivers to deposit red sands, silts, and muds on alluvial plains. These became the Keele and Enville formations. Deposition of the Enville Formation in Early Permian time was followed by coarse-grained angular conglomerates of the Clent Formation. The Clent Formation was deposited as alluvial fans north and east of the Variscan highlands.

Late Permian faulting caused a down-dropped basin west of the South Staffordshire Coalfield. This basin contains the eolian (windblown) Bridgnorth Sandstone, deposited in a desert environment. Continuing uplift and erosion resulted in a widespread unconformity, or erosion surface that truncates rocks ranging from the Upper Carboniferous Keele Formation to the Permian Bridgnorth Sandstone. Above the unconformity, the Triassic Sherwood Sandstone Group consists of red, coarse-grained fluvial and lacustrine conglomerates and sandstones. Triassic strata were later downfaulted both east and west of the boundary faults that define the uplifted coal field block (Powell et al., 1992).

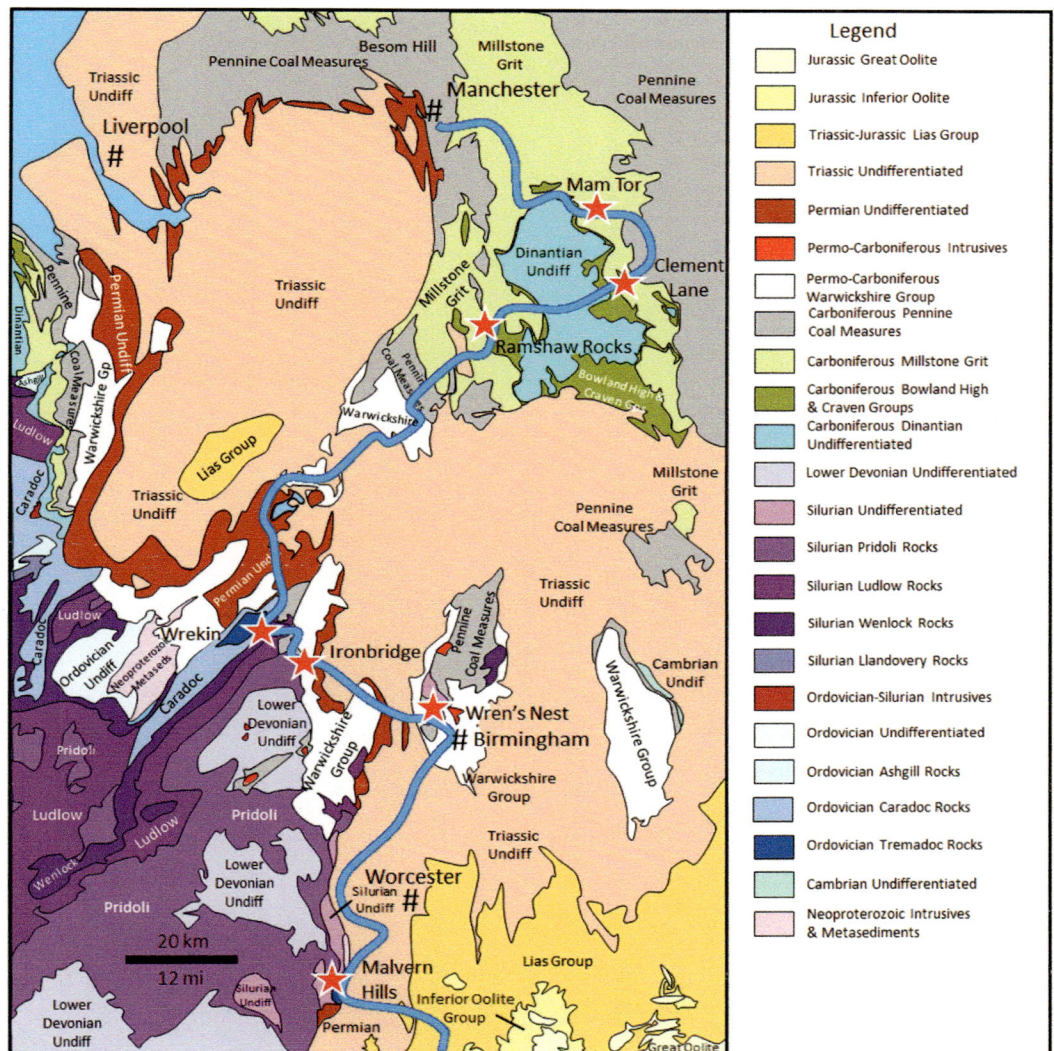

Geologic map of the West Midlands. (Modified after BGS, Geology of Britain Viewer Classic. Contains British Geological Survey materials © UKRI 2022.)

The continental margins around the North Atlantic were for the most part unaffected by deformation, but were subject to widespread post-Triassic uplift and erosion. This history of erosion is far more complex than previously thought in the southwest United Kingdom, with regional kilometer-scale uplift beginning in the Upper Triassic–Lower Jurassic (215–195 Ma) and continuing sporatically until the Neogene (20–10 Ma).

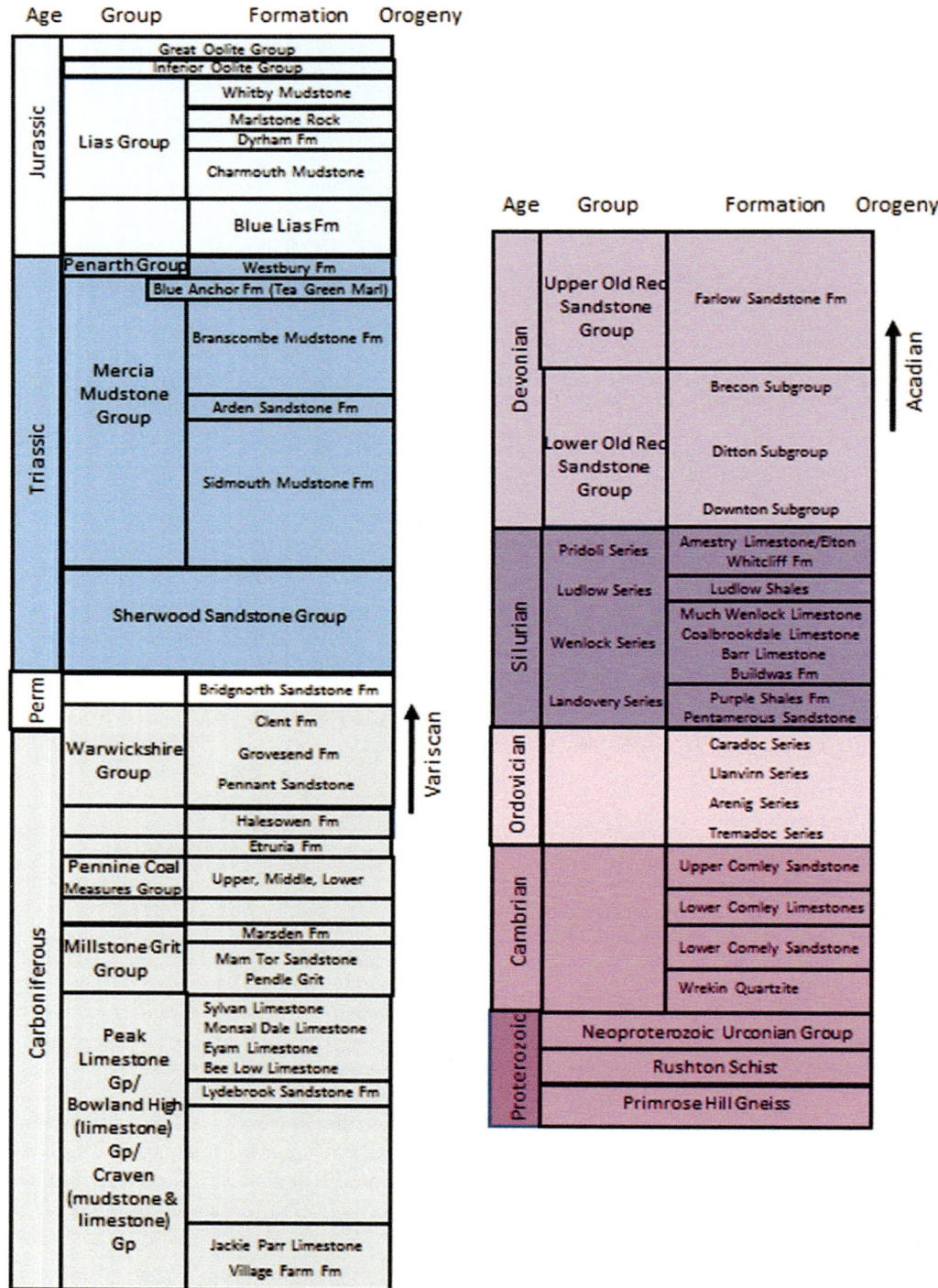

Stratigraphy and tectonic events in the West Midlands. (Modified after Powell et al., 1992; Waters, 2008; BGS, Geology of Britain Viewer Classic.)

Upper Triassic–Lower Jurassic erosion appears confined to uplifted blocks along basin margins and is thought related to the breakup of Pangea. Lower Cretaceous uplift is related to extension and opening of the North Atlantic. Early Paleogene uplift and erosion coincided with an early phase of the Alpine Orogeny. Eocene-Oligocene and Neogene uplift and erosion are distant effects of Pyrenean and Late Alpine compressional events (Kelly, 2010).

MANCHESTER

Manchester, where we begin this tour, is the country's fifth-largest city with 547,627 inhabitants. It is the second most populous metropolitan area in Britain, with a population of 3.3 million as of 2018.

The name Manchester derives from the Brittonic mamm ("breast", referring to a "breast-like hill") or from mamma ("mother", in reference to a local river goddess). Chester is from the Latin castra, a fortified town. Prior to the Roman conquest, it was part of the territory of the Brigantes, a Celtic tribe.

HISTORY

The Roman fort of Mamucium was established around the year 79 near a ford on the River Medlock. The fort was located on a sandstone bluff (Triassic Chester Formation) near the confluence of the rivers Medlock and Irwell. A town, mostly the families of traders and soldiers, grew outside the walls of the fort. Already at that time, one section of town contained an abundance of furnaces for smelting and ironworking (Wikipedia, History of Manchester).

Over time the town was part of several Saxon kingdoms including Northumbria, Mercia, and finally England. By the late 13th century, Manchester contained a fortified manor belonging to the barons Grelleys (or Gresles). The homes and land around the manor were leased to tenants for workshops and gardens. Manchester soon became a market town, gaining the right to hold an annual fair in St Ann's Square. Throughout the Middle Ages, Manchester remained a manorial township.

An influx of Flemish weavers in the 14th century brought the textile trade to Manchester and the town slowly become an important center for wool and linen production. A tradition of textile production developed throughout the region, and the small market town began to grow and prosper. By the late 18th century, the cotton industry was thriving, thanks to the fast-flowing rivers from the Pennines that made it the perfect location for water-powered, and later steam-driven cotton mills.

Cotton imports through the port of Liverpool revolutionized the textile industry in the area. Cotton mills spread rapidly throughout Manchester and the surrounding towns. By the 19th century, Manchester was one of the largest industrialized cities on Earth, even dubbed "Cottonopolis" (New College Group, A Brief History of Manchester). Opening of the Manchester Ship Canal in 1894 created the Port of Manchester, directly linking the city to the Irish Sea 58 km (36 mi) to the west.

Victorian Manchester was the site of one of the world's first passenger railway stations as well as many scientific achievements. Manchester was home to possibly the first telephone exchange in the United Kingdom, and certainly one of the first in Europe, in 1879. Ernest Rutherford first split the atom in 1917 at the University of Manchester, and Fred Williams, Tom Kilburn, and Geoff Tootill developed the world's first stored-program computer there in 1948 (Wikipedia, Manchester).

The post-World War II period saw a decline in Manchester's industrial base that led to depressed social and economic conditions. Investment, gentrification, and rebranding starting in the 1990s have revived Manchester as a center of sporting, broadcasting, and educational institutions (Wikipedia, History of Manchester). In the past 25 years, Manchester has become a cultural, digital, and innovation center in the United Kingdom.

GEOLOGY

Most of Manchester and its suburbs are located on Carboniferous, Permian, and Triassic units. The Carboniferous Pennine Coal Measures underlie the northern and eastern parts of the city. Moving south and west one encounters first the Permian Appleby and Cumbrian Coast sandstones and then the Triassic Sherwood Sandstone. The rocks get younger as you go southwest from Manchester into the Cheshire Basin.

The Pennine Coal Measures Group is a cyclic mudstone–siltstone–sandstone sequence with subordinate coal layers, all deposited in shallow coastal swamps. The mudstone is dark gray to black; sandstones are gray and thin but laterally continuous. The East Pennine Coalfield was the United Kingdom's main coal-producing region: in 1998–1999, it produced 18.1 million tonnes (19.9 million tons) of coal from 14 underground mines and 3 million tonnes (3.3 million tons) from surface mines, nearly 55% of total UK coal production. The coal is high-volatile bituminous and is mostly used to generate electricity in coal-fired power plants (Aitkenhead et al., 2002).

The Permian Appleby Group (Collyhurst Sandstone) lies unconformably on the Coal Measures. The formation consists of red and orange eolian sandstone (sand dunes).

The Permian Cumbrian Coast Group (Manchester Marls) consists of red mudstone, partly marine, with thin limestone and siltstone beds.

The overlying Triassic Sherwood Sandstone Group was deposited by rivers sourced from the Varsican mountains to the south. The upper parts of the unit are eolian. The Sherwood Sandstone Group is a red-brown, sandstone with subordinate red-brown mudstone.

There is a surficial cover of thick glacial till and sand and gravel deposited at the end of the last Ice Age 15,000 to 10,000 years ago.

*Start: Manchester Airport*Take Ringway Road northeast to the A555 and turn left (northwest); drive northwest on A555 to the M56; bear left to merge onto M56/North Cheshire Motorway and drive south and west; continue straight on A494/Bypass Road to the A55/North Wales Expressway; continue straight (northwest) on the A55; after Exit 29, you pass through a large roadcut, after which there is a parking pullout on the left. This is **Stop 1.1, Rualt Roadcut** (53.269390, −3.377078) for a total of 86.8 km (54.0 mi; 57 min).

STOP 1 RHUALT, LLANDUDNO, AND GREAT ORME

Driving west from Manchester we pass through the Cheshire Basin, with primarily Triassic rocks at the surface, and cross into progressively older rocks. We enter Wales on the west side of Chester. There are not many good outcrops in this area known as the Denbighshire Moors but one that you might like to stop at is a massive roadcut just east of Rhualt.

STOP 1.1 RHUALT ROADCUT

The A55 cuts through as much as 28 m (92 ft) of an Early Silurian (Wenlock) mudstone–siltstone– sandstone sequence that dips 15°–20° to the southwest. You can observe the rocks by looking north across the highway or chance walking east about 160 m (500 ft) to the base of the roadcut on the south side of the North Wales Expressway. Please be mindful of the traffic.

Above the Silurian is an erosion surface (unconformity) overlain by Early Carboniferous (Mississippian) limestone. On the west side of Rhualt, a north–south normal fault has dropped the Permian Warwickshire Group sandstones and siltstones so that they are exposed at the surface.

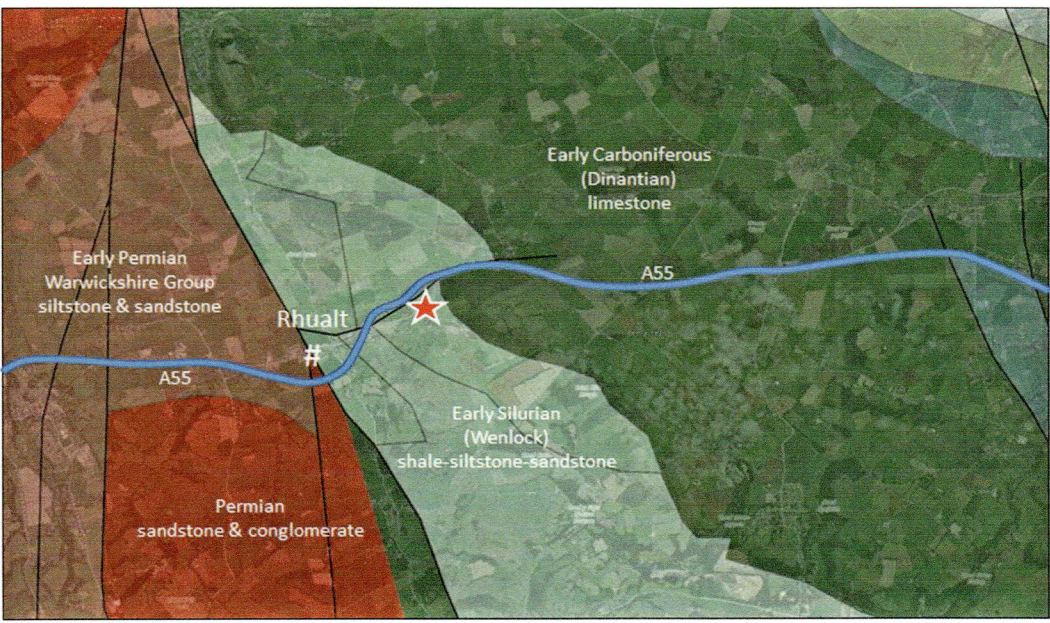

Geologic map of the Rhualt area. (From MacroStrat, https://macrostrat.org/map/#x=-3.389&y=53.265&z=14.)

Rhualt roadcut through Early Silurian mudstones, siltstones, and sandstones. Google Street View northeast.

    Rhualt Roadcut to Llandudno Pier*: Continue west on the A55 to Llandudno; at Junction 20, take the exit to Rhos on Sea/B5115; turn right (west) onto Princess Drive/B5115 and drive to Conway Road/A547; turn right (northwest) onto Conway Road and drive to the roundabout; take the 2nd exit onto Brompton Ave/B5115 and drive northwest; continue straight on Llandudno Road; at roundabout, continue straight onto Penrhyn Hill; continue straight on Colwyn Road/B5115; continue straight onto South Parade; turn right (north) onto North Parade and drive to Llandudno Pier parking. This is* **Stop 1.2, Great Orme from Llandudno Pier** *(53.331604, −3.825163) for a total of 37.7 km (23.4 mi; 30 min). Walk out on the pier for a view of Great Orme; walk 730 m (2,400 ft) north along Marine Drive to examine outcrops.*

STOP 1.2 GREAT ORME FROM LLANDUDNO PIER

Coming upon Llandudno from the east, we see the Great Orme headland. The name Great Orme derives from the Old Norse for a sea serpent. However, "Orme" was not commonly used until the mid-1800s. Previously, Welsh names were used. The peninsula containing Llandudno was known as the Creuddyn, from cwmwd, a medieval Welsh land tract; the headland was called Y Gogarth or Pen y Gogarth (Head of Gogarth; Wikipedia, Great Orme).

    The Great Orme is a limestone headland that rises 207 m (680 ft) above sea level just northwest of Llandudno on the north coast of Wales. The limestones formed during Early Carboniferous (Mississippian; between 339 and 326 Ma) in shallow tropical seas. They are locally dolomitized, and there are minor interbeds of limy mudstone and sandstone that cause the stepped appearance of the cliffs. Common fossils found in the limestone include brachiopods (resemble cockles), crinoids (sea lilies related to starfish), echinoids (sea urchins), bryozoans (moss animals), and corals (Lewis, 1996; Conwy, Discover the Great Orme).

    The Llandudno Pier Dolomite, at the base of the section, is particularly well exposed along the beach near the Pier. These beds were much used as building stone during the last century and were quarried at a number of locations. Diagenetic dolomite is an altered form of limestone; the original rock was altered by magnesium minerals carried in groundwater. Here, the Pier Dolomite provides a good example of the termination of a dolomite body at an ancient reaction front that formed when replacement of the limestone by dolomite stopped as a result of changes in groundwater chemistry (Hollis and Juerges, 2019). Characteristic yellow–brown-weathering limestones occur above the dolomite in the "*Lithostrotian affinephillipsi* beds," named for the coral of the same name (Lewis, 1996).

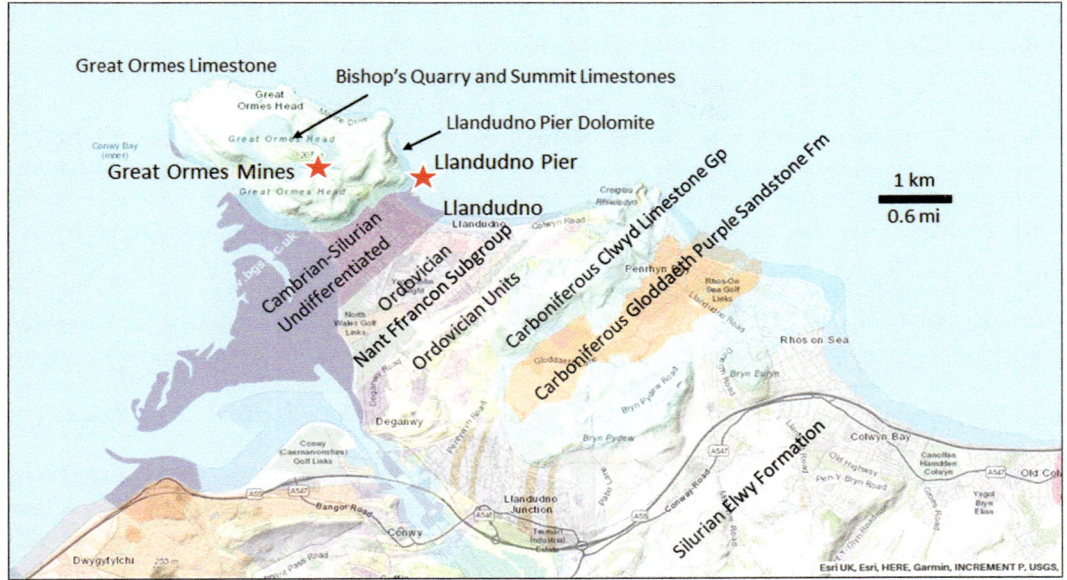

Geologic map of the Llandudno-Great Orme area. (Modified after BGS, Geology of Britain Viewer Classic. Contains British Geological Survey materials © UKRI 2022. Base mapping provided by ESRI.)

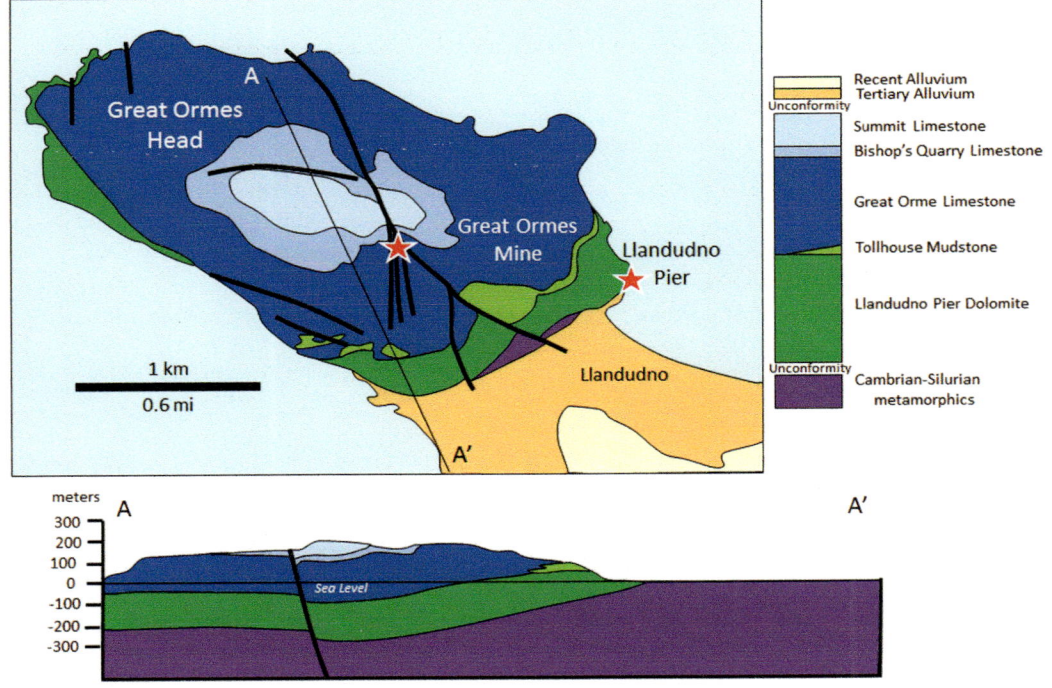

Geologic map and cross section of Great Ormes Head. (Modified after Lewis, 1996.)

Above the Pier Dolomite is the Great Orme Limestone. It formed on an ancient carbonate platform in a shallow, warm equatorial sea that extended across the southern edge of Euramerica, an area covering modern-day Western Europe and North America. Locally, the limestone that outcrops on the Great Orme is part of the North Wales Platform, a carbonate bank that stretched from Anglesey to the Welsh Borders, and formed on the margin of the Wales–Brabant Massif (Lewis, 1996). The Great Orme Limestone shows prominent layering because the strata are capped by red- and yellow-colored mudstone. During the latest Lower Carboniferous, glaciers formed in Gondwana in the southern hemisphere, leading to lower sea levels. Sea levels rose again during interglacial periods (Hollis and Juerges, 2019). The rise and fall of sea levels led to the alternating sequence of mud and limestone at Great Orme.

At the top of Great Orme, the Bishops Quarry and Summit limestones become darker and shalier and are dominated by the brachiopods *Gigantoproductus giganteus*, *Gigantoproductus edelburgensis,* and *Productina margaritacea*. By around 330 Ma, the North Wales Carbonate Platform was gradually covered by suspended clay material derived from the nearby landmass, and drowned by a rising sea level.

In the latest Carboniferous and early Permian, closure of the Rheic Ocean south of the North Wales Platform was caused by collision of the Euramerican plate with Gondwana, the Southern Continent. This collision, the Variscan Orogeny, caused folding and faulting across North Wales and Northern England (Lewis, 1996). The 300+ m (1,000+ ft) of limestone exposed at the headland was deformed into a gentle syncline, and the promontory was cut by north–south near-vertical faults that played a crucial role in allowing groundwater to alter the limestone and introduce ore minerals (Maelgwyn Mineral Services Limited, 2020).

Great Orme as seen from Llandudno pier. View north.

The Great Orme was covered by glaciers during the latest ice ages, from 60,000 to 12,000 years ago. As the ice moved across the headland, it scraped off overlying soil and rock, uncovering the flat to gently inclined limestone. The exposed rock has been subject to weathering since the ice melted. Rainwater entering fractures in the rock slowly dissolved the broken limestone, forming deep vertical channels known as grykes. The hard limestone between the grykes has not weathered as quickly. The bare rock surfaces are called limestone pavements. Evidence of glaciers can also be seen in the large boulders, or glacial erratics, carried by the ice and left scattered over the headland as the glaciers melted (Conwy, Discover the Great Orme).

A view of Llandudno, Llandudno Bay, and the Great Orme. J. Newman & Co., 1855. https://commons.wiki-media.org/wiki/File:Great_Ormes_Head,_north_Wales.jpeg

*Llandudno Pier to Great Orme Mines: Return southeast on Happy Valley Road; continue straight on North Parade; at the roundabout take the 2nd exit onto Church Walks; turn right (north) onto Ty-Gwyn Road and drive to St. Beuno's Road; turn left (southwest) on St. Beuno's Road and drive to Pyllau Road; turn right (north) on Pyllau Road and drive to **Stop 1.3, Great Orme Mines** (53.330147, −3.847586) for a total of 1.8 km (1.1 mi; 6 min). There is abundant free parking. Entering the mines costs £9/adult.*

STOP 1.3 GREAT ORME MINES

History

When first investigated, these workings were thought to have largely originated during the 19th century. However, further study of the mines and artifacts indicates that they are much, much older (Lewis, 1996).

Great Orme Mines.

The copper mines of Llandudno have now been dated to the middle Bronze Age. The oldest confirmed date is around 1,800 BC, and mining continued until around 600 BC. The presently accessible tunnels extend 300 m (1,000 ft) into the Great Orme and sink to 75 m (246 ft) below the surface. A total of about 8 km (4.8 mi) of tunnels have been mapped, making it the most extensive Bronze Age workings surviving in Europe, and perhaps the world (Colley, Great Orme Bronze Age Mine; Maelgwyn Mineral Services Limited, 2020).

The copper ore malachite was mined using stone and bone tools. It is estimated that up to 1,760 tonnes (1,940 tons) of copper was mined during the period, enough to make about 1,810 tonnes (2,000 tons) of bronze. The mine was most productive in the period between 1,700 and 1,400 BC, after which most of the easily reached ore had been mined (Wikipedia, Great Orme). The site became one of Western Europe's major Bronze Age technological centers because it was a rich deposit and because the soft rock enabled easy removal of ore with the simplest tools (Lewis, 1996). Metal traced to this mine by its composition has been found in bronze artifacts in Sweden, France, and Germany (BBC News, 2019).

No ancient metallurgical workings have yet been found, suggesting either that these sites have since been destroyed, or buried, or that the copper was traded elsewhere, probably to Cornwall, where the only deposits of tin are found in Britain – copper and tin being the two metals required for the making of bronze (Maelgwyn Mineral Services Limited, 2020).

At some point in pre-history, the mines were abandoned and weren't worked again until the late 1600s. Modern, industrial-scale mining on the Great Orme ended in the 1850s, although small-scale mining continued until 1881 (Wikipedia, Great Orme). It is estimated that another 45,400 tonnes (50,000 tons) of copper ore was mined here during the 19th century (Maelgwyn Mineral Services Limited, 2020).

In April 1991, the Great Orme Mines site was opened to the public. Pathways and viewing platforms were built to provide access to the surface excavations. A visitor center extension, built in 2014, contains mining tools and bronze axes along with displays about life and death in the Bronze Age, mining, and ancient metallurgy. Also accessible is the 3,500-year-old Great Cavern dug out of the rock with stone and bone hand tools (Wikipedia, Great Orme).

Mineralization

There are strong structural and lithological controls on mineralization and composition of the ores in the Great Orme Mine. Fault and joint planes localized most of the ore minerals as vein or lode ores consisting of sulfide and carbonate minerals, or locally a soft and friable hydroxide-oxide-carbonate mineral assemblage. Ore is mostly confined to north–south fractures and to a lesser extent in the east–west and northwest-southeast fractures. Ore concentrations occur at the intersection of these fractures, forming some of the larger bodies. The bulk of the ore is the result of primary sulfide mineralization followed by intense, widespread, and supergene (near-surface) alteration. The primary minerals are chalcopyrite with minor pyrite and marcasite. The primary sulfide assemblage has been altered to secondary copper sulfides and oxides that include malachite, azurite, and copper-bearing limonite (Ixer, Mineralization at the Great Orme). Minor lead sulfides also occur. The sulfur in the minerals is often oxidized, becoming soluble and combining with calcium to produce selenite, a form of gypsum. Ocher goethite and pale-yellow jarosite (iron minerals) are a result of oxidation of sulfide ores.

Ore is also found as replacement deposits in dolomitized beds. Dolomitization is associated with the north–south and east–west fractures, indicating that these provided conduits for the magnesium-rich fluids responsible for alteration. Dolomitization occurs when calcite, the main mineral of limestone, is replaced by the mineral dolomite. The reaction involves magnesium replacing calcium in the mineral. The process of dolomitization produces a decrease in rock volume of up to 12%, increasing the porosity and cavities in the rock. All of these voids provide sites for later precipitation of copper ore by mineralizing solutions (Lewis, 1996).

Malachite ore, Great Orme Mines.

Many of the dolomites are buff to yellow–brown, a result of oxidized iron replacing magnesium. Dolomitization is favored in cleaner (pure) limestone as opposed to those containing clay or sand. This effect is enhanced if there is a confining layer (impermeable mudstone or sandstone) over a relatively pure limestone. This causes the flow of hot, hydrothermal fluids to be vertically contained, allowing replacement of a limestone layer. Examples can be seen above the tourist entrances where a sandstone bed has confined dolomitization to the beds beneath it. Ore bodies in these dolomites are called bedding-confined or manto replacement deposits.

Chalcopyrite occurs throughout the mine and appears to have been the chief copper ore mined, to depths of 210 m (690 ft) below the surface.

As of 1996, no mineralization age had been determined but analogous lead and zinc deposits in the Llanrwst ore deposit in the Conwy Valley were dated to 280 Ma (Permian). The secondary supergene ores are likely due to the oxidation of primary sulfide ores by percolating groundwater during the Cenozoic (Lewis, 1996). Other estimates suggest that copper ores were formed by fluids circulating into the Great Orme limestone during the Alpine Orogeny in the Late Cretaceous (Hollis and Juerges, 2019).

Detailed work by Ixer (Ixer, Mineralization at the Great Orme) indicates that the galena (lead sulfide) veins are part of Mississippi Valley-type mineralization related to the Northeast Wales Orefield and are probably Permo-Triassic. He believes that the copper-dolomite association is younger, probably Mesozoic-Tertiary in age.

Great Orme Mines to Cwm Idwal*: Turn right (north) toward Bishop's Quarry Road; turn right (east) onto Bishop's Quarry Road; continue straight on Ty-Gwyn Road; turn right (west) onto Church Walks, then immediately turn left (south) onto Mostyn Street; at the roundabout, take the 3rd exit onto Gloddaeth Street/A546 and drive southwest; at the 2nd roundabout, take the 1st exit to Great Ormes Rd/A546 and drive southeast; at the next roundabout, take the 2nd exit to Maesdu Ave/A546 and drive south; continue south onto Ffordd 6G Rd/A546; take the A55 ramp to Bangor/A55 and merge onto the North Wales Expressway heading west; at Junction 11, take the A5 exit to bangor; at the Llys y Gwynt Interchange, take the 1st exit (south) onto the A5; take the A5 southeast to Llyn Ogwen; at the sign for the Youth Hostel, turn right and park in the National Trust parking area. From here, take the mostly paved trail 940 m (3,100 ft) south and west to Llyn Idwal; at the trail junction, take the trail to the right (west) around the lake 900 m (3,000 ft) to **Stop 2.1, Cwm Idwal** (53.118978, −4.025040) for a total of 43.6 km (27.1 mi; 50 min).*

Great Orme Mines to Side Trip 1, Caim Beach*: Turn right (north) toward Bishop's Quarry Road; turn right (east) onto Bishop's Quarry Road; continue straight on Ty-Gwyn Road; turn right (west) onto Church Walks, then immediately turn left (south) onto Mostyn Street; at the roundabout, take the 3rd exit onto Gloddaeth Street/A546 and drive southwest; at the 2nd roundabout, take the 1st exit to Great Ormes Rd/A546 and drive southeast; at the next roundabout, take the 2nd exit to Maesdu Ave/A546 and drive south; continue south onto Ffordd 6G Rd/A546; take the A55 ramp to Bangor/A55 and merge onto the North Wales Expressway heading west; at junction 8A take the A5 exit to Porthaethwy/Menai Bridge/Beaumaris/A545 and turn right (northeast) onto Carreg Bran Interchange/Holyhead Rd/A5; continue straight onto Dale St/B5420 and drive to High St; turn left (north) onto High St/A545; continue straight onto B5109; at the sign for Penmon turn left (north); at the first street turn right (northeast) onto Tyn y Cae Cottages; a trail crosses the road just before the road ends, and there is room for one or two cars on the roadside (53.305453, −4.071624) for a total of 53.3 km (32.6 mi; 59 min). Park and walk 540 m (1770 ft; 15 min) north on the trail to **Side Trip 1.1, Caim Beach** (53.312645, −4.0690537).*

SIDE TRIP 1 ANGLESEY AND GEOMÔN GLOBAL GEOPARK

The entire island of Anglesey and Holy Island, down to mean low tide, is included in Geomôn UNESCO Global Geopark. Known as Ynys Môn in Welsh, the Geopark consists mostly of Precambrian and Cambrian metamorphic, igneous, and sedimentary rocks that are exceptionally well exposed along the coast (European Geoparks, Geomôn). The lowland area consists of rocks as old as 1.8 billion years. In places, the island has been injected with mantle material. The park is noted for its rare Precambrian blueschists in southern Anglesey, Precambrian stromatolitic limestones (860-800 Ma) in northern Anglesey (the oldest fossils in England or Wales), and the largest copper mine in the world between 1780 and 1860, Parys Mountain in Amlwch. The Cambrian-age folds and faults of South Stack, and Carboniferous limestones chock full of fossil corals and bivalve shells, are particularly notable. There are extensive Pleistocene glacial features from the last Ice Age.

Geomôn was admitted to the European Geoparks Network and to the UNESCO-assisted Global Network of National Geoparks in May 2009. GeoMôn covers 720 km² (278 mi²) and is the type locality for a rock type christened "mélange" by Edward Greenly when he mapped the geology of Anglesey in the early 20th century. GeoMôn publishes a series of local trail pamphlets to guide trekkers around the coastal areas of the island along the 200 km (125 mi) long coastal path (Wikipedia, Geomôn).

Contact Information
Tel.:   +44 1248 810287
Email: college@btinternet.com
Address: The Old Watch House, Porth Amlwch, Amlwch, Anglesey, Wales LL68 9DB, UK
Website: http://www.geomon.co.uk/
Official representative and geologist: Dr. Margaret Wood, email: college@btinternet.com

ST1.1 CAIM BEACH

This location is a Site of Special Scientific Interest (SSSI) due to the excellent fossil collecting. You can visit the site and collect fallen rocks and fossils, but hammering or otherwise removing fossils from the outcrop is not permitted.

The road ends at Pentir Cottage. The Wales Coast Path crosses the road just before the road ends. Walk northwest through woods and across fields to the sea cliffs. Be careful descending the steep path to the beach. This site is best visited at low tide.

The rocks are all Carboniferous limestone of the Clwyd Limestone Group. The main part of this beach is Loggerheads Limestone Formation (337–331 Ma); to the east, across a small fault, is the Cefn Mawr Limestone Formation (337–329 Ma; BGS, Geology of Britain Viewer Classic). Some of the limestone has been altered to dolomite.

Corals and brachiopods are common both on the cliffs and in loose pebbles on the beach. Some of the best brachiopods can be seen in limestone at the top of the cliffs. Some of the best corals are in outcrops just above the beach (Cruickshanks, 2017, Caim).

Caim Beach and limestone cliffs. View east.

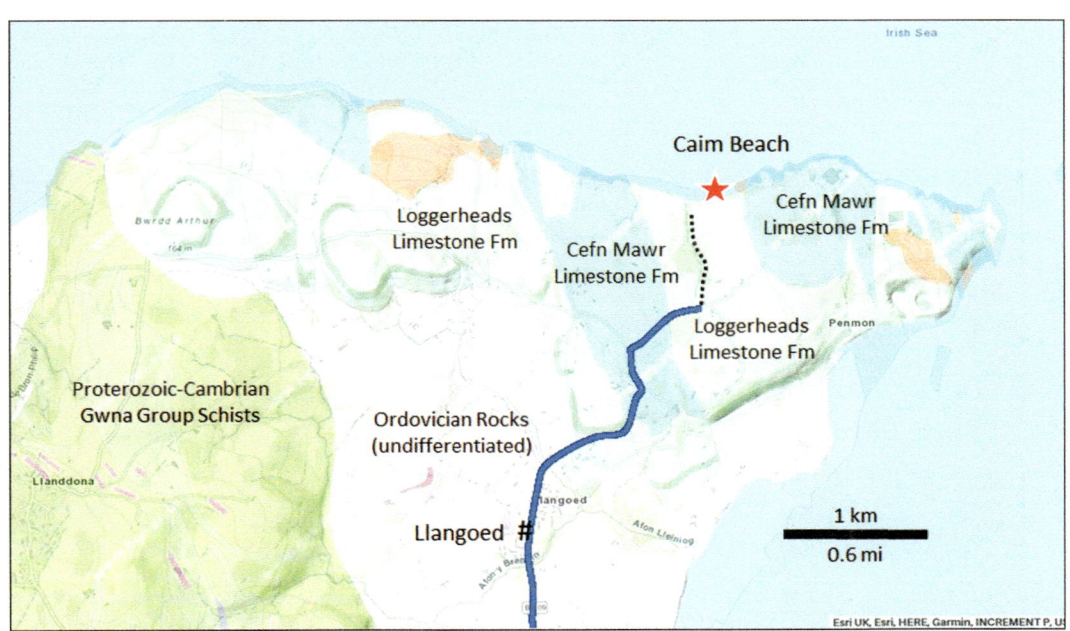

Geologic map of Caim Beach, Anglesey. (Modified after BGS, Geology of Britain Viewer Classic. Contains British Geological Survey materials © UKRI 2022. Base mapping provided by ESRI.)

Side Trip 1: Caim Beach to South Stack: Return southwest on B5109 to Beaumaris and A545 to Menai Bridge; turn right (west) onto Dale St/B5420; at the roundabout take the 2nd exit to Pentraeth

Rd/B5420 driving northwest; at the roundabout take the 1st exit onto A5025 and drive southwest to the North Wales Expressway; turn right (northwest) onto the A55 to Caergybi/Holyhead; at the roundabout in Kingsland continue straight onto Victoria Road/A55; turn left (west) onto Market St; Market St turns left and becomes Thomas St; drive to Alderly Terrace and turn left onto Alderly Terrace/South Stack Road; at the sign for South Stack, turn right (west) to stay on South Stack Road and drive to **Stop ST1.2, South Stack Lighthouse** *parking area (53.307038, −4.697540) for a total of 53.3 km (33.1 mi; 54 min). There are 2 or 3 small parking lots near the end of the road. Parking is free, but entry to the lighthouse will cost £7.50/adult, £3.50/child, or £18.50/family. The steps down to the lighthouse are steep.*

ST1.2 SOUTH STACK

South Stack Lighthouse sits on a sea stack, an erosional remnant of the mainland formed by wave erosion. This particular stack, rising 100 m (328 ft) above the waves, shows spectacular folding and thrusting of the Proterozoic–Cambrian South Stack Formation quartzites and slates. The folding is best seen from the steps leading to the lighthouse. The South Stack Formation was deposited as continental margin marine channel sandstones (Phillips, 1991a).

The overlying Cambrian Holyhead Quartzite forms Holyhead Mountain to the northeast which, at 220 m (720 ft), is the highest point on Holy Island. The Holyhead Formation (maximum age 501 Ma) is part of the South Stack Group, Monian Supergroup. The Monian Supergroup is part of the Cambrian–Lower Ordovician succession found in southern Britain and Ireland whose sediment derived mainly from Avalonia (Collins and Buchan, 2004; Wikipedia, South Stack; Visit Anglesey, Geology).

Early Caledonian deformation of the South Stack Group was accompanied by regional greenschist facies (low temperature, low pressure) metamorphism. Deformation of the Monian Supergroup involved progressive, southeast-directed compression and shortening prior to the Ordovician (Phillips, 1991b).

Deformed Cambro-Ordovician South Stack Formation, South Stack Lighthouse.

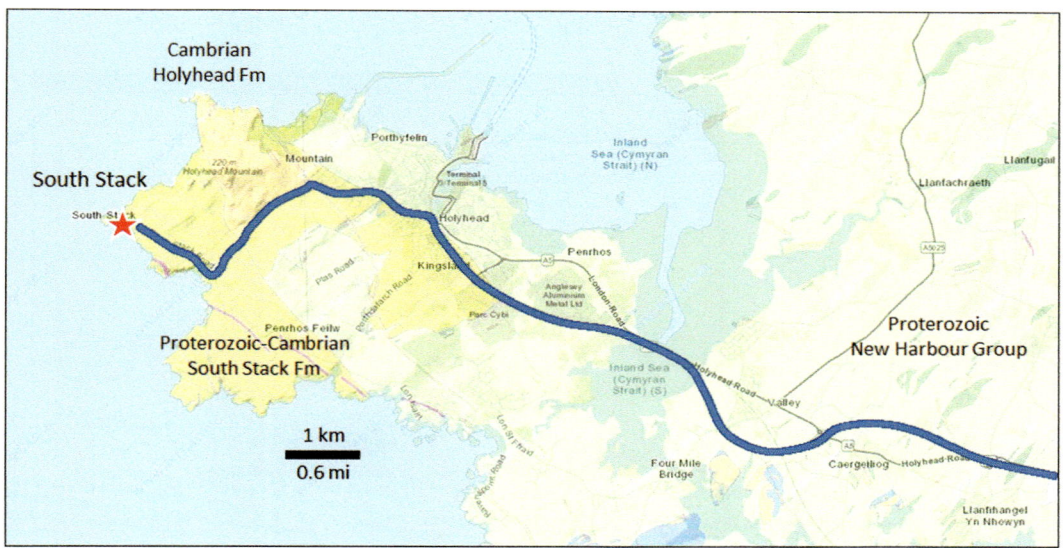

Geologic map of the South Stack area, Anglesey. (Modified after BGS, Geology of Britain Viewer Classic. Contains British Geological Survey materials © UKRI 2022. Base mapping provided by ESRI.)

Side Trip 1- South Stack to Cwm Idwal: Return east on the North Wales Expressway/A55 to the A5; at Junction 11, take the A5 exit to Betws y Coed; at the Llys y Gwynt interchange, take the 3rd exit onto the A5 and drive southeast to Llyn Ogwen; at the sign for the Youth Hostel, turn right and park in the National Trust parking area. From here, take the mostly paved trail 940 m (3,100 ft) south and west to Llyn Idwal; at the trail junction, take the trail to the right (west) around the lake 900 m (3,000 ft) to **Stop 2.1, Cwm Idwal** *(53.118978, −4.025040) for a total of 57.7 km (35.9 mi; 51 min).*

STOP 2 SNOWDONIA NATIONAL PARK

Snowdonia National Park, covering 2,130 km² (823 mi²), lies in a mountainous region in northwest Wales that contains the highest peak in England and Wales (Snowdon, 1,085 m or 3,560 ft) and Llyn Tegid, the largest natural lake in Wales. The park, Wales's first national park and Britain's third, was created in 1951 and contains both public and private lands. Crib Goch, on the north flank of Snowdon Peak, is the wettest place in the United Kingdom with average annual rainfall of 4,473 mm (176 in).

The name Snowdonia derives from the mountain of the same name. In Welsh, the area is known as Eryri, meaning highlands, derived from the Latin *oriri* (to rise).

Rocks in the park are mainly Cambro-Ordovician metasediments and metavolcanics, although there are some Orovician–Silurian igneous intrusive rocks. Folding and metamorphism of these rocks are a result of the Caledonian Orogeny. Metamorphism of mudstones created the slates that the area was once famous for. At one time this region was the most important source of slate in the world. Minor amounts of slate are still produced.

The area was visited by Charles Darwin in the 1840s and contributed to his understanding of the concept of Ice Ages. He recognized that the present landscape is a result of glacial processes, and this realization led to the acceptance of the then-novel concept (Wikipedia, Snowdonia National Park; Snowdonia National Park Authority).

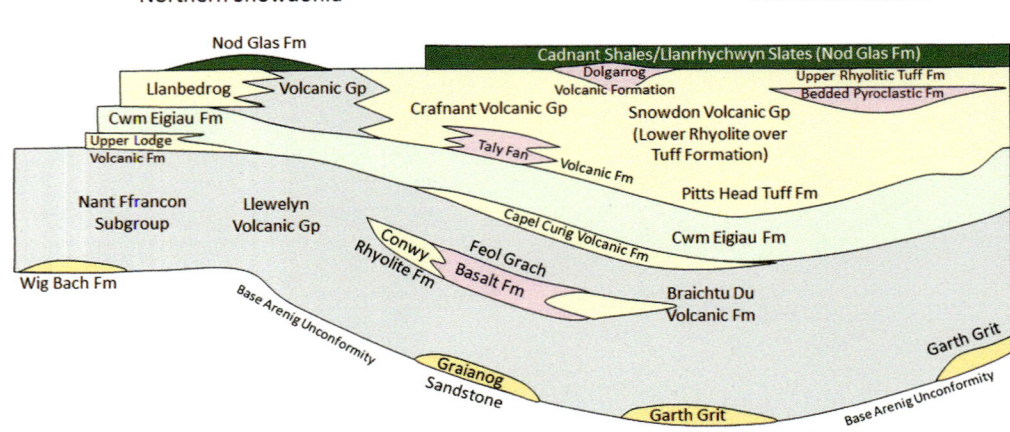

Northern Snowdonia

Central Snowdonia

Ordovician stratigraphic relationships across north and central Snowdonia. (Modified after Howells, 2007.)

2.1 Cwm Idwal

Llyn Idwal, a scenic glacial tarn. Panorama from south to west.

Cwm Idwal is a hanging valley in the Glyderau Mountains of Snowdonia, about 16 km (10 mi) southeast of Bangor. A hanging valley is a side valley that enters a main valley far above the main valley floor as a result of glacial erosion of the main valley. The area became the first Welsh National Nature Reserve in 1954 as a result of its unique landscape, geology, and botany.

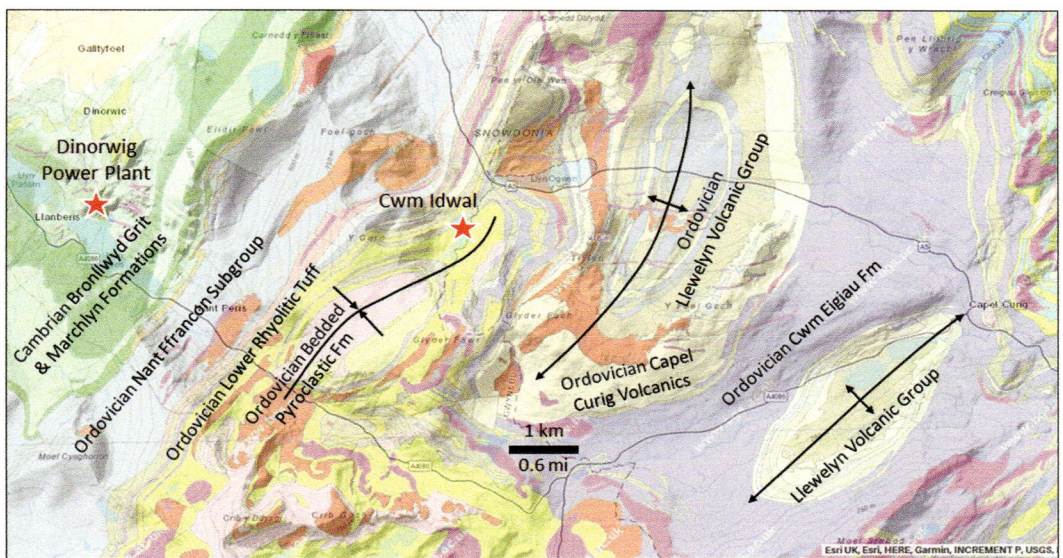

Geologic map of the Cwm Idwal to Dinorwig area, Snowdonia. (Modified after BGS, Geology of Britain Viewer Classic. Contains British Geological Survey materials © UKRI 2022. Base mapping provided by ESRI.)

Ordovician units at Cwm Idwal are deformed into a large northeast-southwest-oriented syncline.

During the Ordovician, North Wales was covered by a shallow sea and experienced intermittent volcanism. Both dark basalts and light-colored rhyolitic rocks erupted from a nearby caldera, or volcanic center. The rhyolitic tuffs (ash) and pyroclastic (volcanic) breccias indicate explosive volcanism, whereas interbedded shelly sandstones and shales record a shallow marine setting.

Sandstones and siltstones of the Cwm Eigiau Formation can be seen in the crags west of Llyn Idwal. They were deposited in a nearshore environment. The overlying Pitts Head Tuff, exposed in the lower western crags along Castell y Geifr to Y Garn, contains flattened pumice fragments that give it a banded appearance. The pumice was deposited as part of a turbulent, fast-moving cloud of hot gas and ash (a *nuée ardente*) and is now a "welded tuff". The area subsided and the sea became deeper. Marine sandstones deposited at this time are exposed in the ridges east of Llyn Ogwen. The next phase of volcanism created the Lower Rhyolitic Tuff, a unit containing ash, marine sediments, rhyolite sills, and pyroclastic breccias (angular blocks of ash and pumice, a frothy gas-rich lava) seen in the Idwal Slabs. The uppermost Bedded Pyroclastic Formation outcrops southwest of Devil's Kitchen and consists of green volcanic sediments deposited in shallow water. This unit is topped by basaltic pillow lavas that erupted underwater.

In 1831, Darwin visited Cwm Idwal and found marine shells in the large, scattered boulders at Llyn Idwal (the lake). He surmised that these rocks must have formed in an ancient ocean and that they had at some later time been uplifted to their present position by mountain-building processes.

At a later date, Darwin returned to this area and looked at it from the perspective of Louis Agassiz, a Swiss geologist that had recently proposed that an Ice Age had shaped the landscapes of northern and central Europe. Darwin saw the steep-sided U-shaped valleys typically formed by mountain glaciers. Where small glaciers merge into a larger glacier, they form "hanging valleys" high above the main valley. Cwm Idwal is just such a glacial hanging valley. As the glaciers melted and retreated around 12,000 years ago, they deposited piles of loose rocks they had been carrying, forming hummocky "moraines."

View north up the glacial valley leading to Cwm Idwal. These U-shaped valleys helped convince Darwin that the Ice Age was real.

*Cwm Idwal to Dinorwic Slate Quarry: Drive east on A5 to Capel Kurig; keep right (southwest) to continue on A4086; at Nant Cynnyd, turn right to continue west on A4086; go over YHA Snowdon Pen-y-Pass and drive to **Stop 2.2, Dinorwic Slate Quarry View** on the southwest shore of Llyn Perris (53.113373, −4.109632) and pull over at driveway on the right for a total of 24.6 km (15.3 mi; 25 min).*

2.2 DINORWIC SLATE QUARRY

The Dinorwic quarry is a large abandoned slate quarry, now home to the Welsh National Slate Museum. Developed in the Cambrian Llanberis Slates Formation (metamorphosed deep marine mudstone deposited 526–508 Ma), it was the second largest slate quarry in Wales, covering over 283 ha (700 ac) in two main sections.

The first commercial attempt at mining here occurred in 1787, but taxes and transportation costs limited development. Business blossomed after construction of a horse-powered tramway to Port Dinorwic in 1824. At its peak, in the late 1800s, the quarry produced 100,000 tonnes (110,000 tons) of slate a year and employed over 3,000 men. Production declined gradually until it ceased altogether in 1969 as the industry went into decline and mining became difficult. Among other things, many of the tips (piles of waste rock) began to slide into the workings.

Today, the quarry is used as a rock climbing and scuba diving venue, and several movies have been filmed there. The quarry's workshop houses the National Slate Museum. Many of the galleries were enlarged and used in construction of the Dinorwig Power Station, a pumped storage hydroelectric scheme (The Geological Society, Dinorwig Power Station and Slate Quarry; Wikipedia, Dinorwic Quarry). The Dinorwig Power Station is a 1,728 MW hydroelectric plant operated as a Short-Term Operating Reserve providing fast response electricity to short-term rapid changes in power demand or a sudden loss of power stations. Water is stored high above the plant in the Marchlyn Mawr reservoir. When power is needed the water is discharged into Llyn Peris through turbines. It is pumped back up to Marchlyn Mawr during off-peak times when energy is cheaper due to low demand.

The plant was built in the abandoned Dinorwic slate quarry in order to preserve the natural beauty of the region. Located deep within Elidir Fawr Mountain, construction began in 1974 and was completed in 1984 at a cost of £425 million. At that time it was the largest civil engineering project ever awarded by the UK government (Electric Mountain, Dinorwig Power Station).

Dinorwic Slate Quarry. View north across LLyn Cwellyn.

*Dinorwic View to National Slate Museum: To visit the power station, continue northwest on A4086 to sign for Padarn County Park/Gorsaf Bwer First Hydro Power station and turn right; at the roundabout, take the 1ˢᵗ exit (north) and follow the signs to **Stop 2.3, National Slate Museum** (53.121912, −4.115159) for a total of 1.9 km (1.2 mi; 4 min). There is ample pay parking for £5.*

2.3 NATIONAL SLATE MUSEUM

The National Slate Museum (previously known as the Welsh Slate Museum and the North Wales Quarrying Museum) consists of the workshops of the now-abandoned Dinorwic quarry. The museum is dedicated to the preservation and display of relics of the slate industry in Wales. The museum is a key component of the European Route of Industrial Heritage (ERIH) and part of the National Museum Wales.

The workshops which served the Dinorwic quarry and its locomotives were built in 1870. Narrow gauge railroads carried the slate from the quarry to Port Dinorwic. The quarry closed in 1969 and the museum opened in May 1972 as the North Wales Quarrying Museum.

After a remodel, the museum has displays featuring Victorian-era slateworkers' cottages that once stood near Blaenau Ffestiniog. They were taken down stone by stone and re-erected on the site. The museum includes the multi-media exhibit *To Steal a Mountain*, showing the lives and work of the men who quarried slate here (Wikipedia, National Slate Museum).

Visit

The museum has exhibits, displays, videos, talks, a shop, and a café.

Hours: 10:00–5:00 daily
Address: Llanberis, Gwynedd LL55 4TY
Phone: 0300 111 2 333
Email: slate@museumwales.ac.uk
Web site: https://museum.wales/slate/

STOP 3 CEREDIGION

Ceredigion is an area on the west coast of Wales, corresponding to the county of Cardiganshire, with a coastline along Cardigan Bay and a mountainous hinterland. There are numerous sandy beaches and excellent outcrops in the sea cliffs along the coast. The bedrock is entirely Ordovician and Silurian marine sedimentary rocks. These rocks were folded and faulted during both the Caledonian and Variscan orogenies. East–west faults appear to have localized lead, zinc, and silver veins. This mineralization was mined perhaps as early as Roman times but certainly during the 17th–19th centuries. There is a widespread glacial till left over from the last Ice Age when the area was covered by an ice sheet moving onshore from the Irish Sea as well as glaciers moving down from the Welsh mountains (Wikipedia, Geology of Ceredigion).

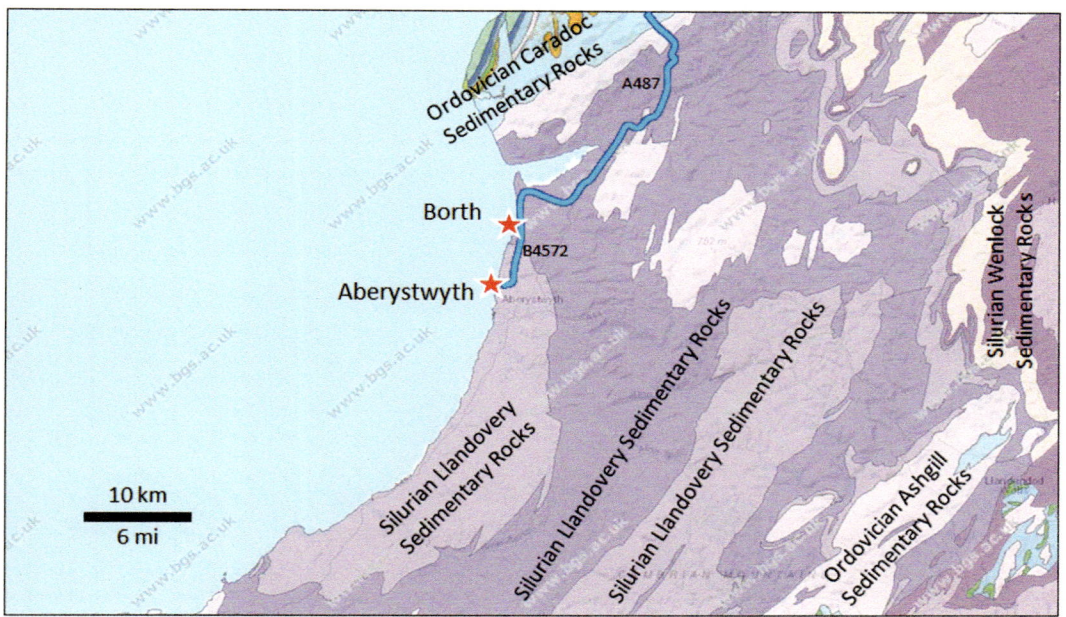

Geologic map of the Borth-Aberystwyth area. (Modified after BGS, Geology of Britain Viewer Classic. Contains British Geological Survey materials © UKRI 2022. Base mapping provided by ESRI.)

*Slate Museum to Borth: Return southeast on A4086 to A498 and turn right (south); drive south on A498 to Beddgelert and turn left (south) to stay on A498; turn left (east and south) onto the A4085 and drive to Garreg; turn left (east) onto B4410 and drive to the A487; turn left (east and south) onto the A487 and drive to Tre' r ddol; turn right (west) on the B4353 and drive to **Stop 3.1, Borth Beach** (52.481656, −4.051090), and park in the car park on the right (west) for a total of 116 km (71.8 mi; 1 h 56 min).*

3.1 BORTH BEACH AND SUBMERGED FOREST

The area around Borth is underlain by the Silurian Borth Mudstone Formation. During the Silurian, this area was part of the Avalonian microcontinent. By Early Silurian, southeastward subduction of the Iapetus oceanic crust beneath Avalonia had ceased, ending volcanic activity in the Welsh Basin. The basin was subsiding and filling with marine turbidite mudstones and sandstones. The Borth Mudstone is a mud-rich submarine fan deposited in shallow to deep water by turbidity currents whose sediments are derived from nearby Avalonia (Geoscience Wales, The Silurian Period; Baker and Baas, 2020). The Borth Mudstone is well exposed in the sea cliffs south of Borth beach.

Borth, however, is well-known for another, much younger geologic feature. When the sand conditions are just right, a 4,500-year-old submerged forest is exposed at low tide. For example, the storms in 2010, 2014, and 2019 removed the sand and allowed large sections of the forest to be examined. Preserved stumps of willow, hazel, oak, pine, and birch indicate that the area was inundated by rising sea levels. The stumps are rooted in peat lying below the beach sand and were preserved by acidic and anaerobic conditions in the continuously waterlogged soil. It was determined that the environment at the time was wet woodlands dissected by a major river channel. Radiocarbon dates from submerged trees at Ynyslas, north of Borth, indicate they were submerged around 4,500 years ago. Animal remains and tracks, including those of auroch, red deer, brown bear, and human prints, have also been found near the tree stumps. The stumps were first observed and commented on by

Gerald of Wales in 1188, and again by Samuel Pepys in 1665. In 1913 the geologist Clement Reid published his *Submerged Forests*, describing the trees scientifically for the first time (BBC News, 2012; Bates, 2014; dyfedarchaeology.org.uk/lostlandscapes).

3,500–5,500-year-old forest exposed at low tide, Borth Beach. View north. (Photo courtesy of Eveengland, https://commons.wikimedia.org/wiki/File:Ancient_sunken_forest_at_sunset_02.jpg.)

These submerged forests are strangely in tune with local legends of Cantre'r Gwaelod, the lost kingdom of Cardigan Bay. Early versions of this legend referred to the lost land as Maes Gwyddno and claim it was submerged when a well maiden named Mererid allowed a well to overflow.

Since the 17th century, the legend refers to the lost land of Cante'r Gwaelod. This mythical land is said to have extended some 32 km (20 mi) west from the current coastline of Cardigan Bay. The land was extremely fertile and contained as many as 16 cities. However, the area depended on a dike to keep the sea out. The dike had sluice gates that were opened at low tide to drain water from the land and closed when the tide returned. Gwyddno Garanhir, the king, had assigned Seithennin, a prince, to operate the sluice gates. After partying particularly hard one night, this normally responsible man fell asleep without closing the gates. As the tide rolled in the kingdom was lost forever beneath the waters of Cardigan Bay. Both Gwyddno Garanhir and Seithennin are referred to in *The Black Book of Carmarthen*, the earliest written collection of Welsh verse, dating from the mid-13th century (Carradice, 2012; Wikipedia, Cantre'r Gwaelod).

The drowned forests along the coast are evidence of rising sea levels since the last Ice Age ended. Bathymetric maps show water depths are less than 50 m (160 ft) as far as 30 km (18 mi) offshore here, and we know that during the peak of the last Ice Age, sea levels were as much as 100 m (330 ft) lower. So the myth may be a folk memory of displacement from low-lying coastal areas by gradually rising sea levels.

Borth to Aberystwyth*: Continue driving south on B4572 to the A487; turn right (west) on A487 and drive to North Road; turn right (northwest) on North Road and drive to Queen's Ave; turn left (west) on Queen's Ave and drive to Victoria Terrace; turn right (north) on Victoria Terrace and drive to **Stop 3.2, Aberystwyth** (52.423034, −4.085435) for a total of 10.0 km (6.2 mi; 14 min). There is free parking on both sides of the street along the seafront. Walk the last 125 m (400 ft) north to the outcrops.*

3.2 ABERYSTWYTH

The sea cliffs north and south of Aberystwyth are mudstones and sandstones in the Silurian Aberystwyth Grits Group. Several hundred meters of this unit were deposited as deep marine turbidites in the Welsh Basin during the Silurian (444–433 Ma). The Group consists of the Mynydd Bach Formation and the overlying Trefechan Formation. Paleocurrent data indicate an overall sediment source to the southwest (Geoscience Wales, The Silurian Period; Wikipedia, Aberystwyth Grits Group).

The rocks dip gently to steeply, depending on the extent of Caledonian folding. The cliffs north of town dip gently to the east.

Excellent views of the headlands to the north and Aberystwyth to the south can be had from Camera Obscura, an overlook perched on the bluffs north of town.

Silurian turbidites, north end of Aberystwyth Beach, looking north.

Aberystwyth to Cemaes Head*: Return south on Victoria Terrace and turn left (east) toward Queen's Road; turn right (south) on Queen's Road; drive south on Queen's Road, and then bear slightly right onto A487/Thespian Street; continue south on A487 to New Town/Cardigan and turn* *right (north) onto B4546; turn left (west) on St Dogmaels Road/B4546 and drive to St Dogmaels; turn right (north) onto Feidr Fawr/B4546 and drive to Poppit Sands Beach; turn left (west) and continue to Allt y Coed Campsite; park and take the Pembrokeshire Coastal Path 765 m (2,500 ft) north to Stop **4.1, Cemaes Head** (52.115351, −4.729262) for a total of 78.3 km (48.6 mi; 1 h 26 min).*

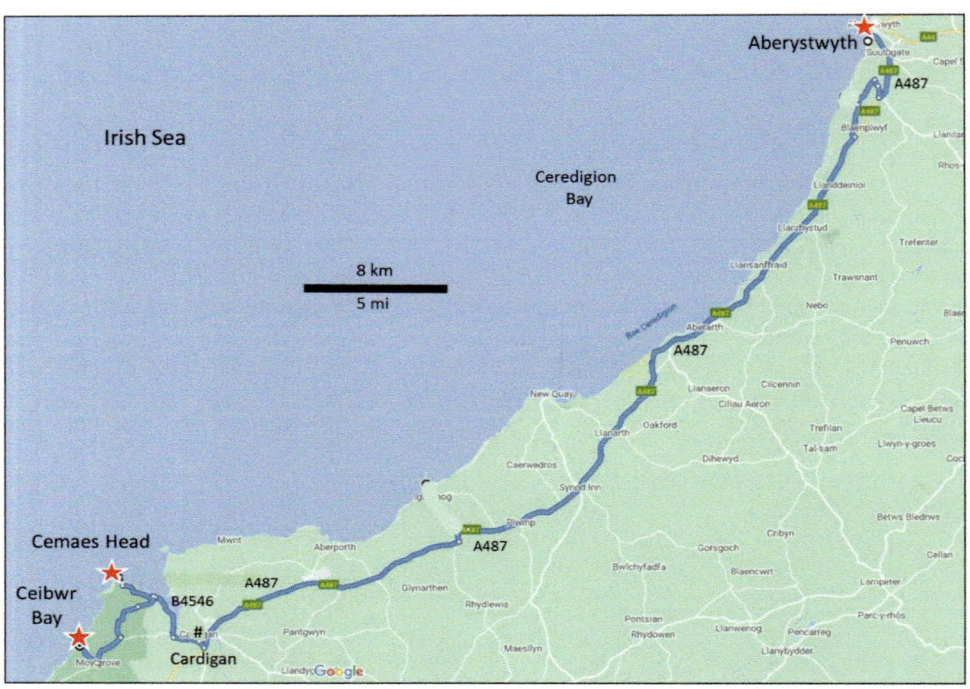

STOP 4 PEMBROKESHIRE COAST NATIONAL PARK

The Pembrokeshire Coast National Park, with an extraordinary variety of rock types and landforms, was designated a National Park in 1952. Most of the park is privately owned, and 22,500 people live in it. It is one of three national parks in Wales. The Pembrokeshire Coast Path, which runs the length of the park along the coast, is a designated National Trail. It was established in 1970 and is 299 km (186 mi) long, much of it running along the top of the sea cliffs. The path begins at Amroth in the south and runs north to St. Dogmaels, where it links with the Ceredigion Coast Path and continues north as part of the 1,400 km (870 mi) Wales Coast Path. Most of our stops in the National Park are on or near the Coast Path.

The 629 km² (243 mi²) park has a varied landscape of rugged cliffs, sandy beaches, wooded estuaries, forested inland hills and valleys, and moors. The National Trust owns several stretches of coastline. Around 40% of the park contains geological features that are protected as Sites of Special Scientific Interest. The ages of the rocks range from late Precambrian to Late Carboniferous (~650–290 Ma).

The Precambrian rocks are exclusively igneous, mostly volcanic ash and intrusions of granite and diorite. They are exposed on the southwestern tip of the St Davids Peninsula and on the north and south-facing cliffs of innermost St Bride's Bay. The Cambrian rocks are mostly sedimentary, a basal conglomerate overlain mainly by sandstones, but some mudstones occur in the middle part of the sequence. They are exposed along the southern and western sides of the St Davids Peninsula and in a section of coastline between Abercastle and Abermawr. The northern coastline of Pembrokeshire, from Ramsey Island to just east of Fishguard, is composed entirely of Ordovician rocks, mainly slates and volcanic rocks. From Dinas Head to the Teifi estuary, the cliffs are mostly turbidite sequences (interbedded sandstones and mudstones). Silurian rocks, both volcanic and sedimentary, form Marloes Peninsula and Skomer Island.

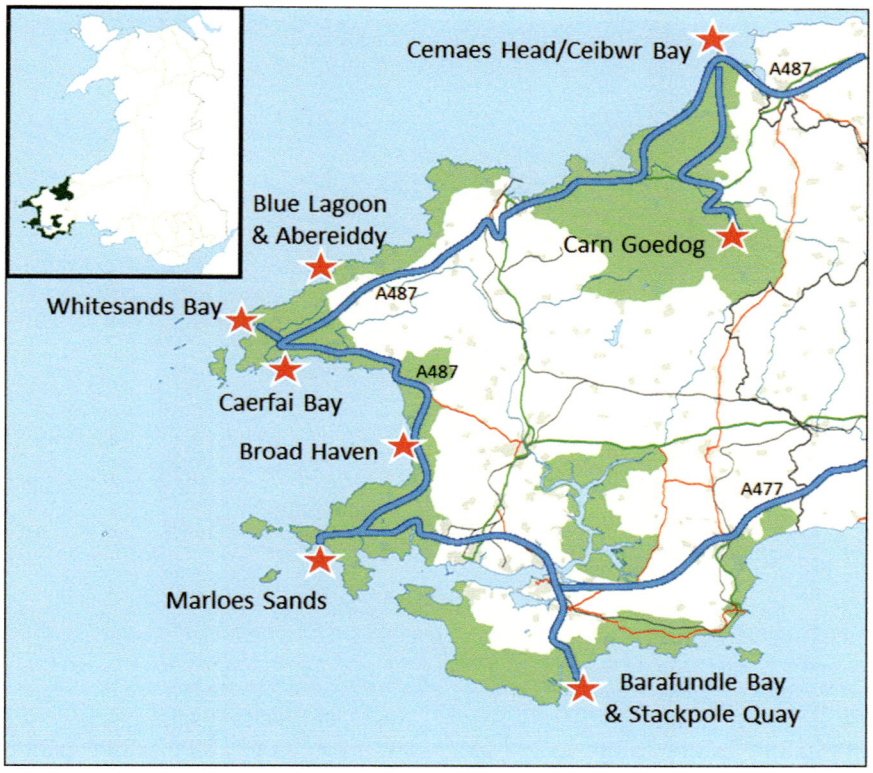

Map of Pembrokeshire Coast National Park, Wales. (Modified after drawing by Nilfanion, https://commons. wikimedia.org/wiki/File:Pembrokeshire_Coast_National_Park_UK_location_map.svg.)

Devonian through Carboniferous rocks outcrop in south Pembrokeshire. The Devonian Old Red Sandstone is mostly red mudstones, siltstones, and sandstones. The Carboniferous Limestone is a dark gray limestone that weathers light gray. Above the limestone, the Carboniferous Millstone Grit and South Wales Coal Measures comprise mainly dark gray or black shales and light gray sandstones (that weather orange-brown). They differ in that the Millstone Grit has a lower and upper coarse sandstone, whereas the Coal Measures include thin seams of anthracite coal.

Remnants of wave-cut platforms, sometimes with cemented pebbles and shells, formed around 125,000 years ago when the sea level was 5 m (16 ft) higher than at present. The wave-cut platforms can be seen at Marloes Sands and Broad Haven, among others.

The park includes many fine examples of natural arches, sea stacks, folds, and sea caves. The south coast has, in addition, the limestone plateau and cliffs of the Castlemartin Peninsula.

The National Park includes 7 Special Areas of Conservation, a Marine Nature Reserve, 6 national nature reserves, 75 Sites of Special Scientific Interest, and numerous Regionally Important Geological Sites. In 2011, *National Geographic Traveler* magazine voted Pembrokeshire the second-best coastal destination in the world for sustainable tourism (Pembrokeshire Coast, Geology; Wikipedia, Pembrokeshire Coast National Park). The area has been reviewed as a potential UNESCO Global Geopark, but the idea has been shelved for now as benefits do not appear to justify costs (Pembrokeshire Coast National Park Authority, Global Geopark Study Results).

		Previous Workers		Nichols, 2019
Silurian	Llandovery			
		Gaerglwyd Fm	Cwmere Fm	Cwmere Fm
				Mottled Mudstone Mbr
Ordovician	Ashgill	Llangranog Fm	Yr Allt Fm	Brynglas Fm
				Drosgol Fm
			Nantmel Mudstone Fm	Nantmel Mudstone Fm
		Tresaith Fm		
	Caradoc		Cwm yr Eglwys Mudstone & Dinas Island Fm	

Stratigraphy of the Pembrokeshire Coast with previous and current naming conventions. (Modified after Nichols, 2019.)

4.1 Cemaes Head

Cemaes Head is a headland and nature reserve in Pembrokeshire Coast National Park. Purchased by the Wildlife Trust of South and West Wales in 1984, and measuring 20 ha (49 ac), it is also a Marine Special Area of Conservation (SAC) and an SSSI. The word "cemaes" is from the old Welsh word "camas," meaning a "bend or loop in a river" or a "bay or inlet of the sea."

Cemaes Head overlooks the mouth of the Teifi estuary and has expansive views of Cardigan Bay. The northwest side, exposed to Atlantic storms, is most dramatic, with cliffs up to 175 m (575 ft). The sheltered eastern side is gentler.

Cemaes Head consists of Ordovician Dinas Island Formation, coarse sandstone–mudstone turbidites up to 1.3 km (4,260 ft) thick. These turbidites accumulated on the north, downthrown side of the west–southwest–east–northeast Ceibwr Bay normal fault in a deep marine environment some 470–440 Ma.

The Ceibwr Bay Fault is the northernmost fault in the Fishguard–Cardigan Fault Belt that crosses the district. This fault zone has a long history of movement and strongly influenced the accumulation of Ordovician sediments. Subsequently, parts of the fault zone underwent major reversals of movement.

The layers exposed along the sea cliffs were deformed, along with the rest of this region, during the Acadian phase of the Caledonian Orogeny, around 400 million years ago. These rocks are folded, buckled, and faulted, and have minor metamorphism and subsequent cleavage. The folds are clearly seen from the top of the bluffs.

The bedrock is covered by a veneer of Quaternary glacial and post-glacial material. Two episodes of glaciation affected this area: early glaciation by Irish Sea ice was followed by the main, Late Devensian glaciation that peaked around 22,000 years ago. Most of the glacial features and deposits derive from this later event.

Cliffs at Cemeas Head. (Photo courtesy of P. Halling, https://commons.wikimedia.org/wiki/File:Cliffs_on_Cemaes_Head,_Pembrokeshire_Coast_-_geograph.org.uk_-_430187.jpg.)

Cemaes Head lies on the Pembrokeshire Coast Path. The section of path over Cemaes Head from Poppit Sands to Ceibwr Bay is about 8.8 km (5.5 mi) long. You can drive part way up Cemaes Head on the lane from Poppit Sand to Allt y Coed campsite. There is a car park available, from which it is a 3.2 km (2 mi) round-trip to the headland. There is also a car park at the hamlet of Cippin, a mile up the lane from Poppit Sands.

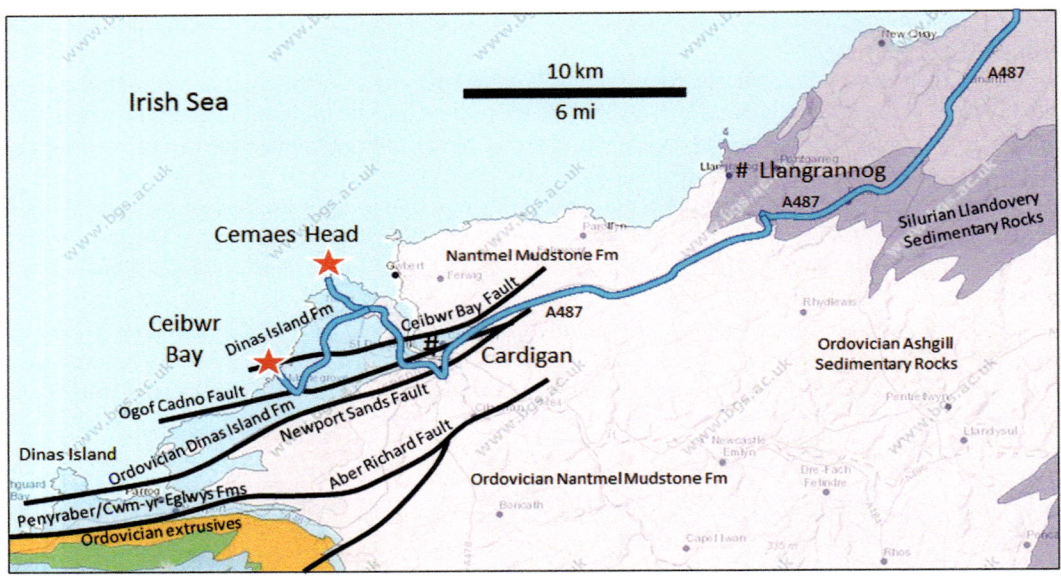

Geologic map of the Cemaes Head area. (Modified after BGS, Geology of Britain Viewer Classic. Contains British Geological Survey materials © UKRI 2022. Base mapping provided by ESRI.)

Cemaes Head to Ceibwr Bay: Return to Poppit Beach and turn right (east) onto B4546 and immediately another right (south) outside the parking lot onto an unnamed single lane road; in 1.3 km (0.8 mi) bear left at the Y and continue driving to the sign for "Trewyddel/Moylegrove 1.5 mi" and turn right (south); continue driving to Trewyddel Moylegrove and in town turn right (west); drive through town to the sign for Ceibwr and turn right (northwest), and drive to the wide spot in the road at the sea cliffs. This is **Stop 4.2, Ceibwr Bay** *(52.077674, −4.762426), for a total of 10.7 km (6.6 mi; 20 min). There is abundant free parking here.*

4.2 CEIBWR BAY

Ceibwr is one of the best-known geological sites in Wales because of the breathtaking vista of sea cliffs and dramatic folding in the Ordovician Dinas Island Formation that extends north to Cemaes Head (see the photo at the start of this chapter).

Sediments were deposited between 458 and 449 Ma by submarine sediment avalanches. Fifty million years or so later, the Caledonian Orogeny raised and crumpled the strata, creating some impressive folding.

The Ceibwr Bay Fault cuts the eastern cliffs of the bay, dropping down rocks that are much younger than those on the west side of the bay. There is about 600 m (2,000 ft) of displacement, juxtaposing the dark, older Ordovician Carreg Bica Mudstone member of the Dinas Island Formation on the south with younger sandstones and mudstones of the Dinas Island Formation to the north.

Ceibwr Bay is a partly-flooded pre-glacial valley. The valley is considered too large to have been eroded by the small river that occupies it. The valley is an Ice Age feature, widened and deepened by glacial meltwater when the sea level was quite a bit lower. The section near the beach is filled with over 20 m (65 ft) of glacial outwash (John, 2019; Wikipedia, Ceibwr Bay).

View of the cliffs at Ceibwr Bay. Note the dramatic folding of the Dinas Island Formation strata.

Ceibwr Bay to Blue Lagoon: *Continue driving west and south 1.7 km (1.1 mi) to the sign for "Trefdraeth/Newport 5" and turn right (west); bear left at the next sign for "Trefdraeth/Newport;" in Newport turn right (west) onto East Street/A487 and drive to Fishguard; at the Rafael Roundabout on the west side of Fishguard take the 1st exit to A40; turn right (west) at the sign for "Manorowen 1.5;" bear right (northwest) onto A4219 and quickly left (southwest) onto A487; continue southwest on A487 to Croesgoch and turn right (west) onto Abereiddy Road; turn right at the sign for "Aberredy 1" and drive to* **Stop 5.1, Blue Lagoon and Abereiddy Bay** *(51.937698, −5.207635), for a total of 41.0 km (25.5 mi; 47 min).*

Ceibwr Bay to Side Trip 2, Carn Goedog: *Continue driving west and south to the sign for "Trefdraeth/Newport 5" and turn left (east) and immediately right (south) at the sign for "Nanhyfer/Nevern 4," drive south to Velindre and turn left (east) onto A487; drive east on A487 to B4329 and turn right (south); take the B4329 south to Croswell and turn left (east) on the unnamed road toward Crymych and Bethabara Chapel; drive just shy of 1 km (0.6 mi) and turn right (south) on an unnamed road and drive 2.2 km (1.4 mi) to a bend in the road and park (51.976784, −4.715653) for a total of 16.1 km (10 mi; 21 min). Parking is free, but space is limited on the side of this road. From here, you hike 1.4 km (4,650 ft; 20 min) south–southwest across sheep pasture to* **Side Trip 2, Carn Goedog bluestone quarry** *(51.965747, −4.725039).*

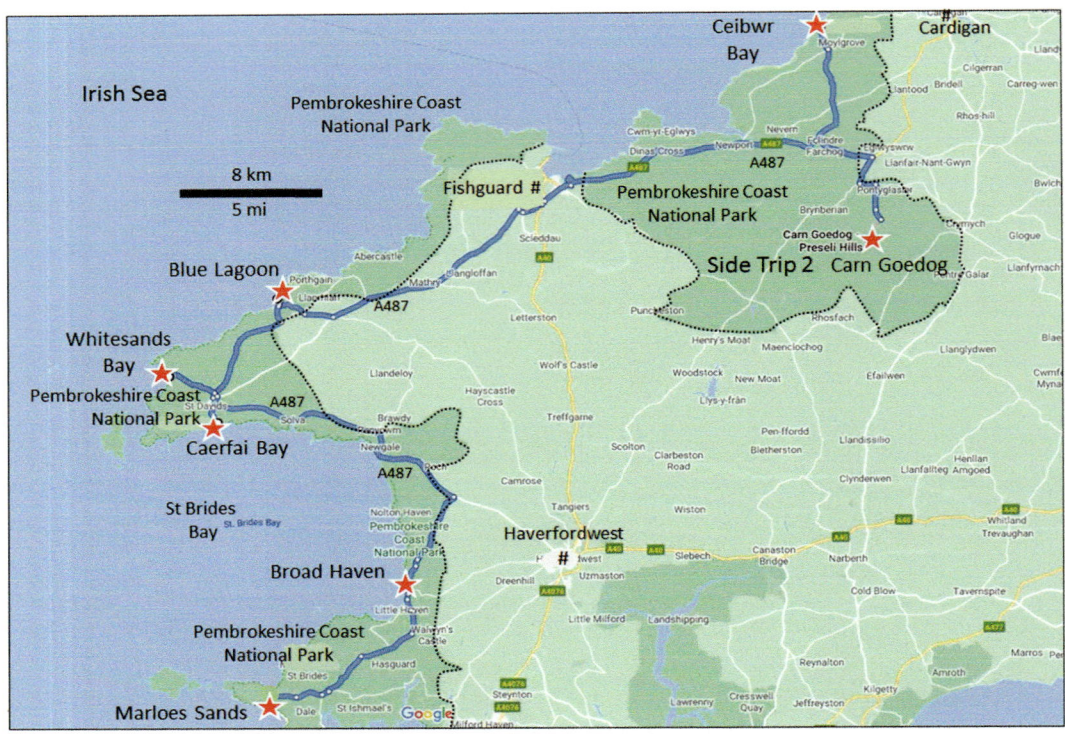

SIDE TRIP 2 CARN GOEDOG BLUESTONES

The discovery of a bluestone quarry at Carn Goedog in 2011, and subsequent archaeological excavations, provide details of ancient quarrying processes along with dates for the extraction of the bluestone monoliths found at Stonehenge. The quarries are in an Ordovician gabbro that intruded marine shale of the Ordovician Aber Mawr Formation (478–461 Ma; BGS, Geology of Britain Viewer Classic).

Excavations by University College London have clearly identified Carn Goedog as the main source of the bluestones (technically a "spotted dolerite" or gabbro) at Stonehenge. The bluestones are named for their slight bluish color. Carn Goedog is an outcrop on the north side of the Preseli Hills where the gabbro occurs as pillar-shaped slabs. Stonehenge bluestones were first sourced to outcrops in Mynydd Preseli in southwest Wales by H.H. Thomas (1923). Identical rock chemistry confirms that the Stonehenge bluestones are from this Neolithic quarry.

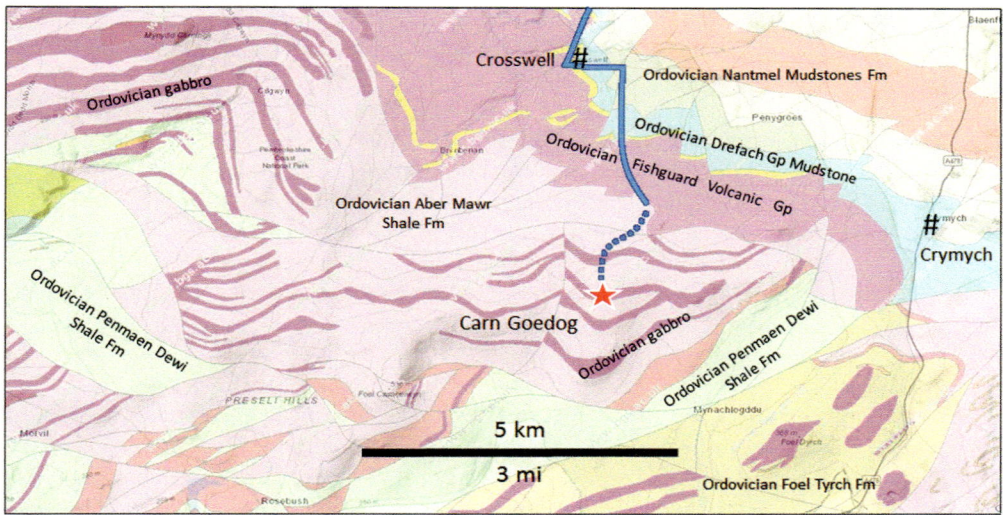

Geologic map of the Carn Goedog bluestone outcrop area. (Modified after BGS, Geology of Britain Viewer Classic. Contains British Geological Survey materials © UKRI 2022. Base mapping provided by ESRI.)

Naturally fractured bluestone slabs near Carn Goedog, Preseli Hills.

The rock's tendency to break into natural pillars would have made it relatively easy to extract slabs with minimal effort and little need for stone tools. Wedge-shaped stone tools made of mudstone or sandstone would have been hammered into V-shaped notches cut into the rock between the naturally fractured slabs, detaching the slabs from the rock face.

Most of the quarry features cannot be dated more closely than the fourth to second millennium BC. Dates of 3,500–3,120 BC are provided by carbonized hazelnut shells from a layer 1.5 m (5 ft) from the slabs. A later phase of quarrying dates to the Early Bronze Age, more than a thousand years after the Middle Neolithic activity. Quarrying of a spotted dolerite pillar at Carn Menyn also dates to the Early Bronze Age (2,160–1,960 BC), suggesting that the Early Bronze Age was a period when numerous standing stones were quarried and erected in Pembrokeshire. This would have been long after the bluestones were taken to Stonehenge.

The bluestones were certainly in place at Stonehenge around 2500 BC and probably were first erected to form a stone circle at Stonehenge around 3000 BC. The quarries' carbon dates are old enough to make it possible that the stones had been erected in a circle in Wales before being taken to Stonehenge (Bevins et al., 2014; Pearson et al., 2015; Archaeology Magazine, 2019). So how did they get to Stonehenge? The latest thinking is that the 1.8 tonne (2 ton) stones were lifted in a framework of poles and carried by up to 60 people at a time eastward along the upper Nevern Valley and the valleys of the Teifi, Tywi, Usk, and Wye, fording the River Severn at Longford, then south to Salisbury Plain. The total distance is around 350 km (220 mi; Pearson et al., 2015; Pearson et al, 2019). That would be an impressive feat even today.

*Side Trip 2, Carn Goedog to Blue Lagoon: Return north to the T-intersection and turn left (west) on an unnamed road to Crosswell; in Crosswell turn right (north) on B4329 and drive to A487; turn left (west) on A487 and drive to Fishguard; at the Rafael Roundabout, west side of Fishguard, take the 1st exit to A40; turn right (west) at the sign for "Manorowen 1.5;" bear right (northwest) onto A4219 and quickly left (southwest) onto A487; continue southwest on A487 to Croesgoch and turn right (west) onto Abereiddy Road; turn right at the sign for "Aberredy 1" and drive to **Stop 5.1, Blue Lagoon and Abereiddy Bay** (51.937698, −5.207635), for a total of 43.7 km (27.1 mi; 47 min).*

STOP 5 SOUTH PEMBROKESHIRE COAST

This section of Pembrokeshire is characterized by dramatic sea cliffs, wave-battered headlands, and remote sandy beaches. The Precambrian to Silurian rocks have been faulted and folded into east–west to east–northeast trends by the Carboniferous–Permian Variscan mountain-building event.

5.1 BLUE LAGOON AND ABEREIDDY BAY

The "Blue Lagoon" is a small slate quarry just north of Abereiddy Bay, worked during the 1800s and abandoned in 1901, that has since been flooded by the sea. The lagoon is 25 m (82 ft) deep. Despite the name, the water is a distinct shade of green owing to minerals in the quarry (Pembrokeshire Coast National Park, Blue Lagoon; Wikipedia, Abereiddy).

The overall structure in the area is an east–west-oriented, faulted syncline overturned to the south, with the youngest unit (undifferentiated Ordovician Caradoc deep marine mudstones) centered on Abereiddy Bay. This Llanrian Syncline formed during the Caledonian deformation of this region.

This is the type section for the Llanvirn Series, the upper part of the Middle Devonian in Wales (467.3–458.4 Ma; Howells, 2007; Michael, 2013; Mitten and Pringle, 2018). The strata are Avalonian marine volcanics and sediments.

The north-dipping Caerhys Shale Formation beds on the south side of Abereiddy Bay contain the graptolite fossil *Didymograptus murchisoni*. On the north side of the bay, the beds dip north but are overturned and inclined 60°–85°. Axial plane cleavage is superimposed on the bedding. The top of the Caerhys Shale is the Llanvirnian/Llandeilan boundary. Above the boundary (an unconformity) lies the Castell Limestone Formation. This is a light-colored shelly and brecciated limestone interbedded with black shale. It outcrops just south of the quarry (Mitten and Pringle, 2018). The Castell Limestone was deposited in shallow seas between 461 and 449 Ma and contains the occasional trilobite.

The Blue Lagoon, an old slate quarry now filled by the sea. (Photo courtesy of A. Schwerdfeger, https://commons.wikimedia.org/wiki/File:Blue_Lagoon,_Abereiddy_2014-09-09.jpg.)

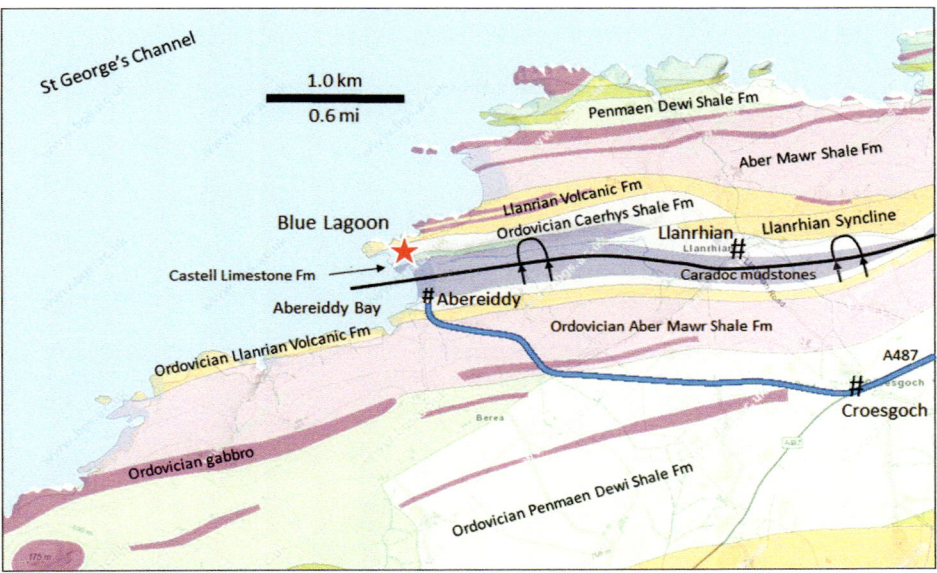

Geologic map of the Blue Lagoon area. (Modified after BGS, Geology of Britain Viewer Classic. Contains British Geological Survey materials © UKRI 2022. Base mapping provided by ESRI.)

The quarry itself is in black slates interbedded with volcanic ash containing dolerite sills. The section is overturned and dips ~55° north. The quarry and adjacent cove are located in *Didymograptus bifidus* beds of the Ordovician Caerhys Shale Formation, deposited ~466–461 Ma in a deep marine setting (BGS, Geology of Britain Viewer Classic; Reading Geological Society, 2009; Wikipedia, Abereiddy).

Fossil collectors look for graptolites in chunks of the Ordovician shale that have fallen from the cliffs onto the beach. The best collecting is done by walking down the beach looking for the easily split shales. Yet Ordovician graptolites can be found in the car park as well (GeoWorld Travel, Pembrokeshire Geology Field Trip).

There is a parking area above the beach. The remains of the Abereiddy quarry are owned by the National Trust. A concrete and stone path approximately 200 m (660 ft) long takes you to an overlook. Ruins of a small group of houses built for quarry workers are near the beach. The quarry is an SSSI for its stratigraphy and fossils and was purchased by the National Trust for its industrial ruins.

Abereiddy Bay is a popular destination for surfing, kayaking, and exploring sea cliffs. In September 2012, the Blue Lagoon was the location for one part of the Red Bull Cliff Diving World Series. Fourteen of the world's best divers dove from a manmade platform, 27 m (88 ft) above the lagoon. The championships returned in September 2013 and 2016 (Wikipedia, Abereiddy).

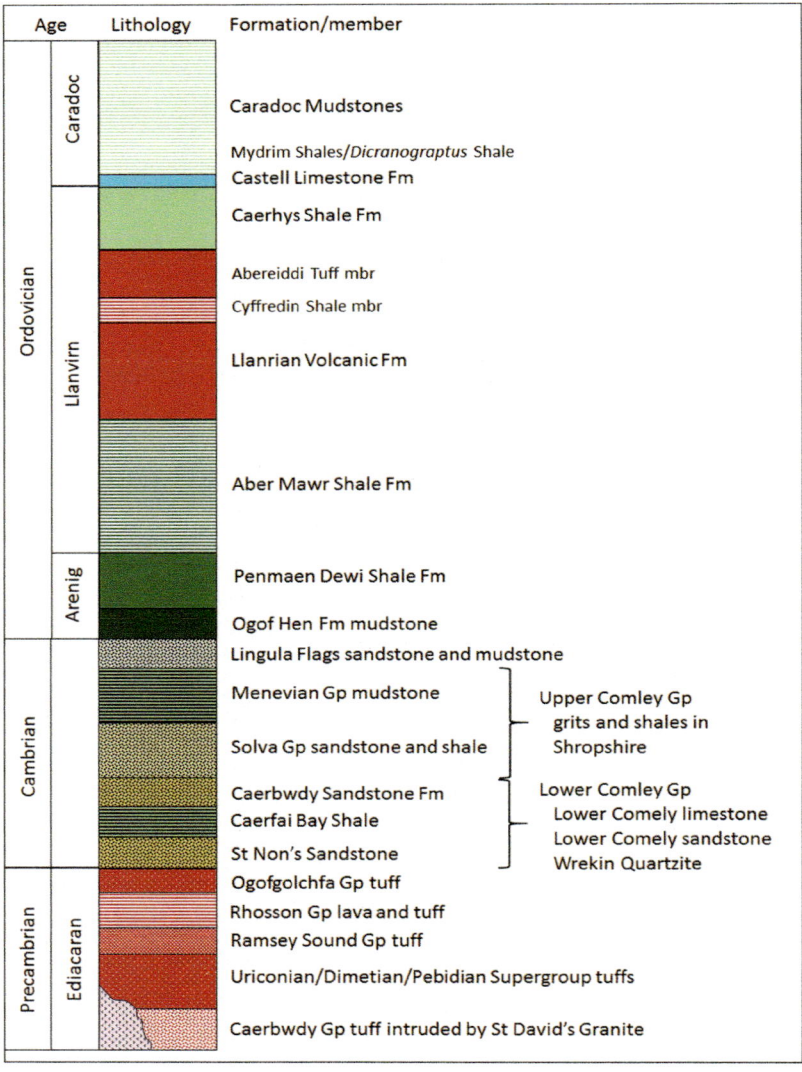

Stratigraphy of southern Pembrokeshire and Shropshire. (Information drawn from Siverter and Williams, 1995; Mitten and Pringle, 2018.)

*Blue Lagoon to Whitesands Bay: Return south on Milon Road to Ildiwch T-intersection and turn right (west) on an unnamed street; drive southwest to B4583 and turn right (northwest); continue northwest on B4583 to **Stop 5.2, Whitesands Bay** (51.896714, −5.295390), for a total of 10.5 km (6.5 mi; 17 min). There is a large car park, £5. It is an easy walk to the beach.*

5.2 WHITESANDS BAY

The units here strike east–northeast and have a more-or-less uniform north dip. South and east of the bay the Precambrian is exposed and consists of 635–541 Ma granites and explosive volcanic tuffs (BGS, Geology of Britain Viewer Classic; Cruickshanks, 2017, Whitesands Bay).

Whitesands Bay. Note the high-tide mark ('bathtub ring') on the sea cliffs. View south.

Beginning in Cambrian time, the Welsh Basin subsided as a result of rifting of Avalonia from Gondwana. The Welsh Basin was a Cambrian to Silurian back-arc basin (that is, it formed behind a volcanic island arc; Rushton and Molyneux, 2011). Cambrian rocks overlying the Precambrian represent a relatively quiet period during which the Welsh Basin was filled with sediments (Berry, 2011). Cliffs at the south end of the beach consist of the Solva Group, interbedded sandstone and shale deposited in shallow seas during the Cambrian approximately 508–499 Ma (BGS, Geology of Britain Viewer Classic). Avalonia completely separated from Gondwana during the Ordovician, the Rheic Ocean was opening, and volcanic arcs were contributing sediments to the basin. During Late Ordovician, Avalonia was approaching Baltica, and their collision would cause the Caledonian Orogeny that peaked in the Devonian.

Most of the south half of the bay is underlain by the Menevian Group, a Cambrian shallow-water mudstone. The north half of Whitesands Bay is underlain by the Lingula Flags, Cambrian shallow-water sandstone, siltstone, and mudstone containing *Lingulella davisii*, a brachiopod (clam-like shellfish). The fossils are usually found in loose blocks of black shale, but they are challenging to find. Rocks with fossils appear to be more abundant in the southern part of this area (Cruickshanks, 2017, Whitesands Bay).

There is an unconformity between the Lingula Flags and the Ogof Hen Formation, an Ordovician (478–466 Ma) deep-water mudstone containing distinctive phosphate nodules. Further north, you cross a fault and are in black, deep-water mudstones of the Ordovician Penmaen Dewi Shale (478–466 Ma) that contains rare trilobites.

Part way down the beach cliffs are dark Cambrian-age gabbro dikes. Partially hidden in the heather at the bottom of the cliffs is an ultrabasic (very dark) half-meter (1.5 ft) boulder believed to be a glacial erratic that originated on the northwest coast of Scotland (Reading Geological Society, 2009). Farther north two gabbro intrusions indicate volcanic activity during the Ordovician (~490 Ma; Berry, 2011).

The low cliffs at this location are composed of near-vertical Cambrian strata. Tilting and folding occurred during the Caledonian Orogeny. A veneer of glacial till covers all these units. A pre-Devensian (pre-last Ice Age) raised beach indicates that either the sea level was higher, or the land has been uplifted since the glaciers melted.

This site is part of the Pembrokeshire Coast National Park and is an SSSI , meaning you can collect loose fossils, but hammering bedrock is not permitted.

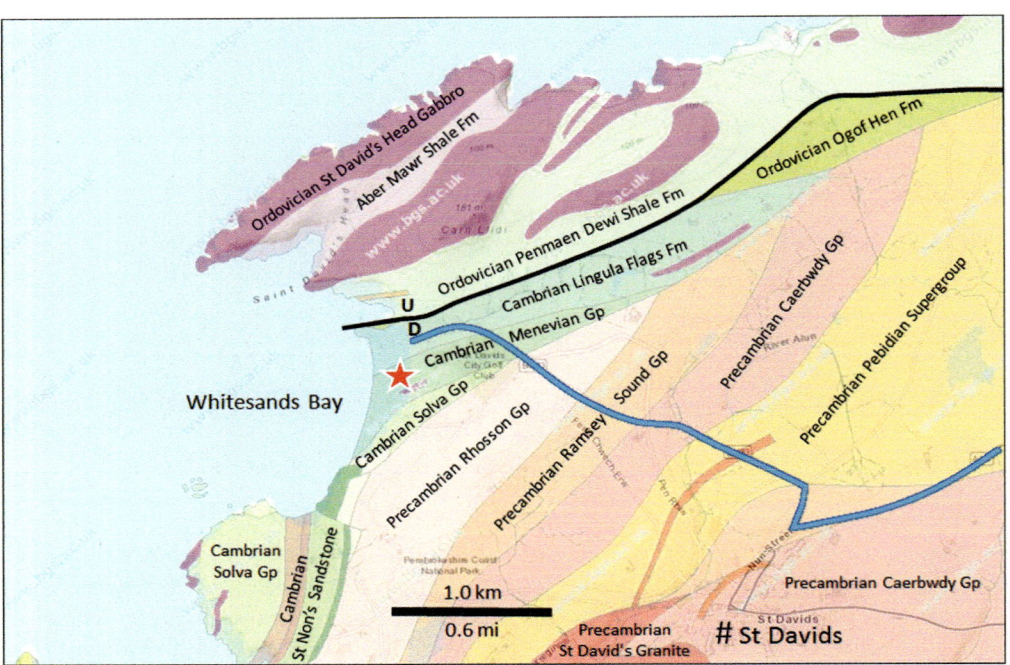

FGeologic map of Whitesands Bay area. (Modified after BGS, Geology of Britain Viewer Classic. Contains British Geological Survey materials © UKRI 2022. Base mapping provided by ESRI.)

Carn Llidi, a rocky outcrop north of the bay that dominates the skyline at 181 m (594 ft), is made up of Carn Llidi Gabbro, another of the dark, igneous dikes injected during the Ordovician roughly 466– 449 Ma (Howells, 2007).

At extreme low tide, and after storms have removed the sand, the remains of an ancient, submerged forest can be seen on the beach. It contains stumps of birch, fir, hazel, and oak trees (Wikipedia, Whitesands Bay).

Layered gabbro exposed on Carn Llidi. View north

The St Davids Head-Carn Llidi area has been occupied since the Stone-age. There are several tombs and the earthworks of an Iron Age fort. A number of megalithic burial chambers, stone hut circles, and Iron Age field enclosures have been found in the vicinity (Reading Geological Society, 2009).

St. Patrick is said to have sailed from this beach in the year 432 to convert Ireland. The site of a Celtic chapel, dedicated to St Patrick, is located under a mound by the car park (Wikipedia, Whitesands Bay).

Today, the Pembrokeshire coastal path passes by Whitesands Bay. In addition to the awesome geology, this bay has been described as the best surfing beach in Pembrokeshire.

*Whitesands Bay to Caerfai Bay: Return south and east on B4583 to the T-intersection to Tdddewi/St Davids (A487) and turn right (south); at Ildiwch turn left (east) onto A487 and drive to Glasfryn Road on the right; turn right (south) onto Glasfryn Road; at the roundabout, continue straight on Ffordd Caerfai to **Stop 5.3, Caerfai Bay** (51.872851, −5.256542) car park (free), at the end of the road for a total of 5.1 km (3.2 mi; 9 min). There is a steep walk to the beach on a good trail.*

5.3 CAERFAI BAY

Caerfai Bay has cliffs of purple Cambrian sandstone and, at low tide, a sandy beach that can be reached by steep steps. Geologically the cliffs are the type section for the Caerfai Epoch – the earliest epoch of the Cambrian Period (Coast and Country, 2019).

Caerfai is on the south flank of the east-northeast-trending St Davids Anticline, with the town of St Davids in the Precambrian core of the structure. As with other folds in this area, the anticline is a result of the Caledonian Orogeny.

Purple Cambrian sandstone at Caerfai Beach. View east.

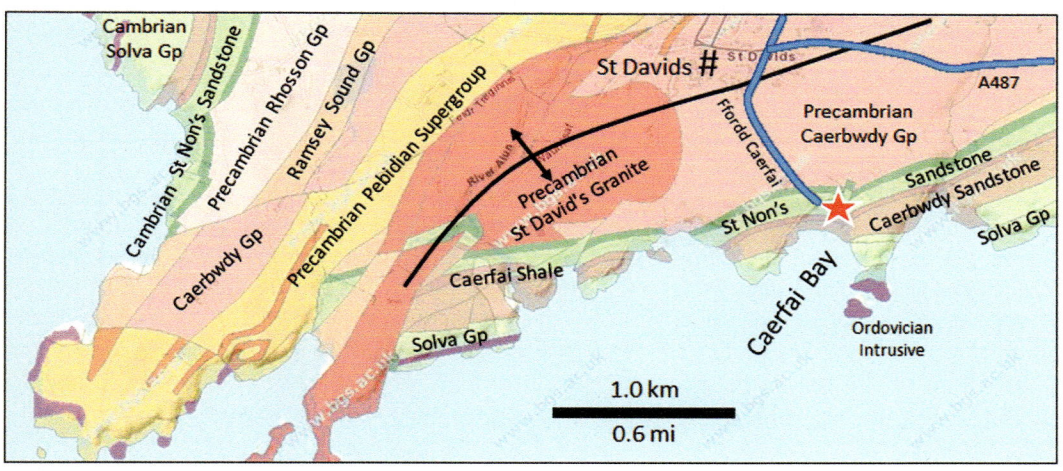

Geologic map of Caerfai Bay area. (Modified after BGS, Geology of Britain Viewer Classic. Contains British Geological Survey materials © UKRI 2022. Base mapping provided by ESRI.)

The St Non's Sandstone Formation basal conglomerate sits unconformably on dark green slates and volcanic tuffs of the Precambrian Caerbwdy Group and dips south. The contact between the green St Non's Sandstone and the overlying red Caerfai Shale crosses the upper beach from west to east. The Caerfai Shale is altered volcanic tuff deposited in a marine setting and contains a few trilobites (Fortey, 2000; Michael, 2013). Quartz veins cross the Caerfai Shale. The Caerbwdy Sandstone Formation lies above the Caerfai Shale and is seen dipping 50°–80° south in the cliffs south of the beach. These are massive beds, up to 3 m (10 ft) thick, of purple fine- to medium-grained sandstone with evidence of burrowing by sea creatures. The distinctive purple-red Caerbwdy Sandstone was quarried nearby for St Non's Chapel and St Davids Cathedral (Michael, 2013; Wikipedia, Caerfai Bay).

Solva Group sandstone and shale form the headlands, and the small offshore island of Penpleidiau consists of an Ordovician felsic intrusion (Smith, 2016; BGS, Geology of Britain Viewer Classic). These units formed on the floor of the nascent Iapetus Sea shortly after the breakup of Pannotia (also known as the Vendian supercontinent, Greater Gondwana, and the Pan-African supercontinent) to form the Laurentian, Siberian, and Baltican microcontinents starting around 540–535 Ma.

Caerfai is the nearest beach to St Davids, and is known as a "suntrap": it is south-facing and surrounded by cliffs. You can enjoy both the sun and the panoramic views. At low tide, the beach is extensive, but it is mostly underwater at high tide. The beach is in Pembrokeshire Coast National Park and on the Pembrokeshire Coast Path. There is free parking above the beach and picnic benches, but there are no public toilets.

Place names here are steeped in history. St Non was born around 475, became a nun, and was raped by Prince Sant of Ceredigion. As a result, she gave birth to a baby boy who later became St David, the patron saint of Wales (Smith, 2016).

*Caerfai Bay to Broad Haven: Return north to St Davids; at the roundabout take the 3rd exit onto A487 and drive east to Simpson Cross; turn right (south) onto Cuffern Road; the road becomes Haroldston Hill; park at the intersection with B4341/Millmore Way (51.782596, −5.101821) and walk 460 m (1,500 ft) north along the beach to **Stop 6, Broad Haven** (51.786389, −5.103889) folds for a total of 24.4 km (15.2 mi; 27 min). The best folding is at the north end of the beach. Parking is on the street and is scarce.*

STOP 6 BROAD HAVEN

Broad Haven is a small seaside resort in the southeast corner of St Bride's Bay. It is best known for UFO sightings in 1977 and is consequently nicknamed the Welsh Triangle, Broad Haven Triangle, or Dyfed Triangle. [The Ministry of Defense was unable to find any evidence of UFOs and speculated that "a local prankster" might have been at work; Wikipedia, Broad Haven.]

The other thing Broad Haven is known for is preservation of classic Variscan thrusted folds developed in Upper Carboniferous (Pennsylvanian) river and delta sandstones of the South Wales Lower Coal Measures Formation (Butler, Folding at Broadhaven). We last saw rocks of Carboniferous age at Great Orme and Anglesey in northern Wales. Here, we return to the Carboniferous and examine deformation related to the Carboniferous–Permian Variscan Orogeny. These strata were deposited in a foreland basin ahead and north of the advancing Variscan deformation front. The sea cliffs between Little Haven and Broad Haven display spectacular examples of thrusted folds. The best place to view them is at Den's Door, 1 km (0.6 mi) north of Broad Haven, where a north-verging fold is cut by north-directed thrusts.

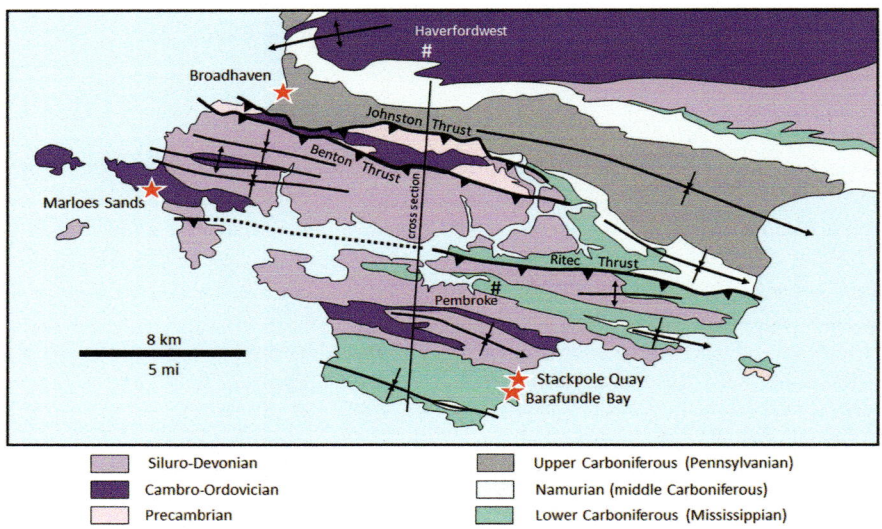

Geologic map of Broad Haven to Stackpole Quay. (Modified after Cawood and Bond, 2020.)

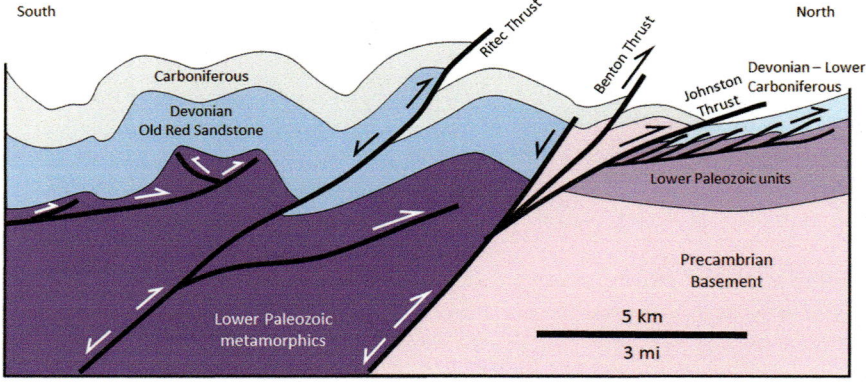

Schematic cross section showing Variscan structures in Pembrokeshire. Early normal faults were reactivated as thrusts during the Variscan Orogeny. The section line is shown on the previous map. (Modified after Howells, 2007.)

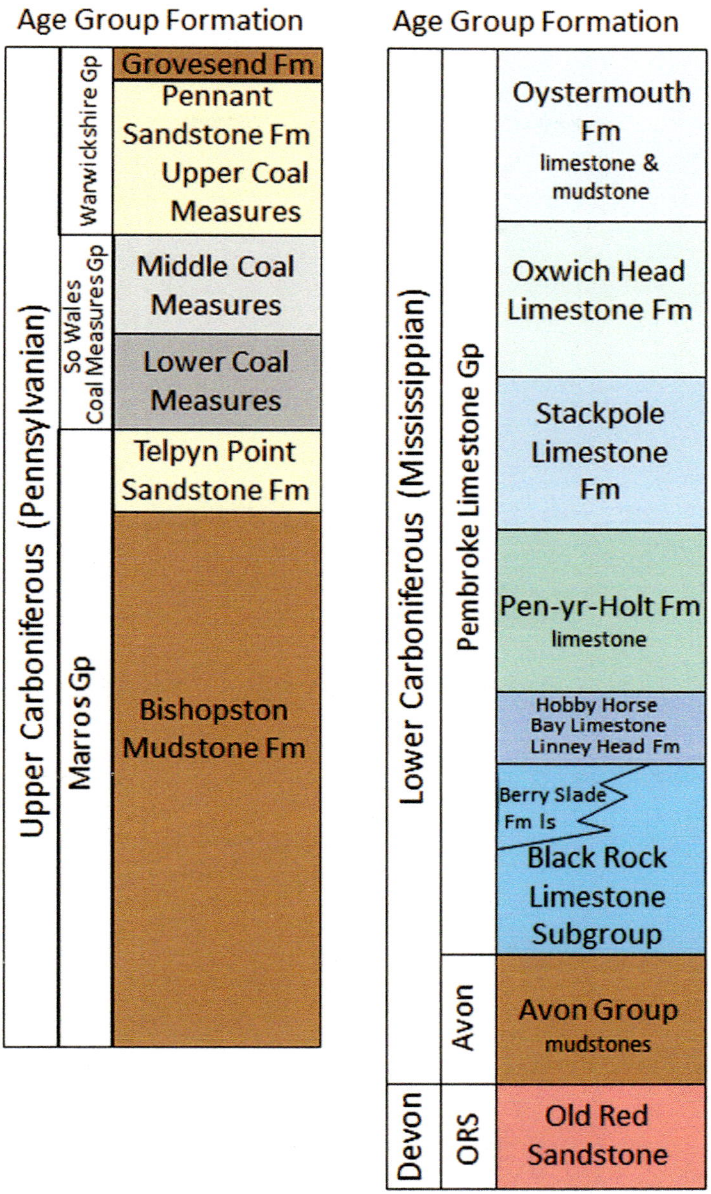

Age Group Formation

Age	Group	Formation
Upper Carboniferous (Pennsylvanian)	Warwickshire Gp	**Grovesend Fm**
		Pennant Sandstone Fm Upper Coal Measures
	So Wales Coal Measures Gp	Middle Coal Measures
		Lower Coal Measures
	Marros Gp	Telpyn Point Sandstone Fm
		Bishopston Mudstone Fm

Age Group Formation

Age	Group	Formation
Lower Carboniferous (Mississippian)	Pembroke Limestone Gp	Oystermouth Fm limestone & mudstone
		Oxwich Head Limestone Fm
		Stackpole Limestone Fm
		Pen-yr-Holt Fm limestone
		Hobby Horse Bay Limestone Linney Head Fm
		Berry Slade Fm ls
		Black Rock Limestone Subgroup
Devon	Avon	Avon Group mudstones
	ORS	Old Red Sandstone

Carboniferous stratigraphy, south Pembrokeshire. (Information drawn from Abbott, 2009; Rowberry et al., 2013.)

This site contains several fault-propagation folds. A fault-propagation fold is one in which the thrust terminates and beds bend around the fault tip. Locally the folds verge to the north and are bounded or cut by north-directed thrusts. This apparently simple structure shows disharmonic folding, with interbedded shales thickening in the axes and thinning on the flanks of fold. Multiple small-displacement thrusts and back thrusts cut the fold. Recent work supports the contention that folding here occurred before a fault formed. Further deformation caused a thrust fault to initiate in the fold; following thrusting, the section continued to deform by movement along the thrust.

Shortening in this area is estimated at 40%, a bit larger than the estimated regional Variscan shortening of 25%–30% (Cawood and Bond, 2020).

Folding in the Carboniferous South Wales Lower Coal Measures. (From Cawood and Bond, 2020.)

Complex Variscan folding and thrusting at Broad Haven.

Broad Haven to Marloes Sands: Continue south on B4341; at the south end of Broad Haven take a slight left onto Walton Road; turn right (south) toward Rosepool and drive to B4327; merge onto B4327 driving southwest; turn right (west) at the sign to "Marloes 1 ¼"; on entering Marloes turn left (west) at the sign to Marloes Sands parking (51.727366, −5.217094); continue on foot ~1 km (3,300 ft) to Stop 7, Marloes Sands (51.723438, −5.214627), for a total of 12.5 km (7.7 mi; 17 min). There is ample pay parking in the National Trust lot, £3 for 3 h. Take the path south down to the beach.

STOP 7 MARLOES SANDS

Marloes Sands is a remote sandy beach southwest of the village of Marloes. The beach is known for The Three Chimneys, three vertical beds of hard Silurian sandstone rising to the full height of the sea cliffs. The area is managed by the National Trust, with a small office at the cliff-top car park. There is a nominal parking charge payable to the National Trust. Information on the geology of the area is available (Discovering Fossils, Marloes Sands; Cruickshanks, 2017, Marloes Sands; Wikipedia, Marloes Sands).

Access to the beach is along a footpath. From the car park, walk about 240 m (800 ft) east along the road to a sign pointing to "Beach 700 m" and take the path south from there.

The beach cliffs at Marloes Sands expose four Silurian–Devonian units. From oldest to youngest they are the Silurian Skomer Volcanic Group, the Silurian Coralliferous Group, the Silurian Gray Sandstone, and the Silurian–Devonian Old Red Sandstone. Head south from the beach access and you will pass through each group in turn, until at the south end of the bay you are in the Old Red Sandstone.

The Skomar Volcanic Group forms the cliffs nearest to the beach access. The Skomer Group, roughly 1,000 m (3,000 ft) thick, consists of subaerial basalt and rhyolite lava flows with volcanic sediments containing breccias, tuffs, conglomerates, quartzites, and red claystone. These formed when a group of volcanic islands produced lava flows, ash, and rock fragments. Sediment was washed down from the volcanoes and spread out to form layers of shallow marine sands and muds. The Skomer Volcanic Group was tilted about 5° before being re-submerged. A few million years later, the Coralliferous Group was deposited unconformably above it.

Tilted Silurian-Early Devonian strata, Marloes Sands. View northwest.

Although the older parts of the Coralliferous Formation contain river-derived sediments, most of the unit was deposited in shallow, warm subtropical seas. The "Three Chimneys" are part of this unit. Mudstones in the unit indicate an estuary environment. The Coralliferous Group, as its name implies, contains most of the fossils in the area. Shallow marine fossils in this unit include brachiopods, gastropods (snails), and crinoids, commonly in the form of beds of skeletal "hash." Burrows, including *Planolites* and *Skolithos*, are common. At times, the water deepened to levels below fair-weather wave base. Shell beds (coquinas) are common, with some layers rich in brachiopods and solitary corals.

The Gray Sandstone Group lies conformably on the Coralliferous Group and is considered middle Silurian. In places, it is out of stratigraphic sequence because of faulting, but it is nicely exposed. It is mainly sandstones, siltstones, and mudstones that were deposited in a shallow marine, near-shore environment. Sand and mud accumulated during periods of high sea level; during periods of low sea level, the emergent region contained estuaries, mudflats, and coastal plains incised by tidal channels. The sands are commonly cross-bedded and ripple-marked. The Gray Sandstone is transitional with the overlying Red Cliff Formation of the Old Red Sandstone (Fitches, 2011; Daniel and Roberts, 2011).

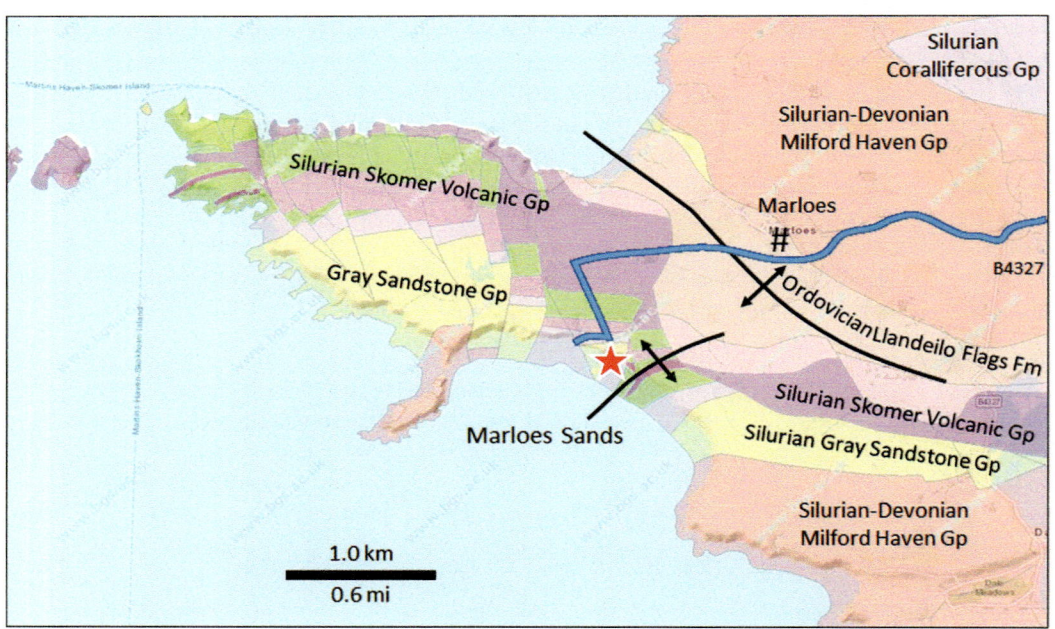

Geologic map of the Marloes Sands area. (Modified after BGS, Geology of Britain Viewer Classic. Contains British Geological Survey materials © UKRI 2022. Base mapping provided by ESRI.)

View southeast at south dipping Silurian units at Marloes Sands. From left to right, the Skomer Volcanic Group, Coralliferous Group, Gray Sandstone Group, and Old Red Sandstone.

At the south end of the bay are outcrops of the Old Red Sandstone. The Lower Old Red Sandstone in Pembrokeshire ranges from Upper Silurian to Lower Devonian in age. Lower parts of the succession seen at Marloes Sands are part of the Milford Haven Group, which contains the Upper Silurian Red Cliff Formation at the base. The Red Cliff consists of mudstones with fossil soil horizons interbedded with sandstones deposited by flash floods on an alluvial plain. The region was fully emergent and semi-arid to arid, and the sediment source was in the rising Variscan Mountains to the south. The overlying Silurian Albion Sandstone Formation is also part of the Milford Haven Group. It is primarily thick-bedded yellow sandstone but has interbedded conglomerate and red mudstone layers, probably deposited by rivers during flash floods (Daniel and Roberts, 2011; Fitches, 2011).

This part of Pembrokeshire was along the northern margin of the Variscan deformation belt, a roughly east–west mountain range across Wales and the United Kingdom and into Europe that was caused by a continental collision in Late Carboniferous to Early Permian time. The Silurian strata exposed here are on the south flank of the large Marloes Anticline. The strata strike roughly east–west, are nearly vertical, and become younger to the south. The Variscan Orogeny imposed a pervasive cleavage on these units (Fitches, 2011).

The best places to find fossils are in loose rock that collects at the base of the cliffs, and the best time is during a falling or low tide. The most common fossils are corals, brachiopods, and other bivalves, but trilobites and cephalopods (mollusks that include the nautilus and squid) have also been found. Corals indicate shallow marine conditions of clear, generally warm water. Corals are ubiquitous, weathering mainly out of the Coralliferous Group. The most productive fossil areas are southeast of the beach access. Head south when you reach the beach and stop when you see another path that descends downwards from the cliff top. A second site is perhaps 300 m (1,000 ft) farther southeast in an area where the layers are near-vertical and consist of a flaky shale (Cruickshanks, 2017, Marloes Sands; Discovering Fossils, Marloes Sands).

This location is an SSSI, so you can collect loose fossils, but you cannot hammer them out of the outcrop.

Marloes Sands to Stackpole Quay: Return east to Marloes and turn right (east) toward Dale/ Hwlffordd/ Haverfordwest (B4327); continue east to B4327 and turn left (east) on B4327 toward Hwlffordd/ Haverfordwest; continue straight on Dale Marloes; continue straight on Dale Road; turn left (east) at sign to Liddeston and drive east on an unnamed road; after the tunnel make a slight left (northeast) onto Cromwell Road; continue east on Thornton Road; continue straight on Neyland Road; continue straight on A477 to Waterloo; turn right (south) on Ferry Lane/A4139 and drive to Pembroke; in Pembroke stay on A4139 as it becomes Main Street, then Well Hill; turn left (south) onto St Daniel's Hill/B4319; continue south on B4319; turn left (east) at the sign to Stackpole; on the outskirts of Stackpole turn left (east) and follow the sign to Stackpole Quay/Barafundle; turn right (south) at the sign to Stackpole Quay/Barafundle, and follow this unnamed road to the parking area at **Stop 8.1, Stackpole Quay** *(51.624634, −4.901038) for a total of 36.4 km (22.6 mi; 43 min). The car park at Stackpole Quay charges £5 per day between March and October (parking is free after 17:30). Take the path about 120 m (400 ft) north over a small rise to see the syncline.*

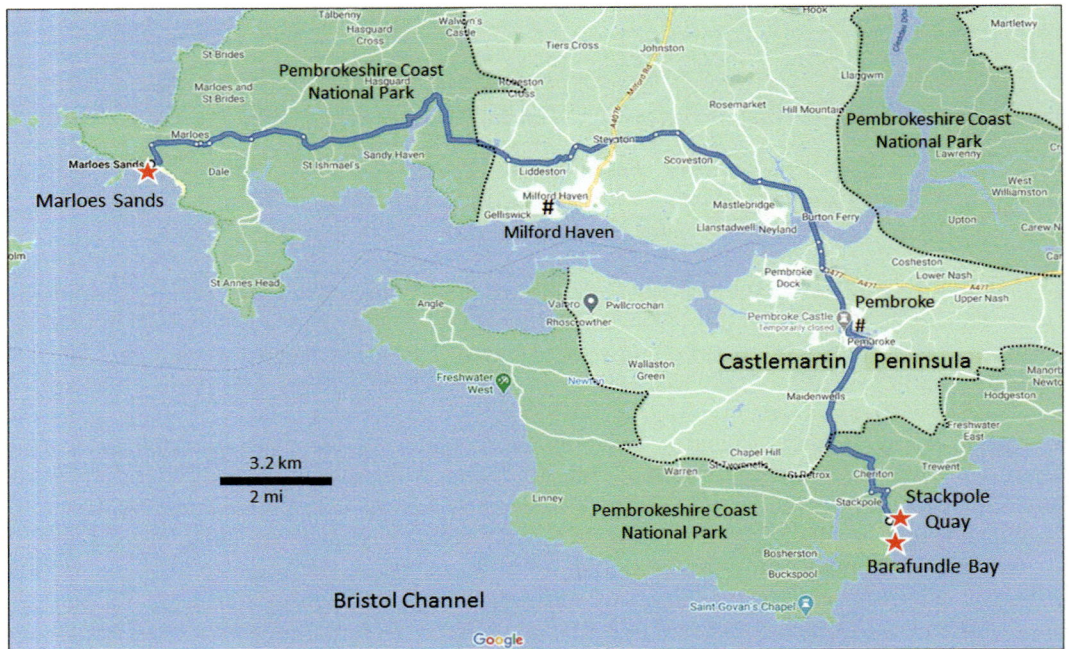

STOP 8 CASTLEMARTIN PENINSULA

With its dramatic limestone cliffs and stunning white-sand beaches, the Castlemartin Peninsula is one of the most scenic areas along the Pembrokeshire Coast.

8.1 STACKPOLE QUAY

Stackpole Quay is on the southwest coast of Wales, along the Pembrokeshire Coast Path, and within Pembrokeshire Coast National Park. The rocks here are folded and faulted Lower Carboniferous (Mississippian) limestones of the Pembroke Limestone Group. The limestone was deposited in a shallow marine setting between 359 and 329 Ma (BGS, Geology of Britain Viewer Classic). This area is within but close to the northern limit of Variscan deformation (Cawood et al., 2017). Fold axes trend west-northwest. The axis of the Stackpole Anticline is located at the slipway; a small islet a short distance to the north contains the axis of the Stackpole Quay Syncline. The Stackpole Fault, an east–west tear fault (a strike-slip fault that cuts a thrust sheet) with about 100 m (328 ft) of right-lateral offset, can be seen in the old quarry face (Earthwatcher, Stackpole Quay).

Stackpole Quay was designated a SSSI in 1977 in order to protect its biological and geological features. The folds and faults exposed here are typical of and provide key insights into the overall structure of the Variscan deformed belt in Pembrokeshire (Wikipedia, Trewent Point).

Small syncline in Pembroke Limestone Group, Stackpole Quay.

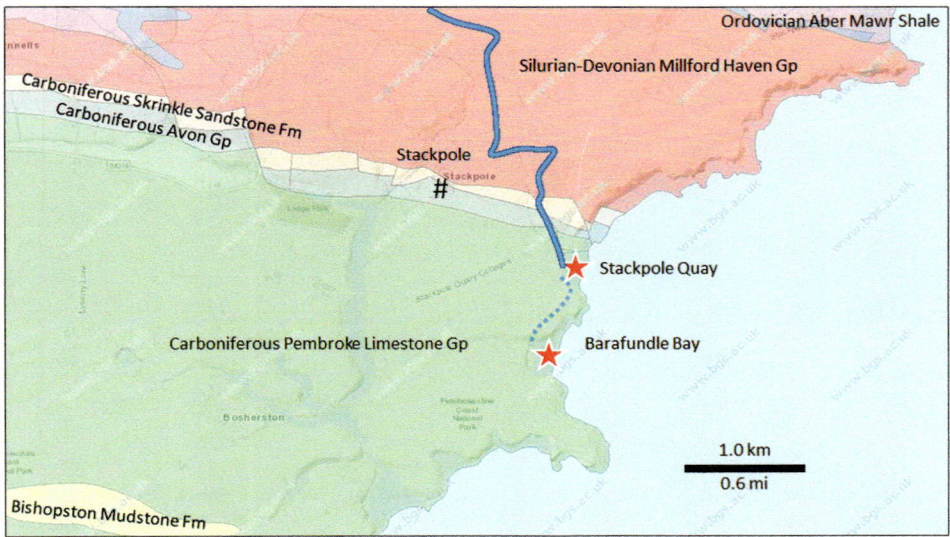

Geologic map of the Stackpole area. (Modified after BGS, Geology of Britain Viewer Classic. Contains British Geological Survey materials © UKRI 2022. Base mapping provided by ESRI.)

*Stackpole Quay to Barafundle Bay: From the Stackpole Quay car park, follow the Pembrokeshire Coast Path 800 m (2,600 ft; 15 min) south along the coast to **Stop 8.2, Barafundle Bay** (51.618060, −4.904144).*

8.2 BARAFUNDLE BAY

Barafundle Bay is a remote sandy beach 0.8 km (0.5 mi) south of Stackpole Quay and 2.1 km (1.3 mi) southeast of Bosherston. Beach access is via the Pembrokeshire Coast Path from Stackpole Quay. Alternatively, you can return to Stackpole, turn south on B4319, and drive to Bosherton and Trefalen car park. The beach is managed by the National Trust (Wikipedia, Barafundle Bay).

Barafundle Beach is set between cliffs of Stackpole Limestone Formation, part of the Pembroke Limestone Group. Breccias on the north wall of the bay include slickensides, polished shear surfaces indicating you are at or near a fault (Butler, Barafundle Bay, and Stackpole Quay).

Stackpole Limestone Formation is a highly-jointed, light gray shelly and oolitic limestone (a limestone made up of small round carbonate grains that precipitated on a shallow limestone bank). The lower boundary is at the base of the first oolitic limestone above the underlying Pen-y-holt Limestone Formation. The upper boundary is above the light gray limy mudstones and below the shell-rich Oxwich Head Limestone Formation. It is 260 m (850 ft) thick at the type locality along the sea cliffs between Stackpole Quay and Barafundle Bay (BGS, Lexicon of Named Rock Units, Stackpole Limestone).

Barafundle Beach. Cliffs are Carboniferous Stackpole Limestone. (Photo courtesy of JKMMX, https://commons.wikimedia.org/wiki/File:BarafundleBeach_StackpoleEstate_WalesUK.jpg.)

In 2004, the magnificent white sands beach at Barafundle Bay was listed among the "Top 12 Beaches in the World" (Wikipedia, Barafundle Bay).

Barafundle Bay to Ogmore-by-Sea: From the Stackpole Quay car park return north to Stackpole and Pembroke; in Pembroke turn right (east) onto Orange Way/A4139; continue on Well Hill/A4139;

*at the roundabout take the 2nd exit onto Holyland Road/A4075 and drive northeast to the A477; turn right (east) onto the A477 and drive to St Clears; at the roundabout, take the 3rd exit onto A40 and drive east to Carmarthen; at the roundabout, take the 2nd exit onto the A48; continue east on the A48 to the M4; at the roundabout, take the 3rd exit to Cardiff/Swansea and drive south on the M4 to North Cornelly; at Junction 37 take the A4229 exit to Pile/Porthcawl; at the roundabout take the 3rd exit onto A48; at the sign for Merthyr Mawr turn right (southeast) and take New Inn Road to Ewenny; in Ewenny turn right (south) onto Ewenny Road/B4265; turn right (west) onto Ogmore Road/B4524 and drive to the car park at roads end. This is **Stop 9.1, Ogmore-by-Sea** (51.463597, −3.640279), for a total of 138 km (85.5 mi; 1 h 47 min). Park free on the street, or £1/1 h in the car park.*

*   **Barafundle Bay to Side Trip 3, Dan-yr-Ogof Caves**: From the Stackpole Quay car park return north to Stackpole and Pembroke; in Pembroke turn right (east) onto Orange Way/A4139; continue on Well Hill/A4139; at the roundabout take the 2nd exit onto Holyland Road/A4075 and drive northeast to the A477; turn right (east) onto the A477 and drive to St Clears; at the roundabout, take the 3rd exit onto A40 and drive east to Carmarthen; at the roundabout, take the 2nd exit onto the A48; continue east on the A48 to the M4; at the roundabout, take the 3rd exit to Cardiff/Swansea and drive south on the M4 to Junction 45; take the A4067 exit to Swansea and drive northeast to the Dan-yr-Ogof turnoff and parking on the left. This is **Sidetrip 2, Dan-yr-Ogof Showcaves** (51.830963, −3.685029) for a total of 124 km (76.9 mi; 1 h 36 min).*

SIDE TRIP 3 DAN-YR-OGOF CAVES AND BRECON BEACONS NATIONAL PARK

Brecon Beacons National Park was established in 1957 and covers 1,340 km² (519 mi²). The Brecon Beacons include the highest peak in South Wales at Pen y Fan at 886 m (2,907 ft). The western half of the national park became the UNESCO Fforest Fawr Global Geopark in 2005. That same year the 161 km (100-mi) Beacons Way was inaugurated. The trail spans the park from Abergavenny in the east to Llangadog in the west. The entire national park became an International Dark Sky

Reserve in 2013. Most of the national park is grassy moorland grazed by mountain ponies and sheep. The relative remoteness and harsh weather of its uplands make this park ideal for Special Forces training (Wikipedia, Brecon Beacons National Park).

Rocks in the national park are a succession of Late Ordovician through Late Carboniferous strata. Most of the peaks consist of the Late Silurian–Devonian Old Red Sandstone. The park sits at the northern edge of the South Wales Coal Basin which accumulated sediments throughout the Carboniferous.

The area was at the southern margin of the Caledonian Orogen and at the northern margin of the later Variscan Orogen. The older units, in the northwest, were folded and faulted during the Caledonian Orogeny. Renewed faulting and folding, mainly in the south, occurred during the Variscan Orogeny.

Glaciers probably covered most of the park at the peak of the last glaciation around 22,000 years ago. Glacial erratics are common, the most obvious being Old Red Sandstone boulders sitting on limestone pavements (Wikipedia, Geology of Brecon Beacons National Park).

Heavy rainfall and soluble limestone have led to a karst landscape. A large number of caves have developed, among them some of the longest in Britain. There are three notable caverns in the cave system here: Dan-yr-Ogof, Cathedral Cave, and Bone Cave.

The river Llynfell emerges from the mountain as you approach the entrance to Dan-yr-Ogof. At this point, it has been underground for more than 6 km (4 mi). Dan-yr-Ogof is a cave system in Brecon Beacons National Park, the third National Park on this tour. It is the main cavern of a show cave complex. Only the first section of the caves is open to the public. In a 2005 poll of Radio Times readers, Dan-yr-Ogof was named as the greatest natural wonder in Britain.

The cave was first explored in 1912 by Edwin, Tommy, and Jeff Morgan. Using candles and primitive equipment they were eventually stopped by a narrow passage known as the Long Crawl. This constriction was eventually passed by Eileen Davies in 1963. Since then, exploration has expanded steadily until a cave system with over 16 km (10 mi) of explored passageways has been mapped.

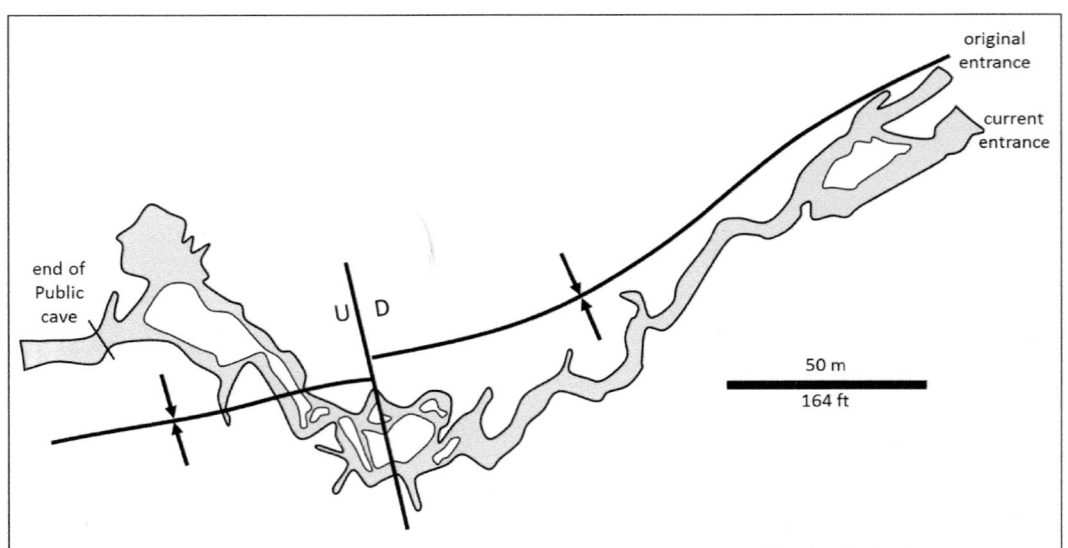

Map of the Dan-yr-Ogof cavern system showing the synclinal axis and fault. (Modified after Ball and Faulkner, 2011.)

Dan-yr-Ogof Showcave is located in the lower part of the Dowlais Limestone Formation (laterally equivalent to the Stackpole Limestone seen at Barafundle Bay). The cavern developed along a faulted, gently southwest-plunging syncline. Part of the Carboniferous Pembroke Limestone Group, the Dowlais Limestone is a thick-bedded, medium to dark gray, bituminous, and fossiliferous limestone with dark shale interbeds. The lowermost beds are dark bituminous limestone containing the coral *Lithostrotion martini* (Ball and Faulkner, 2011; BGS, Lexicon of Named Rock Units, Dowlais Limestone Formation).

The major geological feature in the area is the Cribarth Disturbance (also called the Swansea Valley Disturbance), a lineament comprising a northeast-trending fold and fault system that extends from near the coast at Swansea, along the Tawe Valley, past Cribarth and northeast to Brecon. Where exposed in the coal field to the southwest, movement indicates a left-lateral fault with about 400 m (1,300 ft) displacement. In the Dan-yr-Ogof area, some of the displacement is transferred to folding. The Cribarth Disturbance was active during the Variscan Orogeny. The 50 km (30 mi) long zone is responsible for the Swansea and Tawe valleys, both enhanced by glacial erosion along the pre-existing line of weakness (Wikipedia, Cribarth Disturbance).

Up-slope in younger limestone lies Cathedral Cave. This chamber is developed in well-bedded limestones of the middle beds of the Dowlais Limestone Formation. The main room is 7 m (23 ft) wide. Cathedral Cave is well-lit and has an artificial waterfall. There most likely once was a natural stream flowing over the falls. A 1.5-mi self-guided tour winds through Cathedral Cave, a high-domed chamber with a lake fed by two waterfalls that pour from openings in the rock (Ball and Faulkner, 2011).

Dan-yr-Ogof cavern. (Photo courtesy of Nilfanion, https://commons.wikimedia.org/wiki/File:Dan_yr_Ogof_caves_(8949).jpg.)

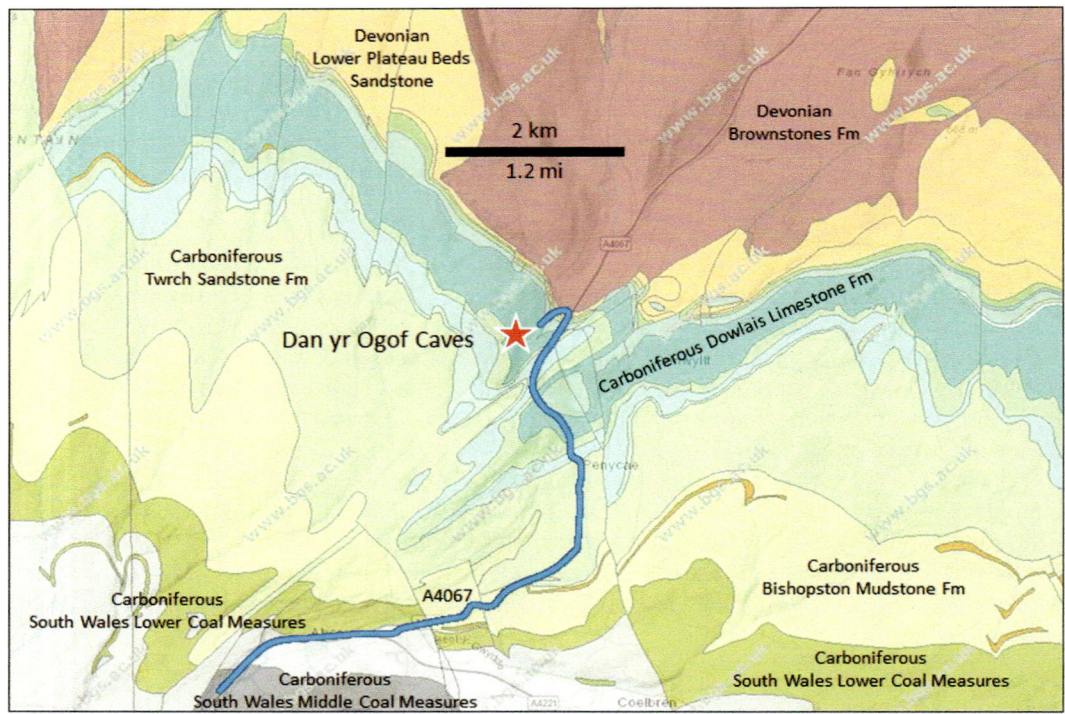

Geologic map of Dan-yr-Ogof caves area. (Modified after BGS, Geology of Britain Viewer Classic. Contains British Geological Survey materials © UKRI 2022. Base mapping provided by ESRI.)

The bones of some 42 Bronze Age humans, as well as numerous animal bones, were found in one of the chambers of the cave system known as Bone Cave (Lonely Planet, Dan-yr-Ogof National Showcaves Centre for Wales; Wikipedia, Dan-yr-Ogof).

Across Tawe Valley to the east is Ogof Ffynnon Ddu (OFD, Cave of the Black Spring). The Cwm Dwr quarry is just below the cave entrance. In the quarry, you can see that folding caused jointing of the limestone that later became preferred locations for dissolution and cave development. There is no connection between OFD and the Dan-yr-Ogof complex, as they developed on opposite sides of the Cribarth Disturbance. OFD is the deepest and third-longest cave system in the United Kingdom (308 m, or 1,000 ft deep, with 48 km, or 30 mi of passages). This cavern is for expert cavers only (Lonely Planet, Dan-yr-Ogof National Showcaves Centre for Wales; Ball and Faulkner, 2011).

Visit

The complex is just off the A4067, north of Abercraf.

The showcave paths are graded and wide. Self-guided tours are available. There is a shop, facilities, a petting zoo, recreated Iron Age farm, life-size dinosaur models, and other non-geologic attractions.

Entrance Fees: Adults £18; Children (ages 3–16) £15; Children aged 2 and under £8.50.
Address: The National Showcaves Centre for Wales, Abercrave, Swansea SA9 1GJ
Web site: https://www.showcaves.co.uk/prices_opening.html
Phone: 01639 730 284
Email: info@showcaves.co.uk
Please contact the Showcaves for hours and entry charges.

Side Trip 2, Dan-yr-Ogof Caves to Ogmore-by-Sea: *Return south on A4067 toward Swansea; at the Ynysforgan Interchange, take the 1st exit left (east) onto the M4 to Cardiff; drive east and south on the M4 to North Cornelly; at Junction 37 take the A4229 exit to Pile/Porthcawl; at the roundabout take the 3rd exit onto A48; at the sign for Merthyr Mawr turn right (southeast) and take New Inn Road to Ewenny; in Ewenny turn right (south) onto Ewenny Road/B4265; turn right (west) onto Ogmore Road/B4524 and drive to the car park at roads end. This is **Stop 9.1, Ogmore-by-Sea** (51.463597, −3.640279) for a total of 66.3 km (41.5 mi; 1 h 9 min). Park free on the street, or £1/1 h in car park.*

STOP 9 GLAMORGAN HERITAGE COAST

The Glamorgan Heritage Coast is a 23 km (14 mi) section of coastline in the Vale of Glamorgan between Ogmore-by-Sea and St Athan. The coast includes the Southerndown Coast SSSI, notable for its sea cliffs of Triassic marl and Carboniferous limestone.

Note the difference between Carboniferous limestone and Carboniferous Limestone. The first indicates the age of a limestone; the second is a Series, or Supergroup that in South Wales includes the Pembroke Limestone Group and Avon Group limestones.

These mostly gently-dipping rocks were subject to low-grade regional metamorphism but were not extensively deformed during Variscan mountain building (Neighbor, 2009; Wikipedia, Glamorgan Heritage Coast).

9.1 Ogmore-by-Sea

At this stop, you can examine Lower Carboniferous limestones, identify different facies of Triassic alluvial fan deposits, and see Jurassic limestones. The beach here is also known as The Flats, or Sutton Flats. As always, access is best at low tide.

The Lower Carboniferous limestones were deposited in warm, shallow seas. The Friars Point Formation is a massive fossiliferous limestone; the overlying Gully Oolite is an oolitic limestone.

The first cliffs south of the beach are red breccias and conglomerates of the Triassic Mercia Mudstone Group. Clearly, then, a large unconformity exists between Carboniferous and Triassic rocks, with much of the Upper Carboniferous and all of the Permian and Lower Triassic sections eroded away. The mainly Permian Variscan Orogeny removed an estimated 4,500 m (14,760 ft) of rock in South Wales. The Triassic breccias are alluvial fans and debris flows deposited around limestone hills in a desert environment. The breccia consists largely of Carboniferous limestone fragments in a red sandy matrix. A few meters further south and east there is a second large alluvial fan breccia composed of smaller limestone fragments in gray Triassic limy mudstone. The fan is extensive and rests on the Lower Carboniferous High Tor Limestone, which dips gently south. Finer-grained sandstones are typical of sediments deposited from sheet floods and are found in the more distal parts of alluvial fans.

Carboniferous limestone beach pavement lies beneath an unconformity, with the red-brown Triassic Mercia Mudstone conglomerate above the unconformity, Ogmore-by-Sea beach.

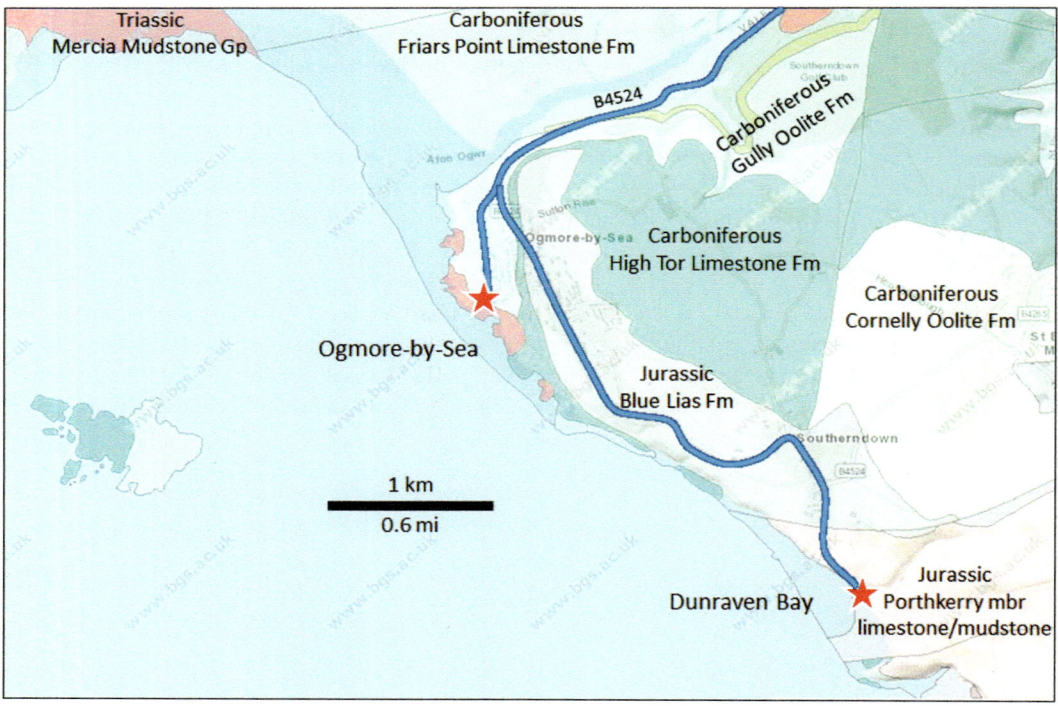

Geologic map of the Ogmore-by-Sea area. (Modified after BGS, Geology of Britain Viewer Classic. Contains British Geological Survey materials © UKRI 2022. Base mapping provided by ESRI.)

Age   Unit and Lithology

Stratigraphy, Ogmore-by-Sea. (Information drawn from Neighbor, 2009.)

Continuing south the High Tor Limestone is a shelly limestone interbedded with thin shales. The limestone was deposited in a nearshore environment: the lower beds contain coral colonies, whereas the upper layers have the solitary coral *Caninia*. Other fossils, including algal mats, brachiopods, and burrows are common.

Farther south, the High Tor Limestone is unconformably overlain by the Jurassic Blue Lias, a shelly limestone and marl also deposited in a warm, shallow sea.

This site is an SSSI because the sediments preserved here give an excellent example of rock character changes within alluvial fan complexes (Neighbor, 2009; Hermolle, 2009).

*Ogmore-by-Sea to Dunraven Bay: Return north to Main Road/B4524 and turn right (southeast); drive 3.1 km (1.9 mi) and turn right (south) onto Beach Road; drive to the parking area at **Stop 9.2, Dunraven Bay** (51.446118, −3.605438) for a total of 4.8 km (3.0 mi; 10 min). There is ample free parking and an easy walk along the beach to see the fault at the south end. If you have time, walk 100 m south of the parking area to see the "walled garden."*

9.2 DUNRAVEN BAY AND HERITAGE COAST CENTRE

Dunraven Bay is bounded on both sides by faults. On the west side of the beach, the Jurassic Southerndown Beds (an oolitic limestone and sandy, limy conglomerate) are interbedded with soft shales. The Southerndown Beds were formed close to the shore during the Early Jurassic and are rich in fossils, including numerous *Gryphaea* (an extinct variety of oysters) in life position. These beds grade upward into the mudstones and shaley limestones of the Blue Lias. The Blue Lias contains corals and brachiopods, the bivalve *Gryphaea arcuata* (Devils Toenail), and some ammonites. The Seamouth Limestone is at the cliff top. The western cliffs are part of a broad ancline that can be seen from the beach.

Dunraven Bay. Jurassic units on the left are separated from Carboniferous units on the right by the Dunraven Fault. View southeast.

Folding along the Dunraven reverse fault. View south.

Bedding on the east side of the bay has been faulted and folded along the Dunraven Fault, a reverse fault that places Carboniferous High Tor Limestone over Jurassic Seamouth Limestone. The Dunraven Fault may have originated during the Late Carboniferous-Early Permian Variscan (290 Ma) Orogeny, but was reactivated and inverted during the Eocene-Miocene Alpine mountain building event (Huff, 2019). Offset and deformation decrease upward in the cliff. The Jurassic lies over the Carboniferous limestone on the north limb of an asymmetrical ancline that can be seen in the headland.

Sea level rise at the beginning of the Jurassic drowned the Triassic desert. Suon Stone was deposited in shallow water while farther out Southerndown Beds accumulated. These units are overlain by alternating limestones, limy muds, and shales of the Blue Lias (Hermolle, 2009; Neighbor, 2009).

Visit

From Dunraven Beach walk or drive 180 m (590 ft) northeast to the Glamorgan Coast Heritage Centre on the right. The Centre has a small geology display and a shop with geology guides. At the time of writing, the Centre is open by appointment only (group visits). Please phone prior to visiting to see if you need an appointment.

Address: Heritage Coast Centre, 2 Beach Rd, Southerndown, Bridgend CF32 0RP, United Kingdom (51.446805, −3.603553)
Phone: 01656 880157
Web site: https://www.visitthevale.com/en/Be-Inspired/History-and-Mystery/Glamorgan-Heritage-Coast-Centre.aspx

*Dunraven Bay to Llantwit Major Beach: Turn right (southeast) onto Beach Road and drive to the B4265; turn right (southeast) onto the B4265 and drive to Llantwit Major; at the roundabout, take the 3rd exit (south) onto Cowbridge Road; at the roundabout, continue straight onto High Street; turn right (west) onto Church Street; turn left (south) onto Colhugh Street; follow the signs to Llantwit Beach parking. This is **Stop 9.3, Llantwit Major Beach** (51.396606, −3.500708) for a total of 13.2 km (8.2 mi; 21 min). There is ample free beach parking.*

9.3 LLANTWIT MAJOR BEACH

Llantwit Major is considered one of the best spots in Wales for Jurassic fossils. Fossils found here include sponges, corals, and echinoids (sea urchin-like creatures). Large 12 cm (5 in) gastropods and 7 cm (3 in) *gryphaea* have been found in the dark shale interbeds. Bones of fish and Ichthyosaurus also occur.

Gryphaea, or Devil's Toenails, is an extinct genus of the oyster family. Found at Llantwit Major beach.

Walk east to find fossils in slabs that have fallen off the cliffs and in boulders tumbled by the waves. The cliffs are too dangerous to collect from, as rock falls are common, so please stay back. The best area begins about 1 km (0.5 mi) east of the café.

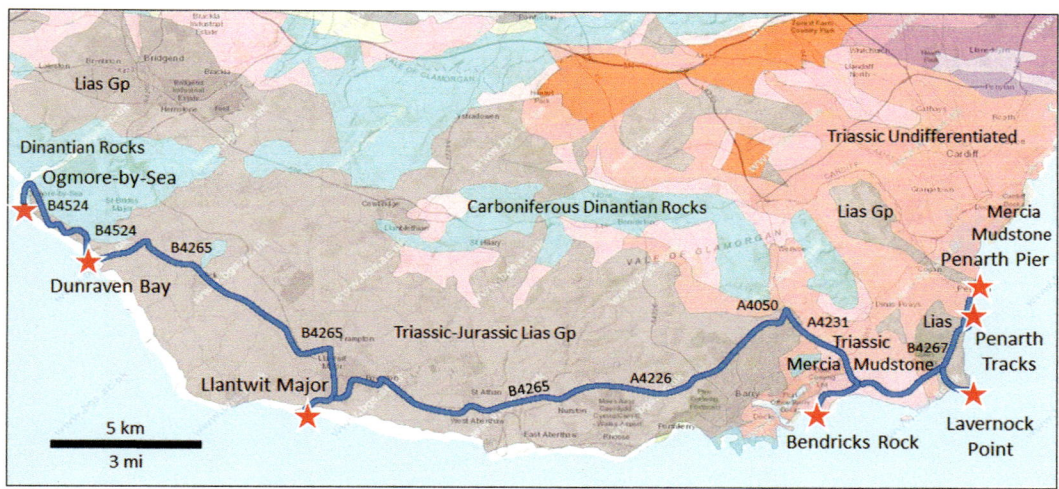

Geologic map of the Llantwit Major to Penarth coast area. (Modified after BGS, Geology of Britain Viewer Classic. Contains British Geological Survey materials © UKRI 2022. Base mapping provided by ESRI.)

Llantwit Major Beach. Cliffs are the Jurassic Porthkerry member of the Blue Lias.

Beach pavement and jointing in Blue Lias limestone, Llantwit Major.

The cliffs consist of alternating layers of limestone and thinner shales. The rocks are part of the Lower Jurassic Porthkerry member of the Blue Lias.

Follow the main road through town heading toward the beach. There is a large car park, café, toilets, and extensive views up and down the coast.

There are no restrictions at this site, so hammering and collecting are permitted (Williams et al., 1996; Rodell, 2007; Cruickshanks, 2017, Llantwit Major; BGS, Geology of Britain Viewer Classic).

Llantwit Major to Bendricks Rock*: Return northeast; turn left onto Colhugh Street; at the roundabout, take the 2nd exit (east) onto Mill Lay Lane; continue straight onto Ham Lane East; turn right (east) onto Boverton Road and drive to B4265; turn right (east) on B4265; at the roundabout, take the 2nd exit (northeast) to A4226; at the roundabout, take the 1st exit (northeast) onto Port Road East/A4050; at the roundabout, take the 2nd exit (southeast) onto A4231; at the roundabout, continue straight onto Sully Moors Road/B4267; at the roundabout, take the 2nd exit (southwest) onto Hayes Road; at the roundabout, take the 1st exit (south) toward Caolfan/ Waste Recycling Centre; drive to **Stop 10.1, Bendricks Rock** (51.395655, −3.246876) parking at the Recycling Centre for a total of 25.5 km (15.8 mi; 30 min). You can reach the tracks by taking a path outside of the security fence around HMS Cambria at Hayes Point or by following the coast path west from the public slip at the Vale of Glamorgan recycling center.*

STOP 10 SEVERN COAST SOUTH OF CARDIFF

Outcrops on either side of the Severn estuary are world-famous sites for hunting Triassic and Jurassic fossils. Shells, corals, and dinosaur bones and teeth can all be found here. Also, because dinosaurs traversed the coastal plain and tidal flats as this area subsided from a continental to a marine environment, dinosaur trackways were quickly buried and preserved.

10.1 BENDRICKS ROCK DINOSAUR TRACKS

Whereas Bendricks Rock is a tidal promontory consisting of Carboniferous Limestone, the tracks that make this site special are found in the Triassic red beds of the overlying Mercia Mudstone Group.

Bendrick Rock is one of the United Kingdom's most important coastal geological localities and is listed both as an SSSI and a Geological Conservation Review Site (GCR). The GCR listing is based on the stratigraphy and the superb early dinosaur tracks. The strata are nearly flat-lying 220 Ma Triassic red beds deposited by flash floods and sheet floods in a desert setting. The exposure consists of outcrops of the Upper Triassic Mercia Mudstone where the strata are mainly fine red mudstones, marls, ripple-marked sandstones, and occasional conglomerates that indicate an environment similar to those found in a modern sabkha (evaporative tidal flats) or playas (evaporating lakes). The mudstones preserve dinosaur prints and trackways.

The tracks were made by several different dinosaurs. Small three-toed footprints were probably made by a 1 m (3 ft) tall theropod dinosaur such as the predator *Coelophysis*. Larger three-toed theropod prints are less common. Theropods walked upright on their hind legs. Prints left by large four-toed dinosaurs are also found; these probably were made by a 7–8 m (23–26 ft) long, plant eater like *Plateosaurus* that walked on all fours. These dinosaurs were early sauropods.

The footprints can be difficult to find. Many are covered at high tide so it is easier to see them at low tide when the tracks contain small puddles of water. It is also easier to spot the footprints when the sun is low in the sky, as longer shadows throw the footprints into relief.

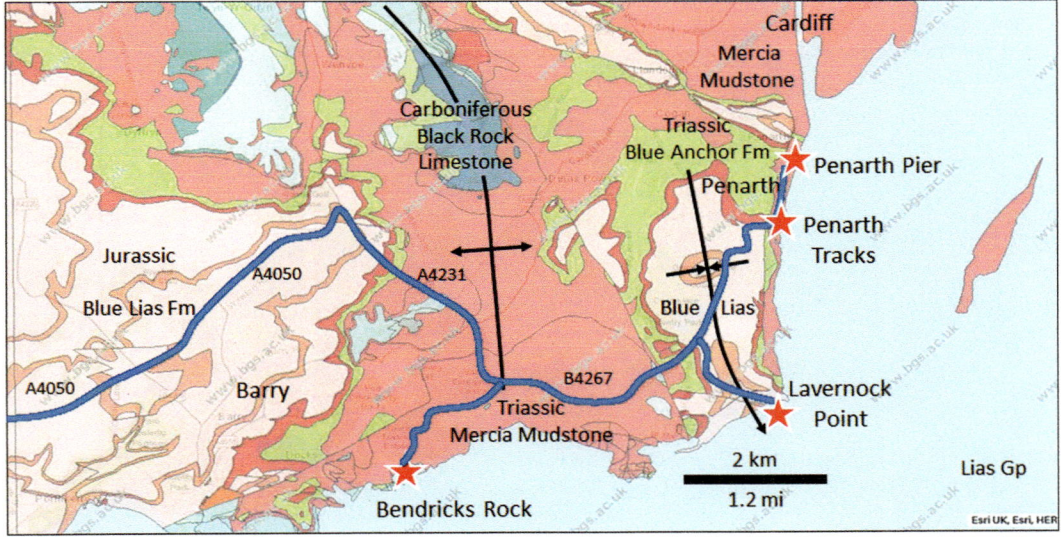

Geologic map of the Bendricks Rock to Penarth Pier area. (Modified after BGS, Geology of Britain Viewer Classic. Contains British Geological Survey materials © UKRI 2022. Base mapping provided by ESRI.)

Dinosaur tracks in Mercia Mudstone at Bendricks Rock. They are damp after a rain.

Possible three-toed theropod print, Bendricks Rock.

These tracks are protected. Do not collect or damage fossil material at this site. In 2005, a large area of the footprint site was dug up and removed illegally. The vandals were caught and prosecuted.

Some of the tracks were excavated and are on display at the National Museum and Galleries of Wales in Cardiff.

You can reach the tracks by a path that follows the outside of a security fence around HMS *Cambria* at Hayes Point, or by following the coast path east from the public slip at the Vale of Glamorgan recycling center (Sharpe, 2007; Evans, Dinosaur Footprints, Bendrick Rock; Wikipedia, The Bendricks).

Bendricks Rock to Lavernock Point: *Return east and north; at the roundabout take the 3rd exit (east) toward Cardiff and the M4; at the roundabout take the 3rd exit (east) onto Hayes Road; at the roundabout take the 2nd exit (east) onto South Road/B4267 and drive to Fort Road; turn right (southeast) onto Fort Road and drive to **Stop 10.2, Lavernock** (51.40686, −3.17026), for a total of 6.9km (4.3 mi; 11 min). There is a small free car park at the end of the road. Take the path to the right (as you face the beach), and take the steps down to the beach.*

10.2 LAVERNOCK POINT

An abundance of fossils makes Lavernock Point a great fossil-hunting site. A mixture of Jurassic and Triassic rocks is seen here. The Point is 330m (1,100ft) east of the north–south axis of a broad, gentle syncline with Jurassic rocks in the core and Triassic rocks on the flanks. At this point, the St Mary's Well Bay member of the Blue Lias Formation is exposed. This comprises interbedded limestone and mudstone deposited approximately 210–199 Ma, a period that spans the Jurassic-Triassic boundary. It was a time when this area was covered in a clear shallow sea conducive to limestone formation, yet periodically had mud washed in from the nearby mainland.

A gentle anticline is evident in the sea cliffs at Lavernock Point. Blue Lias Formation, view northeast.

Cliffs to the south and west, near the synclinal axis, contain the Lower Jurassic Blue Lias Formation, specifically the Porthkerry member (interbedded shallow marine limestone and mudstone) over the Lavernock Shales member (shallow marine mudstone), which in turn lies over the St Mary's Well Bay member (mudstone with interbedded limestone).

Cliffs to the north and east expose the Late Triassic Westbury Formation of the Penarth Group (interbedded shallow marine mudstone and limestone), over the Triassic Blue Anchor Formation (lake-deposited mudstone) over the Triassic Mercia Mudstone Group (desert mudstone; BGS, Geology of Britain Viewer Classic).

Fossils are mostly found in fallen rocks along the shore, but can also be found in the cliff face. The cliffs are generally too dangerous to collect from due to frequent rock falls. As well, this is an SSSI, so although you can collect lots of loose fossils scattered along the shore, hammering the bedrock is not permitted. Many of the best fossils are found where you first enter the beach from the steps.

Jurassic bivalves in beach rocks, Lavernock Point.

In June 2014, the remains of a new dinosaur species, *Dracoraptor*, were discovered at Lavernock Point by the Hanigan brothers. This is believed to be the earliest known Jurassic dinosaur. The skeleton belonged to a dog-sized carnivorous, three-toed theropod.

The Westbury Formation northeast of the point contains sandstone bone beds that include pyrite and the bones and teeth of fish and marine reptiles such as ichthyosaurs and plesiosaurs. Slabs of this unit can usually be found on the beach.

At Lavernock Point, the bedding planes in the limestones contain the giant oyster *Liostrea*, among others. West of the Point, the ammonite *Psiloceras* is quite common. Gastropods, bivalves, and brachiopods are also found (Cruickshanks, 2017, Lavernock).

The National Museum of Wales (www.museumwales.ac.uk) in Cardiff has an "Evolution of Wales" gallery with a reconstruction of South Wales as it was in the Early Jurassic (200 million years ago). They also have on display dinosaur skeletons and footprints, plus specimens of Jurassic marine life that swam in the Jurassic seas of South Wales (Carrier, The 5 Best Dinosaur-Hunting Sites in the United Kingdom).

Lavernock Point to Penarth*: Return northwest on Fort Road to the B4267; turn right (northeast) onto B4267 and drive to Westbourne Road; turn right (north) on Westbourne Road and drive to Forrest Road; turn right (east) onto Forrest Road and drive to The Paddocks/Plymouth Road; turn left (north) onto Plymouth Road and drive to Channel View; turn right (east) on Channel View and drive to **Stop 10.3, Penarth Dinosaur Tracks** (51.425377, −3.171063) parking at the road's end for a total of 3.2 km (2.0 mi; 7 min).*

10.3 PENARTH DINOSAUR TRACKS

Who thought dinosaur tracks could be controversial? Well, they can be. Depressions found on the beach at Penarth in 2009 were noted by French geologists who thought they were dinosaur tracks, but they were not brought to the attention of the public. Around April 2020, these depressions were "rediscovered" by a local woman, Kerry Reese, who sent photos to the Natural History Museum in London. They were described as regularly-spaced and having raised mud rims or "squelch marks," like one would expect if a large sauropod stepped on wet mud or sand.

Yet, some investigators think there may be another cause for these features. According to workers at the National Museum of Wales, these are the type of marks you would expect if concretions formed in layered gypsum, a mineral that precipitates from evaporating seawater.

There is a well-known layer of Penarth alabaster (a translucent white form of gypsum) in the cliffs of Penarth that has been used as a decorative stone. Penarth alabaster contains gypsum nodules. "The nodules look like squashed-rugby-ball-like structures that grow in and displace sediment. This sort of gypsum nodule has never been seen before and, being unprecedented, makes them intriguing to geologists" (Hartley, 2020; Jones, 2020).

At the time of writing the jury was still out on whether dinosaurs or gypsum nodules caused these marks. Check them out and come to your own conclusions.

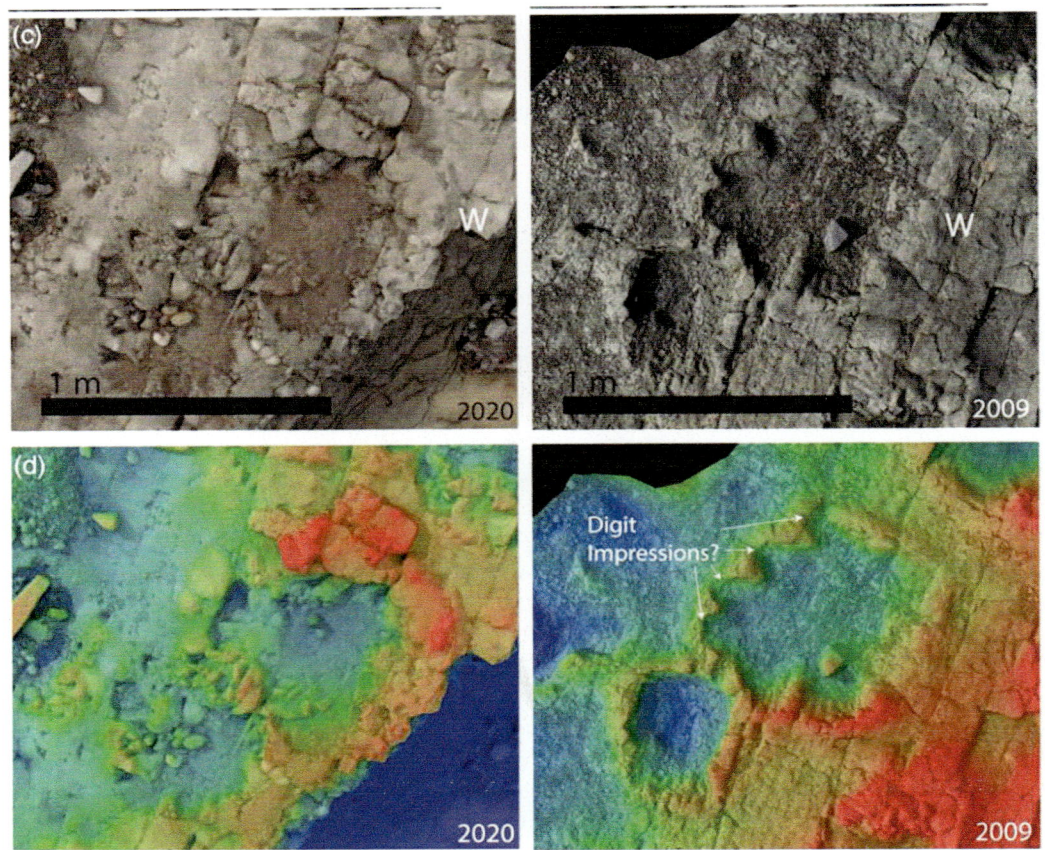

Example of the Penarth Dinosaur tracks as seen in 2009 and 2020. (c) Photo and (d) height-map images of track W as seen in 2009, and again in 2020. Height map blue to red scale indicates 10 cm. (From Falkingham et al., 2021.)

*Penarth Tracks to Penarth Pier: Return west on Channel View to Plymouth Road; turn right (north) on Plymouth Road and drive to Raisdale Road; turn right (northeast) on Raisdale Road and drive to Marine Parade; turn left (north) on Marine Parade; bear right onto Bridgeman Road; turn right (south) onto the Esplanade and park. This is **Stop 10.4, Penarth Pier** (51.434999, −3.167685), for a total of 1.5 km (1.0 mi; 4 min). From here, walk 2.4 km (1.5 mi; 45 min) along the beach to the Cardiff Barrage and back.*

10.4 PENARTH PIER TO CARDIFF BARRAGE

Penarth is the most popular location in Wales for fossil collectors, both because there is an abundance of fossils and because it is easily accessible on the outskirts of Cardiff. It is the Jurassic Coast of Wales.

There is plenty of parking along the Esplanade. Starting at Penarth Pier, walk north along the beach toward Cardiff Barrage. Early Jurassic Blue Lias Formation fossils are found all along this stretch. They are similar to fossils found across the Bristol Channel at Watchet, but there are more and larger gastropods, bivalves, and brachiopods. The ammonites tend to be smaller and less common than at Watchet.

The Triassic red cliffs are mostly the Mercia Mudstone Group and don't have fossils. At the top of this bed and below the Jurassic Lower Lias is the latest Triassic Penarth Group. It consists of mainly marine gray to black mudstone with minor limestone and sandstone beds.

This is an SSSI: you can collect loose fossils, but hammering the bedrock is not allowed (Cruickshanks, 2017, Penarth).

Sea cliffs at Penarth Head looking north to Cardiff Barage. (Photo courtesy of Gareth James, https://commons. wikimedia.org/wiki/File:Penarth_Head_-_geograph.org.uk_-_1905037.jpg.)

Penarth to Aust Cliff: *Head south along Cliff Hill to Marine Parade and turn right (north); continue straight on Bridgeman Road to The Esplanade and turn left (north); continue straight onto Beach Road; continue straight onto Windsor Terrace; at the roundabout take the first exit onto Cogan Hill/A4160; use the right two lanes to turn right (northeast) onto A4055; at the round-about take the 3rd exit onto the A4232/Cardiff Bay Link Road to City Centre; continue on A4232 to the roundabout; take the 2nd exit (east) onto Ocean Way; at the roundabout, take the 1st exit (northeast) onto Rover Way; take Rover Way to roundabout; take the 2nd exit (north) onto A4232; at the roundabout, take the 3rd exit (north) onto the A48; take the A48 to the M4 and merge onto the M4 going east; at Junction 28 take the A48 exit to Newport/Casnewydd; continue east on the A48 to the M4; at Coldra Roundabout, take the 5th exit (east) onto the M4 toward London; at Junction 23, take the M48 exit (east) to Chepstow; at Junction 1 after crossing the Severn take the A403 exit toward Avonmouth; at the Aust Interchange, take the 3rd exit (south) onto the A403; immediately turn right (west) onto New Passage Road; turn right (north) onto Aust Wharf Road and drive to* **Stop 11, Aust Cliff** *(51.602649, −2.629478), for a total of 61.3 km (38.1 mi; 1 h 1 min). Park on the street, and take the path along the shore to the bridge.*

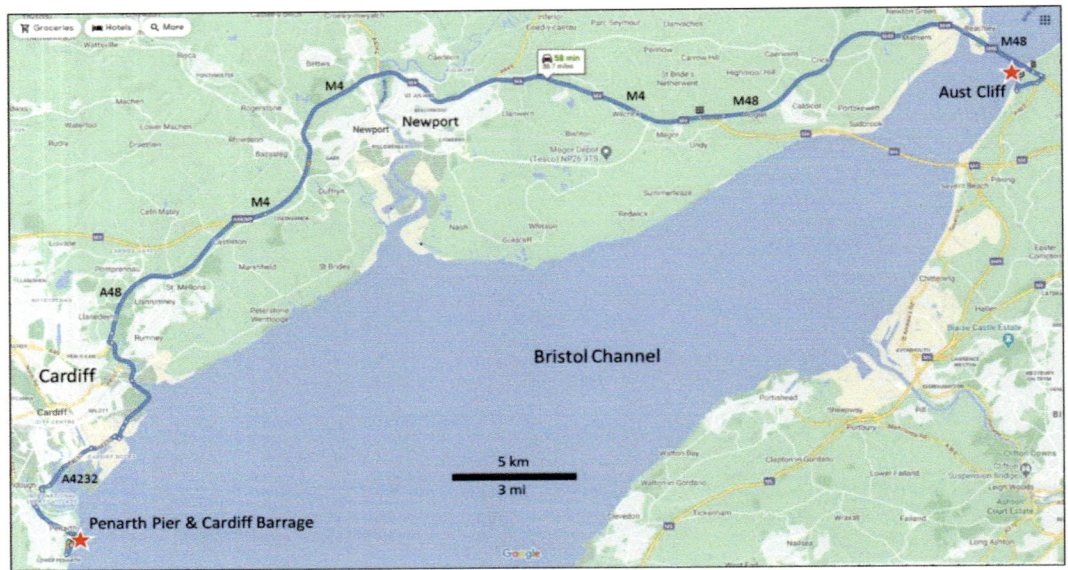

STOP 11 AUST CLIFF

Crossing the River Severn we leave behind Wales and enter South Gloucestershire, England. The road is mostly along the west flank of the Worcester Basin.

The famous red and white cliffs at the east end of the Severn Bridge are another well-known fossil collecting site. Middle and Late Triassic through Early Jurassic strata outcrop just south (downstream) of the Severn Bridge. Access to the cliffs is through a gate onto a concrete path.

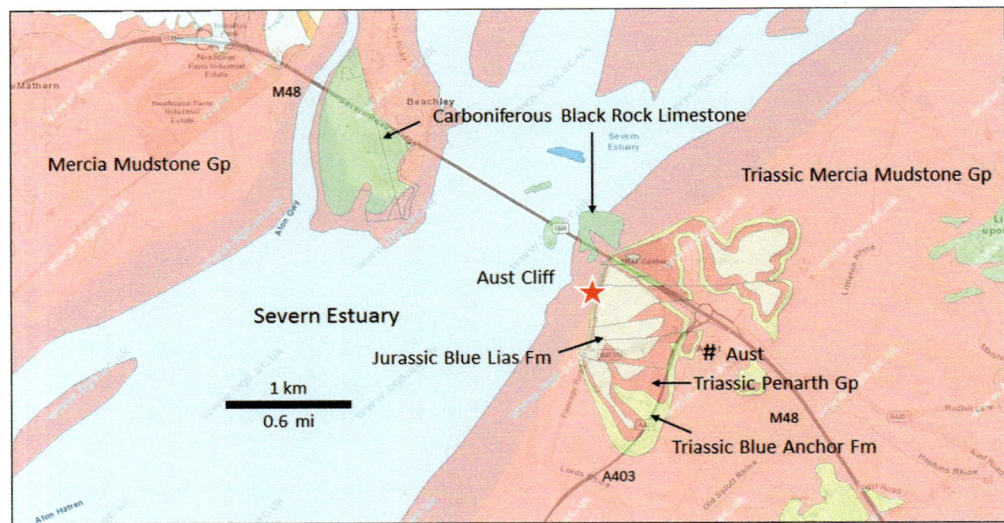

Geologic map of the Aust area. (Modified after BGS, Geology of Britain Viewer Classic. Contains British Geological Survey materials © UKRI 2022. Base mapping provided by ESRI.)

Aust Cliff looking east toward the M48 Severn Bridge.

Cliffs on the east side of the bridge are being actively undercut by the tidal current, causing continuous cliff falls and providing plenty of material to search for fossils. The best locations are in fallen blocks along the shore and areas of shingle.

There are plenty of Carboniferous mollusks eroding from the Black Rock Limestone, actually dolomite here. It was deposited approximately 359–345 Ma in shallow seas. The Black Rock Limestone is exposed below the east end of the Severn Bridge and is partly visible at low tide in outcrops upstream from the bridge.

The Triassic Mercia Mudstone Group red beds (formerly called Keuper Marls) are here part of the fossil-barren Branscombe Mudstone Formation (221–206 Ma), the lower part of the cliffs. Evaporites including gypsum are present and suggest deposition in playas (dry and evaporating lakes). These beds grade upward into green-gray dolomitic mudstones and siltstones of the Triassic Blue Anchor Formation (formerly called the Tea Green Marls). Salt casts indicate hypersaline lakes or tidal flats.

The transition upward from the Blue Anchor Formation to the darker fossil-bearing shale and pyritized, shelly limestone of the Westbury Formation marks a change to a brackish sea or coastal lagoons and estuaries. Above the Westbury Formation is gray marine limestone and mudstone of the Cotham member of the Late Triassic Lilstock Formation, Penarth Group.

At the base of the Westbury Formation is the famous Westbury Bone Bed, a 15 cm (6 in) thick conglomerate consisting of green-gray siltstone, quartz pebbles, and well-preserved vertebrate fragments, all cemented in a sandy matrix. This bed is full of clams, oysters, and bones and teeth from ichthyosaurs and plesiosaurs and fish. It is the most productive Triassic site in the United Kingdom.

At the cliff top are light brown beds of the Early Jurassic Blue Lias Formation (210–195 Ma). Ammonites from this unit indicate an open marine environment.

In summary, the cliffs record the gradual submersion of a desert (Mercia Mudstone, Blue Anchor Formation) beneath a deepening sea (Westbury Formation, Lilstock Formation, Blue Lias) between 221 and 195 Ma.

Structurally the cliffs are part of a gentle anticline that extends northwest-southeast across the Severn. A fault can be seen in the cliff face near the bridge footings. Several normal faults are visible in the cliff from the shore, and springs issue from some of these. Along the concrete causeway, spring water rich in dissolved calcium carbonate (calcite) is precipitating tufa mounds made of travertine.

The best and safest fossil collecting is along the southeastern side of the bridge. The base of the cliffs is dangerous because of falling rocks. Out on the mudflats beware of the tides and the Severn Bore, a large wave created as the tide flows upstream into a narrowing estuary. It can sweep you away. The forecast for the bore can be found at: https://www.severn-bore.co.uk/.

There are no toilet facilities here. An interpretive display at the end of the concrete causeway explains that Aust Cliff is an SSSI. This means you can collect fallen rocks, but hammering bedrock is not permitted (Byles, 2012; Cruickshanks, 2017, Aust Cliff).

*Aust Cliff to Gardiners Quarry: Return south on Aust Wharf Road and turn left (north and east) onto New Passage; turn left (north) onto the A403 and at the Aust Interchange take the 3rd exit (southeast) onto the B4461; merge onto the M48 going southeast; merge onto the M4 going southeast; at Junction 20 take the M5 exit to Midlands/Bristol Airport; continue northeast on the M5 to Junction 12; exit to Gloucester/A38 (north); keep left and merge onto Bath Road/B4008; at the roundabout, take the 3rd exit (north) onto A38; continue straight onto Bristol Road/A430; at the roundabout, take the 3rd exit onto Hempstead Lane/A430; continue north on A430; use the left lanes to bear left onto Over Causeway/A417; continue north on A417; at the roundabout, take the 2nd exit onto Straight Lane and immediately turn left (north) onto Malvern Road/B4208; continue north on B4208 to Welland and turn left (west) onto Marlbank Road/A4104; take A4104 west to Little Malvern and turn left (west) toward A449; turn left onto A449 and drive to Jubilee Drive/B4232; turn right (north) on Jubilee Drive and drive to **Stop 12.1, Gardiners Quarry** (52.076199, −2.342850) parking on the right, for a total of 81.8 km (50.8 mi; 1 h 11 min). There is ample "pay and display" parking.*

STOP 12 ABBERLY AND MALVERN HILLS UNESCO GEOPARK

Driving north from Aust Cliff, we enter the West Midlands as we approach the Malvern Hills. The road runs in or near the west flank of the Worcester Basin. The Worcester Basin is a sedimentary basin filled mainly with Permian and Triassic rocks. It lies between the East Malvern Fault in the west and the Inkberrow Fault in the east and is one of a series of Permo-Triassic basins that extend north–south across England. These basins resulted from regional rifting across northwest Europe and eastern North America related to the breakup of Pangea and opening of the Atlantic that began near the end of Permian time.

The Abberly and Malvern Hills UNESCO Geopark was created specifically to highlight the exceptional geology and scenery of this area. The 1,250 km² (483 mi²) in the Geopark spans the Western Midlands in a north–south swath 83 km (50 mi) long and at most 18 km (11 mi) wide. The unique geology of the area has been recognized for years in the 13 Sites of Special Scientific Interest and 179 Local Geological Sites included within the Geopark.

There are no signs announcing the park, nor any wardens. The park is managed by a partnership of organizations that foster knowledge of geology, forestry, heritage, and landscape protection (Watkins, 2011). The geology spans roughly 700 million years, from Precambrian to Jurassic time. Human history extends from Iron Age hilltop forts to the ruins of industrial coal mining. The park highlights evidence of how geologic and human events have shaped the landscape. Each summer, the Abberley and Malvern Hills Geopark hosts GeoFest, a 3-month program of events and activities. We will visit and describe just two sites within the park that could be considered typical: there are many others worth seeing.

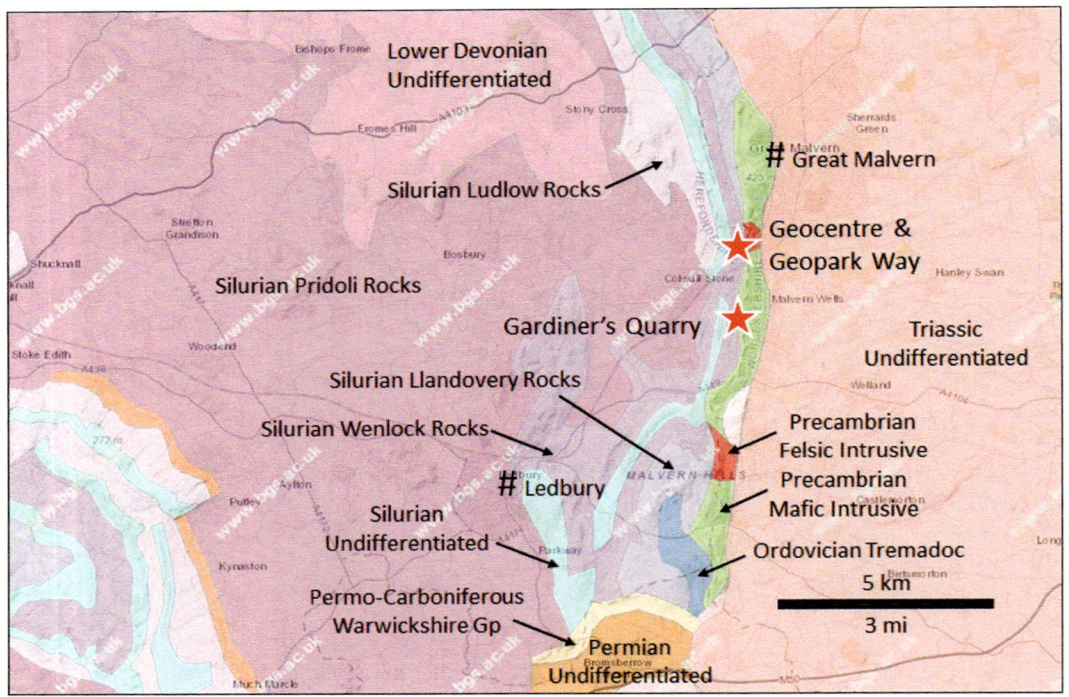

Geologic map of the Abberly and Malvern Hills Geopark. (Modified after BGS, Geology of Britain Viewer Classic. Contains British Geological Survey materials © UKRI 2022. Base mapping provided by ESRI.)

12.1 GARDINERS QUARRY

Quarries are among the best locations to view fresh rock surfaces in otherwise deeply weathered terrain. Gardiners Quarry is one of several quarries scattered throughout the Malvern Hills that were developed for aggregate (crushed rock). This quarry, on the west side of the hills, is currently used as a Malvern Hills Conservators car park. It illustrates multiple intrusions, shearing, and metamorphism seen in the Precambrian of the Geopark.

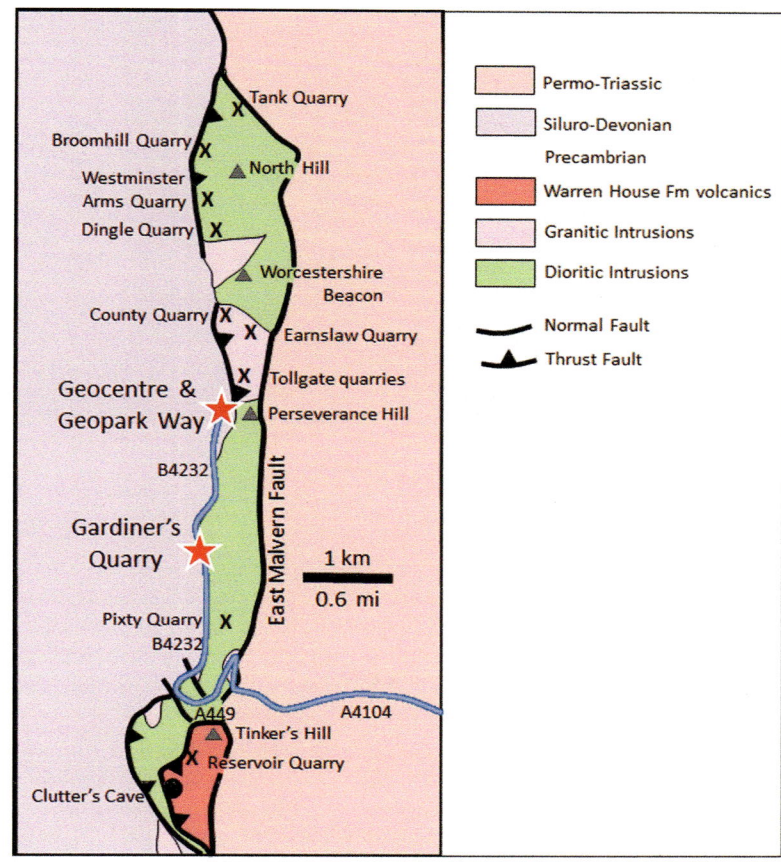

Geologic map of the Malvern Hills. (Modified after Pharaoh, 2019.)

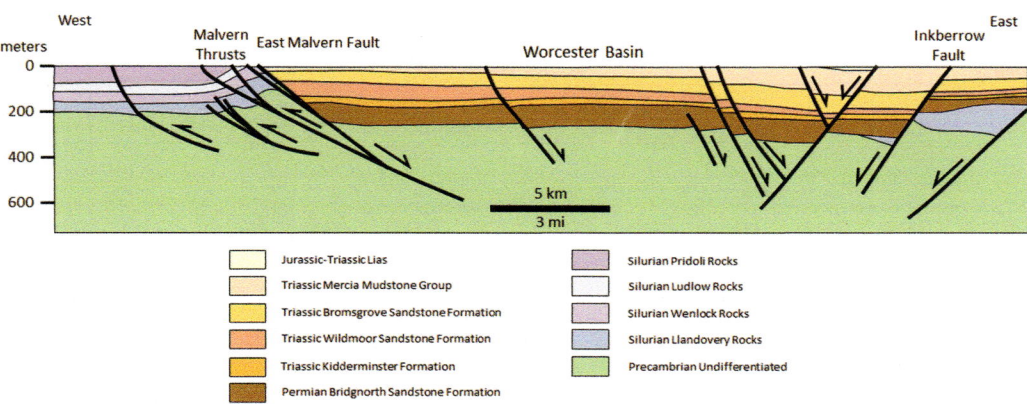

Cross section through the Malvern Hills and Worcester Basin. Variscan (Late Carboniferous) thrusting was followed by Permo-Triassic normal faulting. The East Malvern Fault is an early thrust reactivated as a normal fault. (Modified after Pharaoh, 2019.)

Gardiners Quarry. Diorite (dark) is on the right; granite (pinkish) is on the left.

The dominant rock type here is diorite of the Precambrian Malverns Complex, ~677 Ma (Malvern Spa Association). The diorite is cut by a number of dikes: a dolerite (dark, finely crystalline gabbro) dike is in the center of the quarry, and a series of pegmatite dikes (light, coarsely crystalline granite) emplaced around 610 Ma can be seen throughout the quarry. The dikes intrude the diorite and thus are younger. The quarry contains highly faulted rock as indicated by fractures, mineralized veins, dikes, and slickensides.

A granite intrusion (pink rock) in the quarry has been intensely sheared. The shearing is probably part of the northwest-trending Colwall Fault. Shearing occurred during an early phase of metamorphism that occurred around 652–649 Ma and again at 610–600 Ma (Strachan et al., 1996). The quarry itself is somewhat overgrown, but the diorite (dark) and granite (pink) can also be seen on the surface of the parking area.

Looking west from above the quarry you can see ridges of Silurian Wenlock Limestone running mostly north–south parallel to the Malvern Hills (Earth Heritage Trust, 2011; Abberly and Malvern Hills Geopark, Gardiners Quarry).

North–south ridges of Silurian Wenlock Limestone west of Gardiners Quarry.

Gardiners Quarry to Geocentre: *Continue driving north on Jubilee Drive/B4232 to **Stop 12.2, Malvern Hills Geocentre** (52.089992, −2.339898) on the left, for a total of 1.7 km (1.0 mi; 2 min). There is limited free parking in back of the building.*

12.2 MALVERN HILLS GEOCENTRE AND GEOPARK WAY

The GeoCentre

The Malvern Hills GeoCentre is the official visitor information center for the Geopark Way. The Centre provides information about the Abberley and Malvern Hills Geopark, the Malvern Hills Area of Outstanding Natural Beauty, and the Malvern Hills in general. Wall maps show the Geopark and its geology, the Malvern Hills Area of Outstanding Natural Beauty, and the Malvern Hills. A video wall shows panoramas and videos of the region. There are modest displays of fossils and rocks from the region. The café provides expansive views over Herefordshire toward Wales. A café and shop sells souvenirs, maps, books, and postcards. Facilities include a car park, baby-changing facilities, and free wi-fi. The GeoCentre is open from 10:00 to 4:00 every day except Wednesday, when it is closed.

Dogs are not permitted in the Centre, but they may be tied up outside. There is also an all-terrain scooter available to rent.

Malvern Hills GeoCentre.

Since 2009, the Geopark has developed three geology loop trails near the GeoCentre, each incorporating a section of the Geopark Way. The GeoCentre is about mid-way on the Geopark Way.

The Geopark Way

In 2006, a group of local enthusiasts agreed that a long-distance walking trail was a good idea. By 2009 the Geopark Way trail had been created and a trail guide was published. The result is a 175 km (109 mi) trail winding its way from Bridgnorth in the north to Gloucester in the south.

Starting in Bridgnorth, the trail passes by the Permian sandstone cliffs that separate Bridgnorth into a "High Town" and "Low Town." Crossbedding in the sandstone indicates that the layers were deposited as migrating sand dunes and that this area had been a desert during the Permian.

From Bridgnorth, the trail heads south along the River Severn. The trail guide explains that the Severn is a recent addition to the landscape, formed as the last glaciers melted a mere 12,000 years ago.

Continuing south the trail enters the Wyre Forest Coalfield, an area of Carboniferous (literally "coal-bearing") strata. You see the impact that coal mining had on local communities, including coal mining itself, brick making, stone quarrying, and iron smelting.

Leaving the coal field, the trail passes over a series of Permian to Triassic sedimentary rocks before coming upon Quaternary windblown sand that covers the lower terrace of Hartlebury Common.

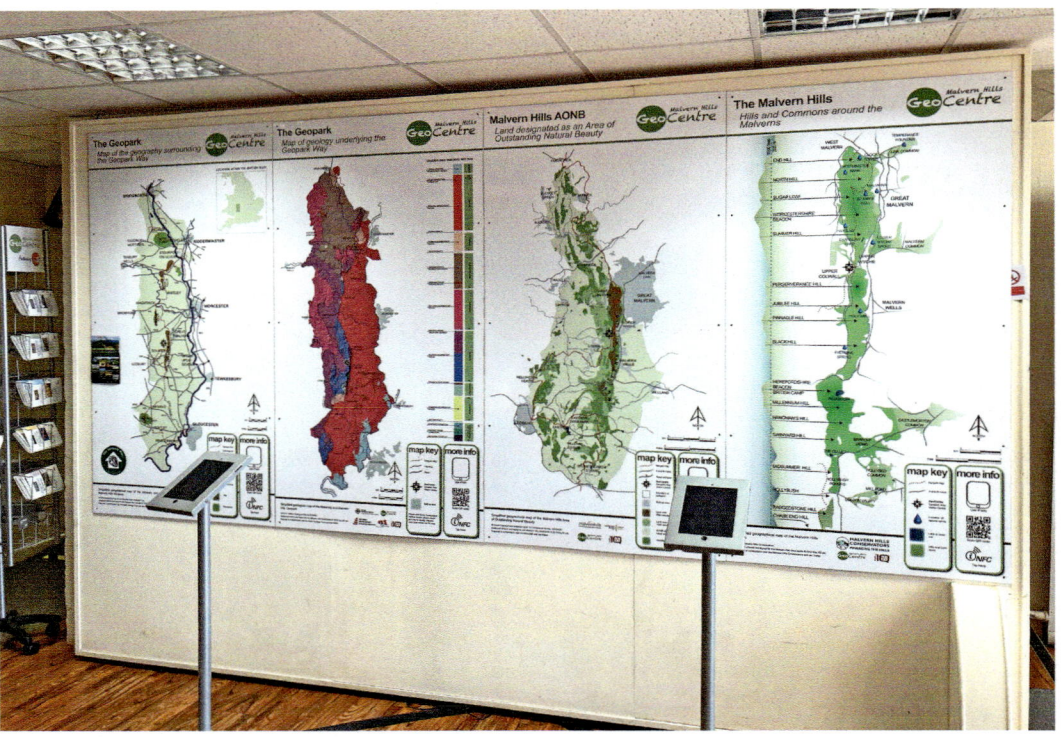

Geocentre displays.

From Hartlebury, the trail heads south and west and crosses the East Malvern Fault. As you cross this down-to-the-east normal fault the topography becomes much more dramatic and the trail enters the uplifted Silurian limestones and shales of the Abberley Hills. The East Malvern Fault has raised up a line of ridges that run north–south through the center of the Geopark.

Malvern Hills. (Photo courtesy of D.M. Hunt, https://commons.wikimedia.org/wiki/File:Malvern_Hills_-_England.jpg.)

Beyond the Silurian hills are the Malvern Hills, a 21 km (13 mi) ridge of resistant Precambrian rocks. The Precambrian includes a number of igneous intrusives, both mafic (dark, iron-rich) and felsic (light, silica-rich). Besides the fascinating outcrops, these hills are famous for their panoramic views.

South of the Malvern Hills the trail enters the Cotswold Hills. The final section of the trail wanders over the relatively low relief plains of Late Triassic and Early Jurassic sediments including the famous Jurassic Painswick oolitic limestone building stones that give the Cotswolds their unique architectural appearance. The trail ends at Gloucester cathedral, built of this white to creamy yellow limestone.

*Geocentre to Wren's Nest: Continue north on B4218 to Wells Road/A449; turn left (north) onto A449; continue straight (northeast) on A449/Worcester Road/Malvern Road; at the roundabout, take the 4th exit (east) onto Temeside Way/A4440; continue driving east on the A4440 to the M4; at the Whittington, Roundabout take the 3rd exit onto Whittington Road; at the roundabout, take the 1st exit onto the M5 north to Birmingham; at Junction 2 in Birmingham take the exit to A4123 toward Dudley and Sandwell; at the Oldbury, Interchange take the 1st exit; at the roundabout, take the 4th exit (northwest) onto Wolverhampton Road/New Birmingham Road/A4123; take A4123 to Priory Road/A4168 and turn left (southwest); turn right (west) onto Mayfield Road and drive to Linwood Road; turn right (north) onto Linwood Road and drive to Wren's Hill Road; turn left (west) onto Wren's Hill Road and drive to **Stop 13, Wren's Nest** (52.527894, −2.095903) parking on the left, for a total of 64.9 km (40.3 mi; 56 min). The car park has 5–6 spaces and is free, but closes at 3:00 p.m. There is plenty of street parking.*

STOP 13 WREN'S NEST AND BLACK COUNTRY GEOPARK

Wren's Nest Hill was established as a National Nature Reserve in 1956 (the first in the United Kingdom for geology) in recognition that it is an exceptional site for Silurian fossils. Over 600 species have been found here, and a third of these were first discovered here. Fossils from this hill are displayed in museums around the world. Wren's Nest was declared an SSSI in 1990. The area was declared a Scheduled Monument in 2004 in recognition of Wren's Nest's contribution to the lime industry and development of the Black Country. Wren's Nest National Nature Reserve is a key site within the Black Country Geopark which, like the Abberly and Malvern Hills Geopark, is acknowledged for its unique geology and landscape. The site was critical to Sir Roderick Murchison, who illustrated 65% of his seminal work on "*The Silurian System*" (1839) with fossils from the Wren's Nest/Dudley area.

The Black Country UNESCO Global Geopark in the West Midlands of England was awarded Global Geopark status in July 2020 (Wikipedia, Black Country Geopark). The Geopark contains rocks of Silurian through Triassic age, mostly sedimentary rocks but some igneous rocks as well. The region has been exploited for its resources of building stone, coal, iron ore, limestone, fireclay, and brick clay for centuries. The Geopark is meant to call attention not only to the distinctive geology but also to the region's industrial heritage based on that geology. Wren's Nest National Nature Reserve (NNR) is a key site within the Geopark. There are a further 16 SSSIs, and 105 sites of Importance for Nature Conservation (SINCs) that are protected within the Geopark.

Steeply inclined Silurian limestone, Wren's Nest.

Wren's Nest is the middle of three adjacent limestone hills (the others are Castle Hill to the southeast and Hurst Hill to the north). They rise above the younger (Carboniferous) rocks of the South Staffordshire Coalfield that surround and separate them. The hills consist of upfaulted and folded Silurian Much Wenlock Limestone Formation (formerly called the Dudley Limestone).

During the Silurian, this part of the Midlands was covered by a warm, shallow sea. The nearest continent was far to the east. The Much Wenlock Limestone, including coral reefs, was deposited about 428 Ma during periods when there was no silt or mud to smother and kill the coral.

Geologists in the United Kingdom classify the rocks of the Silurian Period into four subdivisions; the Llandovery Series (at the base), the Wenlock Series, the Ludlow Series, and the Prídolí (at the top). The rocks at Wren's Nest are from the upper part of the Wenlock Series and the lowest stage of the Ludlow Series.

Toward the end of Silurian time, the area of present-day Wales was uplifted as part of the Caledonian mountain-building event, and the region transitioned from a marine to a continental environment. Any rocks that would have been deposited during Devonian and Lower Carboniferous times were eroded away. As a result, rocks of the Coal Measures, deposited in Upper Carboniferous time, rest unconformably on the Silurian rocks at Wren's Nest. The three limestone hills formed an eroding ridge during the Early Carboniferous; the ridge was later buried by Upper Carboniferous sediments. During the Upper Carboniferous, about 300 Ma, Wren's Nest was near the edge of an extensive tropical swamp that accumulated organic matter that would later become coal.

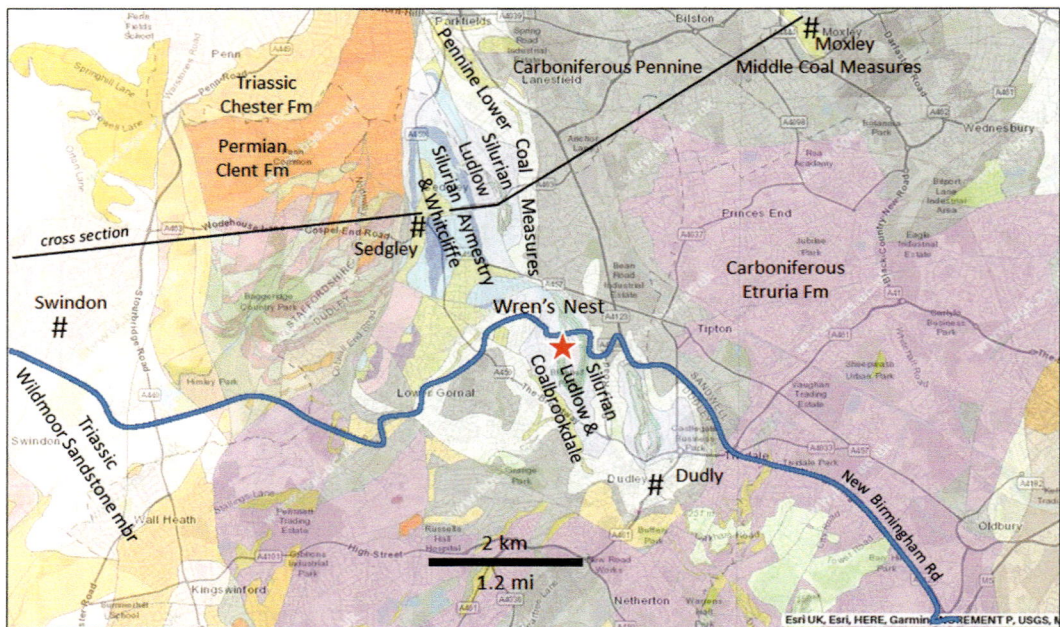

Geologic map of the Wren's Nest area. (Modified after BGS, Geology of Britain Viewer Classic. Contains British Geological Survey materials © UKRI 2022. Base mapping provided by ESRI.)

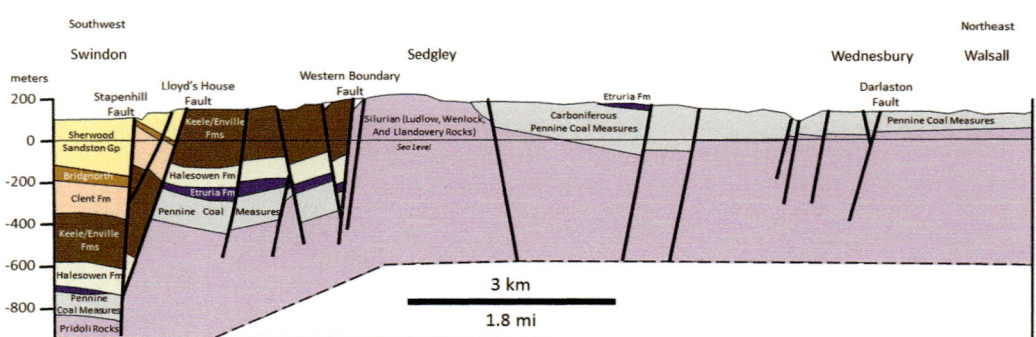

Cross section through the Wren's Nest area. (Modified after Powell et al., 1992.)

The Silurian rocks of Wren's Nest were folded into a north-northwest-trending anticline during the Variscan mountain-building episode at the end of Carboniferous time. As part of the deformation, the structure was faulted along its axis, and the west limb was dropped down. The fold is asymmetric, with a steeper eastern limb dipping up to 80°, and a gentler west limb dipping 50°–60°. The rock is extensively fractured as a result of folding, and it is possible to see slickensides in some exposures.

Pretty much everything that was deposited in the past 300 million years (after the Carboniferous) has been eroded during more recent mountain-building events. The finishing touches, however, were imposed by ice over the last 2.5 million years. The Much Wenlock Limestone at Wren's Nest, being more resistant to the grinding ice than the surrounding Coal Measures, was left as hills when the ice retreated some 12,000 years ago.

This area played a critical role during the Industrial Revolution. The Much Wenlock Limestone had been mined for centuries. The keep of Dudley Castle, built around 1300, consists of local limestone. The earliest reference to underground workings at Wren's Nest is from 1796. The limestone was mined at the surface for building stone and for agricultural lime and mortar. In the late 1700s, there was a demand for limestone as a flux for local iron furnaces, where it helped remove impurities, and underground mining began. During the height of the Industrial Revolution, up to 81,600 tonnes (90,000 tons) of rock were mined annually from quarries around Wren's Nest and used in local blast furnaces. The mines that honeycomb the subsurface around Wren's Nest were finally abandoned in 1924.

A geology/fossil trail extends about 1.8 km (1.1 mi) around Wren's Nest Hill and is well worth the walk. The gravel path takes you through the quarry trenches, past rock faces with ancient reefs, by overlooks with interpretive panels, to collapsed underground workings and room-and-pillar excavations. Fossils found here include bryozoans (moss animals), stromatoporoids (sea sponges), colonial tabulate corals, and compound rugose corals, many in life position. Brachiopods, bivalves, gastropods, and trilobite fragments, together with abundant crinoid (sea lily) heads and stem discs, are common. Trilobites are probably the most famous of the fossils, including the "Dudley Bug" (*Calymene blumenbachii*), nicknamed by quarrymen in the 19th century (Cutler et al., 1990; Black Country GeoPark, 2016; The Geological Society, Wren's Nest).

Visit

The nature reserve has a team of full-time wardens who maintain and care for the site's geological, wildlife, and archaeological features. To book a guided walk or group visit, call the senior warden at 01384 812785.

The Reserve, which attracts thousands of visitors a year, is considered an outdoor teaching laboratory. Hammers are prohibited and collecting is allowed only from the abundant scree.

There is free car parking on site, Monday to Friday from 10 a.m. to 3 p.m. There are no toilet facilities. For that, you should consider visiting the Dudley Museum & Art Gallery a mile (1.6 km) away. The museum has an extensive geological collection and displays of fossils from Wren's Nest, as well as a shop selling books and fossil replicas. There are toilet facilities available. Admission is free.

Address: Wren's Nest National Nature Reserve, Wrens Hill Road, Dudley DY1 3SB
Phone: 01384 812785
Email: wrensnest.country@dudley.gov.uk
Website: www.dudley.gov.uk/wrensnestnnr

Corals at the Wren's Nest. A ten pence coin provides scale.

Wren's Nest to Ironbridge Gorge: Continue west on Wren's Hill to Hillside Road and turn right (northwest); at the roundabout, continue straight onto Maple Green; continue straight onto Elm Green; turn left (south) onto Parkes Hall Road; continue straight (southwest) onto Eve Lane; continue straight onto Jew's Lane/B4175; at the roundabout, take the 2nd exit onto Church Street/B4175; continue southwest on B4175 to Himley Road; turn right (west) on Himley Road/B4176; continue straight on Dudley Road/B4176; continue straight on Bridgnorth Road/4176; stay on B4176 to Telford and Bridgnorth; at the roundabout, continue straight (northwest) onto the A442; turn left (southwest) onto Brockton at the sign for "Broseley 3;" turn right (northwest) onto High Street; continue straight onto Coalport High Street; bear left to stay on The Lloyds; continue straight onto Waterloo Street; turn left onto B4373 and cross the River Severn; turn right (west) onto Ladywood and drive to **Stop 14, Ironbridge Gorge** *(52.626484, −2.484797) car park, on the right, for a total of 34.1 km (21.2 mi; 37 min). There is ample pay parking, £3/2 h.*

STOP 14 IRONBRIDGE GORGE WORLD HERITAGE SITE

Ironbridge Gorge is a World Heritage Site. It is named after the famous Iron Bridge designed by Thomas Pritchard and built by Abraham Darby III. It was the first arch bridge in the world made from cast iron. The bridge was built in 1779 to link the industrial town of Broseley with the mining town of Madeley and the growing industrial area at Coalbrookdale. The deep and wide River Severn provided easy transport of products to the sea.

Known as "the birthplace of the Industrial Revolution," Ironbridge Gorge was cut by the River Severn. When it cut the gorge, between 10,000 and 20,000 years ago, all the ingredients for making high-quality iron (coal, limestone, and iron ore) were exposed in one small area. The juxtaposition of these resources enabled the rapid economic development of the area during the early Industrial Revolution.

There is some debate regarding the cutting of the canyon. Some say that the steep-sided gorge was cut by the erosive action of meltwater flowing beneath an ice sheet that had moved south from the Irish Sea (Fewtrell, Ironbridge Gorge). Others say that this ice sheet dammed a previously north-flowing river, forming glacial Lake Lapworth. When the lake overtopped the hills to the south, the river quickly cut the gorge and established a new course south to the Severn estuary (Wikipedia, Ironbridge Gorge).

The Iron Bridge looking downstream on the River Severn c. 1900. (© IRONBRIDGE GORGE MUSEUM TRUST, image 1972.109.1.)

The Gorge is cut through rocks of Silurian to Upper Carboniferous age: from base to top they are the Silurian Coalbrookdale Mudstone, the Much Wenlock Limestone, and the Lower Ludlow Shale. A major erosional unconformity removed the Devonian and Lower Carboniferous. The upper gorge exposes the Upper Carboniferous Pennine Lower Coal Measures Formation, the Pennine Middle Coal Measures, the Etruria Formation, and the Halesowen Formation.

The valley sides rise 100 m (328 ft) from the river to the plateau above. Rapid downcutting by the river steepened the valley sides such that they became unstable and are known for active landslides. Over 20 landslides are documented in the National Landslide Database in this area. For example, in 1952 a landslide occurred at the village of Jackfield, 2 km (1.2 mi) downstream from the Iron Bridge. It destroyed several houses and disrupted a railway and road. In 1984, ground movement west of the 1952 landslide carried Salthouse Road into the river (BGS, Ironbridge Gorge).

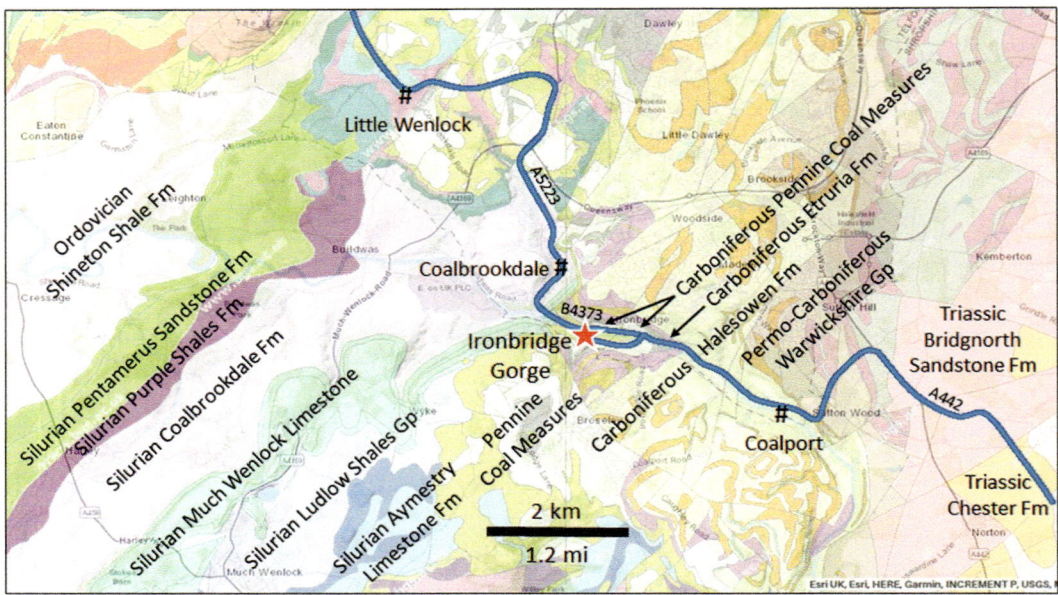

Geologic map of the Ironbridge Gorge area. (Modified after BGS, Geology of Britain Viewer Classic. Contains British Geological Survey materials © UKRI 2022. Base mapping provided by ESRI.)

The car park provides access to views of the gorge and bridge.

View south to the River Severn and Ironbridge Gorge.

Visit

An easy walk (2 km; 1.2 mi) north of the bridge, Ironbridge Gorge Museum on Coach Road (52.639009, −2.493218) provides exhibits about the bridge and the area's industrial history. There is also a shop and tourist information center. See the website for current hours and entrance fees.

Hours: 10:00 a.m. to 5:00 p.m. daily
Entrance fee: £47 Family; £29 Adults; £19 Children and Students
Address: Coach Rd, Coalbrookdale, Telford TF8 7DQ
Phone: +44 1952 435900
Web site: https://www.ironbridge.org.uk/plan/

*Ironbridge Gorge to The Wrekin: Return east on Ladywood; turn left (north) onto B4373; after crossing the River Severn turn left (west) onto Waterloo Street/B43763; at the roundabout, take the 3rd exit (east) onto Madeley Road/B4373; at the roundabout, take the 2nd exit (northeast) onto Parkway/4373; at the roundabout, take the 2nd exit (north) onto Castlefields Way/B4373; at the roundabout take the 1st exit (west) onto A4169; at the roundabout, take the 3rd exit (north) onto A5223; in Coalmoor turn left (west) onto Coalmoor Road; continue straight on Malthouse Bank; in Little Wenlock, turn right (northwest) onto Wellington Road and drive to **Stop 15, The Wrekin** (52.679773, −2.535812) car park on the right for a total of 11.2 km (6.9 mi; 19 min). There is ample pay parking, £1/1 h.*

STOP 15 THE WREKIN

The Wrekin is a hill and prominent landmark in Shropshire. Rising above the Shropshire Plain to a height of 407 m (1,335 ft), it falls in the northern Shropshire Hills Area of Natural Beauty. The hill is open to visitors all the time, and the path is well marked but steep in places. At an average pace, it should take 45 minutes to an hour to climb the hill. The Wrekin provides splendid views of the surrounding landscape (Geogeek1726, 2013; Wikipedia, The Wrekin).

Contrary to popular belief, The Wrekin is not a volcano. However, it does consist of Precambrian Uriconian Volcanics, the remains of lava and ash flows from nearby volcanoes that have since been eroded away or buried. They are mostly rhyolite, tuff (ash), and agglomerate (a coarse accumulation of volcanic material including bombs) that weathers to a dark reddish rock. You can see flow banding in the rhyolite at the inner gate of the Iron Age fort (Geogeek 1726, 2013).

The Wrekin. Google Street View northwest.

The Uriconian is interpreted as having been erupted within a tectonic plate rather than at the margins, most likely as part of a volcanic arc caused by subduction around 566 Ma. A number of dikes intruded these rocks around 563 Ma. The dikes are dark dolerite and can be seen at several locations as you walk up the hill. One of the best places to see a dike is in the center of the main car park at the bottom of the hill. The Uriconian volcanics are also cut by the Late Precambrian Ercall Granophyre, a shallow-emplaced granitic rock dated at 560 Ma. Altogether, these are among the oldest rocks in Shropshire (Wikipedia, Wrekin Terrane).

Urconian outcrop at the Forest Glen Car Park, The Wrekin.

Cambrian and younger rocks lie unconformably on the Precambrian volcanics. These include the Cambrian Wrekin Quartzite, Lower Comley Sandstone, Lower Comley Limestone, and the Permian Bridgnorth Sandstones. Early Carboniferous rocks, including limestone, the Little Wenlock Basalt, and the Lydebrook Sandstone, are found to the southeast (Wikipedia, The Wrekin).

This is the type locality for the Wrekin Terrane. The Wrekin Terrane is one of five inferred fault-bounded terranes that make up the basement rocks of the southern United Kingdom. The terrane is bounded to the west by the Welsh Borderland Fault System and to the east by the Malvern Lineament. The Wrekin, together with the Ercall, forms part of the Church Stretton Complex where different geological terranes meet. The hill is adjacent to the Church Stretton Fault, marking the boundary between the Cymru Terrane to the west and the Charnwood Terrane to the east (Wikipedia, Wrekin Terrane).

The Wrekin is the site of one of the rare meteor falls witnessed in Britain. On the afternoon of April 20, 1876, during a rainstorm a strange rumbling noise was heard, instantly followed by an explosion resembling heavy artillery. In a field about 1.6 km (a mile) north of The Wrekin a man found a hole in the ground which, when probed with a stick, yielded a 3.5 kg (7¾ lb) iron meteorite at a depth of about 46 cm (18 in; Flight, 1881). The meteor is now in the Natural History Museum (Sarah Roberts, oral comm., 25 May 2022).

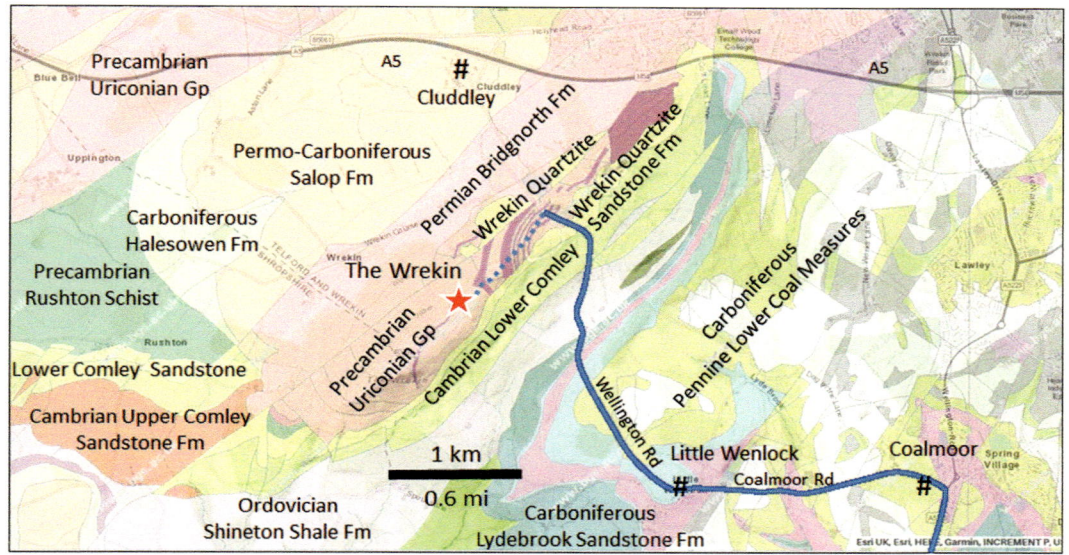

Geologic map of The Wrekin area. (Modified after BGS, Geology of Britain Viewer Classic. Contains British Geological Survey materials © UKRI 2022. Base mapping provided by ESRI.)

An Iron Age fort almost 8 ha (20 ac) in size is located on the summit. First occupied around 3,500 years ago, it appears to have been rebuilt around 450 BC by the Cornovii tribe and may have been their capital. In the year 47, the Romans arrived, stormed the fort, burned it, and moved the defeated tribe elsewhere (Geogeek1726, 2013; Wikipedia, The Wrekin).

There is also a non-volcanic origin story for this mountain. A local legend has it that a giant called Gwendol Wrekin ap Shenkin ap Mynyddmawr had a grudge against the town of Shrewsbury. He decided he was going to destroy the town by flooding it, so he collected an enormous pile of dirt and set off toward the town. Near Wellington, he met a cobbler returning from Shrewsbury market with a large sack of shoes to repair. The giant asked him for directions, adding that he was going to dump his pile of earth in the River Severn and flood the town. "It's a long way to Shrewsbury," replied the quick-thinking shoemaker. "Look at all these shoes I've worn out just walking back from there!" The giant decided it was indeed too far, and dumped the dirt on the ground beside him, thus creating the Wrekin. Scraping the mud off his boots, he created Ercall Hill nearby.

It is said that locals can predict inclement weather in the area based on visibility of the Wrekin. "If you can see the Wrekin, it's going to rain. If you can't see the Wrekin, it's already raining" (Wikipedia, The Wrekin).

The Wrekin to Ramshaw Rocks: *Turn right (north) out of the car park onto Wellington Road; continue straight on Ercall Lane; turn right (east) onto B5061/Holyhead Road; turn left (north) onto Haybridge Road; continue straight (northeast) on Britannia Way; turn left (north) onto Wrekin Road; at the roundabout, take the 2nd exit (east) onto Victoria Road; at the roundabout, take the 1st exit (north) and continue on Victoria Road; at the roundabout, take the 2nd exit (north) onto King Street; continue straight on Whitchurch Road; at the roundabout, take the 2nd exit (east) onto Apley Ave; at the roundabout take the 1st exit (north) onto A5223/Whitchurch Drive; at the roundabout, take the 2nd exit (north) onto A442/Queensway; continue north on A442 to the A53; at the roundabout, take the 3rd exit (northeast) onto A53; stay on the A53 through Stoke-on-Trent and continue toward Buxton to* **Stop 16, Ramshaw Rocks** *(53.152778, −1.97444) for a total of 79.5 km (49.4 mi; 1 h 21 min). Pull into the farm road on the left (west) side of A53.*

STOP 16 RAMSHAW ROCKS

Driving through this gently rolling country you come upon some prominent sandstone cliffs. These outline a gentle, north–south-trending syncline developed in the Carboniferous Millstone Grit (326–315 Ma). The cliffs are often used by climbers.

This and several other rock formations (The Roaches, Hen Cloud) are composed of the Roaches Grit, a subdivision of the Millstone Grit. Nearby the Five Clouds are formed from thin beds of the Five Clouds Sandstone. These grits all originated as delta sands deposited by large rivers draining a landmass to the north. The sandstone beds delineate a syncline known as the Goyt Trough (Wikipedia, The Roaches).

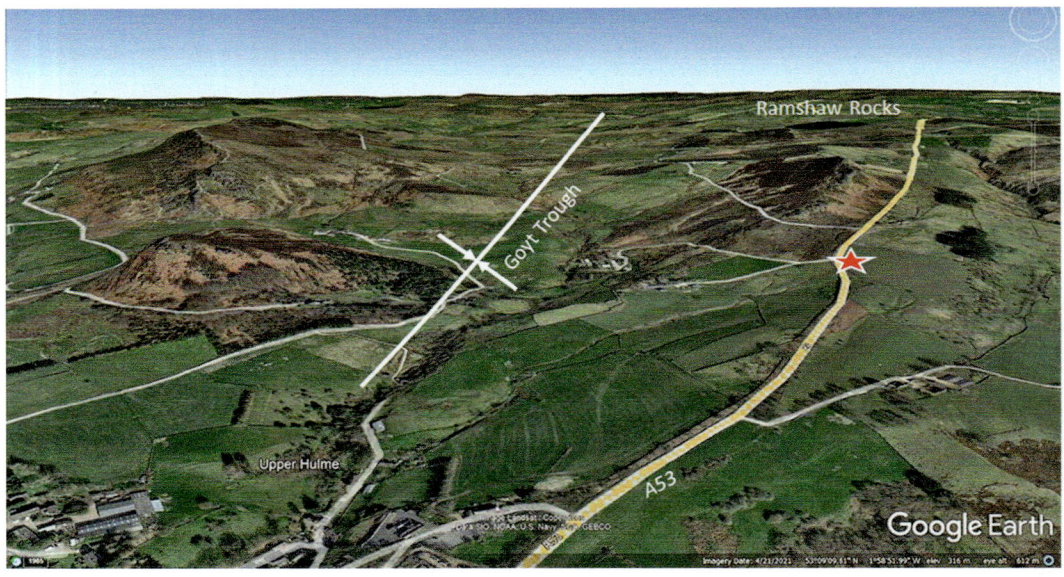

Google Earth oblique view north to a syncline developed in the Millstone Grit at Ramshaw Rocks. (Image Landsat/Copernicus; data SIO, NOAA, U.S. Navy, NGA, GEBCO.)

Ramshaw Rocks syncline, view north.

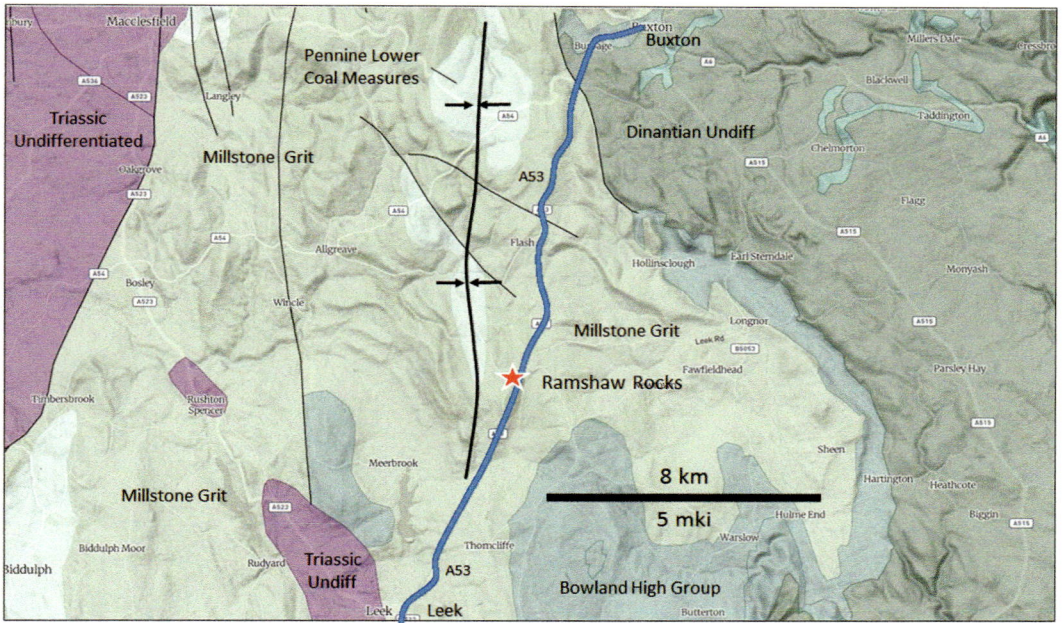

Geologic map of the Leek-Buxton area and Ramshaw Rocks. (From MacroStrat, https://macrostrat.org/map/#x=-1.889&y=53.195&z=11.12.)

Ramshaw Rocks to Clement Lane Quarry: Continue north on A53 to Buxton; at the roundabout, take the 1st exit (east) onto A53/Station Road; at the next roundabout, take the 1st exit (northeast) onto A6/Fairfield Road; turn right (northeast) onto Batham Gate Road; continue straight on Smalldale Road/Batham Gate; at the T-intersection turn left (north) on The Sitch and drive to the A623/Hernstone Lane; turn right (southeast) on A623; on entering Derbyshire Dales turn left (east) on unnamed road toward Bradwell Moor; at T-intersection turn left (north) onto Clement Lane and drive to **Stop 17.1, Clement Lane Quarry** *(53.319455, −1.775935) on the right (south) and pull over, for a total of 27.3 km (17.0 mi; 30 min). Parking is free but limited to one or two vehicles. Walk about 100 m (300 ft) east on the south side of the road.*

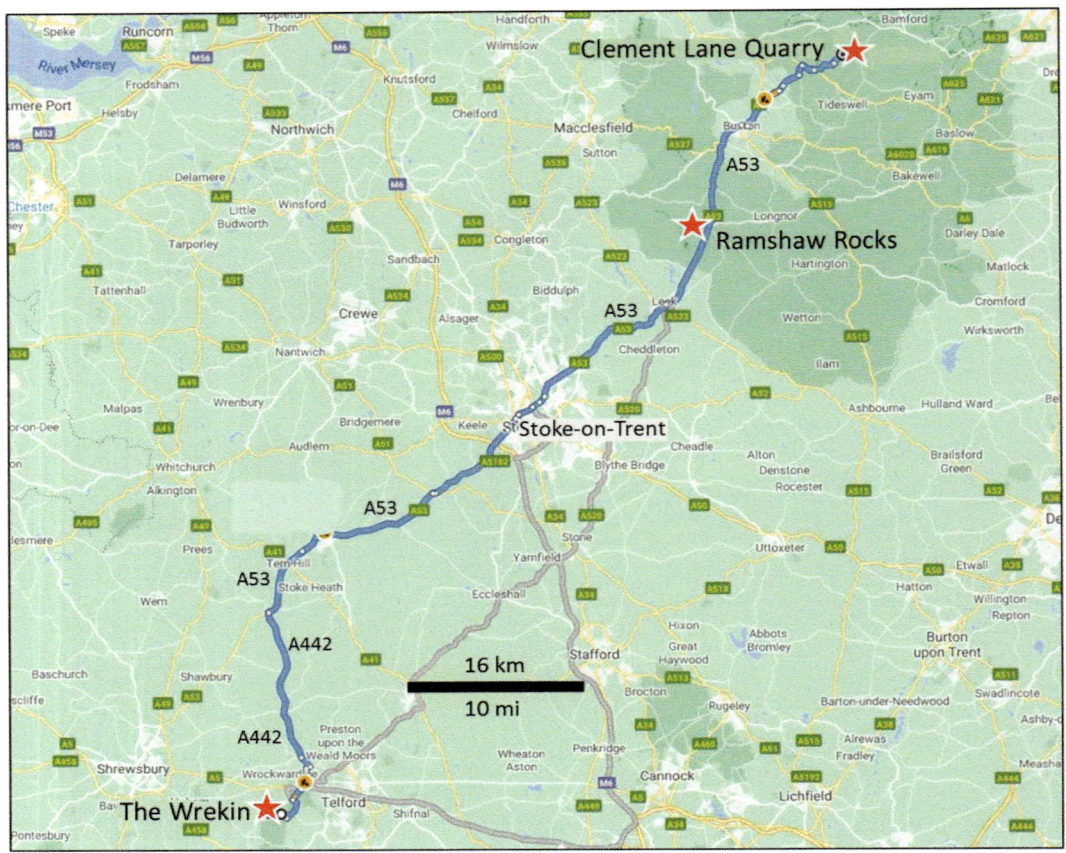

STOP 17 PEAK DISTRICT NATIONAL PARK

Peak District National Park is located in the southernmost part of the Pennine Hills. Much of it is above 300 m (1,000 ft), with the highest point at Kinder Scout (636 m, or 2,087 ft). In 1951 it became the first national park in England and Wales. Largely an agricultural area, mining had been important during the Middle Ages. Cotton mills were located along streams here early in the Industrial Revolution. As mining declined, quarrying grew. Today tourism is a primary industry, thanks to the landscape, spa towns of Buxton and Matlock Bath, and several show caves. Peak District National Park is one of the most visited English national parks. Rambling, cycling, rock climbing, and caving are popular.

The national park covers 1,440 km² (555 mi²), making it the fifth-largest national park in England and Wales. Land within this national park is a mix of public and private ownership. The National Trust owns about 12% of the land in the national park, including Mam Tor.

The National Park is named for its two regions, called the White and Dark Peaks. The White Peak area consists of limestone outcrops that have a distinct light gray color, whereas the Dark Peak, surrounding the White, is higher and consists of dark gritstones, sandstones, and shales. The Peak District is formed almost entirely of Lower Carboniferous (Mississippian, 350–340 Ma) sedimentary rocks. They include limestone at the base of the section, then coarse sandstone (gritstone), limestone, and coal measures at the top. There are rare outcrops of igneous rocks. The rocks here were essentially undeformed until around 60 Ma when the Alpine mountain-building event elevated the units into a broad, north–south-elongated anticline whose western flank is intensely faulted and folded.

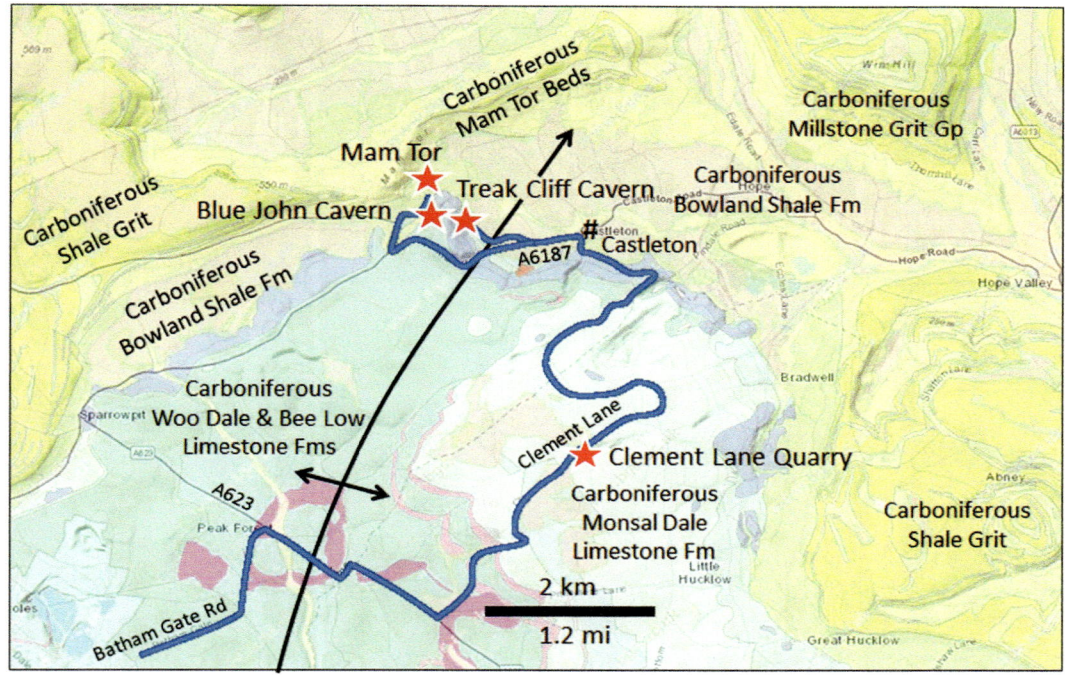

Geologic map of the Mam Tor-Castleton-Clement Lane Quarry area. (Modified after BGS, Geology of Britain Viewer Classic. Contains British Geological Survey materials © UKRI 2022. Base mapping provided by ESRI.)

At some point minerals, mainly lead, were deposited by hot fluids as veins in the limestone. The veins were mined for lead since Roman times.

The Peak District was covered by glaciers during the Anglian Glaciation some 450,000 years ago, as indicated by patches of glacial till and boulder clay found across the area. It was not iced-covered during the most recent Ice Age (Wikipedia, Peak District National Park).

In the central White Peak region, the limestone and wet climate have led to the area's characteristic karst features, including sinkholes, gorges, and caves. These features formed as water percolated through and enlarged existing rock joints by solution weathering (Blackett, 2019).

17.1 Clement Lane Quarry

This stop is an abandoned cement quarry developed in the Carboniferous Monsal Dale Limestone Formation of the Peak Limestone Group. The rocks were deposited between 331 and 329 Ma in warm shallow seas and contain shell fragments and corals from local reefs. The lowest part of the unit is either limestone or lava flows that sit unconformably on the Bee Low Limestone Formation. The Eyam Limestone lies above this unit and is separated from it by an erosional surface (BGS, Geology of Britain Viewer Classic).

Clement Road Quarry. Google Street View southeast.

Carboniferous Crinoids, Monsal Dale Limestone, Clement Lane Quarry. Ten pence coin for scale.

Clement Road Quarry to Treak Cliff Cavern*: Continue northeast on Clement Road to T-intersection; turn left (north) on Washhouse Bottom; bear left to stay on the main road; continue straight onto Siggate; merge onto Pinedale Road; bear slight right onto Back Street; in Castleton turn left (west) onto A6187/Cross Street; bear right (northwest) onto Old Mam Tor Road and drive to* **Stop 17.2, Treak Cliff Cavern** *(53.344292, −1.795729) parking on the left, for a total of 7.4 km (4.6 mi; 13 min).*

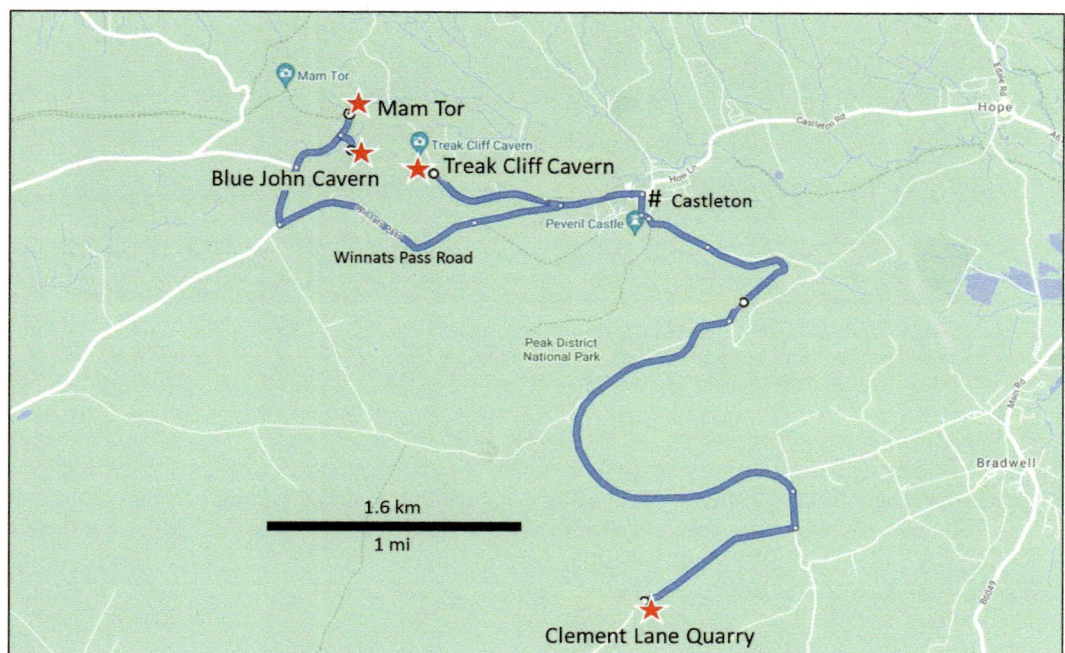

17.2 TREAK CLIFF CAVERN

Treak Cliff Cavern is one of four show caves in the Castleton area and is part of the Castleton SSSI. It is one of only two sites where Blue John is mined (the other is Blue John Cavern). There are two cavern systems at Treak Cliff, the Old Series, discovered by lead miners in the 18th century, and the New Series, discovered by blasting in the 1920s. Blue John, a form of fluorite with purple-blue or yellowish bands, occurs as veins and nodules in cavities within the limestone. The New Series has spectacular flowstone, stalagmites, and stalactites, but only the Old Series contains Blue John.

The Blue John in the Old Series was emplaced around 240 Ma, when the limestone was already 100 million years old. The largest sample of Blue John ever found, The Pillar, weighs about 16 tonnes (17.6 tons) and is located in the Witch's Cave. Another chamber, Aladdin's Cave, was discovered in 1926 by miners exploring for Blue John. Aladdin's Cave leads to a completely different cave system, called the New Series. The Old Series, at the entrance, is more extensive than the New Series, but only three chambers are open to the public.

Raw "Blue John" from Treak Cliff Cavern. (Photo courtesy of Dave Dunford, https://commons.wikimedia.org/wiki/File:Raw_Blue_John.JPG.)

Treak Cliff Cavern lies on the slopes of Treak Cliff, a hill composed of Carboniferous Bee Low Limestone deposited 337 to 331 Ma (Carboniferous) in shallow seas, as evidenced by crinoids, corals, and other fossils. The steep east slope of the hill is thought to have been the edge of a reef on the margin of a tropical lagoon. This limestone was later buried by river and delta sediments of the Millstone Grit (coarse sandstones and shales).

Treak Cliff Cavern. (Photo courtesy of A. Mabbett, https://commons.wikimedia.org/wiki/File:Treak_Cliff_Cavern_-_interior_-_Andy_Mabbett_-_38.JPG.)

There is no modern stream in Treak Cliff Cavern, but dyed groundwater from the cave emerges at the Russet Well beside Peakshole Water near Peak Cavern in Castleton, taking between 13 and 20 h to travel the 1.3 km (4,300 ft).

Stalactites in this cavern grow, on average, about 2.5 cm (1 in) every 1,600 years. Two stalagmite samples from the Aladdin's Cave chamber of Treak Cliff Cavern have been dated at 111,000 years, so the cave itself must be older (Ross, Treak Cliff Cavern; Wikipedia, Treak Cliff Cavern).

Visit

The cavern entrance is located up the hillside from the parking area. You climb up a series of stone steps to the entrance, then wait for one of the tours. The entrance to this underground system is through an old mine tunnel, dug around the year 1750. The tour ends in the Dome of St Paul's, about 50 m (164 ft) below the surface. Dress warm: the temperature in the caves is a constant 10°C (50°F). The cavern floor is wet, so wear non-slip shoes. Tours last 45 min to an hour.

General admission: Adults (including seniors) – £12.50; Youth (16–17 years) – £8.75; Children (5–15 years) – £6.00; Under 5s are free.

Early bird admission: Adults (including seniors) – £10.00; Youth – £7.00; Children – £4.50; Under 5s are free.

Hours: March 1st to October 31st: 9:00 a.m., Early bird; 10:00 a.m. Regular visitors. The last entry is 4:30 p.m.
                  November 1st–February 28th: 10:00 a.m.; last entry 3:00 p.m.

Address: Treak Cliff Cavern Ltd, Buxton Road, Castleton, Hope Valley, Derbyshire, S33 8WP

Phone: 01433 620571 (Monday–Wednesday, 10 a.m.–2 p.m.); 07442 504 762 (all other times)

Email (cavern):   treakcliff@bluejohnstone.com

Email (shop):   lizzy.treakcliff@outlook.com

Website: https://bluejohnstone.com/

*Treak Cliff Cavern to Blue John Cavern: Return south and east on Old Mam Tor Road to Arthur's Way; turn right (west) on Arthur's Way; continue straight on Winnats Pass; turn right onto Old Mam Tor Road and drive to **Stop 17.3, Blue John Cavern** (53.345648, −1.803504) parking, for a total of 3.7 km (2.3 mi; 6 min).*

17.3 BLUE JOHN CAVERN

On the way from Treak Cliff Cavern to Blue John Cavern, you pass through an impressive canyon cut into the Carboniferous Woo Dale and Bee Low Limestone formations as you ascend Winnats Pass. There is no stream in this valley today, so it must have been eroded in wetter, glacial times.

Winnats Pass Road, Google Street View west.

This natural cavern is world famous for Blue John, a semi-precious mineral that is still mined in small amounts. There are eight seams in this cave, and it is found only in Blue John Cavern and the nearby Treak Cliff Cavern. Mining is still done by hand by the cavern guides during the winter months. The mineral veins are thought to be about 240 million years old (Early Triassic).

Blue John is Britain's rarest mineral, probably discovered at Castleton by the Romans almost 2,000 years ago. The world's only known deposits of this stone have been found in the Castleton area. Two vases of Blue John Stone were supposedly unearthed during excavations at Pompeii, evidence that the Romans not only discovered the stone but also appreciated its beauty.

The Carboniferous Bowland Shale is at the surface, but the cavern developed in the underlying Bee Low Limestone Formation (BGS, Geology of Britain Viewer Classic). The caverns, like others in limestone terrain, were formed when underground rivers and groundwater slowly enlarged chambers along fractures in the soluble bedrock.

There are six main natural chambers in Blue John Cavern. Bull Beef, the first chamber, is a working mine that produced some of the largest samples of Blue John ever mined. The Grand Crystallised Cavern, shaped like a cathedral dome, has a high roof and contains more of the Blue John. Waterfall Cavern is completely covered in stalagmites, some of which look like a frozen waterfall. Iron oxide deposits color the roof. The Stalactite Cavern contains the meandering course of the underground river that at one time flowed through the caverns. Lord Mulgrave's Dining Room was formed when two underground rivers merged, creating a whirlpool thought responsible for the circular shape of the chamber. The last chamber open to the public, the Variegated Cavern, is nearly 61 m (200 ft) high. At this point, you are nearly 65 m (214 ft) below the surface (BlueJohn-Cavern. co.uk; Wikipedia, Blue John Cavern).

Entrance to Blue John Cavern.

Visit

The walking route for visitors follows the dry bed of the stream that originally formed Blue John Cavern. The river existed only for short periods during the Ice Age as it was fed by melting ice. The river last flowed about 8,000 years ago.

Today the cavern has a manmade entrance. The original, natural entrance is visible when you reach the first chamber: it is an old sinkhole high in the ceiling from which the first explorers were lowered on a rope. The path is well illuminated, and each tour takes 50 min to an hour. Walking down is the easy part; don't forget you have to walk back up.

Facilities include a café open on weekends. There is a craft shop with a wide variety of Blue John Jewellery set in silver and gold, alongside Blue John ornaments such as bowls, eggs, and goblets.

Fees: Adults £15.00, Senior Citizens £11.00, Students £11.00, Children £10.00;
            Family Ticket (Two adults and two children aged 5–15) £45.00
Hours: 9:30 a.m. (first tour 10:15) to 5:00 p.m. daily.
Address: Cross St, Castleton, Hope Valley S33 8WH
Phone: +44 1433 620638
Website: https://www.bluejohn-cavern.co.uk/

Crinoid stems, Bee Low Limestone, Blue John Cavern. (From Cruickshanks, 2017, Blue John Cavern.)

Blue John Cavern to Mam Tor*: Walk or drive 145 m (480 ft; 5 min) north on Old Mam Tor Road to the end of the road. This is **Stop 17.4, Mam Tor and Landslide** (53.347732, −1.803652). There is ample free parking on this dead-end road.*

17.4 MAM TOR AND THE LANDSLIDE

Mam Tor is a 517 m (1,696 ft) ridge near Castleton, the highest point in the area. The name means "mother hill," because frequent landslides on its slopes have given birth to a number of small bumps on its flanks. The hike from Kinder Scout to Stanage Edge, along which Mam Tor lies, is a popular ridge walk. In fact, Sir Simon Jenkins, a British author, newspaper editor, and former chairman of the National Trust, considers the view from the ridge top to be one of the ten best in England. At the foot of the hill, developed in Carboniferous limestones, are four cave networks: Blue John, Teak Cliff, Speedwell, and Peak caverns.

The hill is crowned by a late Bronze Age hill fort (ditch and rampart) and two Bronze Age barrows (prehistoric tombs covered by earthen mounds). Carbon 14 dating has established an age of around 1,180 BC. The geology, minerals and mining, landscape, and archaeology led to designation of the area as an SSSI.

Mam Tor lies on the southern edge of the Dark Peak (sandstones) and overlooks the White Peak (limestones). The ridge consists of Carboniferous sandstone of the Mam Tor Beds (~322 Ma) over black shales of the Edale Shale Group, Bowland Shale Formation (337–322 Ma). The Mam Tor Sandstone was deposited as repeated fining-upward deep marine turbidite sequences of coarse sand to mud. This represents the first influx of turbidite sands into the Pennine Basin. The underlying

Bowland Shale is a deep marine deposit. Below the Bowland Shale is the Bee Low Limestone, a reef complex that contains abundant fossils including coral, brachiopods, bivalves, crinoids, and goniatites (early ammonites).

The interbedded shales are layers of weakness when saturated, resulting in numerous landslides, including the well-known slide on the southeast slope of Mam Tor. Three more landslides occur on the north slope of the ridge. The limestone below the shale is not involved in slipping.

The still active landslide on the south slope began moving between 3,000 and 4,000 years ago. The slide is in a constant state of creep, with annual movement averaging about 0.25 m (0.8 ft). Movement accelerates when winter rainfall exceeds 210 mm/month and 750 mm in the preceding 6 months (8.3 in/mo and 30 in/6 mo).

The Mam Tor slide has an 80 m (260 ft) scarp and is over 750 m (2,460 ft) long and 300 m (1,000 ft) wide. The slide has caused extensive and spectacular damage to the Manchester-Sheffield road (A625), originally built in 1819, resulting in its abandonment and closure in 1979. The edge of the landslide adjacent to the road has a 2 m (6.5 ft) high compression ridge as a result of downslope movement.

The old road is now closed to vehicles, but from either side, you can see the profile of the landslide, including the prominent scarp, the upper zone of depletion, the lower zone of accumulation, and the toe of the slide.

Mam Tor Peak and landslide. View northwest.

The upper slide zone is a slumped mass that extends south from the scarp. Sandstone layers have been rotated between 30° and 50° into the slide. Walk the road across the main body of the landslide to see where it has been severely damaged. Along the road there are numerous extension cracks, active fissuring, and vertical offsets up to 4 m (13 ft), providing evidence of continued movement.

The back scarp of the slide is an impressive landscape feature. The scarp has an average slope of 45°. Alternating beds of sandstone and siltstone are exposed along the entire face of the scarp.

Ironically, Mam Tor is not a tor. A tor is a rocky hill or outcrop extending many meters above the surrounding landscape. Mam Tor is in fact a sandstone ridge (Donnelly, 2006; Balckett, 2019; BGS, Mam Tor; Wikipedia, Mam Tor).

Mam Tor to Manchester Airport: *Return south on Old Mam Tor Road to Sparrowpit Buxton and turn right (west); take Sparrowpit Buxton/A623.A6; continue straight on Sheffield Road; turn right (north) to merge onto the A6 to Stockport/Manchester; turn left (west) onto A555/Manchester Airport Eastern Link Road; turn left (south) onto Ringway Road to Car Rental Village (end) for a total of 35.1 km (22.1 mi; 35 min).*

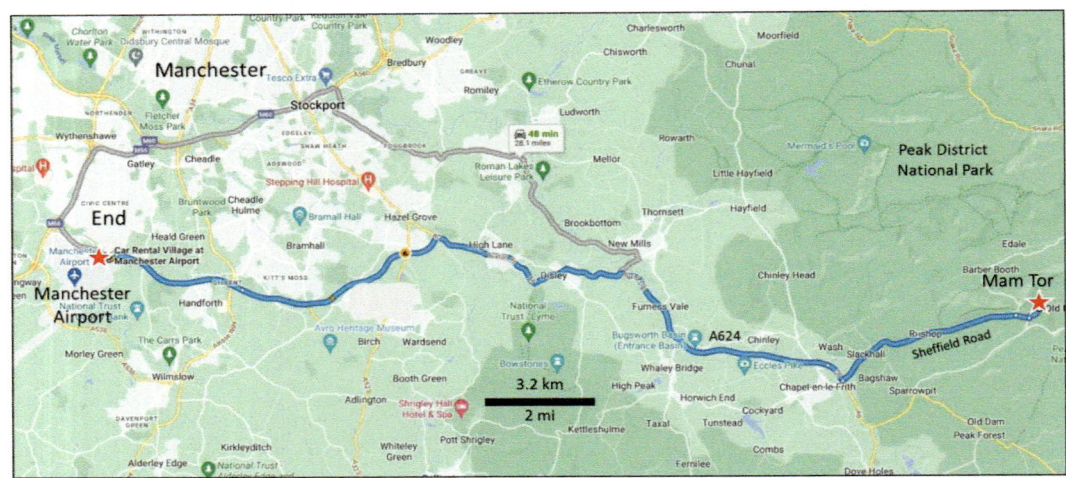

So here we end our tour of the West Midlands and Wales. We have seen the spectacular Pembrokeshire Coast with its breathtaking vistas and deformed sea cliffs, visited fossil sites to rival the classic Jurassic Coast, explored limestone caverns, and toured the birthplace of the Industrial Revolution. Our geo-tour included Snowdonia, Pembrokeshire Coast, Brecon Beacons, and Peak District national parks; Geomôn, Abberly and Malvern Hills, and Black Country Global Geoparks; the Glamorgan Heritage Coast, and Ironbridge World Heritage site. This remarkable sector of British geology is not to be missed.

REFERENCES

Abberley and Malvern Hills Geopark. Geologic Map. Accessed 29 May 2021, http://www.geopark.org.uk/pre2012site/blog/Geology/index.html

Abberly and Malvern Hills Geopark. Gardiners Quarry. Accessed 29 May 2021, http://geopark.org.uk/pub/2015/12/gardiners-quarry/

Abbott, L. 2009. The Depositional Environment of the Pembroke Limestone Group at West Angle Bay, Pembrokeshire. MSc Hons. (Geology), University Manchester, Manchester, 79 p.

Aitkenhead, N., Barclay, W., Brandon, A., Chadwick, R.A., Chisholm, J.I., Cooper, A.H. and Johnson, E.W., 2002. *British Regional Geology: The Pennines and Adjacent Areas* (4th edition). British Geological Survey, Nottingham, 206 p.

Archaeology Magazine. 2019. Neolithic Monolith Quarries Investigated in Wales. Archaeological Inst., America. Accessed 29 May 2021, https://www.archaeology.org/news/7415-190219-stonehenge-carn-goedog#:~:text=LONDON%2C%20ENGLAND%E2%80%94According%20to%20a, side%20of%20the%20Preseli%20Hills.

Baker, M.L., and J.H. Baas. 2020. Mixed sand–mud bedforms produced by transient turbulent flows in the fringe of submarine fans: Indicators of flow transformation. *Sedimentology*, v. 67, pp. 2645–2671.

Ball, K., and T. Faulkner. 2011. Science in South Wales with BCRA. *Speleology*, v. 18, December, pp. 24–25.

Bates, D. 2014. The submerged forest at Borth. Agile Libre. Accessed 29 May 2021, https://www.agilelibre. com/content/the-submerged-forest-at-borth

BBC News. 2012. Ancient Footprints Found in Peat at Borth Beach. Accessed 29 May 2021, https://www.bbc. com/news/uk-wales-mid-wales-17353470

BBC News. 2019. Great Orme Copper Mine 'Traded Widely in Bronze Age.' Accessed 29 May 2021, https:// www.bbc.com/news/uk-wales-50213846#:~:text=North%20Wales%20was%20Britain's%20main, across%20what%20is%20today's%20Europe

Berry, H. 2011. Pembrokeshire Day Two – 8th Oct 2011. Accessed 29 May 2021, https://malvernu3a.org.uk/ geology.malvernu3a.org.uk/walks/2011/Pembrokeshire/day%20two/index.htm

Bevins, R.E., R.A. Ixer, and N.J.G. Pearce. 2014. Carn Goedog is the likely major source of Stonehenge doler- itic bluestones: Evidence based on compatible element geochemistry and Principal Component Analysis. *Journal of Archaeological Science*, v. 42, February, pp. 179–193.

BGS, Geology of Britain Viewer Classic. Accessed 29 May 2021, https://mapapps.bgs.ac.uk/geologyofbritain/ home.html

BGS, Ironbridge Gorge, Shropshire, Landslide Case Study. Accessed 29 May 2021, https://www.bgs. ac.uk/case-studies/ironbridge-gorge-shropshire-landslide-case-study/

BGS, Lexicon of Named Rock Units, Dowlais Limestone Formation. Accessed 29 May 2021, https://webapps. bgs.ac.uk/lexicon/lexicon.cfm?pub=DWL

BGS, Lexicon of Named Rock Units, Stackpole Limestone. Accessed 29 May 2021, https://webapps.bgs.ac.uk/ lexicon/lexicon.cfm?pub=STPL

BGS, Mam Tor, Derbyshire, Landslide Case Study. Accessed 29 May 2021, https://www.bgs.ac.uk/case-studies/ mam-tor-derbyshire-landslide-case-study/

Black Country GeoPark. 2016. Geosite: 2 Wren's Nest. Accessed 29 May 2021, https://dudleypa.files.word- press.com/2016/04/geology-of-wrens-nest-geopark-for-web.pdf

Blackett, M. 2019. The Shaping of England's Peak District National Park. Accessed 29 May 2021, https://serc. carleton.edu/vignettes/collection/68183.html or https://serc.carleton.edu/69110

BlueJohn-Cavern.co.uk. History. Accessed 29 May 2021, https://www.bluejohn-cavern.co.uk/history

Butler, R. Barafundle Bay and Stackpole Quay. The Geological Society. Accessed 29 May 2021, https://www. geolsoc.org.uk/GeositesBarafundle

Butler, R. Folding at Broadhaven. The Geological Society. Accessed 29 May 2021, https://www.geolsoc.org. uk/GeositesBroadhaven

Byles, J. 2012. Aust Cliff. Outcrop, Blog of the Avon RIGS Group. Accessed 29 May 2021, http:// avonrigsoutcrop.blogspot.com/2012/06/rigs-of-month-june-aust-cliff.html

Carradice, P. 2012. The Legend of Cantre'r Gwaelod. Wales History Blog. Accessed 29 May 2021, https:// www.bbc.co.uk/blogs/waleshistory/2012/03/legend_of_cantrer_gwaelod.html

Carrier, R. The 5 Best Dinosaur-Hunting Sites in the UK. Frommer's. Accessed 29 May 2021, https://www. frommers.com/trip-ideas/cultural-immersion/the-5-best-dinosaur-hunting-sites-in-the-uk

Cawood, A.J., C.E. Bond, J.A. Howell, R.W.H. Butler, and Y. Totake. 2017. LiDAR, UAV or compass- clinometer? Accuracy, coverage and the effects on structural models. *Journal Structural Geology*, v. 98, pp. 67–82.

Cawood, A.J., and C.E. Bond. 2020. Broadhaven revisited: A new look at models of fault–fold interaction. In Bond, C.E. and Lebit, H.D. (eds), *Folding and Fracturing of Rocks: 50 Years of Research since the Seminal Text Book of J. G. Ramsay*. Geological Society, London, Special Publications, 487, pp. 105–126.

Coast and Country. 2019. Caerfai Beach – St Davids. Accessed 29 May 2021, https://www.welsh-cottages. co.uk/blog/caerfai

Colley, H. Great Orme Bronze Age Mine. The Geological Society. Accessed 29 May 2021, https://www.geol- soc.org.uk/GeositesGreatOrme

Collins, A., and C. Buchan. 2004. Provenance and age constraints of the South Stack Group, Anglesey, UK: U–Pb SIMS detrital zircon data. *Journal of the Geological Society*, v. 161, pp. 743–746.

Conwy.gov.uk. Discovering the Great Orme. Accessed 29 May 2021, https://www.conwy.gov.uk/en/Resident/ Leisure-sport-and-health/Coast-and-Countryside/Assets/documents/Discover-the-Great-Orme.pdf

Cruickshanks, A. 2017. Aust Cliff. UK Fossil Collecting. Accessed 31 May 2021, https://ukfossils.co.uk/2010/ 03/04/aust-cliff/#:~:text=At%20the%20cliff%20top%20are, and%20White%20cliffs%20at%20Aust

Cruickshanks, A. 2017. Blue John Cavern. UK Fossil Collecting. Accessed 31 May 2021, https://ukfossils. co.uk/2011/03/11/castleton/entrance-to-blue-john-cavern/

Cruickshanks, A. Caim Beach. UK Fossil Collecting. Accessed 31 May 2021, https://ukfossils.co.uk/2008/03/22/caim/#:~:text=FOSSIL%20HUNTING&text=However%2C%20the%20most%20impressive%20fossils, of%20the%20walk%20from%20Caim

Cruickshanks, A. 2017. Lavernock. UK Fossil Collecting. Accessed 31 May 2021, https://ukfossils.co.uk/2002/09/20/lavernock/#:~:text=FOSSIL%20HUNTING, Blue%20Lias%20limestone%20and%20shales

Cruickshanks, A. Llantwit Major. UK Fossil Collecting. Accessed 31 May 2021, https://ukfossils.co.uk/2002/11/22/llantwit-major/#:~:text=Llantwit%20Major%20has%20very%20good,   the%20foreshore%20boulders%20and%20shale

Cruickshanks, A. Marloes Sands. UK Fossil Collecting. Accessed 31 May 2021, https://ukfossils.co.uk/2007/10/26/marloes-sands/#:~:text=FOSSIL%20HUNTING, are%20both%20plentiful%20and%20diverse

Cruickshanks, A. Penarth. UK Fossil Collecting. Accessed 31 May 2021, https://ukfossils.co.uk/2009/08/05/penarth/

Cruickshanks, A. Whitesands Bay. UK Fossil Collecting. Accessed 31 May 2021, https://ukfossils.co.uk/2007/03/01/whitesands-bay/

Cutler, A., P.G. Oliver, and C.G.R. Reid. 1990. *Wren's Nest National Nature Reserve Geological Handbook and Field Guide*. Nature Conservancy Council, Peterborough, 29 p.

Daniel, L., and B. Roberts. 2011. Pembrokeshire Day Three – 9th October 2011. Accessed 29 May 2021, https://malvernu3a.org.uk/geology.malvernu3a.org.uk/walks/2011/Pembrokeshire/day%20three/index.htm

Discovering Fossils. Marloes Sands (Pembrokeshire). Accessed 29 May 2021, http://www.discoveringfossils.co.uk/marloes_sands_fossils.htm#:~:text=Fossils%20are%20not%20abundant%20at,   to%20just%20a%20small%20number

Donnelly, L. 2006. The Mam Tor Landslide, Geology and Mining Legacy around Castleton, Peak District National Park, Derbyshire, UK. IAEG2006 Field Trip Guide Mam Tor and Castleton, The Geological Society of London, 21 p.

Dyfed Archaeological Trust. Submerged Forests. Accessed 29 May 2021, https://dyfedarchaeology.org.uk/lostlandscapes/submergedforests.html

Earth Heritage Trust. 2011. Gardiners Quarry. Accessed 29 May 2021, https://www.earthheritagetrust.org/gardiners-quarry/#:~:text=The%20dominant%20rock%20type%20in,belonging%20to%20the%20Malverns%20Complex.&text=From%20the%20quarry%2C%20the%20landscape,   parallel%20to%20the%20Malvern%20Hills

Earthwatcher. Stackpole Quay, Pembrokeshire. Accessed 29 May 2021, http://www.ipernity.com/doc/earthwatcher/39023152

European Geoparks. Geomôn. Accessed 4 October 2022, https://www.europeangeoparks.org/?page_id=528

Evans, B. Dinosaur Footprints, Bendricks Rock. The Geological Society. Accessed 29 May 2021, https://www.geolsoc.org.uk/GeositesDinosaurBendrick

Falkingham, P.L., S.C.R. Maidment, J.N. Lallensack, J.E. Martin, G. Suan, L. Cherns, C. Howells, and P.M. Barrett. 2021. Late Triassic Dinosaur Tracks from Penarth, South Wales. Geological Magazine, 12 p. Accessed 4 October 2022, https://www.researchgate.net/publication/357404632_Late_Triassic_dinosaur_tracks_from_Penarth_south_Wales

Fewtrell, S. The Ironbridge Gorge. The Geological Society. Accessed 29 May 2021, https://www.geolsoc.org.uk/GeositesIronbridge

Fitches, W.R. 2011. Notes on Stratigraphy and Structure of Marloes Sands and Neighbourhood. Accessed 29 May 2021, University of Wales class notes.

Flight, W. 1881. Report of an Examination of the Meteorites of Cranbourne, Australia; of Rowton, Shropshire; and of Middlesbrough, in Yorkshire. Proceedings of the Royal Society of Lonodon, v. 216–219, pp. 345–346.

Fortey, R. 2000. *Trilobite: Eyewitness to Evolution by Richard Fortey*. Vintage Books, New York, 320 p.

Geogeek1726, 2013. The Wrekin. Accessed 30 May 2021, https://geogeek1726.wordpress.com/2013/12/05/the-wrekin/

Geoscience Wales. 2014. Wall Calendar – April: The Silurian Period. Accessed 30 May 2021, http://geoscience.wales/wall-calendar/2014-calendar/april-the-silurian-period/

GeoWorld Travel. Two Day Trip to Pembrokeshire. Accessed 30 May 2021, https://www.geoworldtravel.com/Geology-Field-Trips.php

Hartley, N. 2020. Penarth 'Dinosaur Footprints' Investigated by Museum. BBC Wales News. Accessed 30 May 2021, https://www.bbc.com/news/uk-wales-53893502

Herbosch, A., and J. Verniers. 2015. Field Guide to the Geology of the Brabant Massif. Memoirs of the Geological Survey of Belgium, Royal Belgium Institute of Natural Sciences, Brussels, 40 p.

Hermolle, M. 2009. Ogmore by Sea. Open University Geological Society. Accessed 30 May 2021, https://ougs. org/southwest/event-reports/144/ogmore-by-sea/

Hollis, C., and A. Juerges. 2019. Geology of the Great Orme, Llandudno. University of Manchester. Accessed 30 May 2021, https://www.greatormemines.info/wp-content/uploads/2019/10/Geology-of-the-Great-Orme-University-of-Manchester.pdf

Howells, M.F. 2007. *Wales: British Regional Geology.* British Geological Survey, Keyworth, Nottingham, 230 p.

Huff, K. 2019. South Wales Field-Trip: Saturday 4th May Locality 2: Dunraven Bay. Dorset GA Group Newsletter Autumn 2019, pp. 1–3. Accessed 3 June 2021, https://dorsetgeologistsassociation.org/wp-content/uploads/2020/02/Autumn-newsletter-2019.pdf

Ixer, Mineralization at the Great Orme Mine. Great Orme Exploration Society Ltd. Accessed 30 May 2021, http://www.goes.org.uk/html/mineralization_on_the_orme.html

John, B. 2019. Ceibwr - Geology and Landscape. Accessed 30 May 2021, https://www.moylgrove.wales/geology-of-ceibwr

Jones, A. 2020. Penarth Dino Footprints? – An Update. Nub News. Accessed 30 May 2021, https://penarth.nub.news/n/penarth-dino-footprints---an-update

Kelly, J.E. 2010. The Post-Triassic Uplift and Erosion History of the Southwestern UK. PhD dissertation, School of Geography, Earth & Environmental Sciences, The University of Birmingham, 619 p.

Lewis, C.A. 1996. Prehistoric Mining at the Great Orme. M.Phil, University of Wales – Bangor, 184 p.

Lonely Planet, Dan-yr-Ogof National Showcaves Centre for Wales. Accessed 31 May 2021, https://www.lonelyplanet.com/wales/fforest-fawr-black-mountain/attractions/dan-yr-ogof-national-showcaves-centre-for-wales/a/poi-sig/1223315/1334391

Maelgwyn Mineral Services Limited. 2020. Mining in Llandudno - North Wales. Accessed 31 May 2021, https://www.maelgwyn.com/who-we-are/mining-in-llandudno-north-wales/

Malvern Spa Association. Geologic History of the Malvern Hills. Accessed 5 June 5, 2021, https://malvernspa.org/geology-malvern-hills/geological-history-of-the-malvern-hills-precambrian/

Michael, J. 2013. The Museum, the Beach and "The Horsebox": Manchester Geological Association's Trip to Pembrokeshire. The North West Geologist, no. 18, pp. 35–42.

Mitten, A., and J.K. Pringle. 2018. *Geology of the Blue Lagoon.* John Wiley & Sons Ltd., The Geologists' Association and The Geological Society of London, *Geology Today,* v. 34, no. 1, January–February 2018, John Wiley & Sons Ltd, Oxford, pp. 35–38.

Murchison, R.I. 1839. The Silurian System, founded on geological researches in the counties of Salop, Hereford, Radnor, Montgomery, Caermarthen, Brecon, Pembroke, Monmouth, Gloucester, Worcester, and Stafford v.1. John Murray, Albemarle Street. Accessed 31 May 2021, https://library.si.edu/digital-library/book/siluriansystemfimurc

Neighbor, G. 2009. Geology Fieldwork Weekend, Saturday March 21st and Sunday March 22nd 2009, Glamorgan Heritage Coast. OUGS (SW) Field Book. Accessed 29 May 2021, https://ougs.org/files/swe/general/south_wales_ougs_fieldbook.pdf

New College Group, A Brief History of Manchester. Accessed 31 May 2021, https://www.newcollegegroup.com/blog/a-brief-history-of-manchester/

Nichols, K. 2019. A Geoconservation Perspective on the Trace Fossil Record Associated with the End – Ordovician Mass Extinction and Glaciation in the Welsh Basin. PhD dissertation, University of Chester – Department of Biological Sciences, 161 p.

Open University. 2011. Geology of the West Midlands. Accessed 31 May 2021, https://ougs.org/westmidlands/local-geology/110/geology-of-the-west-midlands/

Pearson, M.P., R. Bevins, R. Ixer, J. Pollard, C. Richards, K. Welham, B. Chan, K. Edinborough, D. Hamilton, R. Macphail, D. Schlee, J-L Schwenninger, E. Simmons, and M. Smith. 2015. Craig Rhos-y-felin: A Welsh bluestone megalith quarry for Stonehenge. *Antiquity,* v. 89, no. 348, pp. 1331–1352.

Pearson, M.P., J. Pollard, C. Richards, K. Welham, C. Casswell, C. French, D. Schlee, D. Shaw, E. Simmons, A. Stanford, R. Bevins, and R. Ixer. 2019. Megalith quarries for Stonehenge' s bluestones. *Antiquity,* v. 93, no. 367, pp. 45–62.

Pembrokeshire Coast National Park. Abereiddi to Blue Lagoon Adventure Walk. Accessed 31 May 2021, https://www.pembrokeshirecoast.wales/things-to-do/access-for-all/walks-for-all/abereiddi-blue-lagoon/

Pembrokeshire Coast National Park Authority. 2013. Global Geopark Study Results. Accessed 31 May 2021, https://www.pembrokeshirecoast.wales/wp-content/uploads/archive/npa_2013_2013_april_24_18_13_geopark.pdf

Pharaoh, T. 2019. *The Complex Tectonic Evolution of the Malvern Region: Crustal Accretion Followed by Multiple Extensional and Compressional Reactivation*. British Geological Survey, Keyworth, Nottingham, 27 p. Accessed 31 May 2021, http://nora.nerc.ac.uk/id/eprint/522961/1/Complex%20tectonic%20 evolution%20of%20the%20Malvern%20region%20with%20Figs%20v17.pdf

Phillips, E. 1991a. The lithostratigraphy, sedimentology and tectonic setting of the Monian Supergroup, western Anglesey, North Wales. *Journal of the Geological Society*, London, v. 148, pp. 1079–1090.

Phillips, E. 1991b. Progressive deformation of the South Stack and New Harbour Groups, Holy Island, western Anglesey, North Wales. *Journal of the Geological Society*, v. 148, pp. 1091–1100.

Powell, J.H., B.W. Glover, and C.N. Waters. 1992. A Geological Background for Planning and Development in the 'Black Country.' Technical Report WA/92/33, British Geological Survey, Keyworth, Nottingham, 79 p.

Radioactive Waste Management. 2016. Wales – Regional Geology. Curie Avenue Harwell Oxford, 24 p. Accessed 31 May 2021, https://assets.publishing.service.gov.uk/government/uploads/system/uploads/ attachment_data/file/834765/Wales_Regional_Geology_V1.0a.pdf

Reading Geological Society. 2009. Field Trip to Pembroke, September 2009. Reading Geological Society Newsletter 172 November 2009, pp. 3–5.

Rodell, J. 2007. 10 Fossil Sites. The Guardian, Sat 14 Jul 2007 18.46 EDT. Accessed 31 May 2021, https:// www.theguardian.com/travel/2007/jul/14/beach.uk13

Ross, D. Treak Cliff Cavern. Britain Express. Accessed 31 May 2021, https://www.britainexpress.com/ attractions.htm?attraction=2289

Rowberry, M., Y. Battiau-Queney, P. Walsh, B. Blazejowski, V. Bout-Roumazeilles, A. Trentesaux, L. Krizova, and H.M. Griffiths. 2013. The weathered Carboniferous Limestone at Bullslaughter Bay, South Wales: The first example of ghost rock recorded in the British Isles. *Geologica Belgica*, v. 17, no. 1, pp. 33–42.

Rushton, A.W.A., and S.G. Molyneux. 2011. Welsh Basin. In Rushton, A.W.A., Brück, P.M., Molyneux, S.G., Williams, M., and Woodcock, N.H., *A Revised Correlation of the Cambrian Rocks in the British Isles*. Geological Society, London, Special Report 25, pp. 21–27. Accessed 3 June 2021, https://www.research gate.net/publication/279463377_Welsh_Basin

Sharpe, T. 2007. Dinosaur footprints at The Bendricks. Geologists' Association South Wales Group for Cardiff. Accessed 31 May 2021, http://swga.org.uk/wp-content/uploads/2020/01/Bendrick.pdf

Siverter, D.J., and M. Williams. 1995. An early Cambrian assignment for the Caerfai Group of South Wales. *Journal of the Geological Society*, v. 152, pp. 221–224.

Smith, D. 2016. Longer Trip: Pembrokeshire. Open University Geological Society. Accessed 31 May 2021, https://ougs.org/london/event-reports/666/longer-trip-pembrokeshire/

Snowdonia National Park Authority. Accessed 31 May 2021, https://www.snowdonia.gov.wales/home

Strachan, R.A., R.D. Nance, R.D. Dallmeyer, R.S. D'Lemos, J.B. Murphy, and G.R. Watt. 1996. Late Precambrian tectonothermal evolution of the Malverns Complex. *Journal of the Geological Society*, London, v. 153, pp. 589–600.

Talbot, J., and J. Cosgrove. 2011. The Roadside Geology of Wales. The Geologists' Association Guide No. 69, London, 214 p.

The Geological Society. Dinorwig Power Station and Slate Quarry. Accessed 31 May 2021, https://www.geolsoc. org.uk/GeositesDinorwig

The Geological Society. Wren's Nest. Accessed 31 May 2021, https://www.geolsoc.org.uk/GeositesWrensNest

Thomas, H.H. 1923. The source of the stones of Stonehenge. *Antiquaries Journal*, v. 3, pp. 239–260. http://dx. doi.org/10.1017/S0003581500005096

Visit Anglesey, Geology. Accessed 31 May 2021, https://www.visitanglesey.co.uk/en/about-anglesey/geology/

Waters, C.N. 2008. *Stratigraphical Chart of the United Kingdom: Southern Britain*. British Geological Survey and The Geological Society, Keyworth, Nottingham.

Watkins, N. 2011. Walking through Time: The Geopark Way. *GEOExPro*, v. 8, no. 3, 10 p.

Wikipedia, Abereiddy. Accessed 31 May 2021, https://en.wikipedia.org/wiki/Abereiddy

Wikipedia, Aberystwyth Grits Group. Accessed 31 May 2021, https://en.wikipedia.org/wiki/Aberystwyth_ Grits_Group

Wikipedia, Barafundle Bay. Accessed 31 May 2021, https://en.wikipedia.org/wiki/Barafundle_Bay

Wikipedia, Black Country Geopark. Accessed 31 May 2021, https://en.wikipedia.org/wiki/Black_Country_ Geopark

Wikipedia, Blue John Cavern. Accessed 31 May 2021, https://en.wikipedia.org/wiki/Blue_John_Cavern

Wikipedia, Brecon Beacons National Park. Accessed 4 June 2021, https://en.wikipedia.org/wiki/Brecon_ Beacons_National_Park

Wikipedia, Broad Haven. Accessed 31 May 2021, https://en.wikipedia.org/wiki/Broad_Haven

Wikipedia, Caerfai Bay. Accessed 31 May 2021, https://en.wikipedia.org/wiki/Caerfai_Bay

Wikipedia, Cantre'r Gwaelod. Accessed 31 May 2021, https://en.wikipedia.org/wiki/Cantre%27r_Gwaelod

Wikipedia, Ceibwr Bay. Accessed 31 May 2021, https://en.wikipedia.org/wiki/Ceibwr_Bay

Wikipedia, Cribarth Disturbance. Accessed 1 June 2021, https://en.wikipedia.org/wiki/Cribarth_Disturbance

Wikipedia, Dan-yr-Ogof. Accessed 1 June 2021, https://en.wikipedia.org/wiki/Dan_yr_Ogof

Wikipedia, Dinorwic Quarry. Accessed 1 June 2021, https://en.wikipedia.org/wiki/Dinorwic_quarry

Wikipedia, Geology of Brecon Beacons National Park. Accessed 4 June 2021, https://en.wikipedia.org/wiki/Geology_of_Brecon_Beacons_National_Park

Wikipedia, Geology of Ceredigion. Accessed 1 June 2021, https://en.wikipedia.org/wiki/Geology_of_Ceredigion

Wikipedia, Geomôn. Accessed 4 October 2022, https://en.wikipedia.org/wiki/GeoM%C3%B4n

Wikipedia, Glamorgan Heritage Coast. Accessed 1 June 2021, https://en.wikipedia.org/wiki/Glamorgan_Heritage_Coast

Wikipedia, Great Orme. Accessed 1 June 2021, https://en.wikipedia.org/wiki/Great_Orme

Wikipedia, History of Manchester. Accessed 1 June 2021, https://en.wikipedia.org/wiki/History_of_Manchester

Wikipedia, Ironbridge Gorge. Accessed 1 June 2021, https://en.wikipedia.org/wiki/Ironbridge_Gorge

Wikipedia, Mam Tor. Accessed 1 June 2021, https://en.wikipedia.org/wiki/Mam_Tor

Wikipedia, Manchester. Accessed 1 June 2021, https://en.wikipedia.org/wiki/Manchester

Wikipedia, Marloes Sands. Accessed 1 June 2021, https://en.wikipedia.org/wiki/Marloes_Sands

Wikipedia, National Slate Museum. Accessed 4 October 2022, https://en.wikipedia.org/wiki/National_Slate_Museum

Wikipedia, Peak District National Park. Accessed 1 June 2021, https://en.wikipedia.org/wiki/Peak_District

Wikipedia, Pembrokeshire Coast National Park. Accessed 1 June 2021, https://en.wikipedia.org/wiki/Pembrokeshire_Coast_National_Park

Wikipedia, Pennine Basin. Accessed 1 June 2021, https://en.wikipedia.org/wiki/Pennine_Basin

Wikipedia, Snowdonia National Park. Accessed 1 June 2021, https://en.wikipedia.org/wiki/Snowdonia#Snowdonia_National_Park

Wikipedia, South Stack. Accessed 1 June 2021, https://en.wikipedia.org/wiki/South_Stack

Wikipedia, The Bendricks. Accessed 1 June 2021, https://en.wikipedia.org/wiki/The_Bendricks

Wikipedia, The Roaches. Accessed 3 October 2022, https://en.wikipedia.org/wiki/The_Roaches

Wikipedia, The Wrekin. Accessed 1 June 2021, https://en.wikipedia.org/wiki/The_Wrekin

Wikipedia, Treak Cliff Cavern. Accessed 1 June 2021, https://en.wikipedia.org/wiki/Treak_Cliff_Cavern

Wikipedia, Trewent Point. Accessed 1 June 2021, https://en.wikipedia.org/wiki/Trewent_Point

Wikipedia, Wrekin Terrane. Accessed 1 June 2021, https://en.wikipedia.org/wiki/Wrekin_Terrane

Williams, A.T., P. Davies, and A.P. Belov. 1996. Coastal cliff failures at Colhuw beach, Wales, UK. *Transactions on Ecology and the Environment*, v. 9, pp. 75–82.

Woudloper, https://commons.wikimedia.org/wiki/File:Geologic_map_Wales_%26_SW_England_EN.svg

Wrekin Housing Group, https://www.linkedin.com/company/the-wrekin-housing-group-limited/

4 Southern England

White Cliffs of Dover. (Photo courtesy of Johannes Philipp, https://commons.wikimedia.org/wiki/File:Chalkstone_Coast_near_Dover_-_panoramio.jpg.)

OVERVIEW

Southern England is the birthplace of stratigraphy, the geologic map, and paleontology. As one traverses the land from east to west, the rocks range in age from Pleistocene glacial sediments around London to Precambrian granites and metamorphics at Land's End. This geological tour explores canyons and quarries, sea cliffs and cheese caves, tin mines, museums, Roman baths, and prehistoric monoliths. We see rocks deformed by thrusting during the Variscan and Alpine mountain-building events, explain the rare ophiolites of Lizard Point, and visit dinosaur stomping grounds and boneyards along the Jurassic Coast. Fossil collecting sites across the country are described, and the origin of the stunningly brilliant white chalk cliffs of Dover and Southdown National Park is explained. We pass through and examine the geology of 12 Sites of Special Scientific Interest and 5 UNESCO World Heritage Sites.

DOI: 10.1201/9781351165600-4

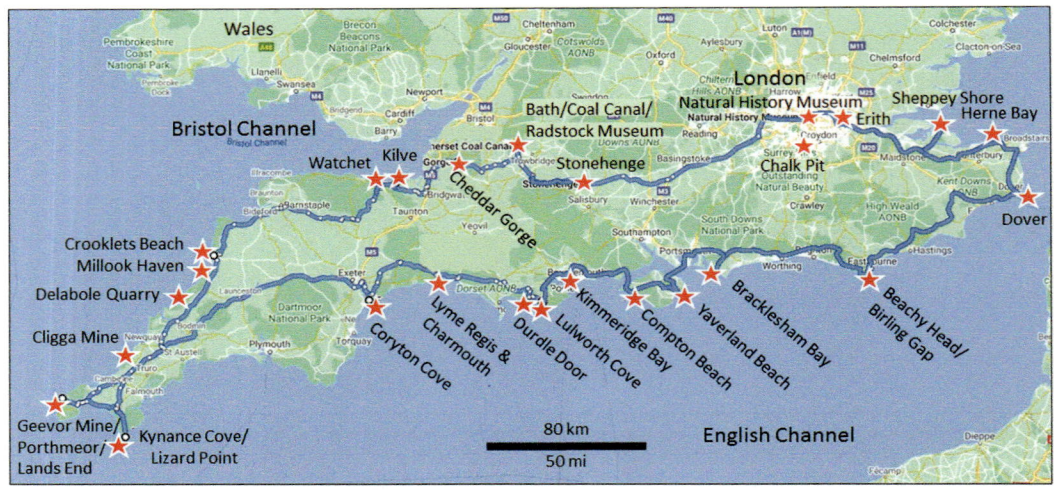

Key stops on the southern England geo-tour.

ITINERARY

Begin –London
Stop 1 Geology of the London Area
      1.1 Natural History Museum
      1.2 Erith Submerged Forest
      1.3 Gilbert's Pit
      1.4 Park Hill Chalk Pit
Stop 2 Stonehenge
Stop 3 Bath Area and the Somerset Coal Canal
      3.1 Somerset Coal Canal
      3.2 Roman Baths
      3.3 Radstock Museum
Stop 4 Cheddar Gorge
      4.1 Horseshoe Bend
      4.2 Landslip Quarry
      4.3 Show Caves
Stop 5 Kilve Beach
Stop 6 Watchet Beach
Stop 7 Crooklets Beach
Stop 8 Millook Haven
Stop 9 Delabole Quarry
Stop 10 Cligga Mine
Stop 11 Penzance Peninsula
      11.1 Porthmeor Cove
      11.2 Geevor Mine and Cornish Miners

START: LONDON

London is the capital of England and the United Kingdom and a true world city. Everyone knows of London, but here are a few things you might not know. The area was originally settled by hunter-gatherers around 6000 BC, and evidence of Bronze Age bridges and Iron Age forts has been uncovered at a ford and narrow point where the River Effra flowed into the Thames (History Channel, London; Wikipedia, History of London).

The Romans founded a port and trading settlement called Londinium in 47 AD, 4 years after invading Britain, and a few years later they built the first wooden bridge across the Thames. However, in 60 AD, the city was sacked by the Iceni, led by Queen Boudica. The city was soon rebuilt, and by the 2nd century, Londinium had become the capital of Roman Britannia. The population was around 60,000. The present City of London, the historic city center, is roughly defined by the Roman walls.

In AD 410, the Romans left and the city went into decline. By the end of the 5th century, it was effectively abandoned. In the early 7th century the area around London area was part of the kingdom of the East Saxons, but it came under Mercian control in the 730s.

London was sacked by Vikings in 842 and again in 851. The "Great Heathen Army" of Danes, which had rampaged across England since 865, took London in 871. The city remained in Danish hands until 886 when it was captured by King Alfred the Great of Wessex and reincorporated into Mercia under Ealdorman Æthelred. After Æthelred died (911) the city became part of Wessex, where it faced competition with Winchester as the political center of the Kingdom of England.

Following King Edmund's defeat at the Battle of Assandun (1016), London was given to the Viking King Cnut along with all of England north of the Thames. Cnut's kingdom dissolved in 1042, and English rule was restored under Edward the Confessor, responsible for founding Westminster Abbey in 1065.

Only a year later, after his victory at Hastings, William the Conqueror became King of England. Willliam and his Normans built the Tower of London, the first stone castle in England. In 1176, the first stone bridge was built across the Thames.

In 1216, during the First Barons' War, London was occupied by Prince Louis of France, who had been called by the barons against King John: he was crowned King of England in St Paul's Cathedral. Following John's death in 1217, Louis's backers withdrew their allegiance, supporting John's son Henry III and forcing Louis to withdraw from England.

Trade increased steadily, and London grew in size and importance as a commercial center. In 1100, London's population was a bit over 15,000; by 1300 it was roughly 80,000. London lost at least half of its population during the Black Death in the 1300s, but its economic and political importance led to a rapid recovery. By the time of King Henry VIII (1509–1547), the population of London was around 100,000.

The Great Plague of 1665 killed about 100,000 Londoners in the city of half a million. A year later much of the city burned during the Great Fire.

The Bank of England, founded in 1694, accelerated London's evolution into an international financial center. By 1840, the city had 2 million people, often living in unsanitary conditions, which helped spread cholera and other epidemics.

During the reign of Queen Victoria, London was the capital of the vast British Empire. The London Underground, the world's first subway system, opened in 1863.

During the London Blitz in World War II, the city was bombed repeatedly by the Luftwaffe. The bombing killed around 30,000 Londoners.

The Great Smog of 1952, caused by coal smoke and stagnant air, killed thousands. Yet the city continues to grow and prosper, establishing itself as one of the great cultural and financial centers of Europe (History Channel, London; Wikipedia, History of London). The 2022 population of the London metropolitan area is 9.5 million.

STOP 1 GEOLOGY OF THE LONDON AREA

London is situated in the London Basin, a broad but shallow, roughly east–west downwarp formed during the Alpine Orogeny, a mountain-building event caused by collision of the African tectonic plate with the European plate. To the north and south of the basin are hills made of Late Cretaceous (roughly 85 Ma) chalk, a porous marine limestone. Within the basin are rare outcrops of Paleocene (~60 Ma) Thanet Formation (a marine sandstone), the Lambeth Group (clay, silt, and sand deposited in swamps, estuaries, and deltas), and the Eocene Harwich Formation (marine siltstone and claystone) and London Clay (a fossil-rich marine clay 56–50 Ma). The London Clay Formation underlies most of the city. The Bagshot Sand, a marine sandstone deposited around 50 Ma (Eocene), forms isolated hills at the top of the sequence. Above this, there are scattered glacial deposits and river gravels less than 450,000 years old, and some windblown glacial silt known as loess.

In early Paleozoic time, the London area was part of the Midlands Microcraton, a small continental mass consisting of Precambrian island arc volcanics with a thin sedimentary cover. These sedimentary rocks consisted of Cambrian to Silurian marine units and Early Devonian river deposits. The section was deformed during the Acadian phase of the Caledonian Orogeny (Early to

Middle Devonian time). During Carboniferous time, this area was part of the Anglo-Brabant massif (sometimes called the Wales-Brabant massif), a topographic and structurally high block that extended from Wales to Belgium. Late Carboniferous extension led to development of foreland basins north of the developing Late Carboniferous-Early Permian Variscan mountain belt. These basins accumulated coal measures (repetitive sequences of sandstone, shale, and coal) and red beds (red sandstone and shale) during Late Carboniferous. During most of the Mesozoic, the region was part of a stable shallow marine area known as the London Platform. Limestone in the form of chalk was deposited on this platform. North–south extension during Permo-Triassic time led to subsidence of the Weald Basin to the south. The London area remained fairly stable during Jurassic and Cretaceous times, covered by a shallow sea with intermittent non-marine sedimentation (Ellison et al., 2004).

Age		Formation	Description
Quaternary		Glacial Till and gravels	
		Norwich Crag Fm *Unconformity*	Marine sandstone & gravels
Eocene	Thames Group	Bagshot Fm	Fine to coarse shoreline/marine sandstone
		Claygate Beds	Shallow marine sandy clay
		London Clay Fm	Gray marine clay
		Harwich Fm	Fine-grained marginal marine sandstone, rounded black flint pebbles
		Blackheath Beds	
Paleocene	Lambeth Group	Woolwich Fm	Lagoonal & estuarine interbedded gray clay and sandstone
		Reading Fm	Red to brown and gray lagoonal clay
		Upnor Fm	Shallow marine sandstone-siltstone
		Thanet Sand Fm *Unconformity*	Shallow marine sandstone
Upper Cretaceous	Chalk Group	Seaford Chalk	White soft chalk (limestone) with nodular flint layers
		Lewes Nodular Chalk	Hard white chalk (limestone) with nodular flint layers
		New Pit Chalk	Hard white chalk (limestone)

Stratigraphy of the London Area. (Information drawn from Clements, 2012; Baker et al., 2016.)

Some uplift and folding occurred in this area between deposition of the chalk and the Thanet Formation. The uplift removed some rock, causing an unconformity (an erosion surface). The deformation occurred during an early phase of the Alpine Orogeny, the mountain-building event that formed the Alps in Late Cretaceous to Eocene time. Maximum local uplift during the Alpine Orogeny, seen in the Weald Anticline south of London, is estimated at about 1,500 m (4,900 ft; Hopson, 2011). A second phase of folding and uplift, north of the main front, started at about 45 Ma and, although apparently tapering off, continues today. The associated uplift caused erosion of pretty much all rocks younger than about 50 million years in this area.

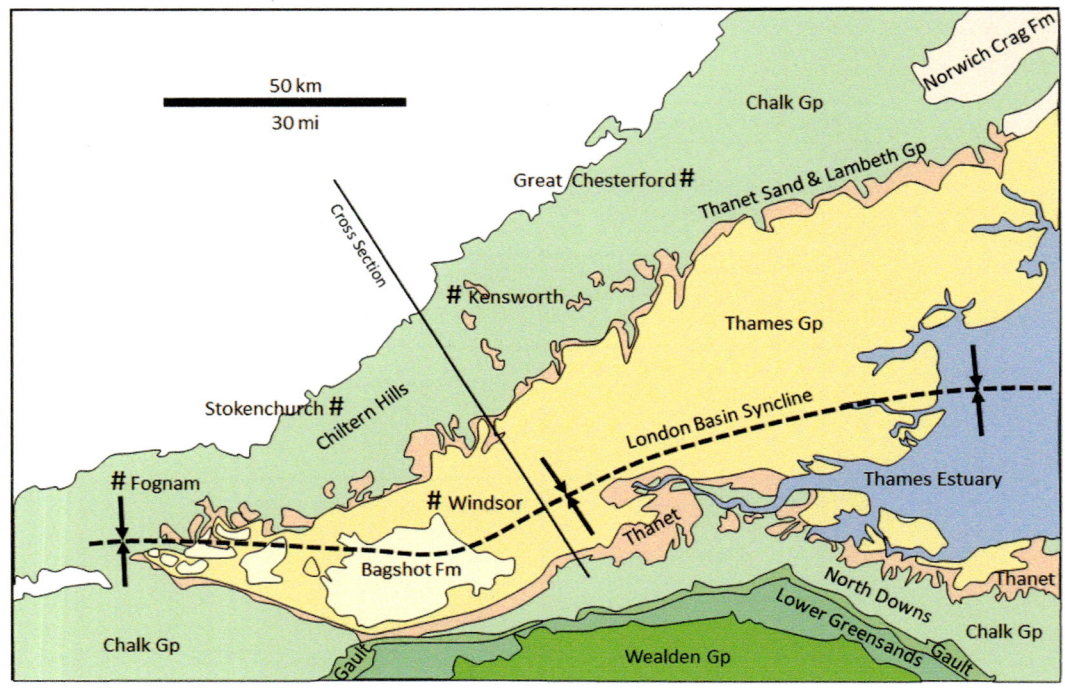

Geologic sketch map of the London area. (Modified after Clements, 2014; with permission of the Geologists' Association.)

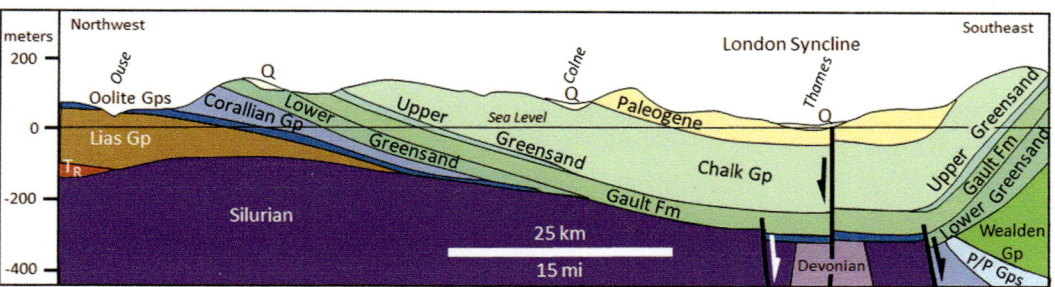

Cross section through the London area. Q, Quaternary; P/P Gps, Portland/Purbeck Groups; T_R, Triassic Units. (Modified after Clements, 2014; with permission of the Geologists' Association.)

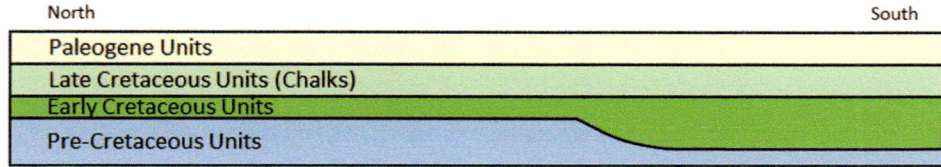

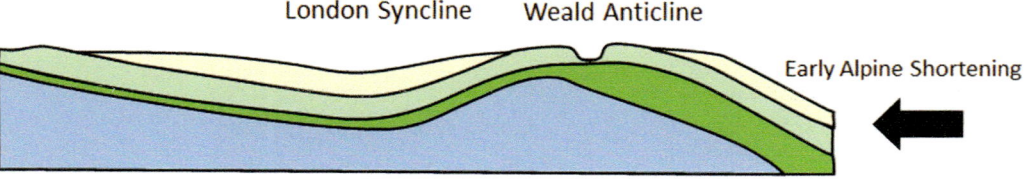

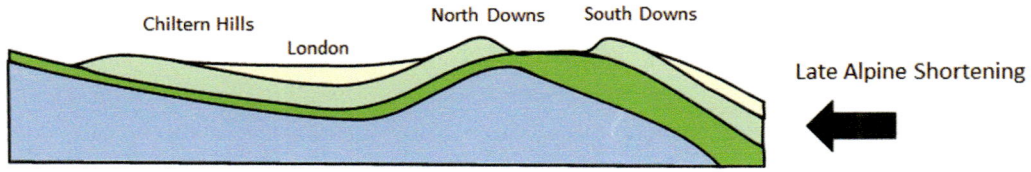

Structural evolution of the London Basin and The Weald. (Adapted from Baker et al., 2016, after Friend, 2008.)

Starting around 450,000 years ago the Anglian Ice Age glaciers reached into the northern districts of London. Large glacial lakes formed in the area and rearranged the local river drainages. Ice pushed the River Thames south to its present course (Clements, 2012; Clements, 2014; Baker et al., 2016).

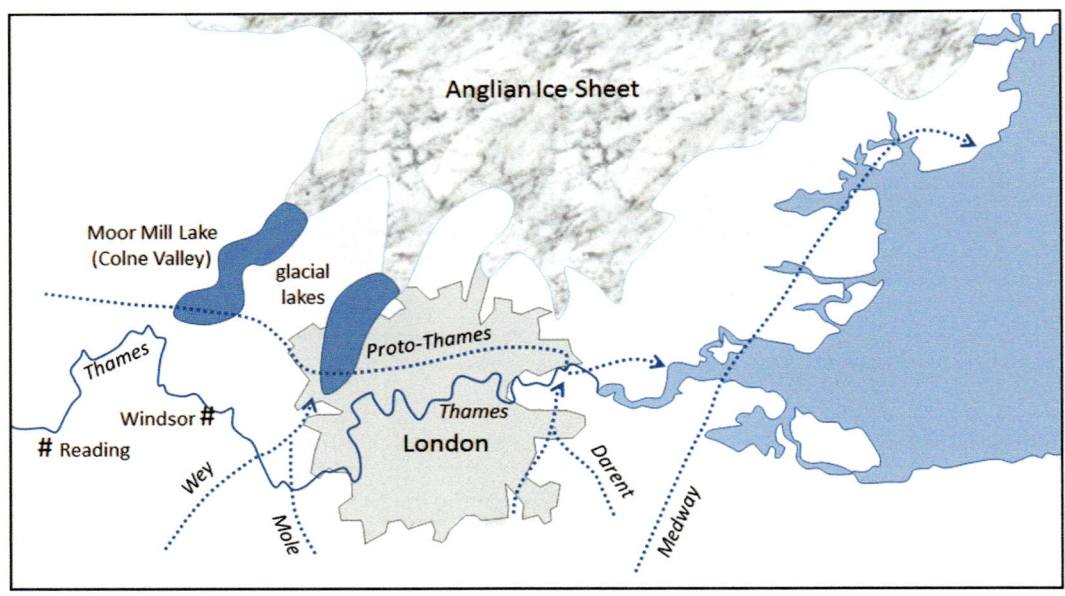

Extent of the Anglian Ice Sheet at the peak of glaciation, along with the prior location of the proto-Thames and tributaries. (Modified after Clements, 2014; with permission of the Geologists' Association.)

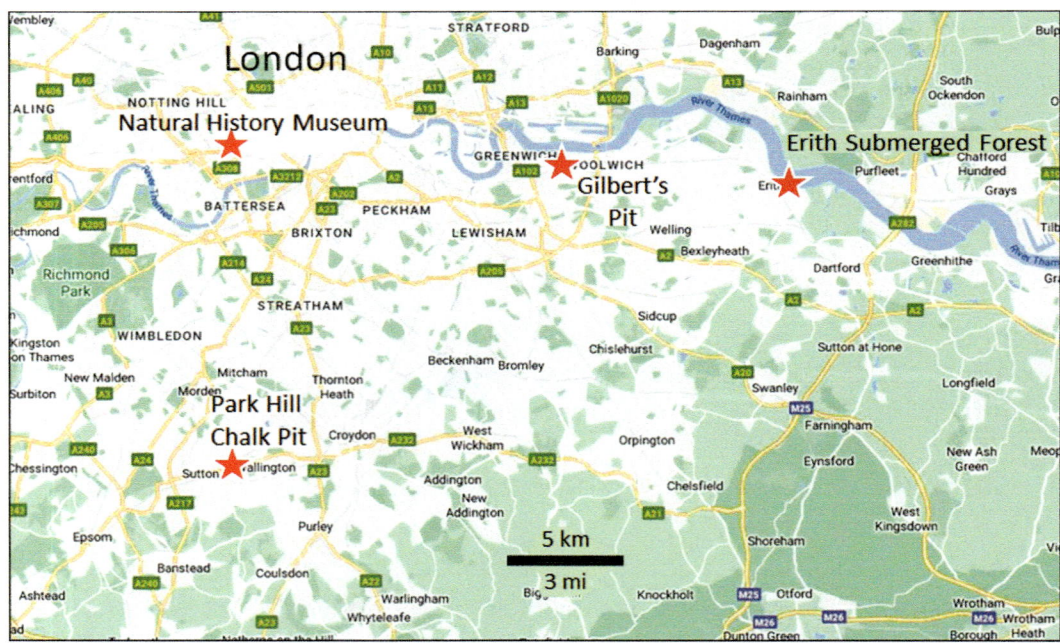

Start: *Natural History Museum, Cromwell Rd, South Kensington (51.496830, −0.175471)*

STOP 1.1 NATURAL HISTORY MUSEUM

Theropod dinosaur, Natural History Museum.

The Natural History Museum is a must-visit when in London. It houses an enormous number of world-class displays, and the collections represent some of the key advances in geology as a science. It is one of three major museums on Exhibition Road in South Kensington, the others being the Science Museum and the Victoria and Albert Museum. The Natural History Museum has one of the world's finest collections of 500,000 rocks, gems, and minerals, including 5,000 individual pieces from 2,000 meteorites. The rock collection consists of approximately 123,000 samples from around the world, including over 15,000 ore and mineral specimens. The museum has the most comprehensive British collection of seabed samples and cores.

More than 80 million earth and life science specimens are contained in five main collections: botany, entomology, mineralogy, paleontology, and zoology. Of great historical value are specimens collected by Charles Darwin. The museum is widely known for its unparalleled dinosaur skeletons.

Pliosaur from the Jurassic Coast, Natural History Museum.

The Natural History Museum originated with the collection of an Ulster doctor, Sir Hans Sloane (1660–1753). Sloane sold his extensive collection to the British Government. A century later, by 1833, the Annual Report stated that, of the 5,500 insects listed in Sloane's catalog, none remained. The inability of the museum to properly conserve its specimens was notorious. Many of those defects were corrected by the renowned paleontologist Richard Owen, appointed Superintendent of the natural history department of the British Museum (as it was then known) in 1856. Owen oversaw the move to a new building in South Kensington in 1880–81. The new museum is housed in a neo-Gothic building built by Alfred Waterhouse. The central axis of the museum is aligned with the tower of Imperial College London and the Royal Albert Hall and Albert Memorial. Together they form part of the complex known as the Albertopolis, an area centered on Exhibition Road and bordered by Cromwell Road to the south and Kensington Road to the north.

The Geological Museum

In 1976, the museum merged with the Geological Museum of the British Geological Survey. The new Geological Museum became world-famous for exhibitions that included an active volcano model and an earthquake machine. The museum's galleries were rebuilt in 1996 as The Earth Galleries. The Natural History Museum's mineralogy displays are purposely left largely unchanged from the 19th century. The central atrium was designed to overcome visitors' reluctance to visit the upper galleries: it draws them up through a model of the Earth. A Stegosaurus skeleton was added in 2015.

The Darwin Centre

The Darwin Centre houses the museum's collection of tens of millions of preserved specimens, contains educational visitor exhibits, and provides a workspace for the museum's staff. Consisting of two buildings adjacent to the main Waterhouse building, the zoological department's "spirit collection" – organisms preserved in alcohol – opened to the public in 2002. Phase Two, housing the entomology and botanical collections – the "dry collections" – opened to the public in 2009. The most famous creature in the center is an 8.62 m (28 ft) long giant squid named "Archie." Archie was captured alive in a fishing net near the Falkland Islands in 2004. The squid is not on general display but is stored in a large acrylic tank in the basement. The creature from the deep is preserved in a formalin and saline solution.

The Natural History Museum is divided into four galleries color-coded according to broad themes.

The Red Zone

This zone is entered from Exhibition Road. The gallery theme is the changing history of the Earth. Sections dealing with geology include Earth Hall (with the Stegosaurus skeleton), Earth's Treasury, Lasting Impressions, Restless Surface, From the Beginning, and Volcanoes and Earthquakes. The "Earth's Treasury" display contains rocks, minerals, and gems. The "Lasting Impressions" exhibit contains specimens that can be touched.

The Green Zone

This zone is accessed from Cromwell Road by way of Hintze Hall. Its theme deals with the evolution of the Earth. Sections of interest to geologists include fossil marine reptiles, fossils from Britain, and minerals.

The Blue Zone

Adjacent to Hintze Hall, the Blue Zone examines the diversity of life on Earth. It includes a section on dinosaurs. Unsurprisingly, the exhibit that attracts the most attention is the dinosaur display with 14 complete skeletons. There is also a full-size robotic Deinonychus eating a Tenontosaurus.

The Orange Zone

This gallery allows the public to see science at work. It is accessible from Queens Gate and contains the Darwin Centre.

Some exceptional mineral specimens, Natural History Museum. (a) Gold. (b) The Ostro Stone, a faceted 9,381 carat treated topaz. (c) Rhodochrosite specimen, Sweet Home Mine, Alma, CO. (d) The Medusa, an emerald from the Kagem Mine, Kafubu, Zambia.

A few of the museum highlights include:

- The Otumpa iron meteorite weighing 635 kg, found in 1783 in Campo del Cielo, Argentina.
- The Latrobe nugget, one of the largest clusters of cubic gold crystals known in the world.
- The Ostro Stone, a 2 kg (9,381 carats) flawless blue topaz, the largest of its kind in the world.
- The first Iguanodon teeth ever discovered (Iguanodon was an Early Cretaceous bulky herbivore).
- Dippy, a replica of the fossilized bones of a *Diplodocus carnegii*.
- The first discovered specimen of Archaeopteryx, generally accepted to be the oldest known bird.
- A rare first edition of Charles Darwin's "On the Origin of Species."

Visit

Address: Cromwell Road, London SW7 5BD
Email: use the enquiries form: https://www.nhm.ac.uk/about-us/contact-enquiries/forms/email
             form.jsp
Phone: If urgent, call +44 (0)20 7942 5000. Lines are open Monday–Sunday 9.00–17.00.
Hours: Monday–Sunday 10.00–17.50 (last entry 17.00). Closed 24–26 December.
Entrance: free, but tickets should be booked in advance to guarantee entry. All visitors (including
             Members and Corporate Supporters) must book a free ticket in advance online for guar-
             anteed entry. Coronavirus restrictions may be in effect at the museum. To keep everyone
             safe, a timed entry system is in place. Only a limited number of people are allowed in the
             galleries.

Areas off limits to the general public can be visited for a fee by booking one of the several Spirit
Collection Tours offered daily.

Subject to capacity on the day, there may be a limited number of walk-up tickets available. These
are not guaranteed to be available.

*To Erith Submerged Forest: It is recommended that you take transit to the Slade Green Rail
Station, then a bus (89, N89, 99) ~1.3 km (0.8 mi) north on Manor Road, then walk 425 m (1,400 ft)
to the Erith Yacht Club. Parking is limited (1 or 2 spots) at the entrance to the Erith Yacht Club.
Walk 150–300 m (500–1,000 ft; 10–20 min) east along the Thames Path to get access to the beach
at low tide. This is* **Stop 1.2, Erith Submerged Forest** *(51.477805, 0.198271).*

STOP 1.2 ERITH SUBMERGED FOREST

Erith submerged forest at low tide. Photo courtesy of Jane Sidell and the London Geodiversity Project.

The best and most extensive place in greater London to view a Neolithic submerged forest is at Erith near the yacht club. Tree trunks, root balls, nuts, and seeds can be seen during low tide in the 3,000- to 5,000-year-old silt and mud. Peat beds are exposed above the high tide level. Although Alder dominates, 15 different tree and shrub species have been identified, much the same as across the river at Rainham. The recent mud containing the forest sits above Cretaceous White Chalk Subgroup (Lewes Nodular Chalk Formation, Seaford Chalk Formation, and Newhaven Chalk Formation).

The site is thought to represent a change from a drier environment, when yew and other "dry" species dominated the forest, to a wetter environment produced by rising sea levels (Brook et al., 2014).

You can reach the Thames shore by a paved path leading about 150 m (500 ft) north from Manor Road toward the Erith Yacht Club. At the Thames Path/Cycle Path head east, then drop down to the shore.

To Gilbert's Pit: It is recommended you take transit to the Charlton Station and walk 1.1 km (3,420 ft) east along A206, or to the Woolwich Dockyard Station and walk 1.4 km (4,700 ft) west along A206 to get to Stop 1.3. If driving, there is limited street parking on Woodland Terrace (51.487878, 0.044605), then walk 150 m (500 ft) north to Stop 1.3, Gilbert's Pit (51.489008, 0.041845).

Stop 1.3 Gilbert's Pit

Gilbert's Pit is an abandoned and mostly overgrown quarry in the Thanet and Woolwich formations. Gray shallow marine sandstone of the Paleocene Thanet Formation (~60 Ma) comprises the lower 2/3 of the face but is mostly covered by scree. The Thanet is overlain by greenish-yellow to yellow-brown shallow marine sandy siltstone of the Paleocene Upnor Formation (also mostly covered), which is in turn overlain by a light-gray shell bed and dark-gray clay of the Eocene Woolwich Formation. The Woolwich Formation accumulated in estuaries and lagoons. The top of the face contains rounded black chert pebbles of the Eocene marine Harwich Formation (~55 Ma), known locally as the Blackheath Beds (Clements, 2014; Baker et al., 2016). Overall, the section documents a transition from fully marine to mudflats, swamps, and lagoons, then subsidence to marine conditions once again.

The old quarry mined the Thanet Sand, an unusually pure quartz sand that was ideal for use as foundry sand (sand used to form the molds for metal casting) at the Woolwich Arsenal. It was also used for making glass bottles. An interpretive panel has been set up next to the eastern face (Brook et al., 2014).

Finding rock outcrops around London is not easy. This is one of the finest and most complete stratigraphic sections in the London area, as well as being the type section for the Woolwich Formation. As such it is considered important enough to be classified as a Site of Special Scientific Interest (SSSI). Although you cannot hammer on the slope, loose pebbles and shells on the slope can be collected.

Access to this site is controlled by Greenwich Parks & Open Spaces. Contact them at 020 8921 2937 or at parks@royalgreenwich.gov.uk, or by calling the Greenwich Council at 020 8856 0100 (Clements, 2012; Hallam-Jones, Gilbert's Pit; Wikipedia, Gilbert's Pit).

Gilbert's Pit quarry face. (Photo courtesy of Mikenorton, https://en.wikipedia.org/wiki/File:Gilbert%27s_Pit_section.jpg.)

To Park Hill Chalk Pit: It is recommended that you take transit to the Carshalton Beeches Train Station and walk 330 m (1,100 ft) north on Park Hill/B278 and turn left (west) onto Bankside Close; walk 78 m (255 ft) west on Bankside Close to Stop 1.4. If you drive there is limited free parking on Bankside Close. This is **Stop 1.4, Park Hill Chalk Pit** *(51.360071, −0.170500).*

STOP 1.4 PARK HILL CHALK PIT

This is one of the few areas within greater London where you can still see the chalk that extends below the city. The former Park Hill Chalk Pit, now filled with residential housing in Bankside Close, still has a chalk face on the north side of the old quarry. This is part of the North Downs chalk escarpment along the south flank of the London Syncline (a broad trough). The quarry was mined for agricultural lime, that is, the crushed limestone was tilled into the clayey soils in the area to make them less acidic. This Cretaceous unit is known as the Upper Chalk or the Seaford Chalk Formation of the Chalk Group. It is a massive chalk with layers of chert nodules and occasional fossils. It is exposed on the nearly vertical 9 m (30 ft) high north face.

You can park on Bankside Close but not in the resident's parking area. The north face is approached through the residential car park. The south face can be reached by going through private open space between houses, but this face appears to be mostly overgrown.

Technically, this is private property, so permission should be obtained.

North face of the former Park Hill Chalk Pit.

Park Hill Chalk Pit to Stonehenge: *Return to Park Hill/B278 and turn right (south); at the next street turn right (southwest) onto Banstead Road; turn right (west) onto Chiltern Road; turn right onto Cotswold Road/B2218; immediately turn left (southwest) onto Brighton Road/B2230; at the roundabout, take the 1st exit to continue straight onto Brighton Road/A217; at the London Orbital Motorway roundabout, take the 4th exit (west) onto the M25 ramp to M3/M4/M40/-Heathrow Airport; merge onto the M25 west to Junction 12; use the left 2 lanes to take the M3 exit to Basingstoke/Southampton; at Junction 8 take the A303 exit to Andover/Salisbury; continue west on A303 to the Countess Roundabout; there take the 2nd exit onto Amesbury Bypass/A303; at the A360 roundabout, take the 3rd exit to the A360 going north; drive to the parking area and Visitor Centre on the right. This is **Stop 2, Stonehenge** (51.186338, −1.859776), for a total of 150 km (93.1 mi; 1 h 43 min).*

STOP 2 STONEHENGE

Perhaps more than any other site, Stonehenge is the iconic symbol of Great Britain. Certainly, it is one of the most important archeological sites, spanning in time from Stone Age humans on the island around 8,000 years ago to around 1500 BC (Bevins, 2012).

First mentioned in Ælfric's 10th century glossary, *henge-cliff* means "precipice"; thus, the *stanenge* or *Stanheng* means "stones supported in the air." Chippindale's *Stonehenge Complete* (2004) states that *Stonehenge* derives from the Old English words *stān,* for "stone," and *hencg* meaning "hinge" (because the lintels hinge on the uprights), although elsewhere in his book, Chippindale refers to them as "suspended stones".

Stonehenge was produced by a pre-literate culture. We still don't know how it was built, and the purpose is still a matter of debate. Proposed uses include as an astronomical observatory and/or as a religious site: the whole monument is oriented toward the rising sun on the summer solstice. Two recent theories suggested that Stonehenge was either a place of healing, such as Lourdes in France, or part of a ritual landscape that included Durrington Walls, where the area around Durrington Walls was a place of the living, whereas Stonehenge was part of the domain of the dead (Wikipedia, Stonehenge; Johns and Jackson, 2012).

Sunset and trilithons, Stonehenge. (Photo courtesy S. Banton, https://commons.wikimedia.org/-wiki/File:November_Sunset_from_inside_Stonehenge_Stone_Circle.jpg.)

DNA extracted from Neolithic human remains found near here reveal that the people who built Stonehenge were farmers from the Eastern Mediterranean with an Aegean ancestry. These Aegean farmers reached Britain about 4000 BC. At that time, Britain was inhabited by groups of hunter-gatherers. DNA studies show that these two groups did not mix much. Instead, the farmers replaced the hunter-gatherers. The next group to inhabit the area are the Bell Beaker people, associated with the Wessex culture, who came from mainland Europe around 2500 BC. They mined tin (a key ingredient in bronze) and traded extensively with Europe. The wealth derived from trade probably permitted the Wessex people to construct the second and third (*megalithic*) phases of Stonehenge.

The monument sits on the Late Cretaceous Seaford Chalk Formation. There are no other rock types within at least 10 km (6 mi) of the site (Johns and Jackson, 2012). We do know that neither of the two main building stones at Stonehenge is from the area. The larger "sarsen stones" are silicified sandstone found only in southern England. The smaller "bluestones" comprise dolerite (gabbro), tuff, rhyolite, and rarely sandstone. The mostly igneous bluestones appear to have originated in the Preseli Hills of southwestern Wales (see Chapter 3).

Stonehenge has been a legally protected Scheduled Ancient Monument since 1882 and became a UNESCO World Heritage Site in 1986. The monument is owned by the Crown and managed by English Heritage; the surrounding land is owned by the National Trust (Wikipedia, Stonehenge).

AGE AND CONSTRUCTION

Local humans appear to have dug post holes in the area starting around 8000 BC when the Salisbury Plain was still covered in forest. Several stone or wooden structures and burial mounds found in the area may date as far back as 4000 BC.

Around 3500 BC, a Stonehenge Cursus (track or path) was built 700 m (2,300 ft) north of the site as the first farmers began to clear the trees and farm the area. The oldest part of the monument is a circular ditch and embankment, of Late Cretaceous Seaford Chalk, measuring 110 m (360 ft) in diameter and surrounding all subsequent structures. This feature is dated to 3100 BC. The main entrance was in the northeast and a lesser entrance was in the south. Within the enclosed area is a circle consisting of 56 pits, each about 1 m (3.3 ft) in diameter. These "Aubrey holes" may have held timbers creating a timber circle, although there is no evidence of them. Recent work suggests that the Aubrey holes may have held a bluestone circle. In 2013 the cremated bone fragments of 63 individuals, originally interred in the Abrey holes, were excavated. There is evidence that the chalk beneath the graves was crushed by a substantial weight, suggesting that the first bluestones brought from Wales may have been used as grave markers. Radiocarbon dating of the remains has put the date of the site between 2400 and 2200 BC. Recent work suggests that the bluestones at Stonehenge had previously been a stone circle at Waun Mawn in the Preseli Hills of Wales. They had been dismantled, moved, and re-erected at Stonehenge. Evidence of a 110-m (360 ft) stone circle at Waun Mawn, which could have held some or all of the stones in Stonehenge, includes a hole that perfectly matches the pentagonal cross section of a Stonehenge bluestone. Each bluestone measures around 2 m (6.6 ft) in height, between 1 and 1.5 m (3.3 and 4.9 ft) wide, and around 0.8 m (2.6 ft) thick. The "Altar Stone" is most likely from part of the Senni Beds (Early Devonian Lower Old Red Sandstone Group) in the Brecon Beacons (Wikipedia, Stonehenge).

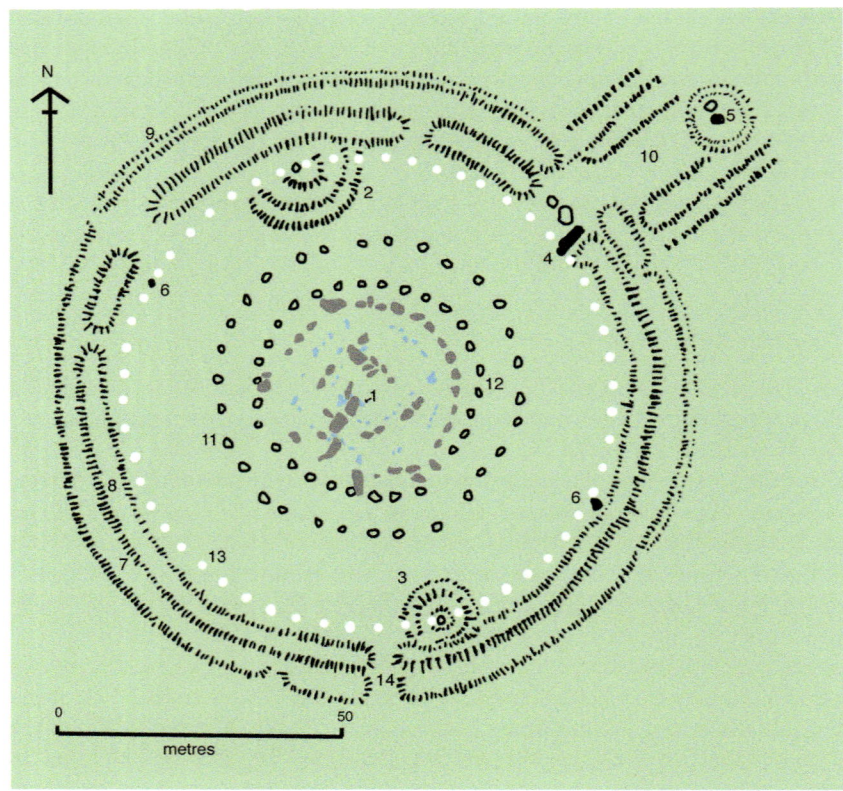

KEY TO THE MAP:

1 = The Altar Stone, a 6-ton monolith of green micaceous sandstone from Wales
2 = barrow without a burial
3 = "Barrows" (without burials)
4 = The fallen Slaughter Stone, 4.9 m long
5 = The Heel Stone
6 = Two of originally four Station Stones
7 = Ditch
8 = Inner bank
9 = Outer bank
10 = The Avenue, a parallel pair of ditches and banks leading 3 km to the River Avon
11 = Ring of 30 pits called the Y Holes
12 = Ring of 29 pits called the Z holes
13 = Circle of 56 pits, known as the Aubrey holes
14 = Smaller southern entrance

The site as of AD 2004. The plan omits the trilithon lintels for clarity. Holes that no longer, or never, contained stones are shown as open circles and stones visible today are shown colored gray for sarsen, and blue for the imported stone, mainly bluestone. (Drawn by Adamsan, https://commons.wikimedia.org/-wiki/File:Stonehenge_plan.jpg.)

There is dated posthole evidence of a wooden structure at Stonehenge from around 3000 BC.

Excavations indicate that around 2600 BC the builders abandoned timber in favor of stone and dug two concentric sets of holes (the Q and R Holes) in the center of the site. The holes held up to 80 standing stones.

The next phase of activity involved quarrying, dressing, transporting, and setting 30 enormous sarsen stones. The Oligocene–Miocene sarsen stones came from a quarry around 25–30 km (15–18 mi) north of Stonehenge, in West Woods and Marlborough Downs, Wiltshire (Bevins, 2012). The stones were fit together with mortise and tenon joints into a 33 m (108-ft) diameter circle of standing stones, with 30 lintel stones on top. The lintels were joined to one another using tongue and groove fittings. Each standing stone was around 4.1 m (13 ft) high, 2.1 m (6.9 ft) wide, 1.1 m (3.6 ft) thick, and weighed around 22.7 tonnes (25 tons). The complete circle would have included 75 stones. The lintel stones are each around 3.2 m (10 ft) long, 1 m (3.3 ft) wide, and 0.8 m (2.6 ft) thick. The top of the lintels is about 4.9 m (16 ft) above the ground.

Within this circle were five trilithons (pi-shaped structures) of sarsen stone arranged in a horseshoe 13.7 m (45 ft) across, with the open end facing northeast. These huge stones weigh up to 45 tonnes (50 tons) each. The trilithons are between 6 and 7.3 m (20–24 f) tall. This phase of construction is radiocarbon dated between 2600 and 2400 BC (Johns and Jackson, 2012; Wikipedia, Stonehenge).

Between 2400 and 2280 BC, the bluestones appear to have been moved within the outer sarsen. Remember that all of this work was being done with Bronze Age tools.

Sometime between 2280 and 1930 BC, the bluestones were rearranged in a circle between the two rings of sarsens and in an oval at the center of the inner ring.

Between 1930 and 1600 BC, the northeastern section of the existing bluestone circle was removed, creating the Bluestone Horseshoe to mirror the shape of the central trilithons. The Y and Z holes are the last known construction at Stonehenge, built about 1600 BC. The monument was probably abandoned during the Iron Age (sometime after 1200 BC; Wikipedia, Stonehenge).

SOURCE OF THE STONES

As previously mentioned, the Oligo-Miocene Sarcen stones, Cretaceous Chilmark Ragstone, and Upper Greensand Sandstone all came from the West Woods and Marlboro Downs of Wiltshire, 25–30 km north of Stonehenge (Green, 1997).

The Altar Stone belongs to the Senni Beds of the Old Red Sandstone formation, which outcrops in many parts of West and South Wales (Johns and Jackson, 2012).

In 1908, geologist Herbert Thomas suggested that the Stonehenge bluestones matched a suite of igneous rocks found in the vicinity of Carn Menyn, a rocky outcrop in the Preseli Hills, 240 km (150 mi) distant in Pembrokeshire, Wales. Detailed petrographic studies later confirmed this match.

The rare clay-like mineral stilpnomelane, found in debris around Stonehenge, matches that from rhyolitic rocks near Pont Saeson, on the northern slopes of the Mynydd Preseli in Wales. This led workers to identify a probable Neolithic quarry site at Craig Rhos-y-felin. Whole rock geochemistry of the Stonehenge spotted bluestones then led to outcrops at Carn Goedog. A Neolithic quarry site was found there in 2011 (Ixer et al., 2020). Excavations by University College London have clearly identified Carn Goedog as the main source of the spotted dolerite (gabbro) at Stonehenge.

It is now commonly accepted that the bluestones came from the Preseli Hills. However, it is still debated whether they were transported by people or by glaciers. Although there is not much evidence of glaciation in this area, one theory is that glaciers flowing south from the ice cap over Wales merged with the Irish Sea Glacier and flowed from west to east up the Bristol Channel. The convergence of these two glaciers acted as a conveyor belt, transporting a train of boulders (glacial erratics) leading straight to Stonehenge (Johns and Jackson, 2012).

Others cite the lack of evidence for Quaternary glaciation in south Wiltshire to refute glacial transport of Welsh rocks to the Salisbury Plain (Green, 1997). Recent work proposes that the 1.8 tonne (2 tons) dressed stones were lifted in a framework of poles and carried by up to 60 people at a time eastward along the upper Nevern Valley and the valleys of the Teifi, Tywi, Usk, and Wye, fording the River Severn at Longford, then south to Salisbury Plain. The total distance would have been around 350 km (220 mi; Pearson et al., 2015).

Stonehenge.

UNRAVELING THE PAST

John Aubrey, a natural philosopher (scientist), was one of the first to examine Stonehenge in 1666. In his survey of the monument, he recorded the pits that now bear his name, the Aubrey holes. William Gowland oversaw the first major restoration of the monument in 1901, involving straightening and setting of sarsen stone 56 in concrete to prevent it from falling over. During the 1920 restoration, William Hawley located the concentric circular Y and Z holes outside the Sarsen Circle. In 1958, three of the standing sarsens were re-erected and set in concrete bases. The last restoration occurred in 1963 after sarsen stone 23 fell over. It was re-erected, and three more stones were set in concrete.

In 2008, an excavation inside the stone circle was able to date the erection of some bluestones to 2300 BC. The restorers discovered organic material from 7000 BC, supporting the contention that the site had been in use at least 4,000 years before Stonehenge was started.

In 2020, a study concluded that the large sarsen stones were "a direct chemical match" to those found at West Woods near Marlborough, Wiltshire, some 25 km (15 mi) north of Stonehenge.

A plan to build a four-lane traffic tunnel below the site was approved in November 2020. It was intended to eliminate the section of the A303 that runs close to the circle and to "restore the landscape to its original setting and improve the experience for visitors." In 2021, archaeologists conducting excavations for the proposed highway tunnel announced they had discovered Neolithic and Bronze Age artifacts including Bronze Age graves and late Stone Age pottery (Wikipedia, Stonehenge).

The first realistic painting of Stonehenge, a watercolor by Lucas de Heere, painted between 1573 and 1575. Corte Beschryvinghe van England, Scotland ende Irland. https://commons.wikimedia.org/wiki/File:Stonehenge_Lucas_de_Heere.jpg

Visit

You will need a ticket to visit this site. Find everything you need on the website. You don't need to book ahead, but if you do you can get the best price and guaranteed entry at the time you want.

Web site: https://www.english-heritage.org.uk/visit/places/stonehenge/
Address: Near Amesbury, Wiltshire, SP4 7DE
Hours: Monday–Sunday 9.30–7, last entry at 5
Entry fee:
   Free to members of English Heritage.
   Non-member adults: £22.00
   Child (5–17 years):   £13.20
   Students/Seniors: £17.60
   Family (2 adults/3 children): £57.20
   Family (1 adult/3 children): £35.20

Other passes are available, and donations are welcome.

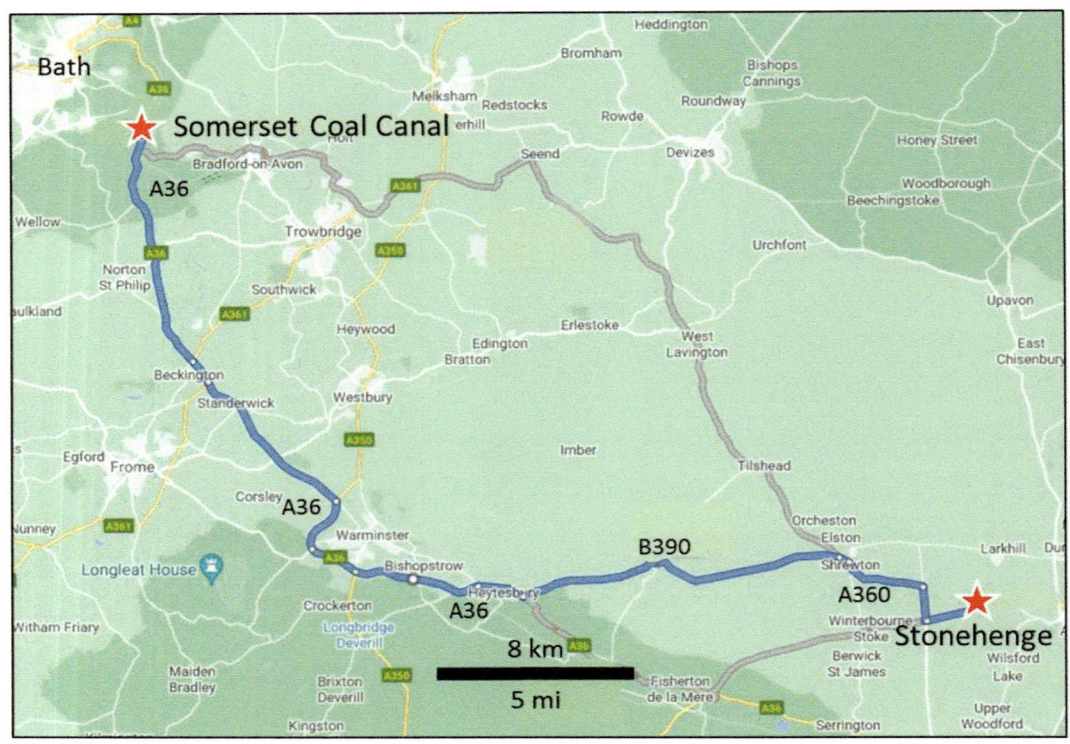

Stonehenge to Somerset Coal Canal: Return to B3086 and turn left (south); immediately turn right (west) onto A360 and drive to Maddington; at the roundabout, take the 1st exit onto A360/- Maddington Street; just past Maddington turn left (west) onto B390/Chitterne Road; at Knook Camp turn right (northwest) onto A36; at a roundabout, take the 4th exit onto B3414; in Warminster turn right (north) onto Imber Road; at the roundabout, take the 1st exit northwest onto Copheap Lane; turn right (northwest) onto Westbury Road; turn left (southwest) onto A350; at the roundabout, take the 3rd exit onto A36; drive northwest on A36 to the outskirts of Bath; turn right (east) onto B3108/Lower Stoke; turn left (north) toward Monkton Senior School, then immediately turn right (east) to the Canal Visitor Centre and park at **Stop 3.1, Somerset Coal Canal** *Brassknocker Basin Car Park (51.357834, −2.313888), for a total of 48.3 km (30.1 mi; 45 min). There is limited parking (£2.20/2 h), toilets, and a café. Walk north along the canal 750 m (0.5 mi) to Dundas Aqueduct.*

STOP 3 BATH AREA AND THE SOMERSET COAL CANAL

Bath is known for its Roman baths, fed by hot springs, that made the village a famous spa town in the 19th century. It is also known as the source of Bath Stone, the warm, honey-colored building stone that faces many buildings in town as well as buildings and monuments across the United Kingdom and, indeed, the globe. Lastly, Bath is renowned among geologists as the home and workplace of William Smith, the surveyor/engineer/geologist who developed the concept of stratigraphic succession and went on to compile the first geologic map and cross section of England, the first such map in the world (Winchester, 2001).

GEOLOGY OF BATH AND THE SOMERSET COALFIELD

The surface rock layers in and near the town of Bath consist of Jurassic limestone, mudstone, and sandstone that are inclined (dip) gently to the east.

The Somerset Coalfield, in northern Somerset, consists of Carboniferous coals that were underground-mined from the 15th century until 1973. The coalfield extends from Cromhall in the north to the Mendip Hills in the south, and from Bath in the east to Nailsea in the west, covering an area of about 622 km² (240 mi²). The field occupies three synclines, or coal basins: the Pensford Syncline in the north, the Radstock Syncline in the south, and farther west the smaller Nailsea Syncline. The Radstock Syncline is cut by a series of east–west thrust faults and north–south normal faults (Wikipedia, Somerset Coalfield). Normal faults usually form during episodes of extension and allow one side to drop down with respect to the other side. Thrust faults are usually nearly horizontal and force older rocks up and over younger rocks.

Bath Stone

The Bath district is best known for the building stone worked from the rocks of the Chalfield Oolite Formation, the so-called "Bath Stone" or "Great Oolite Freestone." Freestones are limestones that can be cut or trimmed in more than one direction and are thus suitable for complex carving and dressing. The best freestones in the Chalfield Oolite are found in the upper part of the Combe Down Oolite member and within the Bath Oolite member, where the rocks consist of fine- to coarse-grained oolitic limestone with a coarsely crystalline cement. Bath Stone formed during the Mid-Jurassic in warm, tropical seas. Waves lapping at white sand beaches rolled small grains or shell fragments in the shallow waters, coating the spherical grains with concentric layers of calcium carbonate until little ooids (egg grains) are formed. Most of Bath is built or faced with this oolitic limestone of the Great Oolite Group. This stone is quarried from the west side of Bath to northeast of Bath near Corsham (Pythian, 2006).

In the Roman and Medieval periods, Bath Stone was extensively used on domestic, ecclesiastical, and civil engineering projects such as bridges.

The freestones of Bath have been worked since Roman times: villas at Box and Bathford used local stone, as did parts of the Great Baths. During medieval times, Bath Stone was used to build Lacock Abbey and Longleat. The stone was also used for the Dundas Aqueduct, which is 150 yards (137.2 m) long and has three arches built of Bath Stone. Much of Bristol Cathedral was built of Bath Stone. In London, the neo-classical Georgian mansion Lancaster House was built from Bath Stone in 1825 for the second son of King George III, as was St Luke's Church, Chelsea in 1824, and several other churches, including Church of Christ the King, Bloomsbury (Wikipedia, Bath Stone). The mined freestones were used in the construction of Buckingham Palace in London and the Royal Pavilion in Brighton and were exported and used to build the Town Hall of Cape Town, South Africa, and Union Station in Washington, D.C. Demand declined, and today only three working sites remain (Barron et al., 2015).

Most of Bath is built or fronted by warm, gold Bath Stone, which gives this World Heritage City its distinctive appearance. Heading down Bath Street, the buildings are faced with Bath Stone. Abbey Church, originally a leper hospital in 1136, was rebuilt mainly using Bath Stone (Pythian, 2006). How did this stone come to be sought after?

Beau Nash was a character who seemed to fail at most of what he did, but he was an opportunist. Becoming Master of Ceremonies in 1704, he embarked on changing Bath from just a resort to a social center with strict building rules.

The Circus, one of several classic, architectural masterpieces in Bath, is faced with Bath Stone.

At the time Ralph Allen was Bath's postmaster. Allen grew wealthy reforming the British Postal system. This allowed him to purchase some of the local stone quarries. He built his mansion at Prior Park using Bath Stone and used it to help market the stone as a building material (Hunt, 2015; Wikipedia, Bath Stone).

During the 17th century, the city square became a major feature of city life. John Wood designed Bath's square in the Palladian style that was later adopted in London in the late 1700s. The larger terraces built in Bath followed a similar design, with an emphasis on order as well as beauty. The curving River Avon and the steep slopes required a novel style. The answer was to build crescents that followed the land's contours. The land in front of the crescents was left without landscaping. Wood designed the first "circus" in England, a ring of townhouses. All this architecture had one thing in common – Bath Stone. There are few towns that can claim such consistent use of stone.

The majority of Bath Stone quarries are underground workings. During World War II, munitions were stored in the mines so as to be out of reach of the Luftwaffe's bombs. When it was realized that the clay layers over the Bath Stone kept out moisture and created an unusually dry environment, various ministries used them for storage. One mine was used as a factory for the Bristol Aeroplane Company to make the Beaufighter, a long-range fighter and torpedo bomber. The Bath Stone industry never recovered from its wartime takeover (Hunt, 2015).

In 1989, a utilities crew unexpectedly broke into part of the Combe Down mines while trenching. This raised concerns about mine collapse among the public, resulting in the Bath City Council commissioning a survey of the mines. It became clear that the mines were dangerous, as about 80% of the mines, which are up to 9 m (30 ft) high, had less than 6 m (20 ft) of cover, and some had as little as 2 m (7 ft). Foamed concrete was chosen to fill the old mine workings. Over 400,000 m^3 (523,180 yd^3) of foamed concrete was injected into the shallowest underground mines (Wikipedia, Bath Stone).

STOP 3.1 SOMERSET COAL CANAL

John Strachey, William Smith, and the Principle of Faunal Succession

Every geology student learns about William "Strata" Smith, the Father of English Geology, but very few have even heard of John Strachey. Turns out that no idea is completely original.

John Strachey (1671–1743) was an English squire living at Sutton Court. Strachey had studied the coal workings at Bishop Sutton, about half a mile southwest of Sutton Court, and at Stanton Drew, about a mile to the north. He had learned from colliers that the same seams as at Sutton were being worked in pits toward Farrington Gurney, about 4 mi away to the southeast of Sutton Court. Strachey's land at Sutton lay between these two known productive areas, and to demonstrate its potential value as a prospect he made a cross-sectional diagram showing his predicted subterranean arrangement of the coal seams, their individual thicknesses, and depths.

In 1719, Strachey drew a diagram illustrating a slice through the geology under his estate and showing how the strata he had seen in nearby coal mines extended into the areas between the workings. He was thus able to predict where to dig for coal. Several of his drawings were published, and are the earliest known cross sections.

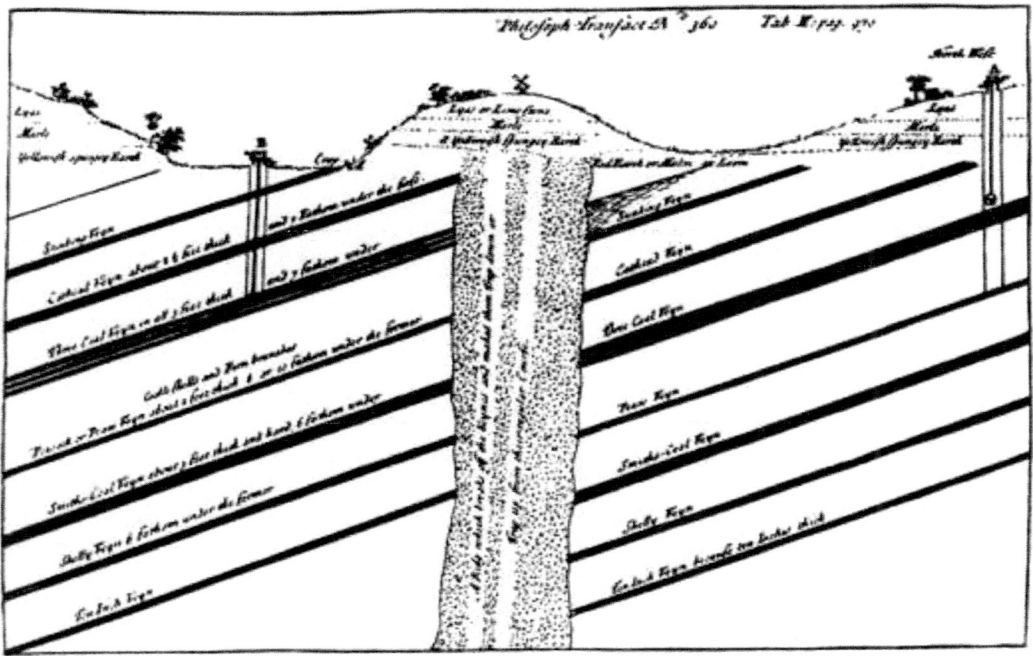

Strachey's 1719 cross section of the Somerset Coalfield.

John Strachey wrote a letter explaining the geology of the local coal industry and the nature of the overlying strata to his friend Robert Welsted, a Fellow of the Royal Society. Welsted read Strachey's letter to the Society at its meeting in May 1719. The Society published the letter along with the cross section, leading to his election as a Fellow of the Society. In a second paper for the Royal Society (1725), and in a booklet published 2 years later, Strachey stated that a distinct group of layers extended northeast across England from the coast of Wessex through Yorkshire and to the coast of Northumberland.

Strachey died in 1743. His niece, Mary Jones, opened a coal pit on her land at High Littleton in 1783. Eight years later she died, and probate on Jones's estate at Stowey (south of Bristol) required a survey and valuation. William Smith did the survey (Fuller, 2006).

William Smith (1769–1839), son of an Oxfordshire blacksmith, was an orphan by age 8. He was sent to work his uncle's farm where the bright kid was noticed by Edward Webb, a land surveyor who took him on as an assistant in 1787. Smith went to live with the Webb family where he learned surveying, measuring, agricultural valuations, and how to improve farm drainage. He began surveying in Somerset when he was 22 years old, mainly in the parishes of Stowey and High Littleton, in the Somerset Coalfield. The Stowey estate where he began work in 1791 was adjacent to lands belonging to the Strachey family of Sutton Court. While working in the area, he came across the geological cross section made by John Strachey illustrating parts of the coalfield. A copy of this section was found among Smith's papers after his death (Fuller, 2006).

Smith's great insight was the observation that fossils succeed one another in time in a distinct order. To that point, fossils had been a curiosity, but for Smith they solved a problem. Many of the layers looked alike, and when he started surveying canals this was a source of confusion. He recalled that early in his career the coal miners had shown him how they could identify the various coal measures by their stratigraphic position as well as by their associated fossils. Smith proposed that to identify any given layer of rock, he could use the unique fossil assemblage in and adjacent to that layer. Taking it one step further, Smith recognized that the rock layers were continuous across the region, always occurred in the same sequence, and contained a unique set of fossils. By plotting the locations of layers on a map, he could reliably forecast what strata the canal would encounter further along its course. He recognized that the canals in the area passed through the same strata in the same order, and he could project the layers from canal to canal and into coal mines (Manz, 2015; Rock Doc Travel – William Smith).

The observation that fossils occur is a specific order through time, that groups of fossils change in an ordered way, and that unique assemblages of fossils are consistently associated with the same rock layers was the basis for Smith's Principle of Faunal Succession. This is what allows geologists to correlate formations from one area to another, to give them relative ages (it is the basis for the Geologic Time Scale), and to make geologic maps.

A geologic map shows the distribution of rock layers. Smith was the first to attempt this over a large area when he traced the boundaries between layers (the formation contacts) across the ground surface (Fuller, 2006). In August 1815 he published the first geologic map of England and Wales along with a cross section and explanation of the different geologic layers. At a scale of 5 mi to the inch (about 1:300,000), it consisted of 15 sheets, each approximately 53 cm (21 in) high by 64 cm (25 in) wide. This was the first geologic map ever made for an entire country. As a result, in 1831 Smith became the first recipient of the Wollaston Medal – The Geological Society's most prestigious award. At the award dinner, he was referred to as "the Father of English Geology" (Manz, 2015). He was elected a Fellow of the Geological Society; he died at Northampton in 1840 at the age of 70.

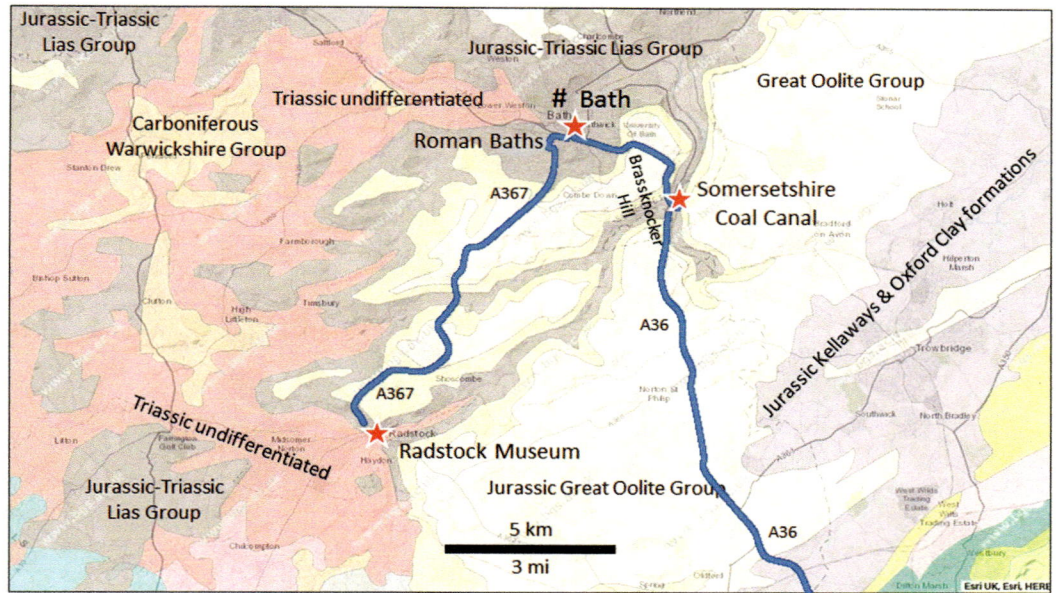

Geologic map of the Bath area. (Modified after BGS, Geology of Britain Viewer. Contains British Geological Survey materials © UKRI 2022. Base mapping provided by ESRI.)

Somerset Coal Canal

Coal was mined in Somerset as long ago as Roman times. By the 1430s, it was being dug in the Kilmersdon district. At first, coal was picked off the surface or dug from shallow pits. Eventually, it was dug from mines. By the end of the 17th century, 9,070 tonnes (10,000 tons) annually were mined in the Somerset Coalfield. Coal was found near Radstock in 1763, but mine development was hindered by the poor roads connecting the mines to markets. In 1792, a canal was planned in South Wales, which would allow Welsh coal to be brought by boat to Bath and Bristol cheaper than it could be delivered by cart or pack horse. Initial surveys were conducted in 1793 by William Jessop and William Smith.

The Somerset Coal Canal (originally known as the Somersetshire Coal Canal) was completed around 1800 from basins at Paulton and Timsbury to Limpley Stoke, where it joined the Kennet and Avon Canal. This gave access from the Somerset Coalfield, which at its peak contained 80 collieries, to London, the ultimate market for the coal (Wikipedia, Somerset Coal Canal).

Somersetshire Coal Canal, Brassknocker Basin, is now used for tourist canal boats.

William Smith was employed to "take the levels." Each section of the canal had to be perfectly horizontal (Somersetshire Coal Canal Society). Digging a canal meant slicing through rock layers across miles of landscape and the cross-sectional views of strata allowed Smith to confirm the order of rock layers across wide distances. As he accumulated observations of strata and fossils in mines and canals, he found that fossils were the key to identifying the layers (Scott, 2008).

Once the canal was opened, the amount of coal it carried increased steadily. By 1854, there was over 16 times as much Somerset coal being sold as there had been before the canal (Somersetshire Coal Canal Society). However, the end of the canal era was already in sight.

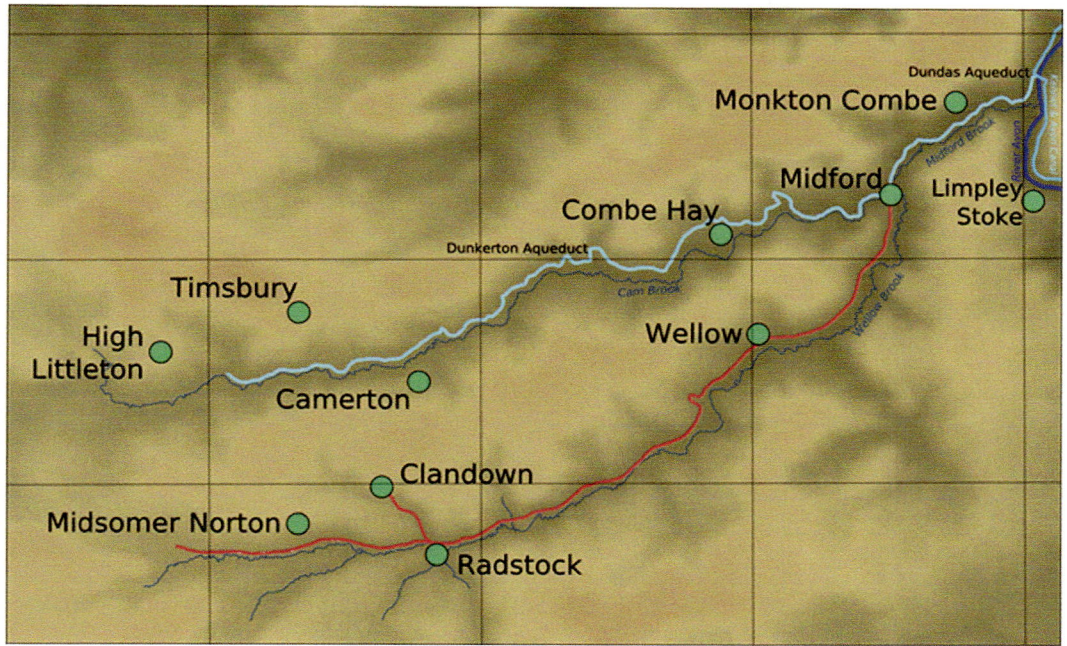

Map of the Somerset Coal Canal. (Topo derived from NASA, PD; detail derived from 1890 OS Mapping, PD by Berne Convention. https://commons.wikimedia.org/wiki/File:SccWip2.png.)

The opening of the Great Western Railway in 1841 was the beginning of the end for the Somerset Coal Canal. It didn't run anywhere near the Somerset Coal Canal, but it took traffic from the Kennet & Avon Canal and the Wilts & Berks Canal. Since the canals could not compete with railways for speed, they had to reduce their tolls in order to maintain market share. The Somerset Coal Canal was forced to charge less to compete. When a branch railway line was opened by the G.W.R. from Frome to Radstock in 1854, there began a slow but steady decline in canal traffic. During the 1870s–1880s, railways were extending into southern England from the north, bringing cheaper Midlands coal and taking over the canal markets. The canals began to lose money. Maintenance was reduced and the canals silted up; the water level fell still further because leaks were not being repaired. The boats could not be fully loaded, so the tolls, which were based on tonnage, brought in even less money. Eventually, the Great Western Railway bought the Somerset Coal Canal and its surrounding land in order to build a railway (Somersetshire Coal Canal Society).

Despite having been closed for 100 years, during the 1980s at Dundas a 400 m (0.25 mi) section of the canal was refilled with water and opened as a mooring for pleasure boats. There is a Visitors' Centre with a detailed map of the whole canal and many interesting artifacts (Wikipedia, Somerset Coal Canal). Any outcrops there might have been along the canal have long since weathered and become overgrown.

The Somersetshire Coal Canal Society arranges regular walks for members and obtains the landowners' permission to visit parts of the canal which are normally inaccessible to the public. Special arrangements can be made for educational trips and the Society can provide guides with a detailed knowledge of the canal. Contact: Derrick Hunt, 43, Greenland Mills, Bradford on Avon BA15 1BL 01225 863066. Website: http://www.coalcanal.org/ (Somersetshire Coal Canal Society).

Somerset Coal Canal to Roman Baths: Return to B3108 and turn right (west); turn right (north) on A36/Warminster Road and quickly left (northwest) onto Brassknocker Hill; at the roundabout, take the 2nd exit (north) onto Claverton Down Road; continue straight (west) on Widcombe Hill Road; turn right (northeast) onto St. Matthew's Place; turn left (west) onto A36/Rossiter Road; at the roundabout, take the 3rd exit to A367/Broad Quay; turn right (east) onto A367/Corn Street; turn right (southeast) onto St James's Parade; continue straight onto Southgate Street; turn left (east) onto A3039 and enter Avon Street Car Park (51.379017, −2.361884; Southgate St, Bath) on the left (£6.40/4 h). Walk 450 m (0.3 mi; 7 min) north to Stop 3.2, Roman Baths (51.380975, −2.359450), for a total of 5.7 km (3.6 mi; 10 min).

Stop 3.2 Roman Baths

Bath is perhaps best known as a spa town with restored Roman baths. Whereas most of the city seen today is Georgian (1714 to c. 1837), there are buildings built of the local stone that actually date back 2,000 years to the Roman occupation (Hunt, 2015). That includes part of the baths.

Archaeological evidence indicates that the site of the baths may have been dedicated to the Celt goddess Sulis and a center of worship. A Roman temple was built here between the years 60–70 during the first years of Roman Britain. The temple led to the development of a small Roman town, *Aquae Sulis*, around the springs. The baths – designed for public bathing – were used until the end of Roman rule in Britain in the 5th century. In 577, the Saxons captured the town at the Battle of Dyrham. According to the *Anglo-Saxon Chronicle*, the Roman baths were in ruins and silted up within a hundred years of being abandoned. In 1066, the Normans conquered Bath along with the rest of England (Greengrove and Davies-Vollum, 2006a). The area around the natural springs was subsequently redeveloped several times during the Early and Late Middle Ages.

The spring is now located in an 18th-century building designed by John Wood the Elder and John Wood the Younger. The visitor entrance is an 1897 concert hall built by J. M. Brydon. It is an eastward continuation of the Grand Pump Room, with a glass-domed center. The Grand Pump Room was begun in 1789 by Thomas Baldwin and completed by John Palmer in 1799. The north colonnade was also designed by Thomas Baldwin. In the 16th century, the city had built a new bath (the Queen's Bath) south of the King's Spring. The modern museum and Queen's Bath, including the "Bridge" spanning York Street to the City Laundry, were designed and built by Charles Davis in 1889. They include an extension to the Grand Pump Room (Wikipedia, Roman Baths – Bath).

The King's Bath is at the heart of the Pump Room complex. It was built over the hot water rising from King's Spring. The King's Bath includes remains that are Roman, medieval, Georgian, Victorian, and 20th century. The King's Bath was built by John of Tours around its hot spring in the 12th century on the foundations of the earlier Roman baths. The King's Bath provided niches for bathers to sit in, immersed up to their necks in the hot water. The Bath is supervised by a statue of King Bladud, the mythical discoverer of the hot waters and founder of the City of Bath (Heritage Services – Bath & NE Somerset Council, 2018).

The town of Bath was given responsibility for the hot springs in a Royal Charter granted by Elizabeth I in 1591. That responsibility has now passed to the Bath and North East Somerset Council, who monitor pressure, temperature, water quality, and flow rates. The thermal waters contain sodium, calcium, chloride, and sulfate ions in concentrations that many consider curative.

The restored Roman bath at Bath. Bath Abbey is in the background.

A tragedy in 1978 forced the ancient baths to close. A young girl swimming in the restored Roman Bath contracted meningitis and died. Tests showed *Naegleria fowleri*, a deadly pathogen, in the water. There are, however, two recent bathing spas, the newly built Thermae Bath Spa and the refurbished Cross Bath. Both are nearby, and allow bathers to experience the hot springs with fresh water from recently drilled boreholes (Wikipedia, Roman Baths – Bath).

Geologically, the spring water that emerges at the baths ultimately derives from rain that falls on the Mendip Hills to the southwest. It percolates down through limestone to depths between 2,700 and 4,300 m (8,900 and 14,100 ft). The Earth's heat warms the water to between 69°C and 96°C (156.2°F and 204.8°F). The heated water then rises along fissures and the Pennyquick Fault that cuts the limestone.

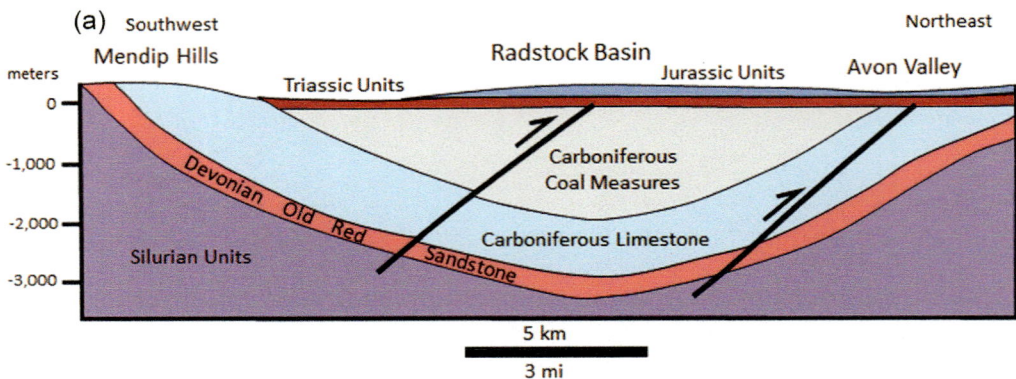

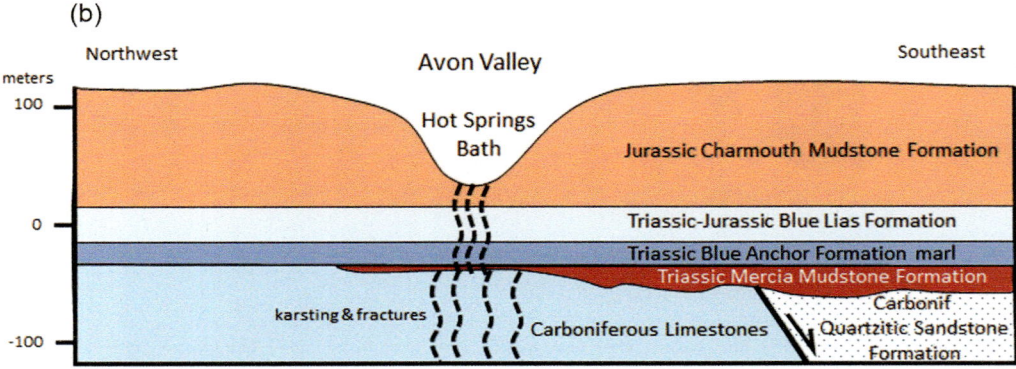

(a) Cross section from the Mendip Hills to Bath. (b) Cross section through the Avon Valley showing a possible origin for the Bath hot springs. (Modified after Gallois, 2006.)

In 1810, the hot springs failed. William Smith excavated the Hot Bath spring: at the bottom, he found it was not the spring that had failed, but rather it was clogged with minerals and had flowed into a new fracture system. Smith restored the water to its original course.

Water at a temperature of 46°C (114.8°F) flows into the pools at 1,170,000 l (257,364 imp gal) per day. In 1982, a borehole was sunk to provide a clean, safe supply of spa water for drinking in the Pump Room (Wikipedia, Roman Baths – Bath).

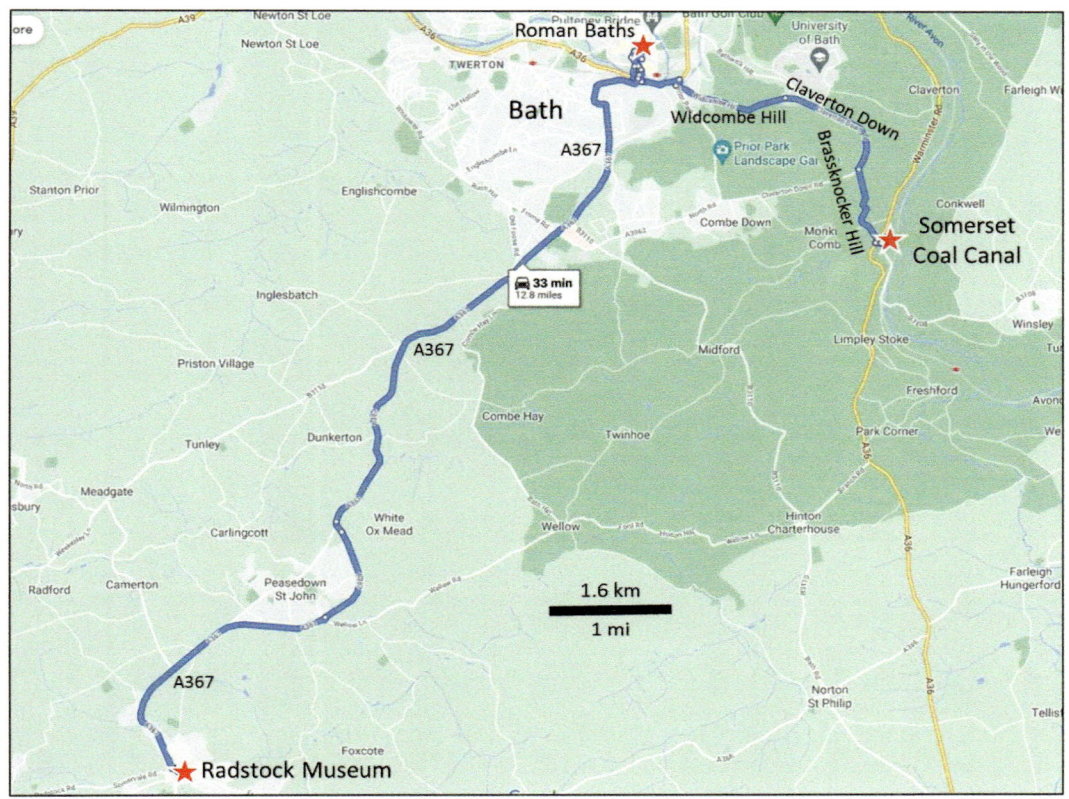

Roman Baths to Radstock Museum: *Head west on A3039/Dorchester Street toward Southgate Street; turn left (south) onto A367/Broad Quay/Churchill Bridge; at the roundabout, take the 2nd exit onto A367/Wells Road; at Odd Down Roundabout, take the 3rd exit onto A367/Roman Road; continue on A367 to Radeng and turn left (east) onto Waterloo Road.* **Stop 3.3, Radstock Museum** *(51.293051, −2.446691; Radstock BA3 3AP) is on the left, for a total of 14.1 km (8.8 mi; 15 min).*

STOP 3.3 RADSTOCK MUSEUM

Radstock Museum has a range of exhibits which offer an insight into North Somerset life since the 19th century. The museum opened in 1989 and moved to its current site in a restored and converted Victorian Market Hall built in 1897. Many of the exhibits describe the Somerset Coalfield and local geology, including exhibits of William Smith's fossils used in developing the Principle of Faunal Succession and correlating formations across Somerset and England. Permanent displays include Coal Mining, Geology and William "Strata" Smith, Somerset and Dorset Railway, and Great Western Railway artifacts and memorabilia.

The Radstock Museum.

Visit

There is free parking nearby on Waterloo Road and limited pay parking across the street from the museum. All areas of the museum are accessible by disabled visitors and there are two parking spaces for blue badge holders immediately outside the main entrance

Address: Radstock Museum, Waterloo Road Radstock BA3 3EP
Phone: +44 (0)1761 437722
Website: http://radstockmuseum.co.uk/
Email: info@radstockmuseum.co.uk
Hours: The museum is open from February to November.

   Tuesday–Friday: 2 p.m.–5 p.m.
   Saturday: 11 a.m.–5 p.m.
   Sunday: 2 p.m.–5 p.m.
   Closed all of December and January

Admission: Adults: £6.00

   Children 6–15: £2.50
   Senior Citizens/Students 16+: £5.00
   Free admission for children aged 5 and under and helpers for the disabled
   Family Ticket: £13.00 (2 adults and 2 children)

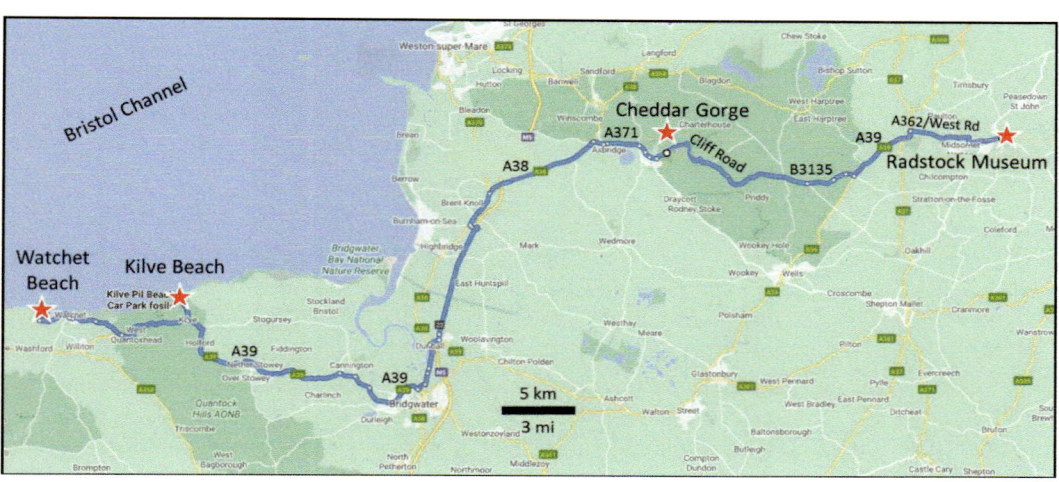

Radstock Museum to Horseshoe Bend, Cheddar Gorge: Head west on Waterloo Road to A367; turn left (south) on A367; at the 2nd roundabout, take the 2nd exit (west) onto A362/Somervale Road; at the roundabout, take the 3rd exit (north) onto A362/Station Road; stay on A362 as it becomes West Road and Thicket Mead; turn left (south) onto A37/A39/Rush Hill; in Bathway turn right (west) onto Cheddar Road, then slight left onto Broad Street; continue straight onto Torhole Bottom; continue straight onto B3135/Plummers Lane; continue straight on B3135/Cliff Road to Stop 4.1, Horseshoe Bend (51.285105, −2.759398) and pull into the parking area on the left for a total of 28.3 km (17.7 mi; 35 min).

STOP 4 CHEDDAR GORGE

This is where Cheddar cheese was invented. It is also the location of a rare English gorge with caves in Early Carboniferous (Mississippian, ~340 Ma) limestone. We are on the southwest flank of the Mendip Hills, a low limestone range that extends from Weston-Super-Mare on the west to near Frome on the east.

As a result of the Variscan mountain-building, the Mendip Hills have been folded into a series of east–west anticlines. These folds expose Silurian volcanics and Devonian sandstones in their cores. Cheddar Gorge incises the southern flank of the Winscombe-Beacon Batch Anticline (informal name). The Mendips were probably part of an impressive mountain range 290 million years ago, perhaps similar to the Alps today. Erosion has taken its toll, and today it is a karsted landscape of hills with gorges, caves, and limestone pavements typical of karst landscapes. The 200 m (650 ft) deep gorge is likely a result of dissolution along fractures in the limestone terrane, probably accelerated by runoff during the melting of glaciers over the past half million years.

The units in the gorge are inclined about 20° south. The Carboniferous limestone is about a km (half a mile) thick, although only the upper part is exposed in the gorge. Fossil-rich layers occur in the limestone, full of brachiopods (sea shells) and corals, but are difficult to see because of extensive weathering (West, 2019). The formations exposed in the gorge are part of the Carboniferous Limestone Group and were deposited on a continental shelf in clear and warm tropical waters. Oolite horizons indicate high-wave-energy shallow water conditions.

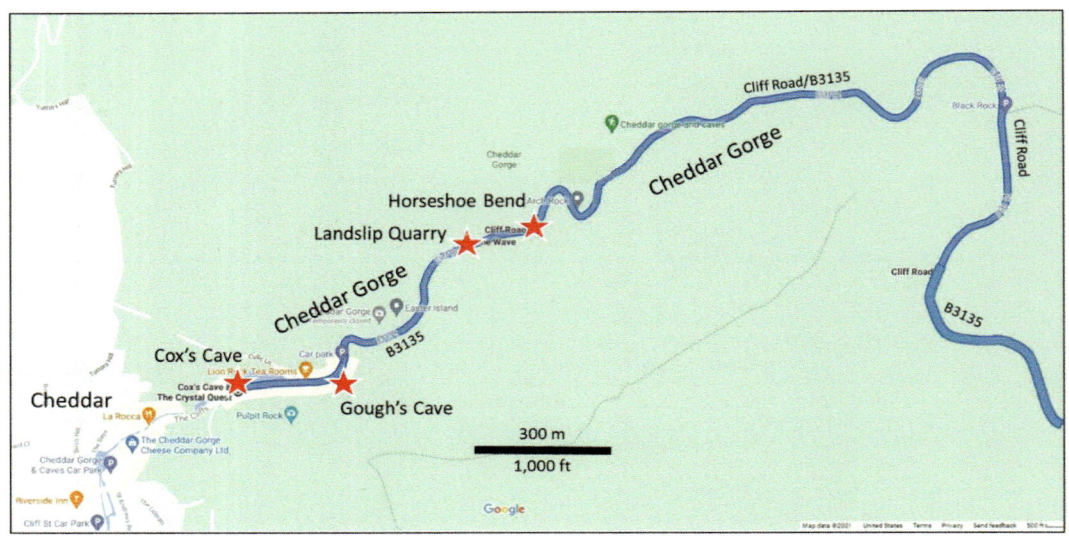

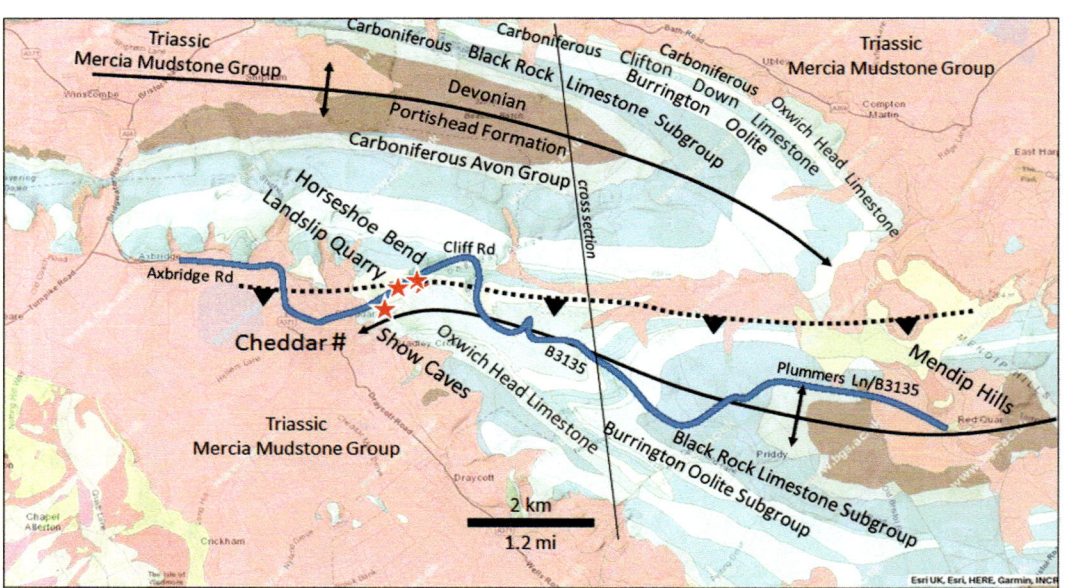

Geologic map of the Cheddar Gorge area. Dotted line is the subsurface thrust front. The cross section is shown in the next figure. (Modified after BGS, Geology of Britain Viewer. Contains British Geological Survey materials © UKRI 2022. Base mapping provided by ESRI.)

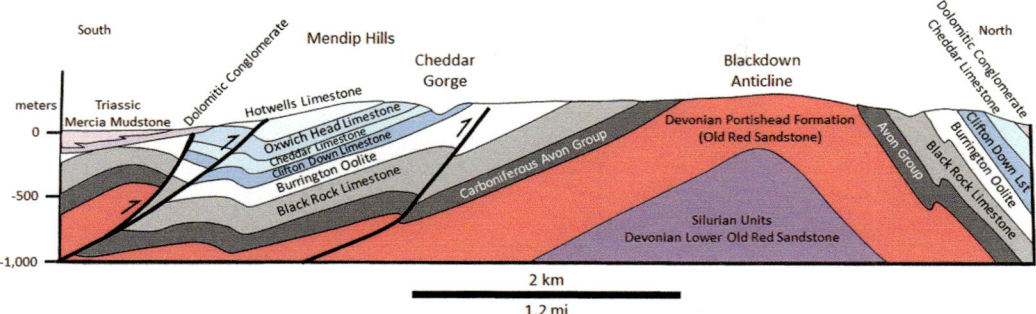

Cross section through Cheddar Gorge and Mendip Hills. Location is shown in the previous figure. (Information derived from the previous figure, from West, 2019, and Discovering Black Down, Geology.)

Age	Unit	Description
Early Carboniferous (Mississippian)	Oxwich Head Limestone	Carboniferous crinoidal limestone
	Clifton Down Limestone	Carboniferous limestone; Caves Stromatolite beds Lithostrotion beds Chert beds Horseshoe beds
	Cheddar Oolite	Carboniferous oolitic limestone
	Cheddar Lst	Crinoidal limestone
	Burrington Oolite	Carboniferous oolitic and crinoidal limestone

Stratigraphy of Cheddar Gorge and the Mendip Hills. (Information derived from West, 2019, and BGS, Cheddar Gorge.)

View east up Cheddar Gorge from Show Caves parking area.

STOP 4.1 HORSESHOE BEND

The oldest unit in the gorge is the Cheddar Oolite. It occurs in the upper part of the gorge and is generally not well exposed, but there is a good outcrop at Horseshoe Bend, where it is a pale gray porous oolitic limestone. Look for brachiopod fossils in the fallen rocks.

Street View west toward Horseshoe Bend, Cheddar Gorge.

*Horseshoe Bend to Landslip Quarry: Continue west on B3135/Cliff Road for 170 m (0.1 mi; 1 min) and use the pullout on the right (north) side of the road. This is **Stop 4.2, Landslip Quarry** (51.284642, −2.761673).*

STOP 4.2 LANDSLIP QUARRY

This abandoned quarry is a great place to examine the Clifton Down Limestone Formation, the main unit forming the canyon walls. The Clifton Down Limestone contains chert (flint) beds, stromatolite layers (thin layered bacterial mats), and Lithostrotion beds full of the coral *Lithostrotion martini*. These corals are quite common in the Landslip Quarry. Also seen are *Composita ficoidea*, a smooth-shelled brachiopod. The Clifton Down was deposited in a shallow shelf (crinoidal limestone) to lagoon setting (Lithostrotion beds). Stromatolites, oolites, and calcitic mudstones appear to be associated with times when sea level was dropping. Rising sea levels are most often associated with shelly limestone. There are lots of fossils, but few species, indicating a stressed environment (high salinity, brackish, too warm). In most places, the Clifton Down is a dark, hard, finely crystalline, and low porosity limestone, although it does contain dark-gray dolomite and light-gray oolitic layers.

Landslip Quarry (left), Cheddar Gorge. Street View north.

Above the Clifton Down Limestone is the Oxwich Head Limestone Formation (formerly called the Hotwells Limestone). This is a thick-bedded fine- to coarse-grained, shelly limestone with some oolitic layers.

Quarrying within the gorge has now ceased. The quarries produced mainly Clifton Down Limestone, which was crushed and sent to South Wales as lime used in the manufacture of steel (West, 2019).

*Landslip Quarry to Show Caves: Continue west on B3135/Cliff Road for 450 m (0.3 mi; 1 min) and use the parking spaces on the right (north) side of the road. This is **Stop 4.3, Gough Cave** (51.281975, −2.765867). **Cox's Cave** (51.281801, −2.768791) is another 250 m west.*

Stop 4.3 Show Caves

Cheddar Gorge is famous for its caves. The two largest, Gough's Cave and Cox's Cave, are both in the lower, western part of the gorge. Cox's Cave is the smaller cave but has excellent stalactites and stalagmites. Gough's Cave is larger and better known. The caves are a result of surface and groundwater dissolving the limestone along fractures.

Cox's Cave was discovered by George Cox in 1837 during quarrying. It was opened to the public later that year. The cave tour takes about 10 min.

Gough Cave, originally a low tunnel extending to a choke point, was dug out from 1890 to 1898 and opened to the public by R.C. Gough and his sons in 1898. "Cheddar Man," a Stone Age skeleton, was discovered in 1903 when the entrance to Skeleton Pit was being cleared. An archaeological dig during 1927–1930, found hundreds of latest Paleolithic flints in cave fill material. This cave walk can be done in about 40 min.

Cave formations, Gough Cave, Cheddar Gorge. Red colors are related to iron oxide leaching out of the limestone.

A short distance above Horseshoe Bend, there are two small rectangular caves on the south side of the gorge. They are a result of water moving along and dissolving bedding planes in the limestone.

The car park 200 m (660 ft) above Gough's Cave contains Cooper's Hole, a cave entrance. Great Oone's Hole is at the base of the Oxwich Head Limestone Formation, about 49 m (160 ft) above the car park at Gough Cave. Most people are not aware of the cave at Great Oone's Hole, as it is mostly hidden from below. It is owned by Cheddar Caves Ltd., and there is occasional access (West, 2019).

Cheddar Cheese

No visit to this spot would be complete without a few words about Cheddar cheese. According to the United States Department of Agriculture, Cheddar cheese is the world's most popular variety of cheese. The cheese originated in the village of Cheddar. Cheddar Gorge, on the edge of the village, has a number of caves with the ideal humidity and a constant cool temperature for maturing the cheese.

Cheddar has been produced here since at least the 12th century. In fact, the Romans may have originally brought the recipe to Britain from the Cantal region of France. King Henry II in 1170 records the purchase of 10,240 lb (4,640 kg) of this cheese. Charles I (1600–1649) also bought cheese from Cheddar. Joseph Harding (1805–1876), "the father of Cheddar cheese," introduced technical innovations and dairy hygiene, and taught modern cheese-making techniques.

During the manufacture of Cheddar cheese, the curds and whey are separated using rennet, an enzyme normally produced in the stomachs of newborn calves. "Cheddaring" refers to an additional step in the production of Cheddar where, after heating, the curd is kneaded with salt, cut into cubes to drain the whey, and then stacked and turned. Strong, extra-mature Cheddar is cured for 15 months or more.

Cheddar cheese being aged in Gough Cave.

The ideal quality of the original Somerset Cheddar was described by Joseph Harding in 1864 as "close and firm in texture, yet mellow in character or quality; it is rich with a tendency to melt in the mouth, the flavor full and fine, approaching to that of a hazelnut". Cheddar made in the classic way tends to have a sharp, pungent flavor, often slightly earthy (Wikipedia, Cheddar Cheese).

Visit

If the roadside car parks in the gorge are full, park in the Cheddar Park-and-Ride (Prudden, 2004). Both sides and the base of Cheddar Gorge are on private land, so be courteous. Many of the best exposures are in quarries, some abandoned. Be mindful of falling rocks. The roads are narrow, so watch for cars.

Black Rock Nature Trail is a path on the southeast side of the gorge that leads to the Somerset Wildlife Trust Reserve. This is a quiet oasis at the head of the gorge. There is also a 3 mi marked trail around the gorge.

The Cheddar Showcaves are open to the public and include a museum with a collection of Paleolithic tools. Wear warm clothes and be able to climb up and down several dozen steps.

Address: Cheddar Caves and Gorge, Cheddar, Somerset BS27 3QF.
Phone: 01934 742343.
Website: www.cheddarcaves.co.uk.
Gough Cave: 51.282026, −2.765584 The two caves are 220 m (720 ft) apart
Cox's Cave: 51.281707, −2.769358
Hours: Contact Cheddar Caves for hours.
Admission: Caves & Gorge Explorer Ticket: Adults £20.95, Children (5-15) £15.70, Children
               (0–5) free.

For those interested in a cheesy experience, check out the Cheddar Cheese Company, open daily from 11:30 a.m. to 3:00 p.m.

Address: Cheddar Gorge, Somerset, UK. BS27 3QA
Phone: 01934 742810
Email: info@cheddar.ltd
Website: https://www.cheddaronline.co.uk/

*Show Caves to Kilve Beach: Continue west on B3135/Cliff Road to Cheddar; at the roundabout, take the 2nd exit (west) onto B3135/Tweentown; continue straight on B3135/The Barrows; continue straight on B3135/Axbridge Road; turn left (west) on Shipham Road and immediately turn right (northwest) onto A371/Axbridge Road; bear left (west) toward Taunton; continue straight onto Cross Lane; turn left (southwest) onto A38/Bridgewater Road; continue straight onto A38/Turnpike Road; continue straight on A38/Bristol Road; at the East Brent Roundabout, take the 1st exit onto A38/Bristol Road; at the Edithmead Roundabout, take the 1st exit onto the M5 (south) ramp to Taunton/Exeter; at Junction 23 take the A39 exit to Bridgewater; at the Dunball Interchange take the 3rd exit onto A39; at the Dunball Roundabout, take the 1st exit onto A38/Bristol Road; turn right (west) onto Wylds Road; turn right (northwest) onto A39/Western Way; continue straight on Homberg Way; continue straight on Quantock Road; at the roundabout, take the 2nd exit onto A39/New Road; continue west on A39 to Kilve; in Kilve turn right (north) onto Sea Lane; drive to the car park on the left. This is **Stop 5, Kilve Beach** (51.191451, −3.225267), for a total of 51 km (31.7 mi; 54 min). Walk 200 m (650 ft) north to the beach at low tide. There is ample pay-and-display parking (£2/2 h) and facilities.*

STOP 5 KILVE BEACH

This famous fossil beach has exposures of the Jurassic Blue Lias Formation both on the horizontal shore surface and in the near-vertical sea cliffs, allowing a three-dimensional view of the structure. As indicated in the oblique satellite view, the bedding is extensively folded and faulted. The details of faults and folds seen here are studied by students and professional geologists as analogs for structures seen in the subsurface elsewhere (Butler, Kilve).

Google Earth oblique view south over deformed Jurassic Blue Lias, Kilve Beach. Dashed lines show trace of folded bedding on the shore. (Image courtesy of Landsat/Copernicus, SIO, NOAA, US Navy, NGA, GEBCO; © 2021 Google.)

The outcrops are mainly gray bituminous oil shales interbedded with thin, hard limestones. They are the same rocks found offshore in the Bristol Channel Basin and across the channel in South Wales. They are primarily Lower Jurassic Blue Lias Formation, time-equivalent to rocks exposed at Lyme Regis on the Jurassic Coast. Paleozoic rocks are exposed south of here in the Quantocks. The oil shales are thought to represent the remains of planktonic blooms deposited in an anoxic environment. Dewatering of the shales has led to extensive compaction and flattening of fossils.

A Global Boundary Stratotype Section and Point (GSSP) marker occurs at Kilve Beach. A GSSP is an international reference point that defines a stage, in this case, the Sinemurian. Here the boundary is defined by an ammonite species (a coiled mollusk related to the nautilus and squid).

The limestones are characterized by regular orthogonal (2 sets at right angles) jointing, probably a result of extensional tectonics. Jurassic extension formed the Bristol Channel by down-dropping the area along east–west normal faults. The basin was later partially inverted (uplifted) by Alpine north-directed shortening and compression during the Late Cretaceous and Tertiary. Relay ramps, overlap zones between normal faults, allow displacement on one fault to transfer to another fault. Relay ramps also develop among strike-slip faults (faults with horizontal rather than vertical offset). Relay ramps are important because they represent pathways for sediment transport and can be migration routes for hydrocarbons. Normal faults, relay ramps, and strike-slip faults can all be found on Kilve Beach.

On this beach, strike-slip faults displace the east-west-oriented normal faults. A dextral strike-slip fault extends into the cliffs. With a dextral, or right-lateral strike-slip fault, the opposite side moves to the right with respect to the observer.

The Quantocks Head Fault, west of Kilve, is a classic example of a structure that began as a normal fault and later, during the Alpine Orogeny, was reactivated as a reverse fault. Reverse faults usually form during episodes of compression and cause one side to move up and over the other side of the fault. The offset on the fault is about 70 m (230 ft). Some beds in the hanging wall (in this case the upthrown side) are offset by small thrust faults. Vertical slickensides (polished surfaces with scratch marks indicating a fault) indicate normal offset, whereas oblique slickensides reveal reverse movement, and horizontal slickensides represent dextral displacement. Normal faults usually dip ~60°. The steepness of the Quantocks Head Fault implies that it began as a normal fault. Folding of the strata in the hanging wall indicate later reactivation as a reverse fault.

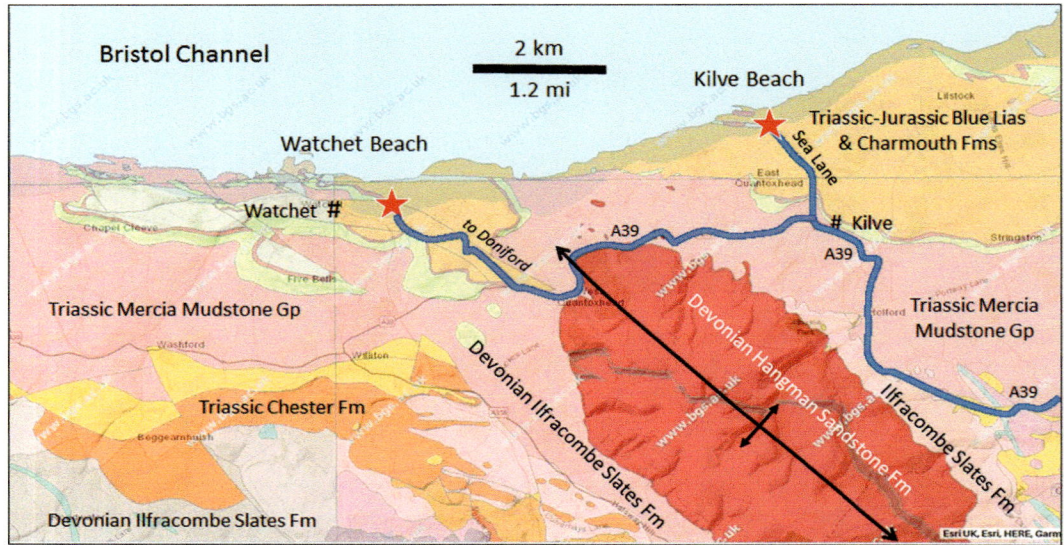

Geologic map of the Kilve-Watchet area. (Modified after BGS, Geology of Britain Viewer. Contains British Geological Survey materials © UKRI 2022. Base mapping provided by ESRI.)

The fossils here are similar to those found across the Bristol Channel at Penarth, but the ammonites are larger and the gastropods (snails), bivalves (sea shells like clams), and brachiopods tend to be smaller. Ammonites are found in shale or in concretions. Marine reptile bones of ichthyosaurs and plesiosaurs have been found here.

Around 1920 it was determined that the Blue Lias shale in the sea cliffs contained high organic carbon content, up to 12 or even 18%. In 1924, a retort was built by the Shaline Company to heat the shale and extract the oil. Alas, more energy was used than was recovered, and the project was never economic. The abandoned ruins can be seen near the Kilve Beach car park (Webster, 2017).

This beach is part of the Blue Anchor to Lilstock Coast SSSI, mainly due to the fossils found in outcrops (Wikipedia, Kilve). Hammering and collecting are *not* allowed. The area is best visited at low tide.

Folded and faulted Jurassic Blue Lias Formation offset along faults, Kilve Beach.

__Kilve Beach to Watchet Beach__: Return south to Kilve and turn right (west) on A39/Lagger Hill; turn right (northwest) to Doniford and continue on Doniford Road; continue straight onto South Road; turn right (north) onto B3191/Swain Street; turn left (west) on Market Street and drive to the parking lot on the right. This is __Stop 6, Watchet Beach__ (51.182961, −3.331744), for a total of 11.4 km (7.1 mi; 17 min).

STOP 6 WATCHET BEACH

The car park is near a small museum by the harbor. The museum has local fossils (including an ichthyosaur skeleton) and books illustrating the fossils and local geology.

Take the footpath next to the steam railway east onto the top of the seacliff. After about half a km you come to steps that lead onto the beach. Walk east from the lighthouse, as the rocks around the lighthouse are all Triassic and lack fossils (UK Fossil Collecting, Watchet).

The tides are rapidly eroding north-dipping red marls and green siltstones just east of the harbor. In the foreshore at the base of the steps the east–west-trending Doniford Bay Fault is exposed at low tide. South of the fault is downthrown Jurassic Blue Lias. The Jurassic is juxtaposed against mudstones and marls of the Triassic Mercia Mudstone Group (formerly known as Keuper Marls) on the north side of the fault. The Mercia Mudstone at Watchet has horizontal gypsum veins and alabaster nodules parallel to the bedding. The gypsum veins are thought to have formed by hydraulic fracturing of the marl in an overpressured environment (Webster, 2017).

Over 150 m (500 ft) of the Lias is exposed, extending up to the semicostatum zone (*Arnioceras semicostatum*, a Lower Jurassic ammonite that serves as a zone marker; UK Fossil Collecting).

Red Triassic sea cliffs at Watchet. View west.

Both extensional and compressional faults are seen in the cliff face between the steps and the harbor. The Quantocks Head and the Warren Bay faults are both east–west normal faults reactivated by Alpine compression. These faults dip south, making them susceptible to north-directed shortening and inversion. The Lias is intensely folded in the hanging wall of the Warren Bay Fault (Webster, 2017). Further east, in Doniford Bay, the Lower Lias shales and limestones are deformed into both large and small-scale folds (Prudden, 2004).

The cliffs just west of Watchet are Mercia Mudstone Group, barren of fossils. However, reptile bones are found west of a major fault where the Lias starts again. The Triassic Blue Anchor bone bed contains fish, reptiles, and shark fossils. Although bones can be found anywhere on the foreshore, the best locations are (1) from the caravan camp west to the red Triassic beds, and (2) from the large fault undercut by the sea west to just before Blue Anchor. The bones erode out of the upper part of the cliff and can be picked up along the shore. This is a good location for plesiosaur and ichthyosaur bones, especially after cliff falls.

Rich fossil collecting occurs in Jurassic units including, from base to top, the Blue Anchor Formation, the Westbury Formation, and the Blue Lias Formation. Bivalves, including scallops, oysters, and Gryphaea are common, as are crinoids. Ammonites, mainly broken, are found scattered on the foreshore. Occasionally you can find a complete ammonite that has recently broken out of its nodule. Otherwise, look for nodules or concretions to crack open. Ammonites are also found in the shingle near Blue Anchor. Slabs containing crinoids are found on the foreshore between Warren Bay and Blue Anchor.

Ammonite found on Watchet Beach. The fossil is approximately 34 cm (14 in) across.

For this area, park at the Warren Bay Caravan Park (51.180645, −3.350121) and follow the footpath north to Warren Bay.

The best time to collect is at low tide when the foreshore has been scoured. Most fossils are exposed during tidal scouring (UK Fossil Collecting, Watchet). This site is an SSSI: you can collect loose rocks, but hammering bedrock is not permitted.

Watchet Beach to Crooklets Beach: *Head east on B3191/Market Street; turn left (south) on Swain Street; turn right (southwest) onto B3191/Brendon Road; continue straight on B3191/-Washford Hill; at Raleigh's Cross turn right (west) to stay on B3191; in Bampton turn right (west) onto B3227/Castle Street; at Newton turn left (west) onto A361; in Barnstaple at the roundabout, take the 1st exit to stay on A361; at the next roundabout, take the 1st exit onto A39; in Bush, on* *the outskirts of Bude, turn right (west) onto Stonehill; turn right (west) onto Poughill Road; turn right (west) onto Oceanview Road; turn left (south) onto Crooklets Road; turn right (west) to **Stop 7, Crooklets Beach** car park (50.835685, −4.552192), for a total of 128 km (79.2 mi; 2 h 3 min). There is ample pay-and-display parking (1 h £0.90; 1–2 h – £2.70). Walk up to 550 m (1,800 ft) north along the beach.*

STOP 7 CROOKLETS BEACH

Start at First Cove, north of Crooklets Beach, and walk south to The Breakwater. From there continue to Compass Point, a total of about 1.5 km (4,800 ft). Observe the structures developed in the Carboniferous Bude Formation in the sea cliffs and on the surface of the shore.

The coast at Bude is an SSSI due to its unique geology. Whereas most of the coast of Cornwall consists of Devonian slate, granite, and Precambrian metamorphics, Bude Formation sandstone and shale are exposed in the cliffs and wave-cut platform along the shore. The layers were faulted and folded during the Late Carboniferous-Early Permian Variscan Orogeny. Chevron folds plunge west and appear as "V" patterns on the beach. The folds are cut by a left-lateral strike-slip fault that extends northwest from the Bude Sea Pool (Higgs, 2015; Wikipedia, Bude).

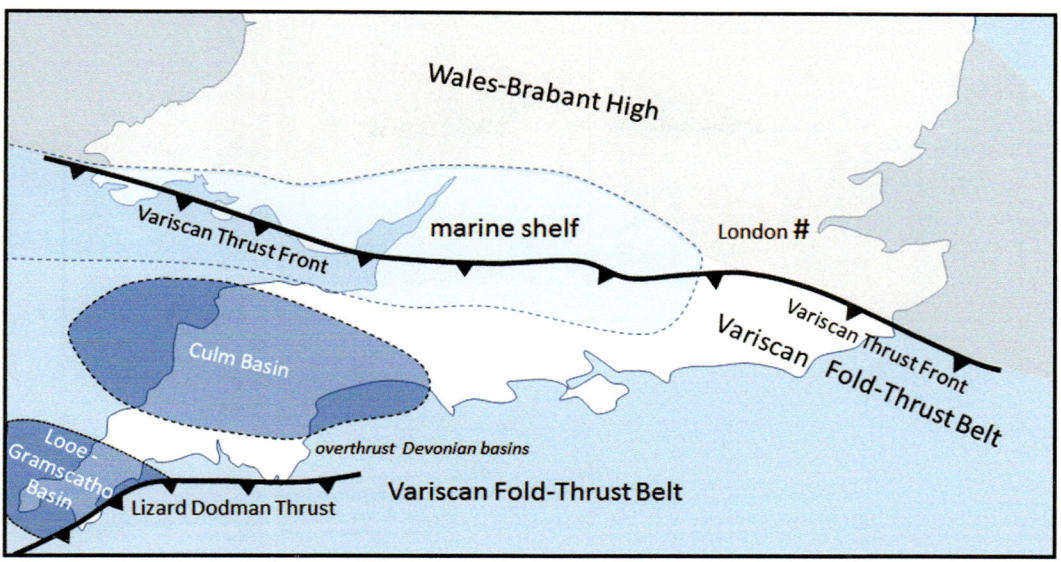

Location of the Culm and Looe-Gramscatho basins, Variscan Fold-Thrust Belt, and Wales-Brabant High. (Modified after Jackson, 2015.)

The Bude Formation consists of up to 1.3 km (4,260 ft) of alternating sandstone and shale deposited in the Culm Basin south of the Bristol Channel. These beds may have been deposited as turbidites or a submarine fan in a tropical sea or, possibly, in shallow "Lake Bude" with occasional marine incursions. A lack of fossils and burrows, and the presence of a unique fish, *Cornuboniscus budensis* (displayed in Bude Castle museum), suggest a lake environment. Slumped beds toward the top of the Bude Formation show an overall south-directed fold vergence (the folds lean to the south), suggesting soft-sediment deformation on a south-sloping surface (Anderson and Morris, 2004).

The Culm Basin developed as a foreland basin north of the advancing Variscan deformation front (Anderson and Morris, 2004). Thrust loading and downward flexing of the basin were a result of Upper Carboniferous crustal shortening (Hecht, 1992). In some cases, folding/slumping affected the Bude Formation while it was still soft. The basin was shortened and inverted during the latest Variscan Orogeny. Bedding was deformed into spectacular chevron folds by the collision of Euramerica and Gondwana that formed the Variscan mountain belt across central Europe (Higgs, 2019). Variscan thrusting and folding produced over 50% shortening of the Culm Basin (it was only half as wide as it originally was). Early bed-parallel thrusts are folded by east–west-trending chevron folds in all parts of the basin (Anderson and Morris, 2004). The intensity of deformation and low-grade metamorphism generally increases from Millook Haven southward (Hecht, 1992).

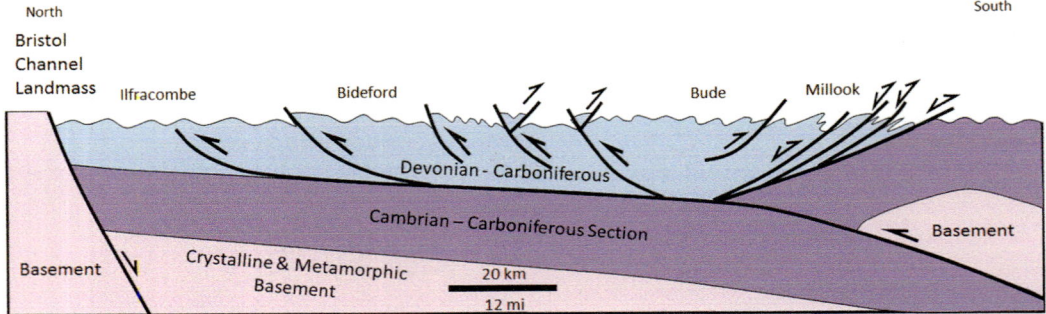

Schematic cross section of the Culm Basin showing Upper Carboniferous Variscan deformation. The sense of vergence changes on either side of Bude, and intensity of deformation and metamorphism increases south of Millook Haven. (Modified after Hecht, 1992.)

Widespread normal faults cut the earlier thrusts to produce the present distribution of units and associated folds. These are the latest structures seen in the area and probably developed during a phase of extension in the Late Carboniferous and Early Permian (Anderson and Morris, 2004).

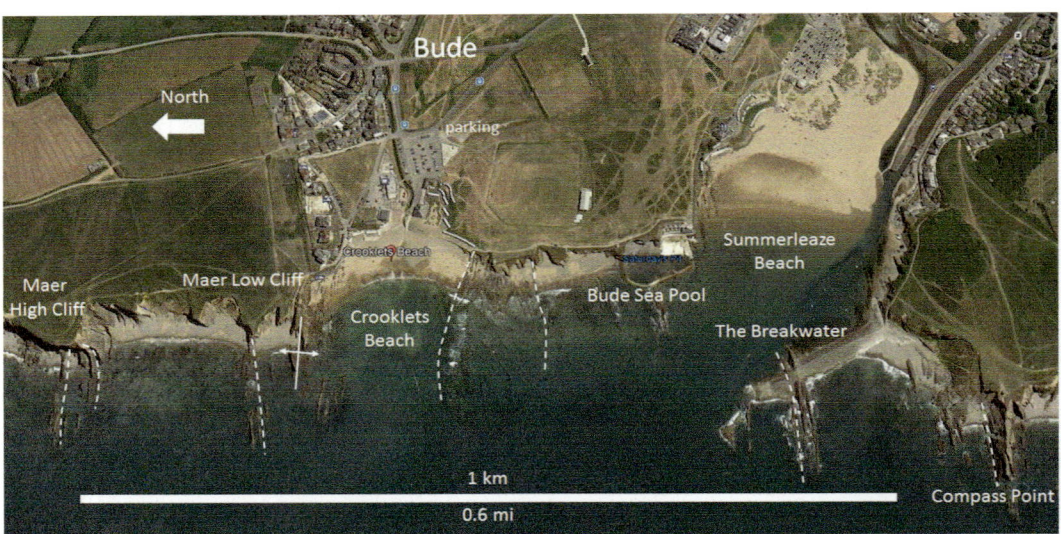

Google Earth vertical image of Bude and Crooklets Beach. North is to the left. (Image courtesy of SIO, NOAA, US Navy, NGA, GEBCO, ©2022 TerraMetrics.)

Variscan-age folding on Crooklets Beach.

Variscan chevron folds in the Bude Formation on Crooklets Beach. View north from Maer Low Cliff. White dashed lines outline the bedding.

Crooklets Beach to Millook Haven: Drive south on Crooklets Road; turn left (northeast) on Golf House Road; turn right (southeast) on Burn View; continue straight on Lansdown Road; turn left (south) onto The Strand; at the roundabout, take the 2nd exit onto The Crescent; continue straight on Vicarage Road; continue straight on Lynstone Road; continue straight on Marine Drive; turn right to park at Millook Water. This is **Stop 8, Millook Haven** *(50.771423, −4.576016) for a total of 9.6 km (5.9 mi; 16 min). There are 3 or 4 free parking spaces on the roadside.*

STOP 8 MILLOOK HAVEN

The sea cliffs at Millook Haven are well known for their spectacular folding, so much so, that The Geological Society voted it Number One in their top ten sites for "folding and faulting" locations in the United Kingdom (The Geological Society, Millook Haven).

The cliffs comprise Carboniferous Crackington Formation sandstone with thin shale interbeds. This unit was originally deposited as turbidites in a deep marine setting (Geology Page, 2018).

The sandstones were deformed into recumbent chevron folds (V-shaped folds lying on their sides) during the Variscan Orogeny. Chevron folds are characterized by long, straight limbs and sharp fold hinges, and this style of folding is a result of the mechanical contrast between the stiff sandstone layers and more ductile shale interbeds (Kim, 2015). The folds resemble nothing so much as a giant rock accordian.

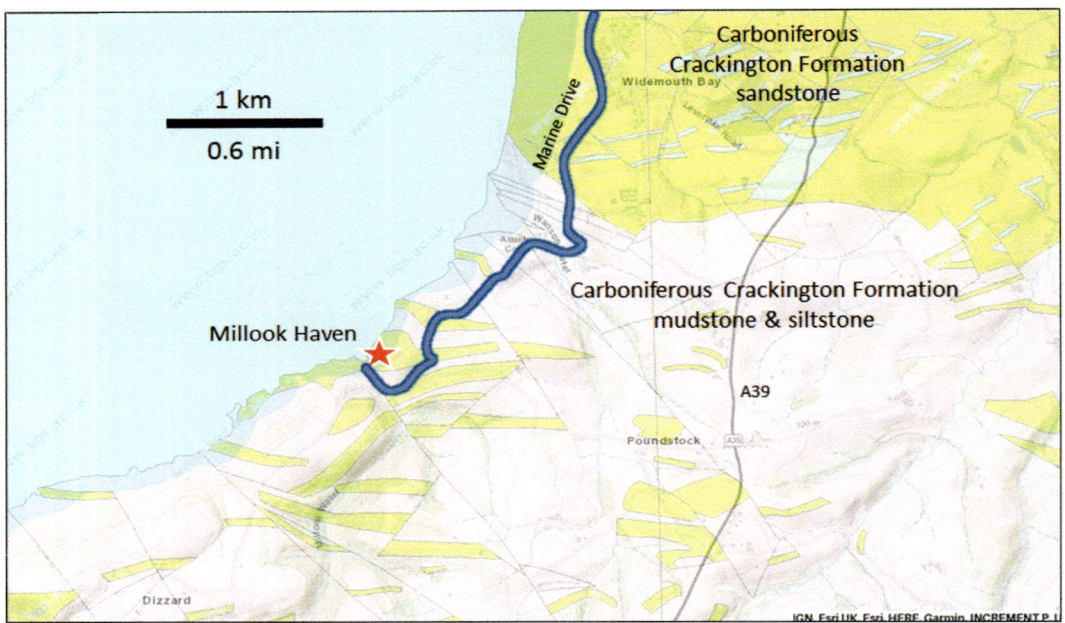

Geologic map of the Millook Haven area. (Modified from BGS, Geology of Britain Viewer.) Contains British Geological Survey materials © UKRI 2022. Base mapping provided by ESRI.

Recumbent chevron folds in the Crackington Formation, Millook Haven.

Millook Haven to Delabole Quarry: *Continue south on unnamed roads to the A39 at Wainhouse Corner; turn right (south) on A39 and drive to B3314; turn right (west) on B3314 toward Port Isaac and Delabole; drive to Pengelly in Delabole and turn left (east); continue to the quarry parking area. This is* **Stop 9, Delabole Quarry** *(50.620314, −4.724215), for a total of 25.4 km (15.8 mi; 31 min).*

STOP 9 DELABOLE QUARRY

Delabole Slate Quarry produces high-quality roofing slate (Bristow, 2014). The quarry is said to be the deepest man-made hole in Britain at 150 m (490 ft) deep (Blue Suede, 2010). The quarry is approximately 633 × 435 m (2,100 × 1,427 ft).

Delabole Slate has been used as a building material for perhaps as much as 800 years and has been quarried continuously since the early 17th century. It is the oldest continuously working slate quarry in England. During the reign of Elizabeth I, there were five quarries in the vicinity of the present pit. They produced slate used throughout Britain and exported it to Brittany and the Netherlands (Delabole Slate, 2013). In 1841, the five quarries were combined to make the Old Delabole Slate Quarry (Wikipedia, Delabole). The Old Delabole Slate Quarry Ltd was liquidated in 1977 by the company's bankers. It was run in receivership by Rio Tinto Zinc until 1999 when a local group bought it out. In 2005, the quarry was acquired by the Hamilton family (Walsh, 2011).

Delabole Slate Quarry. Note the heavy equipment for scale. (Photo courtesy of Martin Bodman / St Teath: Delabole Slate Quarry/CC BY-SA 2.0, https://commons.wikimedia.org/wiki/File:St_Teath, _Delabole_Slate_Quarry_-_geograph.org.uk_-_934450.jpg.)

In 1859 about 1,000 men were employed at the mines (Havaej et al., 2015). In 1910, 500 people were employed at the quarry, but as of 2007 only five quarrymen work the pit and the total workforce is no more than forty (Wikipedia, Delabole). Today the mine produces an average of 120 tonnes (132 tons) of slate daily. Using diamond wire saws, 600 tonne (661 ton) blocks are cut from the quarry face (Delabole Slate, 2013). Current production at the quarry is limited to the north–east face, where blocks are cut from 4 to 10 m (13–33 ft) high vertical benches (Havaej et al., 2015).

The village of Delabole was established in the early 1900s and is named after the Delabole Quarry. The town and quarry lie in the Cornwall Area of Outstanding Natural Beauty (Wikipedia, Delabole).

The rock being mined is the Delabole member of the Tredorn Slate Formation (BGS, Geology of Britain Viewer). Delabole slate is noted for its distinctive silvery gray color (Cornwall Council, 2018). The unit is a fine-grained, quartz-chlorite-sericite slate with small pyrite blebs. The Delabole member is a lenticular body with an arcuate outcrop around the Davidstow Anticline.

The slate was deposited approximately 372–359 Ma (Late Devonian) as open marine mud in extensional grabens (fault-bounded basins) on the southern margin of Laurasia/eastern Avalonia. After burial, they were metamorphosed during the Variscan Orogeny. A ductile folding episode was followed by thrusting, which was in turn overprinted by extensional faulting. Metamorphism in north Cornwall (the Tintagel High Strain Zone) was of the low temperature-high pressure variety as well as contact metamorphism related to the Bodmin Moor granite intrusion to the southeast (Havaej et al., 2015; The Geological Society, Slate Cornwall).

Visit

Visitors can come for a presentation, a short walking tour, or a long walking tour.

Presentations: One daily at 2:00 p.m. lasting 30 mins. Cost: £2.50 per person (there is no charge for children under 5 years). The presentation consists of a visit to the showroom where you receive an introduction to the company and a video presentation of Delabole Slate's operation. You get a short talk on the history, working methods, and geology, and can observe the quarry from the pit viewing platform.

Short Walking Tours: There is one daily at 2:00 p.m. lasting about 75 mins. Cost: Adults - £8.00, Students & Seniors – £6.00, Family (two adults & two Children) – £22.50. The short walk consists of a visit to the showroom (introduction to the company, a video presentation of Delabole Slate's operation, and a short talk on the history, working methods, and geology). A tour of the surface works includes observing the sawing of large slate blocks as well as the traditional hand-splitting of Delabole Roofing Slates.

Long Walkabout Tours, By Reservation Only: A minimum of ten persons is required. The long tour lasts about 2 h. Cost: Adults – £12.50, Students & Senior Citizens – £10.00. The long walk consists of a visit to the showroom (introduction to the company, a video presentation of Delabole Slate's operation, and a short talk on the history, working methods, and geology) followed by a tour of the pit. A tour of the surface works includes observing the sawing of large slate blocks as well as the traditional hand-splitting of Delabole Roofing Slates. See the diamond wire sawing of 600 tonne slate blocks (if in operation at the time) and observe the working faces within the pit.

Presentations & short tours are available at 2:00 p.m. from 1 May to 31 August, Monday to Friday, excluding bank holidays. During all other months, or for a long tour, please phone for a booking.

Pit Hours: 9:00 a.m.–4:30 p.m., Monday–Friday
Address: The Delabole Slate Company Ltd, Pengelly, Delabole, Cornwall PL33 9AZ, UK
Phone: 01840 212242
Email: sales@delaboleslate.co.uk
Website: http://www.delaboleslate.co.uk/quarry-tours.asp

For all of those readers who fell in love with the fictional town of Portwenn in the TV series Doc Martin, you are only 10 km (6 mi) from the actual town of Port Isaac where the program was filmed. If you chose to visit this quaint and scenic spot, you will be traveling over Middle Devonian mudstone, siltstone, and sandstones.

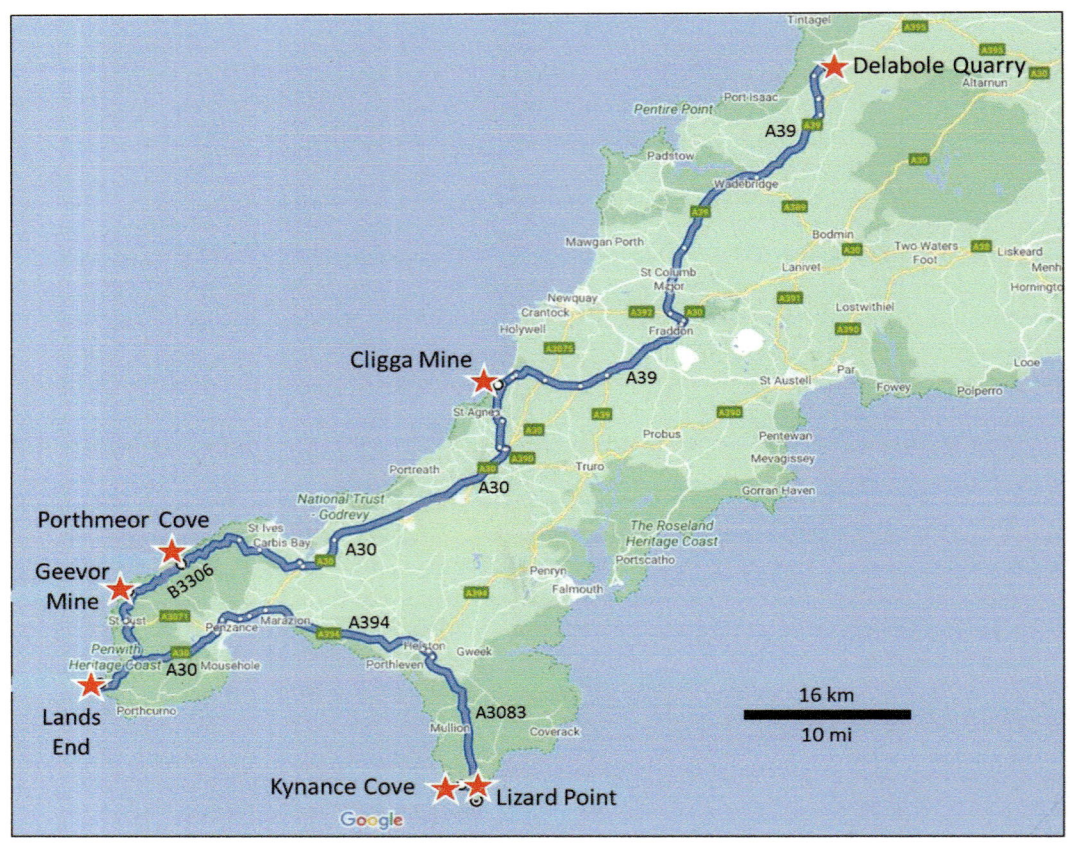

Delabole Quarry to Cligga Mine: *Return to B3314/High Street and turn left (southwest); turn left (south) onto B3267/West Downs and drive to St. Teath; turn right (south) at The Square, then left (south) toward Trehannick; turn right (south) on A39 and drive to the A30 at Toldish/Indian Queens; at Highgate Hill Roundabout, take the 4th exit onto the A30 toward Redruth/Truro/Fraddon; drive southwest on A30 to the B3285; turn right (west) onto B3285 and drive to Perranporth; continue south on B3285 to the sign for the Cligga Head Industrial Estate on the right; turn right (west) and drive to parking in front of the road gate (50.337597, −5.171926). You can walk or drive the remaining 615 m (2,000 ft) to **Stop 10, Cligga Mine** (50.339304, −5.178784), for a total of 57 km (35.4 mi; 58 min). There is ample free parking, but no facilities.*

STOP 10 CLIGGA MINE

Cornwall has more than 2,000 years of mining history, including tin, copper, lead, silver, and other metals, from Phoenician traders to modern times. The key product of the region has been and is tin. The Cligga Head area has been mined for at least several hundred years. Extensive wolframite (iron-manganese-tungsten oxide) mining and explosives factories have been in the area since before World War I. In the same neighborhood as the Cligga Head Mine are the Perran St George, Good Fortune, and Wheal Prudence mines, all abandoned (Themes, 2020).

The tin mine at Cligga Head closed in 1945, after contributing both tungsten (for machine tools) and tin (for solder) to the war effort (Truro School, 2020).

Just east of the headland is the site of a former explosives works. First established in 1889 by the British and Colonial Explosives Company, in 1893, the Nobel Explosives Company took over. By 1905, the works had shut down, with a short revival during World War I (Themes, 2020).

View of old mine adits (tunnel entrances) on the cliffs of Cligga Head. It is hard to imagine how workers got to these openings.

Parts of the Cligga Head and mine area are an SSSI. In these areas, no hammering or collecting is allowed (Variscan Coast, 2021). Today the site has a collection of unimpressive industrial ruins. On the other hand, the cliffs are quite impressive both for their geologic and scenic value.

While the clifftop is fairly safe, beware of the cliff edges as they are constantly eroding and breaking away (Themes, 2020). On windy days, the gusts can be quite powerful. Large waves smashing into the base of the cliffs can send spray over the cliff tops. On such days, the path to the beach can be treacherous.

In Truro check out the Royal Cornwall Museum, which has an impressive collection of Cornish minerals (Evans, 2014). First opened in 1818, it has local mining and mineral exhibits, among others.

Address: Royal Cornwall Museum, 25 River St, Truro TR1 2SJ
Phone: +44 1872 272205
Website: https://www.royalcornwallmuseum.org.uk/
Email: enquiries@royalcornwallmuseum.org.uk

GEOLOGY

The Cligga Granite, a Permian (~295–270 Ma) muscovite-rich granite, was intruded into Devonian Porthtowan Formation, a hornfels or "clay slate." This granite is a cupola, or high point, extending above the regional Cornubian granite batholith that formed during the Variscan Orogeny (Wilson, 2014).

The Porthtowan Formation was altered by the heat and fluids of the intruding granite, changing it from a mudstone to a hornfels. Minerals in the mudstone recrystallized to form biotite and andalusite. The metamorphic aureole, or area of country rock altered by the intruding granite, extends several kilometers beyond the granite (Variscan Coast, 2021).

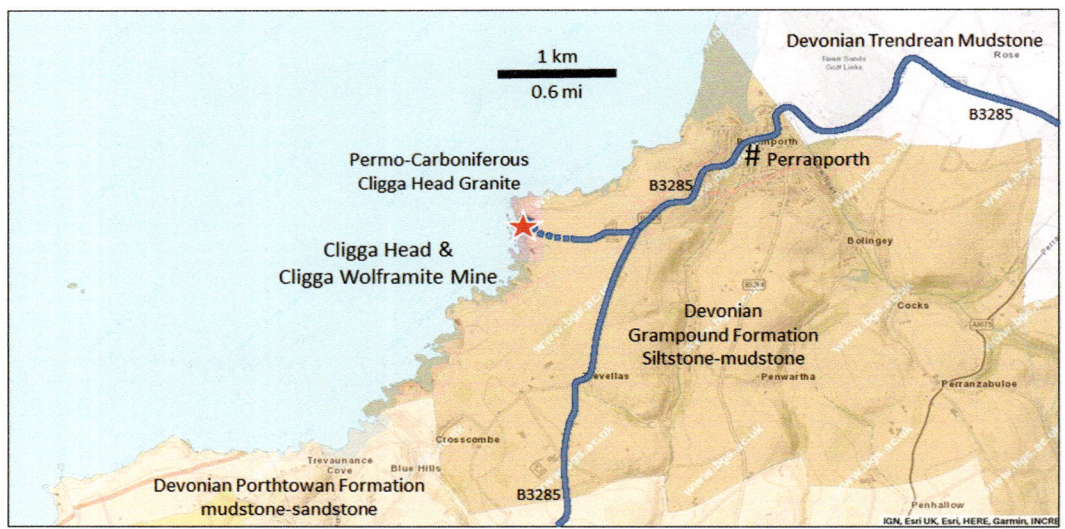

Geological map of the Cligga Head-Perranporth area. (Modified after BGS Geology of Britain Viewer. Contains British Geological Survey materials © UKRI 2022. Base mapping provided by ESRI.)

The intrusion may have taken advantage of a fault that lies at the contact between the intrusion and the country rock. Cooling joints in the granite were filled by late-stage volatiles and hydrothermal fluids that precipitated minerals in a "sheeted tin-tungsten vein complex" (Variscan Coast, 2021). Wolframite is the main tungsten mineral in the greisen-bordered veins; there is also minor cassiterite (a tin oxide mineral; Variscan Coast, 2021). Greisen is an old Saxon mining term for granitic rock with tin ore and virtually no feldspar; later it referred to altered granite consisting mainly of quartz and mica (Wilson, 2014). This is also known as "phyllic" alteration (quartz-sericite-pyrite with local tourmaline). The granite would have been between 4 and 5 km (2.4–3 mi) below the surface at the time of intrusion. Emplacement of the granite was followed by intrusion of rhyolite dikes. This was followed by several episodes of mineralization.

The Cligga Granite was extensively altered to a greisen. Greisen is a result of boron-bearing fluids circulating through fractures in the granite and precipitating tourmaline near the fractures. Look for mineral specimens in the dumps below the mine workings. A quarry below the mine has an excellent exposure of greisen-bordered sheeted veins in the granite. The quarry was designated an SSSI because of this exposure (Nottle, 2006).

Greisen granite on Cligga Head.

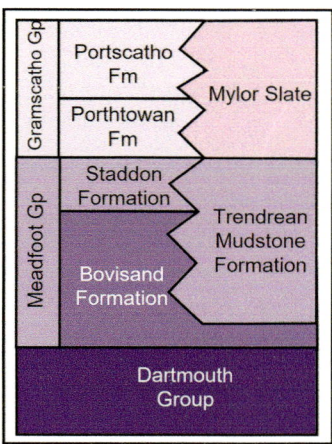

Devonian units in the Gramscatho and Looe basins. (Information drawn from Hollick et al., 2006.)

Kaolinization (also called "argillic" alteration, where feldspars degrade to quartz and kaolinite clay) also affected the core and southern part of the granite (Moore and Jackson, 1977; Variscan Coast, 2021). The granite became soft and easily eroded because of the kaolinite clay (Nottle, 2006).

To the south, there is a chalcopyrite-rich elvan (copper-iron sulfide mineral in quartz porphyry) which also intruded into the metasedimentary rocks (Variscan Coast, 2021).

MINERALIZATION

A widespread Permian-aged vein system extends across the cliff. Near-vertical east–west veins contain wolframite with cassiterite. Old mine workings, mainly from the 19th century, can be seen on the cliff. The veins have a greisen border. The wolframite-cassiterite veins are overprinted by later Permian-aged sulfide veins. These are all crosscut by north–south-trending, steeply dipping Jurassic lead-zinc veins (Variscan Coast, 2021).

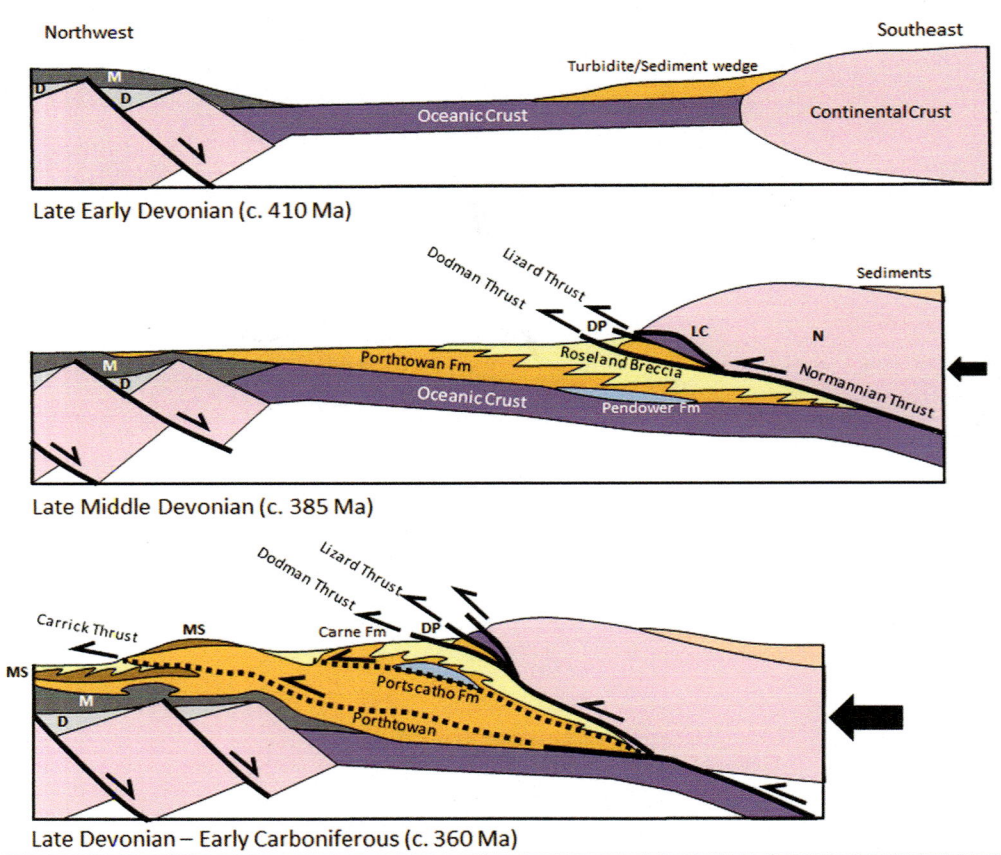

Tectonic development of the Gramscatho Basin during Devonian to Early Carboniferous. M, Meadfoot Group; D, Dartmouth Group; N, Normannian Nappe; LC, Lizard Complex; DP, Dodman Phyllites; MS, Mylor Slate. (Modified after Leveridge and Shail, 2011.)

A series of pressure release and cooling episodes resulted in deposition of minerals in the fractures as the ability of the fluids to hold minerals in solution decreased. Later fractures were filled with sulfide-rich fluids derived from the surrounding thermally altered country rock (killas). Tin, tungsten, and arsenic mineralization extends into fractures in the surrounding Porthtowan Formation country rock. At the north end of Perranporth Beach, there is evidence of mining of a low-grade iron deposit (Nottle, 2006).

Quartz-tourmaline-topaz veins have been found. Even native gold has been noted at Cligga Head, in iron-rich metasediments.

Cornish mines were among the first where concentric zoning of mineralization, from the granite core out into the host rocks, was observed. Most tin, tungsten, copper, and arsenic production was located in the granitic core, with the western end (including Cligga Head) enriched in tin, copper, and zinc. In contrast, most iron, arsenic, antimony, fluorine, barium, and kaolin production has been to the east. Lead and silver production was concentrated along the south flank of the intrusion (Wilson, 2014).

Cligga Mine ruins, view north.

MINERAL COLLECTING

At the end of the road, past the industrial estate, a footpath leads to the ruins of some of the old mine buildings, mainly the mill. The views from here are breathtaking. Below the mine ruins you come upon greisen sheet vein stockworks. Collecting is permitted at the old mine dumps. Look for cassiterite crystals (a silvery tin oxide) on the surface of vein quartz. Collectors have also found wolframite, scorodite (iron arsenate), arsenopyrite (iron-arsenic sulfide), chalcopyrite, pyrite (iron sulfide), stannite (tin-copper-iron sulfide), molybdenite (molybdenum sulfide), chalcocite (copper sulfide), sphalerite (zinc sulfide), and tourmaline (Nottle, 2006; Evans, 2014; Wilson, 2014).

STOP 11 PENZANCE PENINSULA

The South West Coast Path National Trail winds 1,000 km (600 mi) along the coast of Cornwall from Somerset to Dorset. It is a great way to see geology. If you don't have time to do the walk, you can drive and stop at a few choice locations.

Most of the rocks, with the exception of the Lizard Peninsula, are Middle and Upper Devonian Gramscatho Group metamorphosed marine shales, siltstones, and sandstones. The Gramscatho consists of the Portscatho Formation interbedded meta-sandstone and slate (384–376 Ma) overlying the Porthtowan Formation interbedded slaty mudstone and subordinate turbidite sandstone (392–384 Ma) (BGS Lexicon, https://webapps.bgs.ac.uk/lexicon/lexicon.cfm?pub=POAN; Hollick et al., 2006).

In addition to the sandstone-rich Gramscatho Group, there is the laterally equivalent and mudstone-rich Mylor Slate. The Gramscatho Group was deposited in deep basin and continental rise settings, whereas the Upper Devonian Mylor Slate was deposited as deep marine and continental slope channel deposits and proximal (nearer-to-shore) turbidites grading upward into distal (deep marine) turbidites (Leveridge and Shail, 2011).

These units were deformed and perhaps lightly metamorphosed during the Variscan Orogeny. Several million years later they were intruded by Early Permian granite, the Cornubian batholith, that can be seen today at Bodmin Moor, Land's End, the Isles of Scilly, St. Michael's Mount, and Porthmeor Cove. As the magma intruded the sedimentary cover it created a "metamorphic aureole," or ring of altered and metamorphosed rock around the granite.

Granites in Cornwall in many cases contain economic quantities of tin, tungsten, copper, and minor amounts of zinc, lead, iron, and silver. These precipitated in fissures in the granite and the country rock, forming vein or lode-type deposits. Mining was an important part of the economy of this otherwise poor agricultural area.

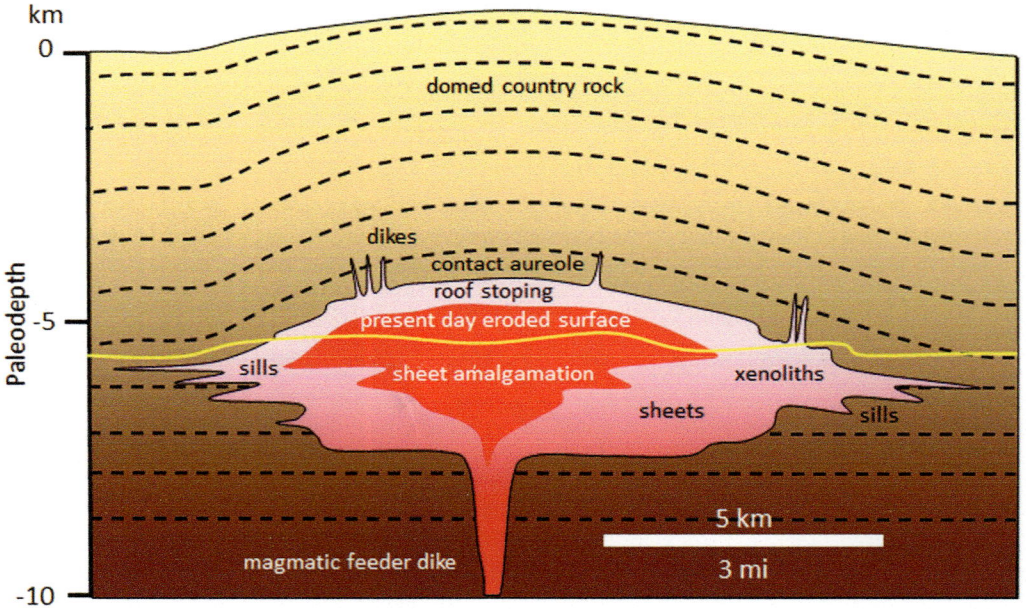

Schematic cross section, Land's End laccolith. A laccolith is a dome-shaped sill-like magma intrusion (Modified after Pownall et al., 2012.)

Tin placer deposits in streams were used in prehistoric times to produce bronze… metal traders from the Mediterranean called Britain the "Cassiterides," or "Tin Islands." In the 1500s, miners followed the deposits to their underground sources. Cornwall's copper mining, for a short time in the 1830s, dominated world copper production. As the mines got deeper, they were often flooded by groundwater. The invention of steam-powered pumps in the early 1700s allowed mining to continue and flourish. The last operating tin mine in Cornwall closed in 1998.

The landscape of Cornwall has been affected by changes in sea level during and after the last Ice Age. Wave-cut platforms and several levels of raised beaches can be seen along the coast, as well as drowned-river valleys or "rias" (Whaley, 2010).

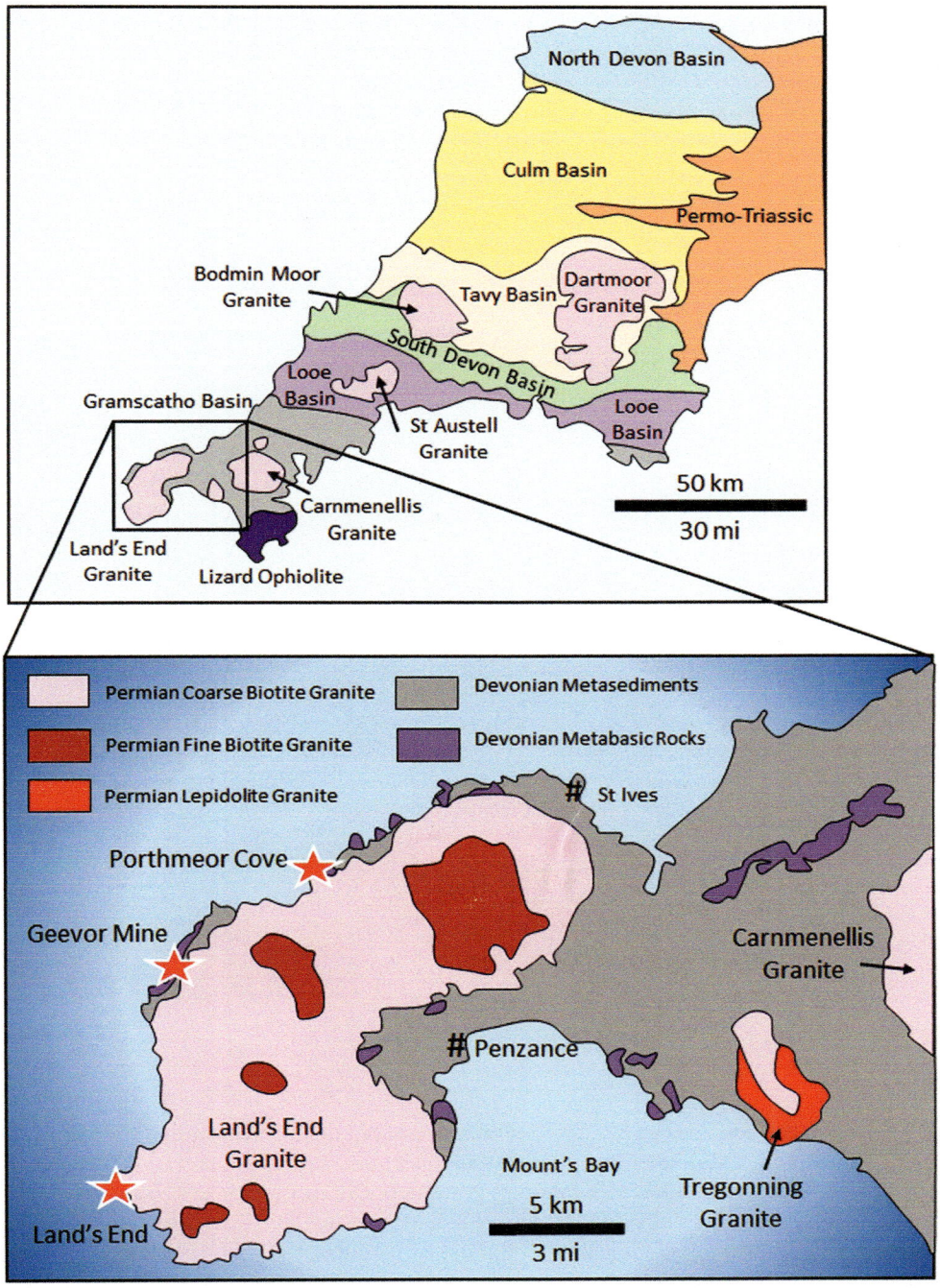

Geologic setting of granites in the Variscan fold-thrust belt of Europe during the Early Carboniferous. (Modified after Pownall et al., 2012.)

*Cligga Mine to Porthmeor Cove: Return to B3285 and turn right (south); at Barkla Shop turn left (south); turn left (southeast) on an unnamed road to Truro/Redruth/Silverwell and drive to Mithian Downs; in Mithian Downs turn right (south) on the unnamed road to Truro/Redruth/Porthtowan; turn left (southeast) onto B3277 and drive to Three Burrows; at Chiverton Cross take the 4th exit (southwest) onto A30/Blackwater Bypass; follow A30 to St Erth; at the St Erth Roundabout, take the 2nd exit (west) onto A3074/Nut Lane; at the roundabout, take the 1st exit onto Mill Hill; turn left (south) onto B3311 and immediately turn right (west) toward Towednack; turn left (west) onto B3306 and drive toward Porthmeor; 460 m (1,520 ft) south of Porthmeor pull over on the right and park next to a trail leading northwest to the coast path (50.175722, −5.600504) for a total of 48.4 km (30.1 mi; 48 min). Parking is free, but there only room for maybe five cars. Access is somewhat overgrown, immediately south of the cattle grid. No facilities. This is a moderately steep walk about 950 m (3,130 ft) northwest to **Stop 11.1, Porthmeor Cove** (50.181156, −5.606857).*

STOP 11.1 PORTHMEOR COVE

View north over Porthmeor Cove toward the West Penwith Granite.

Porthmeor Cove has been designated an SSSI because of the unique exposure of the West Penwith Ganite (a light-colored, coarse biotite granite) and its contact with the darker, overlying Devonian slates (iWalk Cornwall, Circular walks visiting Porthmeor Cove). Porthmeor Cove is the only place in southwest England with a fully-exposed granite "cupola," an upward extension of a vast, deep granitic intrusion. The roof of the granite, the contact with the overlying rocks, is exposed in the sea cliffs and shows fine- and coarse-grained dikes extending into the surrounding rocks, which are extensively veined and altered (Whaley, 2010).

Beneath the nearly horizontal contact is a layered pegmatite (coarsely crystalline) zone up to 3 m (10 ft) thick that consists of alternating fine- to medium-grained tourmaline granite. Three generations of dikes are recognized. The oldest is the only known albite-rich granite in the Land's End Pluton (a "pluton" is an old, crystallized magma chamber). The youngest dikes are fine-grained tourmaline-bearing granite (Muller et al., 2006).

This is a place of solitude. There are no lifeguards. Dogs are allowed all year. It is a nudist/naturist beach. There is roadside parking, but there is only enough space for 2 or 3 cars (Cornwall Beaches, 2021).

There are two small cupolas here. The north cupola intruded exclusively into slate, is exposed mainly in the sea cliffs as a tall granite dome with sharp, angular contacts. A network of aplite (finely crystalline) sills connects it to the smaller but more accessible south cupola, which intruded both slates and metagabbro. The roof of the south cupola granite, beneath a slightly domed contact, consists of a banded light-colored granite-pegmatite sequence (Pownall et al., 2012).

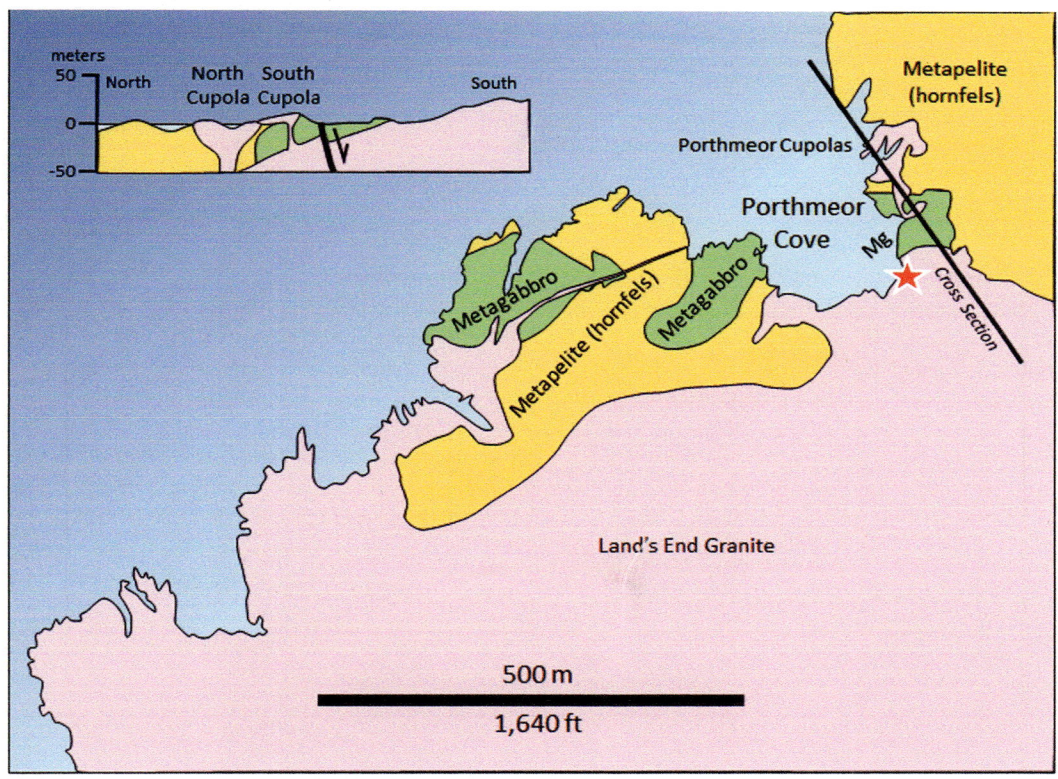

Geologic map and cross section of the Porthmeor Cove area. (Modified after Pownall et al., 2012.)

Porthmeor Cove to Geevor Mine: Continue south on B3306 to Trewellard/Pendeen/Boscaswell; turn right (northwest) at signs for the Geevor Tin Mine and drive to the parking area. This is **Stop 11.2, Geevor Tin Mine** *(50.151681, −5.676536) for a total of 7.1 km (4.4 mi; 12 min).*

Stop 11.2 Geevor Mine and Cornish Miners

Geevor Tin Mine is in the St Just Mining District (Geevor Mine Home Page). It is the largest preserved mining site in England and is part of the Cornish Mining World Heritage Site. The site includes the Holman Collection, a collection of engineering artifacts (The Geological Society, Geevor Tin Mine). It also lies on the South West Coast Path.

Tin has been mined in Cornwall since prehistory. Yet, the greatest era of mining in Cornwall was the early 19th century when for a short time Cornwall became the world's leading supplier of copper. By around 1860, foreign sources became dominant, and tin became Cornwall's most important export (The Geological Society, Geevor Tin Mine). The Geevor Mine finally closed in 1990 when the ore became uneconomic.

Lode ore is located in the outer zone of a Carboniferous biotite granite that intruded the surrounding Devonian metasediments and metavolcanics, known locally as "killas." The metamorphic country rock is visible northwest of the mine at the sea cliffs. As you approach the granite, the country rock acquires the mineral cordierite, then mica-rich phyllite, and ultimately becomes hornfels and tourmaline schist at the contact.

The tin ore is primarily cassiterite in northwest-southeast veins accompanied by quartz, tourmaline, metal sulfides, and fluorite. Cassiterite is around 1% of the vein material (Wikipedia, Geevor Tin Mine).

It is worth noting that Cornwall played a leading role in the spread of mining skills (The Geological Society, Geevor Tin Mine). Cornish miners were forced to take jobs overseas due to a lack of work at home during the 18th and 19th centuries. Known as "Cousin Jacks," miners migrated to mining districts around the world in search of work. It is estimated that as many as 250,000 migrated to the United States, Canada, Australia, Argentina, New Zealand, Mexico, Panama, South Africa, and Brazil just in the years from 1861 and 1901 (Wikipedia, Cornish Diaspora).

Victory Shaft headframe, Geevor Tin Mine. (Photo courtesy of Ahwmay, https://commons.wikimedia.org/-wiki/File:Geevor_tin_mine_headframe.jpg.)

The Geevor Mine only shows visitors a fraction of the total workings. A huge network of tunnels runs from the top of the site down below the road to the sea and far out below sea level.

Visit

The underground workings were part of the 18th-century Wheal Mexico Mine. Starting in 1919 miners got to the subsurface using a 9-man cage that went down the Victory Shaft, while ore was raised in 2 tonne (2.2 ton) skips. The shaft extended to a total depth of 480 m (1,575 ft) below the surface in 1975. In 1975, an inclined shaft was begun from near the bottom of Victory Shaft to access workings in the nearby Levant Mine due west on the coast. The shaft eventually reached a position of 650 m (2,130 ft) below the surface, putting the tunnels well below sea level.

Surface facilities include the mill, where the rock was crushed to produce tin concentrate, The Dry, the changing room for the miners, and the Hard Rock Museum. The changing room is preserved just as the miners left it on their last day in the mines in 1990. The museum has exhibits about metal mining in Cornwall, interactive activities, and a gallery with a collection of minerals and mining artifacts.

Family activities include panning for gold and gems, neither of which were found in the mine.

The shop at Geevor carries many items made in Cornwall. In particular, the shop has a large selection of Cornish books, minerals and gems, jewelry, clothing, and accessories.

For more information on the shop:

T: 01736 788006
E: shop@geevor.com

The shop at Geevor is open Sunday–Friday 10:00 a.m.–4:30 p.m.

There is also a café with stunning views of the Cornish coast.

Address: Pendeen, Penzance TR19 7EW, United Kingdom
Hours: Open 9:00 a.m. till 5:00 p.m.; last entry at 4:00 p.m. Sunday through Thursday.
Website: geevor.com
Phone: +44 1736 788662
Prices: Adult (full admission)   £17.70
    Adult (museum only)   £14.25
    Child (4 years and over)   £9.90
    Child (under 4 years old)   Free
    Student   £9.90
    Senior Citizen   £15.12
Family (two adults and up to three children) £54.75

You must use the online booking system to reserve tickets.

STOP 12 PENWITH AND LIZARD HERITAGE COASTS

The Penwith Heritage Coast extends from Penzance, around Land's End, to the resort town of St. Ives on the Atlantic coast. It has the mildest climate in England.

According to Cornish legend, the Giant Myen du lived at Maen Castle near Land's End. Faeries known as "Coopers" would bang stones together to alert residents of St. Ives that rich shoals of fish were approaching. At Zennor, a mermaid is said to have lured the local squire's son to his death, and some say that her song can still be heard if you listen carefully. Legend has it that if you climb through the hole at Men an Tol nine times against the setting sun, all your injuries will be cured.

Numerous prehistoric monuments lie near the coast, including the Merry Maidens Stone Circle and Tregiffian Bronze Age grave, near Lamorna.

Abandoned mines lie scattered across the peninsula, reminders of Cornwall's mining heritage.

On the cliffs near Porthcurno, a white pyramid marks the spot where the first transatlantic cable came ashore in 1880.

The Lizard Heritage Coast just to the east stretches from Porthleven to Enys Head and includes The Lizard, the southernmost point of the English mainland: it too has a mild climate, a result of the warm Gulf Stream. This allows subtropical vegetation to flourish and is home to one of Britain's rarest birds, Cirl's Bunting. The long-distance South West Coast Path passes through both heritage coasts.

Whereas Land's End is granitic, the Lizard Peninsula contains serpentine, unique in Western Europe. Serpentine is derived from ancient ocean floor that was uplifted and metamorphosed when

Gondwana banged into Laurasia during the Variscan Orogeny between 270 and 320 Ma (Ross, Penwith Heritage Coast; Ross, Lizard Heritage Coast).

*Geevor Mine to Land's End: Return to B3306 and turn right (south); in St Just continue straight onto A3071/Market Square; on the south side of town turn right (south) onto B3306 and drive south to the A30; turn right (west) onto the A30 and drive southwest to **Stop 12.1, Land's End** (50.068276, −5.715332), for a total of 13.8 km (8.6 mi; 23 min).*

12.1 LAND'S END

One of four Permian granite domes that together form the backbone of Cornwall, was emplaced around 270 million years ago at what is now Land's End.

The domes extend above a regional granite intrusion, the Cornubian batholith, and forced their way through the overlying rocks. The cliffs around Land's End contain two varieties of granite. Below the Land's End Hotel, the granite is coarse and contains large crystals; the granite below the First & Last House is finer and has smaller crystals. You can tell the difference between the two because of the smoother weathering of the finer rock (Coastal Landscape, 2019). The cliffs at Land's End are between 61 and 122 m (200–400 ft) high.

The Lower Permian Land's End Granite intruded Upper Devonian metasedimentary and metavolcanic rocks of the Mylor Slate Formation that had been previously deformed and regionally metamorphosed during the Variscan Orogeny. Granite generation and emplacement occurred in response to regional north–northwest–south–southeast crustal extension. Anisotropy of magnetic susceptibility work reveals an igneous fabric that reflects late magmatic deformation, that is, there is a rock texture or pattern caused by flow of the large, early-forming crystals during magma emplacement (Kratinová et al., 2003, 2010).

Land's End Granite looking north from Land's End.

Land's End, looking west. The offshore rocks used to be onshore, but the sea is relentlessly wearing away at England.

Land's End to Kynance Cove: Return east on A30 to the Newtown Roundabout, and take the 2nd exit onto A394; continue east on A394 to Helston; at the roundabout, south of town take the 3rd exit onto A3083 and drive south to the sign for Kynance Cove; turn right (west) and drive to **Stop 12.2, Kynance Cove** *(49.974452, −5.225111), for a total of 56.0 km (34.8 mi; 1 h 6 min). There is ample National Trust parking for £2/h, with facilities.*

12.2 KYNANCE COVE

The Lizard Peninsula is unique, and important in geology. It contains the largest outcrops in Britain of serpentine and ophiolite (Thomas, 2019). The Lizard Complex is Britain's most complete example of an ophiolite, ancient ocean floor that was thrust onto the continent during the Variscan Orogeny (Leveridge and Shail, 2011). Much of the peninsula consists of dark green and red serpentinite, which is not always obvious, but nice samples can be found. This rock breaks down into the poor soil that underlies the flat and marshy heaths of Goonhilly Downs (Wikipedia, Geology of Cornwall).

Serpentinite cobbles at Kynance Cove.

The name The Lizard potentially derives from a couple of expressions in the Old Cornish dialect. "Lezou" translates as headland, and "Lis-ardh", means fortress or headland and is derived from "Lis", place, and "Ardh" for high area (Whaley, 2010; Thomas, 2019).

Another local place name has had great historical repercussions. Predannack Downs are on the west side of the Lizard Peninsula, extending from Lizard Point to Mullion. The Greek explorer and geographer Pytheas in the 4th century BC referred to *Prettanike* or *Brettaniai* as a group of islands off the coast of northwest Europe. In the 1st century BC, Diodorus Siculus called the island *Pretannia*, after the *Pretani* people believed by the Greeks to live there (Wikipedia, Britannia; Thomas, 2019).

Marconi chose The Lizard to establish his radio station that sent the first transatlantic radio message (Thomas, 2019).

The Lizard Peninsula is partly owned and cared for by Natural England, the National Trust, and Cornwall Wildlife Trust. It is an Area of Outstanding Natural Beauty, and there are eight Sites of Special Scientific Interest to protect both wildlife and geology (Thomas, 2019).

Kynance Cove looking south toward Lizard Point.

The simplest geology has often been compared to a layered cake, with the oldest layers on the bottom and getting progressively younger toward the top. Now imagine that the cake has been tilted and planed off such that the bottom layers are exposed in one corner, and the overlying layers are exposed in succession across the cake. With a few wrinkles, this might be a good analogy to what we see in England. In southwest England, the oldest rocks are exposed, and as we travel east across England we see progressively younger strata, from Devonian (416–359 Ma) on the Lizard Peninsula to Cretaceous (roughly 67 Ma) at Dover. This pattern was revealed on one of the first comprehensive geologic maps in the world, *A Delineation of the Strata of England and Wales with part of Scotland*, published by William Smith and The Geological Society in 1815 (Winchester, 2001).

The oldest rock in Cornwall is the 500 Ma Man of War Gneiss, part of the pre-Devonian, pre-rifting continental basement. It is exposed on small islands off Lizard Point. Ophiolites form when two continents collide and a section of the intervening oceanic crust is thrust over continental crust. The thrusted material lies over the Old Lizard Head Series, mica schist, and quartzite derived from metamorphosed mudstone, sandstone, and volcanics that are also part of the pre-rift basement. One of the only places they can be clearly seen is at Lizard Point.

A classic ophiolite consists, from the bottom up, of ultramafic (very low silica and high iron and magnesium) mantle rocks overlain by layered gabbros and volcanic dikes, with pillow lavas and marine sediments at the top. Most of these occur on The Lizard. Typically, these rocks form at mid-ocean spreading centers. Structurally, The Lizard consists of several thrust fault-bounded ophiolite sheets (Whaley, 2010).

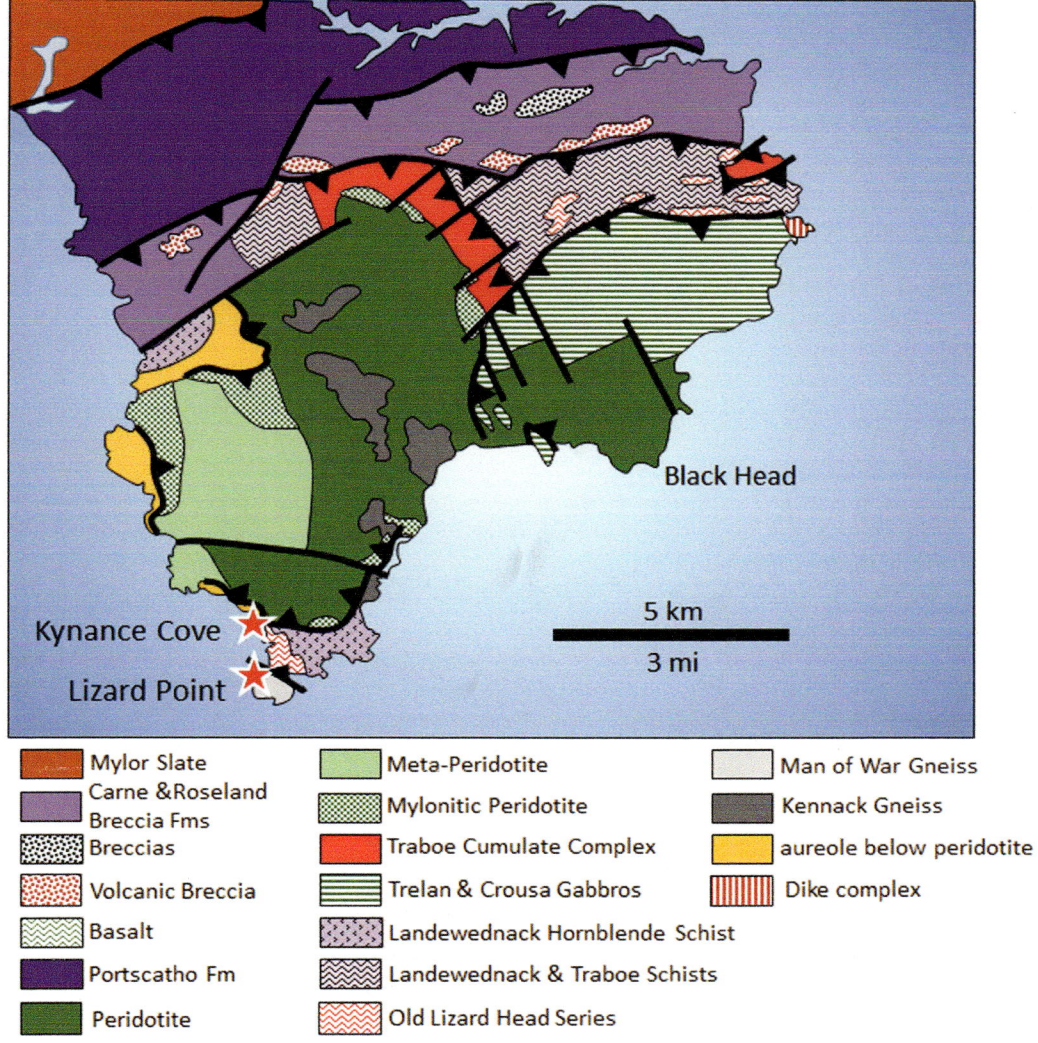

Geologic map of the Lizard Complex. (Modified after Leveridge and Shail, 2011.)

The Lizard Complex comprises a pre-rift continental basement (Man of War Gneiss, Old Lizard Head Formation); an ophiolite that includes mantle peridotites exposed by rifting; igneous rocks generated during rifting (Traboe Cumulate Complex, Trelan and Crousa gabbros, basalt dikes, and Landewednack Hornblende Schist); and magma (Kennack Gneiss) intruded during the initial emplacement of thrusted ocean and mantle slivers on the continent. The ophiolite was thrust onto Devonian metasediments (Bromley, 1991).

The Man of War Gneiss is metamorphosed Late Cambrian gabbro that was altered to amphibolite facies in Mid-Late Devonian. Amphibolite facies is a mineral assemblage that is the result of medium pressure and average to high temperatures. The upper contact of the Man of War Gneiss is a thrust below amphibolite facies metasediments of the Old Lizard Head Formation, a garnet schist derived from metamorphosed mudstones, sandstones, and tuffs (Leveridge and Shail, 2011).

The Traboe Cumulate Complex contains metamorphosed gabbros, dunites , pyroxenites, and anorthosites formed during initial continental rifting. The most recent hypothesis is that the Traboe Cumulate Complex formed by intrusion of mantle-derived magma during rifting. The Crousa Gabbro and Trelan Gabbro overlie these mantle-derived rocks.

The gabbros are cut by northwest-southeast dolerite–basalt dikes similar to those found today at Mid-Ocean Ridges. This sheeted dike complex is thought to have formed in a slow-spreading oceanic ridge. The Landewednack Hornblende Schists consist of metamorphosed mid-ocean ridge volcanics and intrusives.

The Kennack Gneiss is derived from metamorphosed mid-ocean ridge basalts and granites. It intruded around 376 Ma (Middle Devonian) and was metamorphosed to amphibolite facies about 370 Ma (Late Devonian).

The oceanic crust of the Lizard formed either on a mid-ocean ridge or in a small pull-apart basin that developed along the southern margin of the Variscan Orogen (Leveridge and Shail, 2011). The resulting rift and its sediment infill are known as the Gramscatho Basin (Pownall et al., 2012). The Lizard ophiolite formed during continental rifting that commenced in the Early Devonian (Leveridge and Shail, 2011).

Early Devonian marine sediments of the Meadfoot Group were deposited in the east–west Gramscatho Basin; they came mainly from the southern continental margin, which was actively uplifting during the Devonian.

Early Devonian to Early Carboniferous sediments accumulated on a passive margin, that is, a rifting continental margin. These sediments were affected by Late Devonian-Carboniferous inversion and crustal shortening involving folding and north–northwest-directed thrusting. The European Variscides (Variscan mountain belt) was formed by collision of Laurussia and Gondwana after closure of the Rheic Ocean between Late Devonian and Early Carboniferous time. Remnants of this once mighty mountain chain are found in several large massifs from Iberia to Bohemia. Relics of Rheic oceanic crust and mantle are locally preserved in, for example, the Mid-Devonian Lizard Complex (Nance et al., 2010; Pownall et al., 2012).

Three thrust sheets occur on the peninsula: from the base, they are the Carrick, Veryan, and the Dodman nappes, all dipping southeast. The Carrick Nappe (thrust sheet) contains the Portscatho Formation, a Late Devonian graywacke sandstone and mudstone turbidite.

The Veryan Nappe consists of the Middle to Late Devonian Pendower Formation and the Carne and Roseland Breccia formations. The Pendower Formation contains mudstones, mid-ocean ridge basalt, and chert, an assemblage usually associated with spreading centers. The Carne Formation sediments are interpreted as continental slope deposits and turbidites. They pass upward into the Roseland Breccia Formation, a major olistostrome, or unit formed by slumping and gravity flows. It consists of silty mudstone with large and small fragments of sedimentary, metamorphic, volcanic, and intrusive rocks. .

The uppermost Dodman Nappe contains greywacke sandstone and mudstone. In the English Channel, it is overthrust by the Normannian Nappe consisting of Eddystone Rock, a garnet gneiss. The Lizard Ophiolite lies between the Dodman and Normannian nappes.

Thrust sheet deformation is characterized by tight folds that indicate north–northwest compression and shortening. Metamorphism, related to early deformation, occurred in Mid-Devonian to Early Carboniferous time (385–355 Ma) and with time progressed north–northwest across the thrust sheets.

Thrusting of the Dodman Nappe and Lizard Ophiolite, and filling of the rifted marine basin, occurred in Mid-Devonian. Continued shortening transferred displacement to the Veryan Thrust, and then to the Carrick Thrust. This last thrust event marked the closure of the Gramscatho Basin in southwest England (Leveridge and Shail, 2011).

Kynance Cove is surrounded by cliffs and islands of red and black serpentinite altered from the original olivine-rich peridotite. The alteration is due to retrograde metamorphism that results from a decrease in temperature and pressure on the rock, usually during uplift. Serpentinite is named for its resemblance to snake skin: the rock here is banded and streaked with veins and polishes to an attractive red, green, and black rock.

A steep path leads down the sea cliffs and suddenly opens onto a small bay where the cliffs shine red and green where they have been polished by the sea.

Kynance Cove to Lizard Point: Return east to A3083 and turn right (south); drive to Lizard and continue straight on Lighthouse Road; bear left to the National Trust Lizard Point car park. This is Stop 12.3, Lizard Point (49.960850, −5.204136), for a total of 3.3 km (2.1 mi; 10 min). There is ample National Trust parking and facilities. Walk about 250 m (820 ft) southwest on a moderately steep trail to the overlook.

12.3 Lizard Point

This is the southernmost point on the UK mainland. The thrusted Lizard Complex lies over the Old Lizard Head Series. One of the only places this series can be clearly seen is at Lizard Point (Whaley, 2010).

The Old Lizard Head Formation at Lizard Point is a garnet mica schist and quartzite derived from metamorphosed mudstones, sandstones, and tuffs (Leveridge and Shail, 2011). It lies in thrust contact with the Man of War Gneiss below, and its upper contact is the Carrick Thrust that carries the Late Devonian Portscatho Formation.

Old Lizard Head Formation crenulated mica schist at Lizard Point.

View northwest from Lizard Point, the southernmost tip of Great Britain.

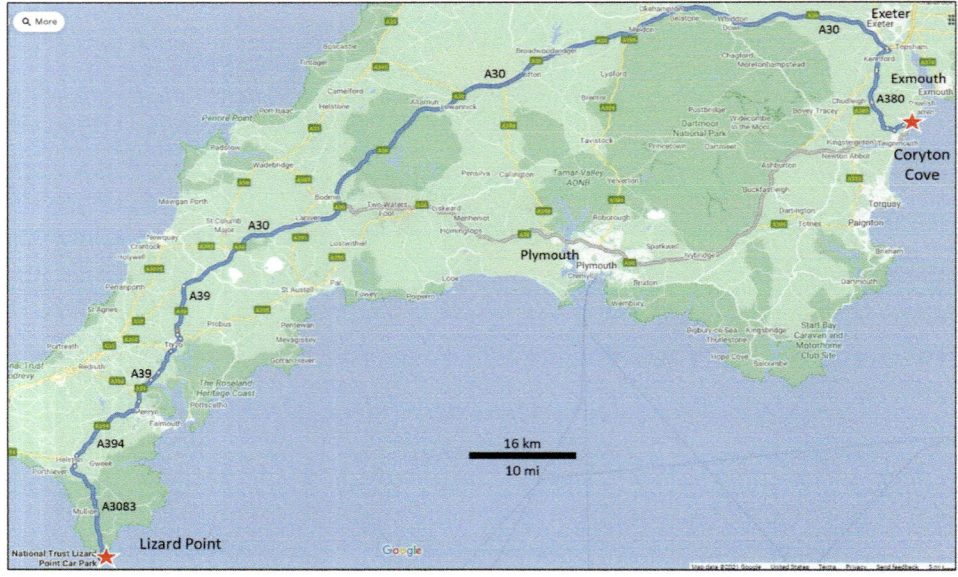

Lizard Point to Coryton Cove: *Return north on A3083 to Helston/Turnpike Cross; continue straight on A394/Clodgey Lane; at the roundabout, take the 1st exit (north) onto B3297/Clodgey Lane;*

*continue north to Redruth and turn right (south) onto B3300/Southgate Street; at the roundabout, take the 1st exit (north) onto A393/Sandy Lane; continue north on A393 to Scorrier; at the roundabout, on the west side of Scorrier take the 1st exit (north) onto A3047; at the roundabout, take the A30 ramp to Bodmin (east) and merge onto the A30/Blackwater Bypass; continue driving northeast on the A30 to Exeter; south of Exeter follow the signs for A38/A380/Plymouth/Torquay (south); continue south on the A380 to Torquay; take the B3192 exit toward Teignmouth; continue straight on B3192/Higher Exeter Road; turn left (northeast) to stay on Holcomb Down Road; continue straight onto Oak Hill Road; turn right (east) onto John Nash Drive; turn left (north) onto A379/Teignmouth Road; turn right (south) onto Marine Parade and drive to **Stop 13, Coryton Cove** parking area (50.578264, −3.467178), for a total of 206 km (128 mi; 2 h 36 min). Walk 560 m (1,830 ft) south along the Coast Path.*

STOP 13 CORYTON COVE

This is the westernmost part of the Dorset and East Devon Coast (Jurassic Coast) World Heritage Site.

The sea cliffs at Coryton Cove provide excellent exposures of the Permian New Red Sandstone, an eolian (windblown) sandstone that preserves dune-type crossbedding. This unit, around 3,000 m (10,000 ft) thick, outcrops between Torquay and Lyme Regis (Gallois, 2014). It was deposited in a desert setting between 269 and 265 million years ago. The red color derives from iron oxides coating the sand grains. Interbedded pebble zones indicate flash flood deposits.

Start at the south end of Coryton Cove and work your way north along the sea wall. The term New Red Sandstone encompasses several formations. In Coryton Cove, the rocks are Alphington Breccia and Heavitree Breccia formations (undifferentiated) on the BGS map, or the laterally equivalent Coryton and Tiegenmouth Breccias (Stevenson and Colley, 2012; Gallois, 2014). These sandstones were deposited in environments dominated by rivers (BGS, Geology of Britain Viewer). The term "breccia" refers to the angular pebbles that indicate alluvial material and flash flood deposits. As you walk north you cross a northwest-oriented fault that juxtaposes the breccia formations against the younger Dawlish Sandstone Formation.

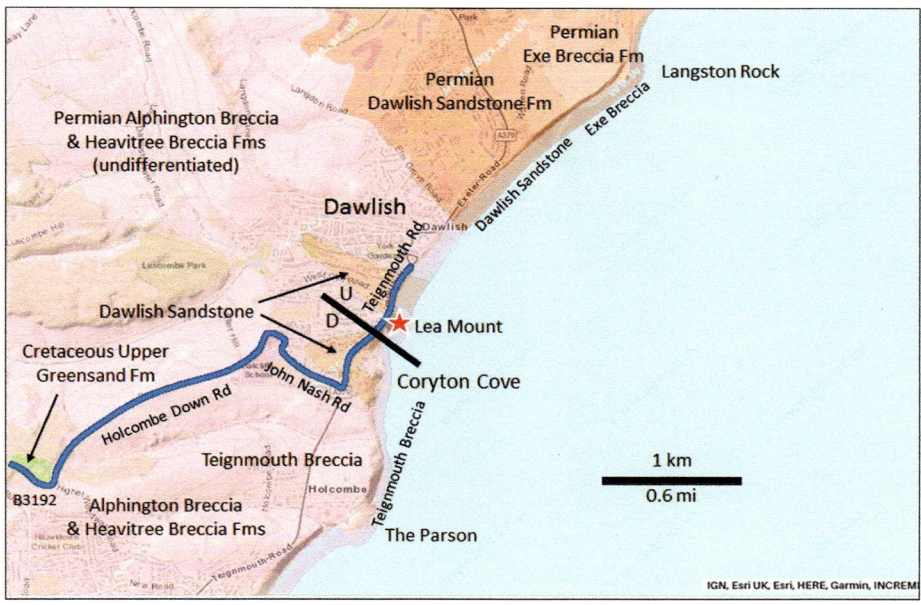

Geologic map of the Coryton Cove area. (Modified after BGS, Geology of Britain Viewer; and West, 2016a. Contains British Geological Survey materials © UKRI 2022. Base mapping provided by ESRI.)

The Teignmouth Breccia-Dawlish Sandstone contact is transitional. You can see it in the cliffs near the Coast Guard station, where dominantly breccias with minor interbeds of river and windblown sandstone pass up into dominantly eolian sandstones with minor breccia interbeds (Gallois, 2014). The units dip around 20° northeast here. The Dawlish Sandstone Formation is mainly a cross-bedded eolian sandstone with some fluvial (river) conglomerates. The conglomerates grade upward from small angular fragments at the base to finely-layered sand at the top. Each graded unit is thought to be the result of a single, rapid event, probably a flash flood. The Dawlish Sandstone is up to 120 m (390 ft) thick (West, 2016a).

If you make it all the way to Langstone Rock (about 2.4 km) you are in the Exe Breccia Formation, dominated by coarse alluvial fan deposits. This is the type section for the Exe Breccia, that is, where the unit was originally described (Stevenson and Colley, 2012). The transitional contact between the Dawlish Sandstone and the Exe Breccia is well exposed in Langstone Cliffs. Windblown sandstone of the Dawlish Sandstone is overlain by river sandstones with local breccia lenses that pass up into interbedded breccias and river sandstones. The base of the Exe Breccia is arbitrarily set where the breccia becomes dominant (Gallois, 2014).

About 17 faults are mapped along the cliffs between Teignmouth and Dawlish Warren Beach. The faults mostly trend northwest and are not well exposed except at Coryton Cove. There, on the south side of Lea Mount at the entrance to the Kennaway railroad tunnel, the down-to-the-south Coryton normal fault is exposed (West, 2016a).

Permian Teignmouth Breccia/Dawlish Sandstone red beds, Coryton Cove.

Age	Group	Lithology	Formation	Description
Jura	Penarth		Blue Lias	marine mudstone & limestone
Triassic	Penarth		White Lias	marine limestone
			Cotham	green lagoon mudstone
			Westbury Mudstone	gray lagoon mudstone
	Mercia Mudstone		Blue Anchor	green & gray mudstone & thin limestone
			Branscombe Mudstone	red-brown mudstone & evaporites
			Dunscombe Mudstone	green-gray-purple mudstone & evaporites
			Sidmouth Mudstone	red-brown mudstone & evaporites
	Sherwood Sandstone		Otter Sandstone	river channel sandstone
			Budleigh Salterton Pebble Beds	braided stream gravels
Permian	Aylesbeare		Littleham Mudstone	red-brown mudstone & evaporites
			Exmouth Mudstone & Sandstone	red-brown mudstone & sandstone
	Exeter		Exe Breccia	breccia & braided stream and dune sandstone
			Dawlish Sandstone	dune sandstone
			Teignmouth Breccia	alluvial fan breccia and sandstone
			Oddicombe Breccia	alluvial fan breccia and sandstone
			Watcombe Formation	alluvial fan breccia and sandstone
			Torbay Breccia	alluvial fan breccia and sandstone
			Devonian-Carboniferous	

Permo-Triassic stratigraphy of the Coryton Cove area. (Information drawn from Gallois, 2014; Gallois, 2019.)

STOP 14 JURASSIC COAST

The Jurassic Coast refers to a 154 km (96 mi) stretch of coast from East Devon to East Dorset (Kovacs, 2019; Hand Luggage Only, 2020). It was designated a UNESCO World Heritage Site in 2001, primarily for its unique abundance of Jurassic fossils (Trenchard, 2020). In reality, the formations range in age from Triassic to Cretaceous, but it is the Jurassic Blue Lias that has the most spectacular fossils. With every high tide or winter storm the cliffs are eroded back a little farther. Winter is the best time to hunt for fossils, as landslides and rock falls occur during the rough weather and the waves winnow the fossils. The beaches also have fewer people in winter (Jurassic Coast Trust, Jurassic Coast). The safest time to hunt fossils is during a falling or low tide.

The beaches between Charmouth and Lyme Regis are considered the best for fossil collecting. Pay attention to collecting rules: you are not permitted to excavate fossils, but if you find one lying on the beach, you may collect it. The West Dorset Fossil Collecting Code of Conduct applies to a 17-km stretch of coast from Lyme Regis east to Burton Bradstock and also applies in East Devon. It states that collectors may take home small and common fossils, but can only keep more significant finds – "category one" fossils – after reporting them. That includes many dinosaur bones. Museums must be offered category one fossils before private buyers if a fossil goes for sale, and the sale must be reported (Trenchard, 2020).

In some places, fossil collecting is not allowed without specific permission (Jurassic Coast Trust, Jurassic Coast).

Pyritized (fool's gold) ammonites can be found. They look like a coiled nautilus shell. Belemnites are the internal hard parts of a squid-like animal. Some are found intact, but most are broken into segments (Jurassic Coast Trust, Jurassic Coast). Scattered backbones of ichthyosaurs, a type of giant marine reptile, are relatively common around Charmouth and Lyme Regis. Plesiosaur bones are less common. Sea shells found along the Jurassic Coast include pelecypods (relatives of oysters), clams, and scallops (Jurassic Coast Trust, Jurassic Coast).

Both the Charmouth Heritage Coast Centre and Lyme Regis Museum provide guided fossil hunting walks throughout the year.

The Charmouth Heritage Coast Centre has excellent displays of local fossils. Entry is free, and the visitor center has interactive displays and information on fossils and fossil hunting. They provide a video microscope to examine your finds and have a warden or volunteer to help identify fossils. The exhibits include the "Charmouth Dinosaur" Scelidosaurus, an armored herbivorous dinosaur that lived on islands in the tropical Jurassic seas. The fossil cast in the museum was made by combining the best parts of the eight Scelidosaurus skeletons found in this area.

VISIT – CHARMOUTH HERITAGE COAST CENTRE

Address: Lower Sea Lane, Charmouth, Dorset DT6 6LL
Phone: 01297 560772 (+44 from overseas)
Email: info@jurassiccoast.org
Website: www.charmouth.org/chcc and https://jurassiccoast.org/visit/attractions/charmouth-heritage-coast-centre/
Entry: free

Lyme Regis lies within an Area of Outstanding Natural Beauty. It is also the birthplace of Mary Anning, one of the most important early fossil collectors and paleontologists. The Lyme Regis Museum, built on the site of Mary Anning's family home, has a collection of local fossils, displays, and also provides regular fossil hunting walks (Jurassic Coast Trust, Jurassic Coast).

Visit – Lyme Regis Museum

Address: Bridge Street, Lyme Regis, Dorset DT7 3QA
Phone: 01297 443370 (+44 from overseas)
Email: www.lymeregismuseum.co.uk
Website: https://jurassiccoast.org/visit/attractions/lyme-regis-museum/

Mary Anning, an Original Geologist

Mary Anning was a fossil collector, fossil dealer, and early paleontologist who became famous for her discovery of Jurassic marine reptile fossils in the sea cliffs at Lyme Regis along the English Channel. The cliffs consisted of Blue Lias (shallow marine limestone and shale) and Charmouth Mudstone. Lyme Regis was, and continues to be, one of the richest fossil localities in Britain. Among her discoveries are the first ichthyosaur skeleton; the first two nearly complete plesiosaur skeletons; the first pterosaur skeleton found outside Germany; and many fossil fish.

Not a member of the Church of England and, worse, a woman, Anning was effectively barred from the establishment scientific community of 19th-century Britain. As a woman, she was not eligible to join The Geological Society of London, and rarely got credit for her discoveries. Through perseverance and her innate intelligence, she became well known in geological circles in Britain, Europe, and America (Wikipedia, Mary Anning).

Fossil collecting was in vogue in the late 18th and early 19th centuries among English gentlemen. It began as a means of collecting "curiosities" to display and gradually transformed into science as the importance of fossils became known due to men like William Smith, who used them to correlate formations and put them in their proper time sequence.

Mary's father Richard was a cabinetmaker and occasional fossil collector. He often took her and her brother fossil hunting to supplement the family's modest income. They sold their discoveries to tourists. When her father died at age 44, he left the family in debt. The family relied on the sale of fossils to survive. Mary made her first major discovery in 1811 at age 12 (Trenchard, 2020; UCMP Berkeley, Mary Anning). She and her brother found an ichthyosaur skull. A few months later she found the rest of the skeleton. Henry Henley of Sandringham House in Sandringham, lord of Colway manor near Lyme Regis, paid the family £23 for the fossil before reselling it to William Bullock, a collector who displayed it in London. In 1819 it was sold to the British Museum as a "Crocodile in a Fossil State." Charles Konig of the museum renamed it *Ichthyosaurus,* meaning "fish lizard." In 1821 William Conybeare and Henry De la Beche of The Geological Society of London analyzed the fossils found by Anning and concluded that ichthyosaurs were a previously unknown marine reptile.

Thomas Birch was a wealthy collector and became a family friend. Concerned about the impoverished family, he began to auction fossils he had bought from them and provide them with the proceeds and credit for their discoveries. Anning's reputation spread as she made new discoveries: in 1823, she found the first complete *Plesiosaurus* ("near lizard"); in 1828 she found the first British pterosaur, called a "flying dragon" when it was displayed at the British Museum; in 1829 she found a *Squaloraja* fish skeleton; in 1830, Anning discovered the skeleton of a new type of plesiosaur. Her shop in Lyme Regis, Anning's Fossil Depot, displayed many fine fossils and attracted collectors from across Europe and America. George Featherstonhaugh bought Anning's fossils for the newly opened New York Lyceum of Natural History, now the Academy of Sciences, in 1827. Swiss paleontologist and glaciologist Louis Agassiz visited Lyme Regis in 1834 and worked with Anning to find fish fossils…. he acknowledged and thanked her in his book, *Recherches sur les Poissons Fossiles* (Studies of Fossil Fish, 1833–1843). King Frederick Augustus II of Saxony purchased an ichthyosaur skeleton for his natural history collection in 1844 (Wikipedia, Mary Anning).

Anning must have been a sympathetic character. Frozen out of the scientific establishment, she knew more about fossils and geology than many of the wealthy collectors she sold them to, yet they were the ones that published scientific descriptions of the specimens, usually without crediting her contribution. One who did was Henry De la Beche. Friends since teenagers, they stayed in touch as he became one of Britain's leading geologists. William Buckland, who taught geology at Oxford, often visited Lyme Regis on his Christmas holidays and went fossil hunting with Anning. In fact, she suggested to him that the objects they found known as "bezoar stones" were really the fossil feces of ichthyosaurs and plesiosaurs. These stones, found in the abdomen of ichthyosaur skeletons, often contained fish bones and scales when broken open. Buckland named them coprolites. Another leading geologist, Roderick Murchison, did some of his early fieldwork around Lyme Regis. He arranged with Mary to teach his wife Charlotte how to hunt fossils. The two women became fast friends. Charlotte Murchison used her husband's connections to help Anning build a network of customers throughout Europe. Anning corresponded with Charles Lyell, who asked her opinion on how the sea affected the cliffs around Lyme Regis. She also communicated with Adam Sedgwick, who taught geology at Cambridge (Charles Darwin was one of his students).

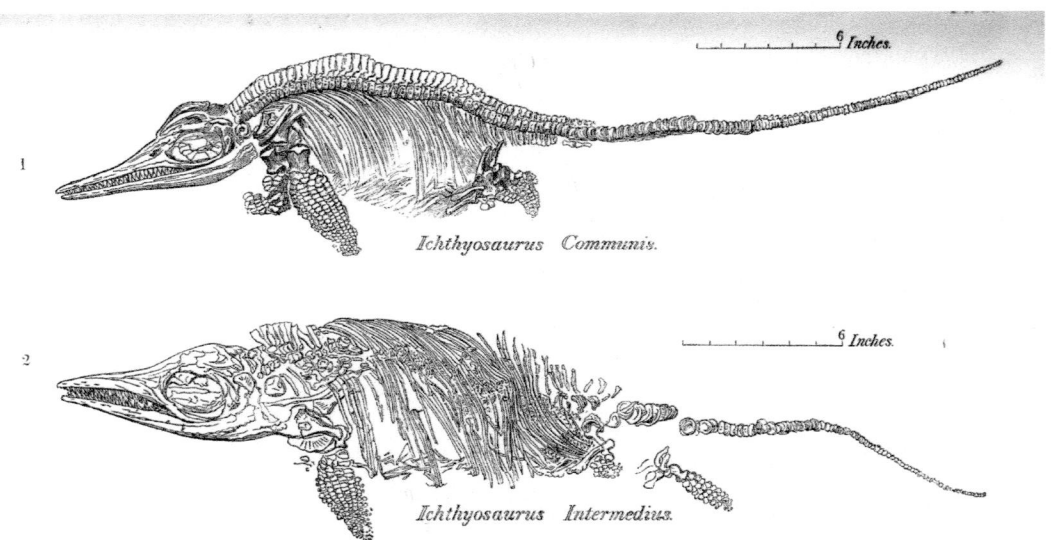

Ichthyosaurs sketched by William Buckland in 1837.

In 1835, Anning became destitute after making a bad investment. Her friend William Buckland persuaded the British Association for the Advancement of Science and the British government to award her a £25 annual pension in return for her many contributions to geology.

Mary died from breast cancer in 1847: she was a mere 47 years old. After Anning's death, De la Beche, president of The Geological Society, wrote a eulogy that he read to a meeting of the society and published in its Transactions, the first ever eulogy for a woman and non-member.

Anning's work and discoveries led to the acceptance of the concept of extinction. The French geologist Georges Cuvier argued in the late 1790s that extinction was a reality, but until the early 1820s most scientists still did not believe it: extinction implied that God's creation had been imperfect. Yet evidence, much of it from Anning's work, was mounting that Earth had once been inhabited by creatures much different from those living today and that there had been an "Age of Reptiles" when reptiles had been the dominant form of animal life.

In 2010, 163 years after her death, the Royal Society included Anning in a list of the ten most influential British women in the history of science (Wikipedia, Mary Anning).

Coryton Cove to Orcombe Point: Head northeast on Marine Parade on A379 toward Teignmouth Hill; continue on A379/Exeter Road; continue straight on A379/The Strand/South Town/Sannerville Way; at roundabout, take the 2nd exit (northeast) onto A379/Bridge Road; at the roundabout, take the 3rd exit (southeast) onto Topsham Road; continue straight onto Exeter Road; at the roundabout, take the 1st exit (east) onto Station Road; at the roundabout, take the 2nd exit (southeast) onto Elm Grove Road; bear left (east) onto Bridge Hill Road; continue straight on Topsham Road; at the roundabout, take the 2nd exit (south) onto A376/Exmouth Road; on the northern outskirts of Exmouth turn left (south) onto Gipsy Lane; at the roundabout, take the 2nd exit (south) onto Withycombe Village Road; at the roundabout, take the 1st exit (south) onto Marpool Hill; continue straight on Claremont Grove; turn right (southwest) onto B3178/Salterton Road; turn left (south) onto Cyprus Road; turn left (east) onto Douglas Ave; turn right (south) on Maer Road; at the roundabout, take the 3rd exit (west) onto Queens Drive, then a right turn (south) onto Marine Drive and drive to parking area; from the end of the parking lot walk 630 m (2,070 ft) southeast to **Stop 14.1, Orcombe Point and Geoneedle** *(50.607118, −3.385170), for a total of 32.5 km (20.2 mi; 40 min). There is ample pay parking along the beach, as well as toilets.*

STOP 14.1 ORCOMBE POINT AND GEONEEDLE

The sea cliffs at Orcombe Point are colorful orange and red and brown Triassic sandstone of the Exmouth Mudstone and Sandstone Formation of the Aylesbeare Mudstone Group. These rocks were deposited in desert environments that included sand dunes, estuaries, evaporite lakes, and river channels and floodplains.

The Geoneedle is a monument built from local stone that represents all of the formations found along the Jurassic Coast. The sequence, from oldest to youngest, includes the New Red Sandstone, White Lias, Blue Lias, Ham Hill Stone, Forest Marble, Portland Stone, and Purbeck Group. The monument was dedicated in 2002 (Greengrove and Davies-Vollum, 2006b).

The Geoneedle at Orcombe Point. (Photo courtesy of Geof Sheppard, https://commons.wikimedia.org/-wiki/File:Orcombe_Point_Geoneedle.jpg.)

Orcombe Point to Lyme Regis: *Return northwest on Marine Drive to Queens Drive and turn right (east); at the roundabout, take the 1st exit (northwest) onto Maer Road; turn right (east) on Douglas Ave; turn right (north) onto Crawford Ave; turn right (east) onto Littleham Road; at the roundabout, take the 1st exit (north) onto The Broadway; turn right (east) onto B3178/Salterton Road; continue on B3178 to Newton/Poppleford; at the roundabout, take the 2nd exit (east) onto A3052/High Street/Four Elms Hill; continue east on A3052/Sidmouth Road to Lyme Regis; turn right (south) onto Cobb Road; turn right (west) onto Ozone Terrace; turn left (south) and drive to Monmouth Beach Car Park and walk west 750 m (2,450 ft) to the Ammonite Pavement. This is* **Stop 14.2, Lyme Regis** *(50.720016, −2.941157), for a total of 44.2 km (27.5 mi; 53 min). There is ample pay-and-display parking for £1.50/h, and facilities.*

STOP 14.2 LYME REGIS AND MONMOUTH BEACH

Lyme Regis, the "Pearl of Dorset," is world famous for the fossils found on its beaches and sea cliffs. The town began as a fishing village around the mouth of the River Lim (or Lym). The population was estimated to be 4,805 in 2019.

Lyme is mentioned in the Domesday Book of 1086. There has been a church on this site since 1145, and Sherborne Abbey (a Saxon church since the year 705) had salt-boiling rights near the River Lym. In the 13th century, it became a major British port: the 13th century harbor wall, The Cobb on the west side of town, may be the oldest breakwater in the country. A Royal Charter was granted by King Edward I in 1284, and "Regis" was added to the town's name (Wikipedia, Lyme Regis). The town is famous for residents Mary Anning and Henry Thomas de la Beche. Anning, whom we already met, was a skilled paleontologist and fossil collector, and de la Beche, founder of the British Geological Survey, in 1826, provided the first complete descriptions of the local rocks and fossil successions (Gallois and Davis, 2001). The town holds an annual Mary Anning Day and Lyme Regis Fossil Festival, usually in the May to July period.

Lyme Regis is important to geoscientists for its contribution to the early development of geology and paleontology. This historical contribution and the unusually rich source of Early Jurassic fossils are the main reasons it is part of the Jurassic Coast World Heritage Site. The sea cliffs expose a continuous sequence of Triassic, Jurassic, and Cretaceous rock formations that span 185 million years of Earth's history (Wikipedia, Lyme Regis).

The town of Lyme Regis lies in a valley between low ridges of Upper Greensand Formation and Chalk Formation. The town and river valley are located in Early Jurassic Charmouth Mudstone. The beach cliffs south of town are Early Jurassic Blue Lias limestone (Greengrove and Davies-Vollum, 2006d).

The rocks are gently inclined to the east such that the oldest layers are exposed in the west and the youngest to the east. The sea cliffs are Blue Lias and Charmouth Mudstone unconformably overlain by Cretaceous sandstones of the Gault and Upper Greensand formations (Lymeregis.com, Geology Lyme Regis). The Jurassic rocks consist of various shales, thin limestones, sandstones, and marls over the Blue Lias and White Lias limestone-mudstone successions. A peculiar-sounding unit, at least to North American ears, is the Shales with Beef, a mudstone just above the Blue Lias. This is an organic-rich shale with fibrous calcite veins. "Beef" is an old quarryman's term for fibrous calcite. This unit is a significant oil source rock in the Wessex Basin to the east and south (West and Gallois, 2019).

These Jurassic rocks were deposited in relatively deep tropical seas teeming with marine life. Fossils are common. Some, such as belemnites and ammonites, are easy to find. Others, like ichthyosaurs and fish, are rare (Lymeregis.com, Geology Lyme Regis).

Thin clay-rich layers in the Charmouth Mudstone serve as bedding-failure surfaces that cause landslides near town, especially where they are inclined seaward (Gallois and Davis, 2001). The Gault outcrops also cause extensive landslides. Large landslides almost always occur during or just after rainy periods when water lubricates the bedding planes.

In 2005, work began on a £16 million project to stabilize the cliffs and protect the town from coastal erosion (Wikipedia, Lyme Regis). As a result of this project a seaward-plunging, fault-bounded syncline in the Blue Lias was discovered and mapped beneath the central part of Lyme Regis. Around 30 normal faults were identified, almost all of which have offsets less than 2 m (Gallois and Davis, 2001). Small elongate anticlines in the Blue Lias limestones along the shore below Ware Cliff are a geologic curiosity. Similar ridges, about a foot high, had been observed at Whitlands Cliff (Humble Point) by Mary Anning. They are speculated to be pressure ridges that develop at the toe of slope ahead of a landslide (West and Gallois, 2019).

Ammonites large and small in a beach cobble, Monmouth Beach, Lyme Regis. A centimeter scale is shown.

Monmouth Beach, Lyme Regis. View west. Some of the best fossils are at the base of the cliffs near the west end of the beach.

Landslides, especially during the wet winter months, cause large mudflows onto the beaches. Storm waves then wash away the mud, leaving fossils scattered in the sand. The best and safest time to collect fossils is at low tide after such a storm. The cliffs are dangerous and prone to rock falls and mudflows. The best place to search is on the beach below a landslide scar (Lymeregis.com, Geology Lyme Regis).

Since the completion of the sea wall in 2014, the area fronting it has been almost entirely sediment-free, exposing the geology better than ever before. The new sea wall provides easy and safe access to the intertidal area for fossil collectors.

Cars can park at the large Holmbush car park on the west side of the town. This car park is less expensive than the Monmouth Beach car park near the Cobb.

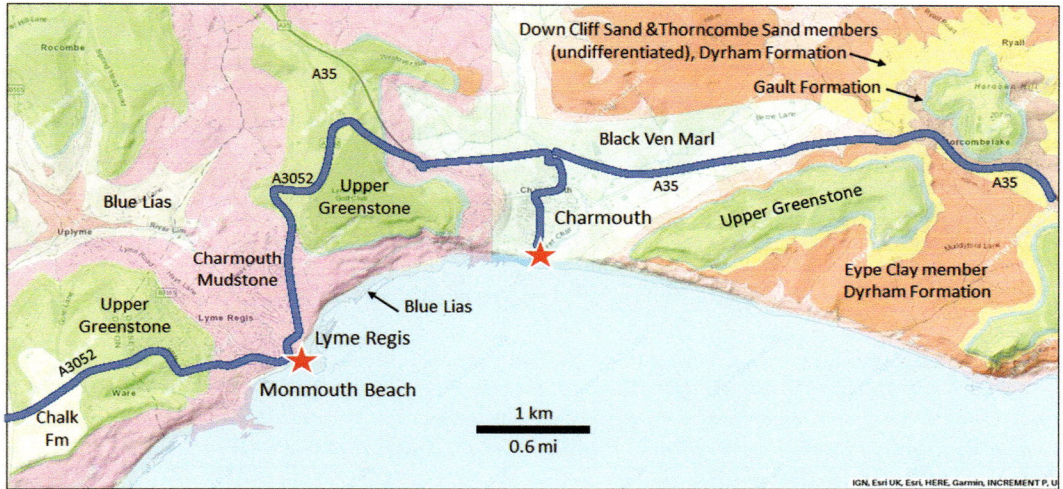

Geologic map of the Lyme Regis-Charmouth area. (Modified after BGS, Geology of Britain Viewer. Contains British Geological Survey materials © UKRI 2022. Base mapping provided by ESRI.)

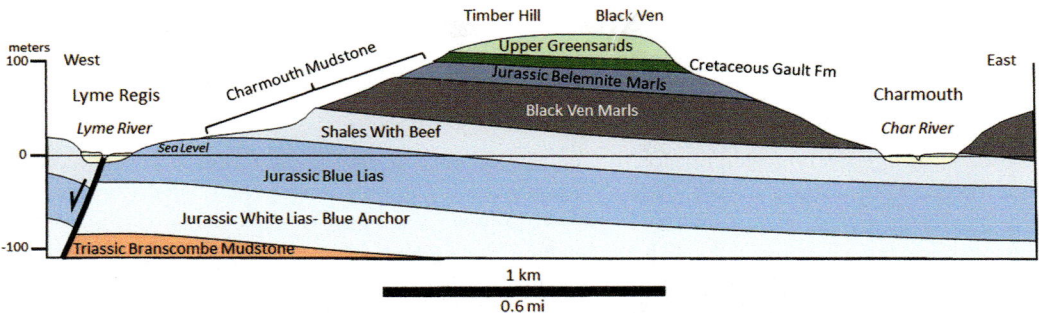

Cross section from Lyme Regis to Charmouth showing the various units.

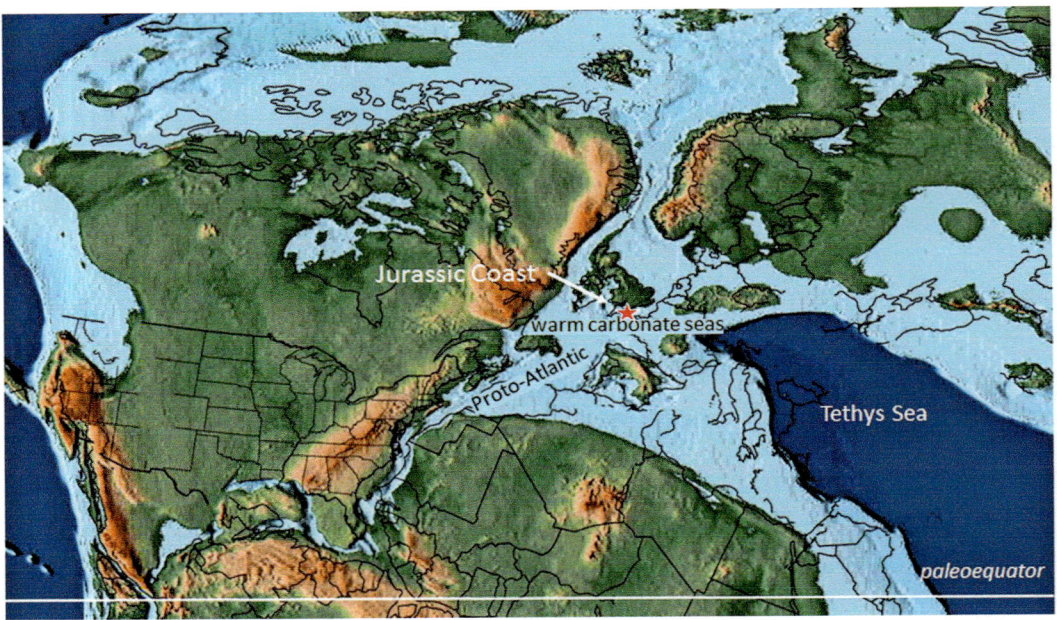

Geography during deposition of the Lower Jurassic. (Modified after Scotese, 2013.)

*Lyme Regis to Charmouth Beach: Turn left (north) onto Ozone Terrace, then again left (north) onto Cobb Road; turn right (east) on A3052/Pound Street; at the roundabout, take the 4th exit (south) onto Axminster Road; continue straight on The Street; turn right (south) onto Lower Sea Lane and drive to Charmouth Beach parking on the left; walk west up to 800 m (2,625 ft), or east up to 500 m (1,640 ft). This is **Stop 14.3, Charmouth Beach** (50.734224, −2.900141), for a total of 6.6 km (4.1 mi; 13 min). The Charmouth Heritage Coast Centre is across from the parking lot. Pay-and-display parking is £3 for 2 h; £5 all day. There are no facilities.*

Stop 14.3 Charmouth Beach

Charmouth Beach and the coast both west and east of it are among the best places to go fossil hunting along the Jurassic Coast. The pyritized ('fool's gold') ammonites found there are famous Charmouth fossils that are easily located after storms. Scattered vertebrae (backbones) of ichthyosaurs are relatively common around Charmouth. Marine reptile teeth and bones, belemnites, nautiloids, and crinoids are also found. Nautiloids and ammonites are both cephalopod mollusks (Chaffee, 2012; Cook, 2018; Jurassic Coast Trust; UKAF, East Beach (Stonebarrow), Charmouth).

Fossils are mostly found at the tide line, but can also be found at the bottom of scree slopes, landslides, and occasionally in rock falls. Hammering outcrops is not permitted. Look for loose nodules ("flatstones"), as these often contain excellent fossils. You will need a hammer: hit them on the side to split the nodule along the bedding plane.

By far the most common fossils at Charmouth are ammonites. The larger, pyritized ammonites are found near Stonebarrow about 750 m (2,460 ft) east of the River Char. Several complete fish fossils have been found in these nodules. The Black Ven Marl member of the Charmouth Mudstone Formation is famous for ichthyosaur bones that are washed out of the landslide clay (Cruickshanks, 2018).

The cliffs and foreshore between Charmouth and Seatown are mainly Early Jurassic outcrops (190–185 Ma). During this time, a mostly shallow tropical sea (<100 m, or 330 ft deep) covered the area, accumulating alternating layers of clay and limestone (Shepherd, 2021b).

Rocks exposed between Charmouth Beach and Golden Cap to the east include the Lower Jurassic Blue Lias, overlying Charmouth Mudstone Formation, and younger Cretaceous layers above an erosional unconformity. The Blue Lias limestone is rich in marine fossils including ammonites, bivalves, brachiopods, and marine reptiles (ichthyosaurs and plesiosaurs).

Golden Cap as seen from Charmouth. Note the massive landslide and escarpments. (Photo courtesy of Nigel Chadwick, https://commons.wikimedia.org/wiki/File:Golden_Cap_from_Charmouth_-_geograph.org.uk_-_1184579.jpg.)

The Charmouth Mudstone is a thick pile of shales and mudstones. It was in Charmouth Mudstone that the first remains of Scelidosaurus were found in 1858 (Chaffee, 2012). The Black Ven Marl near the base of the Charmouth Mudstone is the lowest and oldest unit and is the first encountered traveling east toward Golden Cap. The Black Ven Marl is broken into the lower "Shales with Beef," characterized by shales with bands of fibrous calcite (the "Beef"), and the upper Black Ven Marls, shale with calcareous concretions containing ammonites. You can also find fish, crinoids, and insects in the shales (Cruickshanks, 2018). The Black Ven Marl is best seen along the beach beneath Golden Cap (Shepherd, 2021b).

The Stonebarrow Marl member (formerly the Belemnite Marls) is near the top of the sea cliffs but can be easily examined on the beach beneath the east side of Golden Cap. The Belemnite Marls are famous for belemnites, even though ammonites are also common (Cruickshanks, 2018). The Stonebarrow Marl is easily identified by its alternating pale and dark layers.

The Seatown Marl member (formerly called the Green Ammonite Beds) lies above the Stonebarrow Marl. These are gray mudstones with occasional limestone layers, all rich in ammonites. The Seatown Marl is best seen from the base of Golden Cap eastward toward Seatown (Shepherd, 2021b).

Overlying the Jurassic sediments are younger Cretaceous deposits, including the Gault Clay and gold-colored Upper Greensand, deposited around 106–102 Ma (UKAF, East Beach (Stonebarrow), Charmouth).

Marine fossils can be found along the entire coast between Charmouth and Seatown, although the greatest volumes seem to be within the first 1 km (0.6 mi) of the beach access. At extreme low tide, traces of a submerged forest (including bones of mammoth and red deer) have been seen (Cruickshanks, 2018; Shepherd, 2021b).

In general, high tides and storms wash away the soft clay surrounding the more resistant fossils. A 7-day tide forecast is available on the BBC website [for Lyme Regis https://www.bbc.com/weather/coast-and-sea/tide-tables/10/28; in general https://www.bbc.com/weather/coast-and-sea/tide-tables]. As elsewhere, the best fossil collecting is during a falling or low tide after winter storms. At other times of the year, huge numbers of commercial collectors battle it out for the best specimens.

This beach is part of the Jurassic World Heritage Coast, is an SSSI, and is partly private land. The car park at Charmouth is very close to the beach and nearby are toilets, a café, and the Charmouth Fossil Heritage Information Centre.

STOP 15 LULWORTH AREA & WESSEX BASIN

In Dorset, the flat-lying chalk that is so common along the south coast of England is deformed into a ridge, with layers inclined to the north, that extends west from Ballard Downs near Swanage to Lulworth Cove. This is the Purbeck Monocline. The chalk, Portland Stone (Jurassic limestone), is part of the north limb of the offshore Kimmeridge-Purbeck Anticline (an arch-shaped fold) beneath the English Channel south of Dorset.

The Wessex Basin (or Channel Basin) covers an area of over 40,000 km² (15,440 mi²) in Dorset and extends beneath the English Channel. It contains Permian to Tertiary sediments that locally are over 3 km (9,840 ft) thick. The Permian red beds (mudstones, sandstones, and basal conglomerates) were deposited on deformed Devonian-Carboniferous (Variscan) basement. Triassic strata lie over the Permian: they are red bed sandstones and conglomerates in the lower part and mudstones with salt in the upper part. The Jurassic consists of a sequence of mudstone and limestone/marl, with minor sandstones. Early Cretaceous strata are preserved only in a few areas of the basin. Overall, the Permian to Early Cretaceous units dip east beneath an unconformity capped by Late Cretaceous and Tertiary strata (Harvey and Gray, 2013).

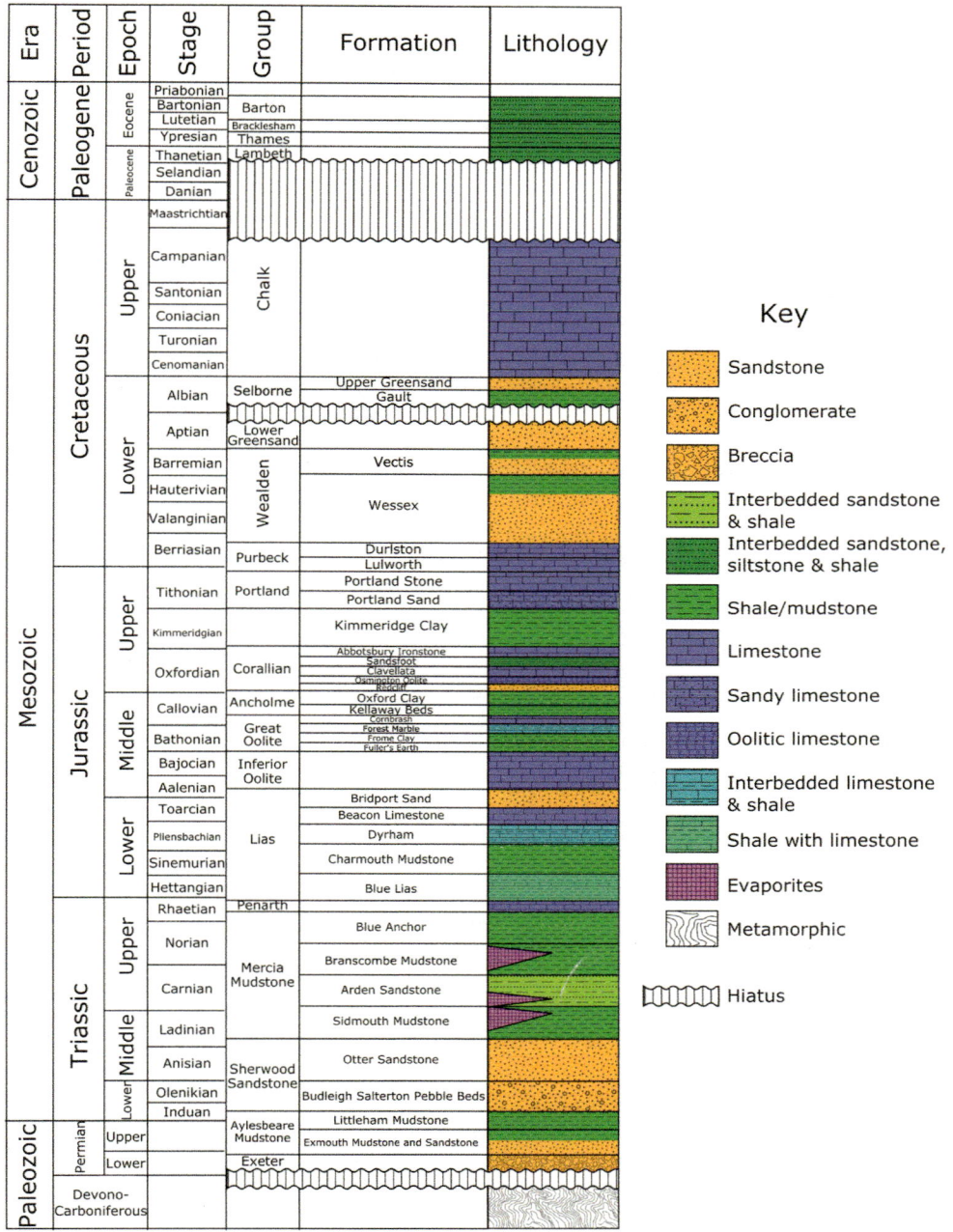

Era	Period	Epoch	Stage	Group	Formation	Lithology
Cenozoic	Paleogene	Eocene	Priabonian			
			Bartonian	Barton		
			Lutetian	Bracklesham		
			Ypresian	Thames		
		Paleocene	Thanetian	Lambeth		
			Selandian			
			Danian			
Mesozoic	Cretaceous	Upper	Maastrichtian	Chalk		
			Campanian			
			Santonian			
			Coniacian			
			Turonian			
			Cenomanian			
		Lower	Albian	Selborne	Upper Greensand	
					Gault	
			Aptian	Lower Greensand		
			Barremian	Wealden	Vectis	
			Hauterivian		Wessex	
			Valanginian			
			Berriasian	Purbeck	Durlston	
					Lulworth	
	Jurassic	Upper	Tithonian	Portland	Portland Stone	
					Portland Sand	
			Kimmeridgian	Corallian	Kimmeridge Clay	
			Oxfordian		Abbotsbury Ironstone	
					Sandsfoot	
					Clavellata	
					Osmington Oolite	
					Redcliff	
		Middle	Callovian	Ancholme	Oxford Clay	
					Kellaway Beds	
			Bathonian	Great Oolite	Cornbrash	
					Forest Marble	
					Frome Clay	
					Fuller's Earth	
			Bajocian	Inferior Oolite		
			Aalenian			
		Lower	Toarcian	Lias	Bridport Sand	
					Beacon Limestone	
			Pliensbachian		Dyrham	
			Sinemurian		Charmouth Mudstone	
			Hettangian		Blue Lias	
	Triassic	Upper	Rhaetian	Penarth	Blue Anchor	
			Norian	Mercia Mudstone	Branscombe Mudstone	
			Carnian		Arden Sandstone	
					Sidmouth Mudstone	
		Middle	Ladinian			
			Anisian	Sherwood Sandstone	Otter Sandstone	
		Lower	Olenikian		Budleigh Salterton Pebble Beds	
			Induan	Aylesbeare Mudstone	Littleham Mudstone	
Paleozoic	Permian	Upper			Exmouth Mudstone and Sandstone	
		Lower		Exeter		
	Devono-Carboniferous					

Key

	Sandstone
	Conglomerate
	Breccia
	Interbedded sandstone & shale
	Interbedded sandstone, siltstone & shale
	Shale/mudstone
	Limestone
	Sandy limestone
	Oolitic limestone
	Interbedded limestone & shale
	Shale with limestone
	Evaporites
	Metamorphic
	Hiatus

Stratigraphy of the Wessex Basin. (Diagram courtesy of Mikenorton, https://commons.wikimedia.org/-wiki/File:Wessex_basin_lithostratigraphy.png.)

The Wessex Basin resulted from Permian through Early Cretaceous crustal extension, rifting, and subsidence related to the opening of the Atlantic. By mid-Cretaceous time, the extension had ceased, but the region continued to subside due to thermal relaxation and cooling of the crust.

Structurally, the Wessex Basin consists of four north-inclined and roughly east–west elongated sub-basins (tilted half-grabens) filled with sediments that thicken to the north. All four sub-basins are bounded by south-dipping normal faults. The southernmost Channel, or Portland-Wight sub-basin, is located along and just south of the line of hills and spectacular folds along the Dorset coast. It is bounded by the Purbeck-Isle of Wight (or Portland-Wight) Fault that was inverted by Alpine compression. To date, this sub-basin is the only one with a petroleum system (Harvey and Gray, 2013).

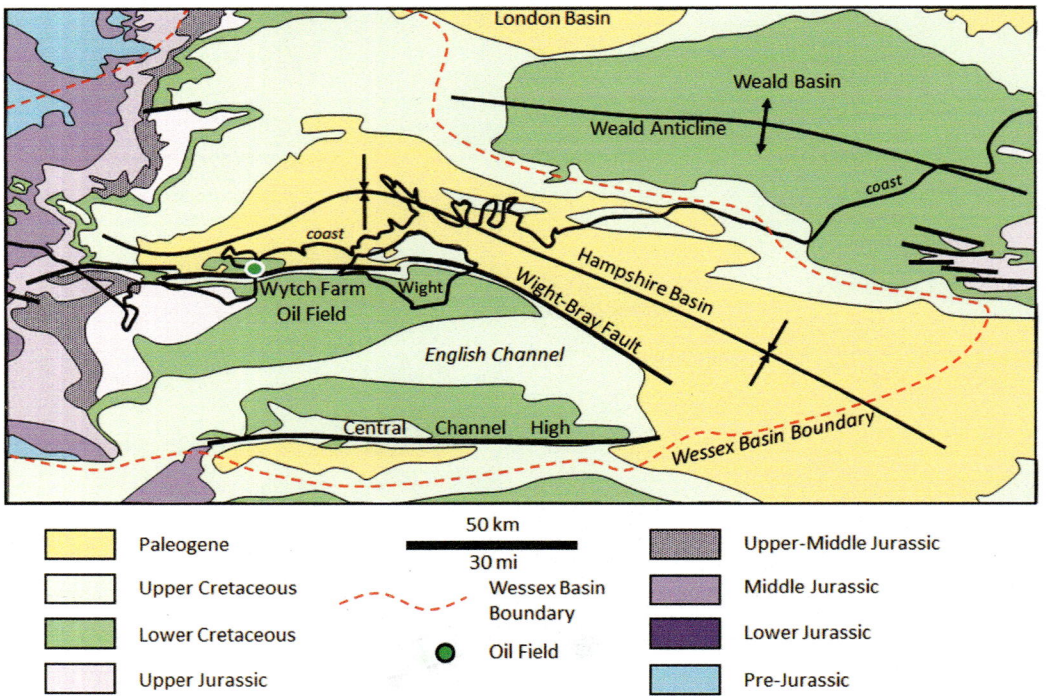

Geologic map showing the Weald and Wessex basins and sub-basins. (Modified after Underhill and Stoneley, 1998.)

Hampshire-Dieppe High
Porchfield Anticline
Bournemouth
#
Hampshire Basin
Sandown Fault
Wytch Farm Field
Needles Fault
Sandown Anticline
Purbeck Fault
Compton Beach
Yaverland Beach
Brighstone Anticline
Lulworth Cove
Portland-Wight Basin
Isle of Wight
10 km
6 mi
Winspit Quarry
Portland-Wight High

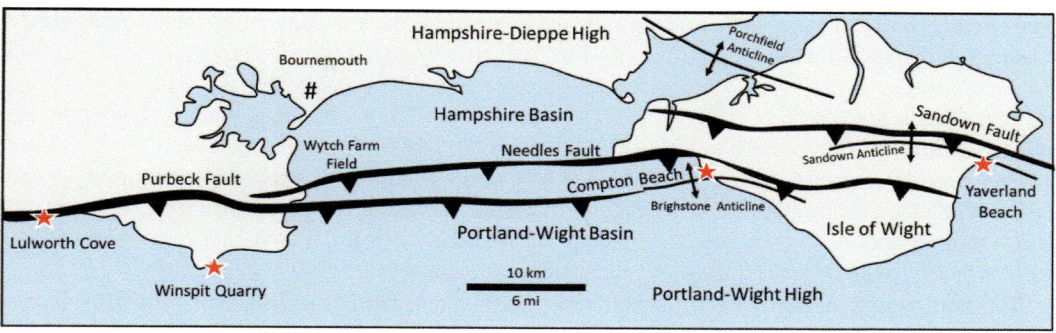

Structure of the Hampshire-Dieppe High and Portland-Wight Basin. Thrust faults were originally down-to-the-south normal faults. (Information drawn from Mortimore, 2018; Sweetman and Martill, 2019.)

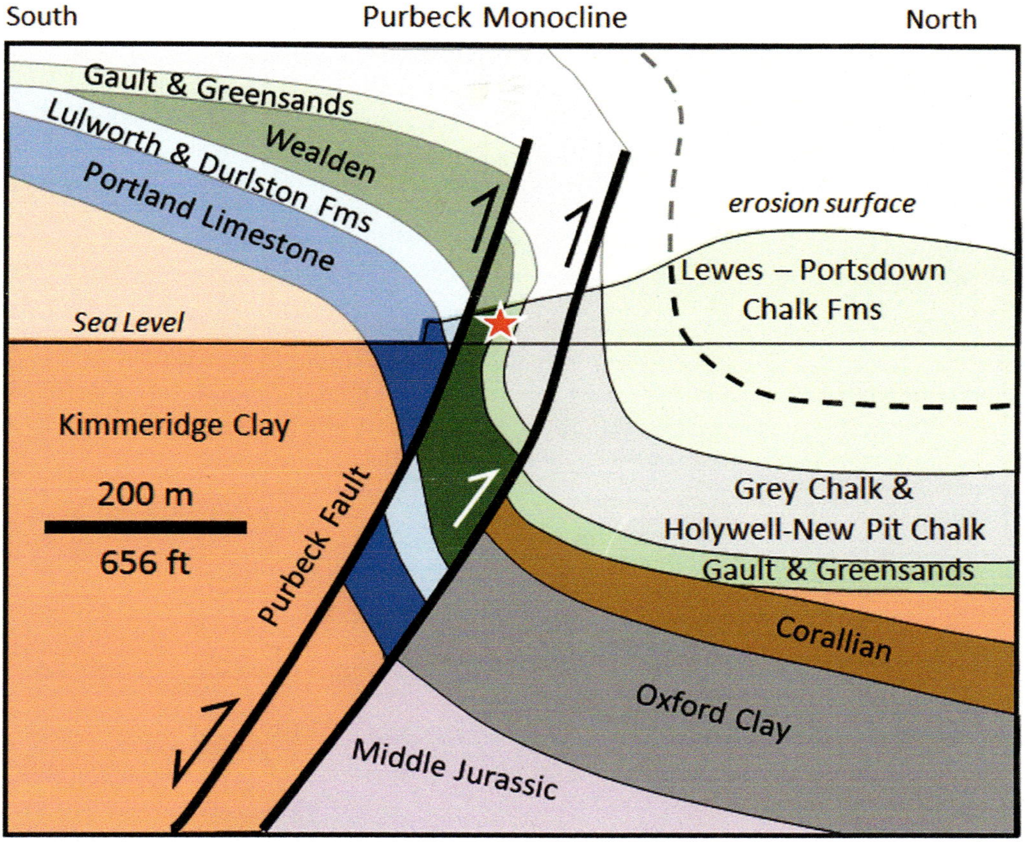

South
Purbeck Monocline
North
Gault & Greensands
Wealden
Lulworth & Durlston Fms
Portland Limestone
erosion surface
Lewes – Portsdown
Chalk Fms
Sea Level
Kimmeridge Clay
200 m
656 ft
Purbeck Fault
Grey Chalk &
Holywell-New Pit Chalk
Gault & Greensands
Corallian
Oxford Clay
Middle Jurassic

Cross section through the Purbeck Fault/Purbeck Monocline at Durdle Door. (Modified after Mortimore, 2019.)

The Kimmeridge-Purbeck Anticline was formed by tectonic inversion. North-directed compression caused the Purbeck Fault, originally a normal fault that bounded the sub-basin, to be reactivated as a thrust. Alpine compression and inversion were caused by convergence of the African and European tectonic plates over the past 50 million years (The Geological Society, Lulworth and Wessex Basin; Wikipedia, Wessex Basin).

The overall effect of Alpine compression has been to convert former Mesozoic depocenters (areas with thick accumulations of sediment) into uplifted areas. Inversion structures such as anticlines and monoclines lie above the main basin-bounding normal faults (Harvey and Gray, 2013). The Purbeck-Isle of Wight Fault is one such inversion structure.

You can recognize inversion by noting that a layer that thickens on the downthrown side of a fault is now on the upthrown side. Compression across the Mesozoic normal (extensional) faults reactivated them as thrust faults (Wikipedia, Wessex Basin).

The Wessex Basin is a petroleum basin, and inversion played a role in the accumulation of the hydrocarbons. Oil and gas were generated in source rocks south of the Purbeck Fault as a result of deep burial and cooking of organic-rich shales. The source of oil is thought to be black shale in the Early Jurassic Blue Lias and "Shales with Beef," as seen in outcrops at Lyme Regis and Charmouth (Simmons, 2016). Other potential source rocks occur in the Oxford Clay and Kimmeridge Clay, both Jurassic (Harvey and Gray, 2013). The hydrocarbons migrated upward by buoyancy along porous and permeable layers and along faults until encountering reservoir rocks. Excellent oil reservoir (rock with good porosity, lots of voids between the grains that can be filled with oil) can be seen in outcrops of the Triassic Sherwood Sandstone at Ladram Bay; a secondary reservoir is Early Jurassic Bridport Sands, as seen at West Bay. The hydrocarbons were trapped along faults and in anticlinal folds. Faulting seen at Seaton Hole and deformation seen at Stair Hole formed at the same time as oil-trapping structures in the subsurface (Simmons, 2016). The result is northwest Europe's biggest onshore oil field, Wytch Farm, along with the smaller Kimmeridge and Wareham oil fields (The Geological Society, Lulworth and Wessex Basin).

The Wytch Farm field is well known among petroleum engineers and drillers as having some of the first and longest horizontal-reach wells in the world. The M-11 well, drilled by BP in 1997, broke the 10 km (6 mi) horizontal distance record at the time it was drilled. The well produces from the Triassic Sherwood Sandstone at 1,605 m (5,266 ft) true vertical depth. These wells were drilled from onshore to tap into offshore reservoirs with minimal environmental disturbance.

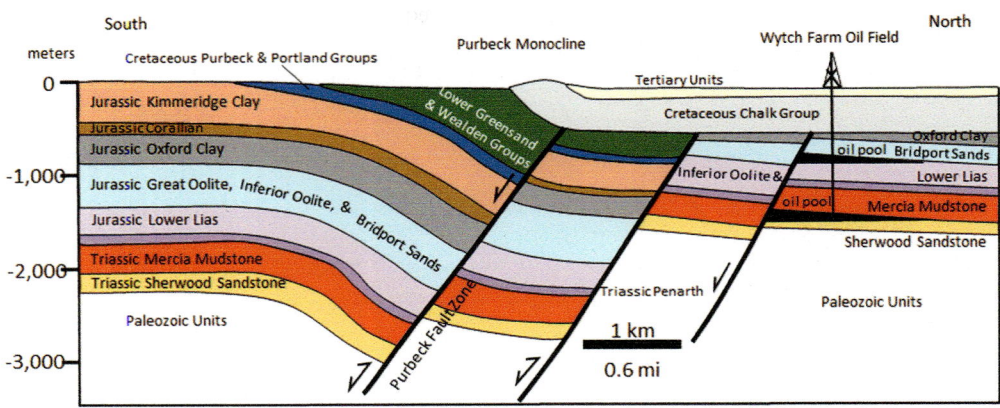

Schematic cross section through the Wytch Farm oil field. (Modified after Underhill and Stoneley, 1998.)

The Bencliffe Grit member of the Late Jurassic Corallian Group at Osmington Mills is a silty sandstone with oil staining. The oil stains were first noted in 1937 by d'Arcy Petroleum. It turns out there are other oil stains or seeps along the Dorset coast: in the Early Cretaceous Wealden Sands at Lulworth Cove, and in Wealden Sands at Mupe Bay (Simmons, 2016). Exploration was first carried out in 1936–1937 at Broadbench, in Kimmeridge Bay, with traces of oil noted on joints in gray sandstones in the Corallian Group, Sandsfoot Grit. All early wells were non-economic and abandoned.

In 1958 traces of oil were found in Early Jurassic sandstones in a well at Radipole, near Weymouth. Three more wells were drilled. Broadbench 2 (renamed Kimmeridge 1) was drilled in 1959 and encountered oil at a depth of 570 m (1,880 ft) in the Middle Jurassic Cornbrash Limestone. Production tests yielded between 4.8 and 684 cubic meters of oil per day (m³/d; 30–4,300 barrels of oil/day, or bopd). Two other wells were drilled to define the extent of the oilfield. The Kimmeridge Field began producing in 1961.

Continued exploration led to the discovery, in 1973, of the Wytch Farm oil field south of Poole Harbor. This field was discovered by drilling into an anticline. The field has recoverable reserves estimated at around 79.5 million cubic m (mm m³; 500 million barrels; Harvey and Gray, 2013). Peak production at Wytch Farm was 15,900 m³/d (100,000 bopd); production in 2016 was around 2,385 m³/d (15,000 bopd; Simmons, 2016).

Charmouth Beach to Durdle Door: Return north on Sea Lane; turn right (east) on The Street; turn right (east) onto A35; at The Crown Roundabout, take the 2nd exit (north) onto A35/Sea Road South; at the next roundabout, take the 3rd exit (east) onto A35/East Road; stay on A35 to the east side of Dorchester; turn right (southeast) onto the A352; stay on A352 to Water Lane; turn right (south) on Water Lane; turn left (south) on the unnamed road to West Lulworth and Durdle Door and drive to Durdle Door parking on the right; walk west 880 m (2,880 ft) to cliff lookout; take stairs to the beach. This is **Stop 15.1, Durdle Door** *(50.623792, −2.268449), for a total of 56.5 km (35.1 mi; 55 min).*

STOP 15.1 DURDLE DOOR

Durdle Door is a natural arch caused by marine erosion in near-vertical beds of Late Jurassic limestone. The Jurassic Portland Stone Formation, forming cliffs up to 150 m (490 ft) high, is part of the north flank of the offshore Kimmeridge-Purbeck Anticline. The east–west ridge is mapped as the Purbeck Monocline. The Miocene monocline developed over the Purbeck Thrust during the Alpine Orogeny when south-dipping normal faults were reactivated as thrusts in a north-directed compressional regime. The Durdle Cove Thrust can be seen just above beach level in the Late Cretaceous Lewes Nodular Chalk Formation. It is characterized by a series of small caves eroded along the thrust plane (Mortimore, 2018).

The Purbeck Monocline runs from Durdle Door east to Swanage. North of the main ridge of Portland Stone, erosion has worn away the softer Late Cretaceous Purbeck Group limestones and Wealden Group sandstones and siltstones (Colley, Lulworth). Still further north is a stronger, thicker band of Cretaceous chalk that forms the Purbeck Hills (West, 2014; Wikipedia, Durdle Door).

Durdle Door, a natural arch in Late Jurassic Portland Stone Formation limestone. (Photo courtesy of Saffron Blaze, via http://www.mackenzie.co, https://en.wikipedia.org/wiki/File:Durdle_Door_Overview.jpg.)

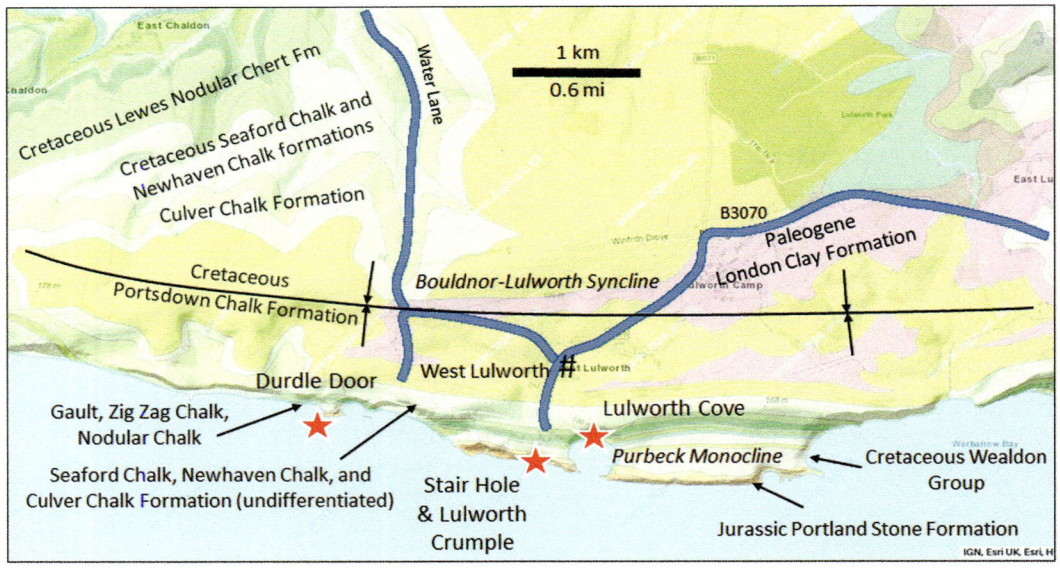

Geologic map of the Lulworth area. (Modified after BGS, Geology of Britain Viewer. Contains British Geological Survey materials © UKRI 2022. Base mapping provided by ESRI.)

Some of the beds have a rich Cretaceous fauna that includes *inoceramus* bivalves, ammonites, echinoids (sea urchins), and brachiopods (Mortimore, 2018).

Age	Group	Lithology	Formation
Upper Cretaceous	Chalk Group — White Chalk		Portsdown Chalk Formation — 100 m / 328 ft
			Culver Chalk Formation
			Newhaven Chalk Formation
			Seaford Chalk Formation
			Lewes Nodular Chalk Formation
			New Pit Chalk Formation
			Holywell Nodular Chalk Formation
	Grey Chalk		Zig Zag Chalk Formation
			West Melbury Marly Chalk Formation
Lower Cretaceous	Selborne Group		Upper Greensand Formation sandstone
			Gault Clay Formation
	Lower Greensands Group		Monk's Bay Sandstone
			Sandrock Formation sandstone
			Ferruginous Sands Formation
			Atherfield Clay Formation
	Wealden Group		Vectis Formation siltstone and mudstone
			Wessex Formation mudstone and sandstone

Cretaceous stratigraphy of the Lulworth-Isle of Wight area. (Information drawn from Hopson, 2011; Mortimore, 2018.)

It is worth noting that the Isle of Portland, between Durdle Door and Charmouth and 17 km (11 mi) south of Dorchester, contains several quarries of the latest Jurassic Portland Stone, a much-used building stone. This is the source of the name Portland Cement, invented by Joseph Aspdin in 1824, and so named because of its resemblance, when set, to Portland Stone.

*Durdle Door to Lulworth Cove: Return north to Church Road and turn right (east); drive to West Lulworth and bear right (south) on B3070/Main Road; turn right (southwest) to Lulworth Cove parking. This is **Stop 15.2, Lulworth Cove** (50.619908, −2.253309) for a total of 2.6 km (1.6 mi; 7 min). There is ample "pay-and-display" parking at £5/4 h, with toilets at the Visitor Centre.*

*Walk 170 m (550 ft) south to **Stairhole View and Lulworth Crumple** (50.618230, −2.253993); walk 300 m (1,000 ft) east to **Lulworth Cove** (50.618924, −2.249562); walk 1.4 km (4,500 ft) east and south to the **Fossil Forest** (50.616466, −2.241817).*

STOP 15.2 LULWORTH COVE, STAIR HOLE, LULWORTH CRUMPLE, AND FOSSIL FOREST

Lulworth Cove and the surrounding area contain some of the most visited geologic sites in the United Kingdom, with estimates of 500,000–1,000,000 visitors each year (Google, Lulworth Cove; West, 2020). The cove is a semicircular bay on the Jurassic Coast. It is protected from the open sea by near-vertical to steeply north-dipping strata of massive Jurassic Portland Limestone Beds. This barrier was breached by the sea. The breach is thought to be where an Ice Age stream eroded downward to create a water gap along the trace of an ancient fault.

Lulworth Cove panorama from east (left) to southeast (right).

The Portland Limestone was deposited in a shallow clear sea toward the end of the Jurassic Period. Beneath the bay are softer and more easily eroded claystones, sandstones, and limestones of the Late Cretaceous Purbeck and Wealden formations. The thinly bedded Purbeck Limestone was deposited in freshwater and brackish lagoons around 146 Ma. The Wealden Formation is mainly river-deposited sandstone and lagoonal mudstone. Above the Wealden are the deep marine Gault Clay and Upper Greensand marine sandstone deposited between about 119 and 98 million years ago. These beds are frequently covered by landslides. Further north, the sea cliffs consist of resistant Late Cretaceous Chalk Formation which was deposited in clear seas far from land around 73 Ma (Nowell, 1998).

The Stair Hole, view southeast.

The east–west vertical ridges are, as at Durdle Door, part of the Miocene Purbeck Monocline that defines the north flank of the offshore Kimmeridge-Purbeck Anticline.

The Stair Hole, an easy walk (220 m; 720 ft) west of Lulworth Cove, is a small inlet where the sea has breached the Portland Limestone by means of natural arches and gaps to form a sea-filled depression in which the softer Purbeck and Wealden sediments are being eroded. The east wall of Stair Hole exposes the Lulworth Crumple, tight folding in the thin-bedded limestone and interbedded shale of the Purbeck Formation. The localized folding, like the regional anticline and monocline, formed during the Alpine Orogeny. Whereas massive, rigid layers such as the Portland Limestone and Chalk Formation merely bend under compression, the more ductile layers sandwiched between them were free to deform into complex folds by flow and slip along bedding planes.

North of the Purbeck strata in Stair Hole are easily eroded pink, gray, and yellow claystones, marls, and sandstones of the Wealden Group. A naturally oil-stained sandstone occurs in the Wealden Formation just above the Purbeck Group contact (West, 2017).

Lulworth Crumple looking east.

The Fossil Forest is an easy walk (800 m; 2,600 ft) east of Lulworth Cove. Fossil trees of the ancient Cypress (*Protocupressinoxylon*) are rooted in ancient soil, the Great Dirt Bed. The trees are roughly 140 Ma. Most are casts, but some are silicified. Above the trees is a microbial-mounded limestone, or thrombolite. Thrombolites are the result of precipitation of calcium carbonate around microbial mounds in hypersaline water (West, 2016b; West, 2017).

The Fossil Forest is in a military firing range. The location is open most but not all weekends. Call the Army Range Control (01929–462721, ext. 819) for entry information. Hammering outcrops is not permitted. The area is most easily reached by walking east around Lulworth Cove (West, 2016b).

The Stair Hole. Oil painting by J.M.W. Turner, 1811.

Lulworth Cove to Kimmeridge Bay: Return north on B3070/Main Road to Burngate and turn right (east) to stay on B3070; in East Lulworth continue straight on an unnamed road toward Tyneham and Wareham; continue straight toward Tyneham Village; bear right (southeast) on the unnamed road toward Steeple; turn right (south) on the unnamed road toward Etches Fossil Collection and Kimmeridge Bay Access; turn right (north) and follow the sign "To The Sea" as enter Kimmeridge; pass the Etches Collection and take Kimmeridge Bay Toll Road to the shore parking lot. This is Stop 16, Kimmeridge Bay (50.611427, −2.130290), for a total of 14.4 km (9.0 mi; 24 min).

STOP 16 KIMMERIDGE BAY

The cliffs at Kimmeridge Bay provide excellent exposures of, and are the type locality for the Kimmeridge Clay Formation and the Kimmeridge Stage, a roughly 5-million-year period of time (157–152 Ma) during the Late Jurassic (Gallois, 2004). It is part of the Jurassic Coast World Heritage Site, is an SSSI, and contains the Purbeck Marine Wildlife Reserve. The Dorset Wildlife Trust operates the Fine Foundation Wild Sea Centre in the bay, with exhibits, an aquarium, and various events. Nearby is the Etches Collection Museum of Jurassic Marine Life. The museum contains an extraordinary collection of Kimmeridgian fossils collected by Steve Etches (Colley, Kimmeridge Bay). Steve Etches, a local fossil collector and expert, has spent over 30 years collecting and researching over 2,000 Late Jurassic specimens (Jurassic Coast Trust, Jurassic Coast).

VISIT – ETCHES COLLECTION, MUSEUM OF JURASSIC MARINE LIFE, KIMMERIDGE

Address: Kimmeridge, Wareham BH20 5PE, United Kingdom
Hours: 10:00 a.m.–5:00 p.m. daily
Phone: +44 1929 270000
Website: https://www.theetchescollection.org/

Exquisite specimen of *Thrissops* from the Etches Collection. The fossil was found in the Kimmeridge Clay at Kimmeridge Bay. (Photo courtesy of DaCaTaraptor, https://commons.wikimedia. org/wiki/File:Thrissops_KC.jpg.)

Hammering outcrops is not permitted, but you can collect loose samples on the beach. Visiting the bay at low tide is best, although the tidal range is small and there is little danger of being caught out by an incoming tide. Charnel is at the border of the firing range: when the range walks are closed and the red flag is flying, you must not go west beyond the danger notice at Charnel (West, 2018).

Cliffs of Kimmeridge Bay, view east. Note the east dip of bedding. (Photo courtesy of Lies Thru a Lens, https://commons.wikimedia.org/wiki/File:Kimmeridge_Bay_-_Single_Shot_(20367044755).jpg.)

The cliffs of Kimmeridge Bay contain a rhythmic succession of shales, claystones, and yellow-brown dolomites ("stone bands") apparently deposited in repeating cycles. The cycles are recognized because hard bituminous shales form ledges while the softer mudstone beds are recessed. The Kimmeridge bituminous shales have high organic content and are considered to be the main source rocks for oil in the Wessex and North Sea basins (Greengrove and Davies-Vollum, 2006c;

Colley, Kimmeridge Bay). Deposited on the floor of a shallow (50–200 m; 164–656 ft) tropical sea, ammonites are abundant in some of the shale horizons, and bivalves and brachiopods are also common. The Kimmeridge Clay Formation is almost entirely exposed at the surface at Kimmeridge Bay. Seismic acquired by British Petroleum indicates that the full thickness of the unit is between 535 and 585 m (1,755–1,919 ft) in this area (Gallois, 2004; West, 2018).

The cyclicity of the Kimmeridge Clay sequence has been the subject of much speculation. Most recently the cycles are considered to be a result of Milankovitch orbital cycles. Measurements of the rock cycles at Kimmeridge Bay show an apparent 38,000-year periodicity. "Ecliptic" is the plane of the Earth's orbit around the sun. The Obliquity of the Ecliptic, also known as Axial Tilt, is the inclination of the plane of the ecliptic relative to the plane of the Earth's equator. The Axial Tilt may have varied in a cyclical manner in the past, and the change in the tilt of the ecliptic results in tropical zones expanding and contracting periodically with resulting climatic, sea level, and depositional changes. For comparison, obliquity cycles at present are ~41,000 years.

Almost all the faults seen in the cliffs around Kimmeridge Bay are north–south extensional normal faults with displacements of a few meters (West, 2018).

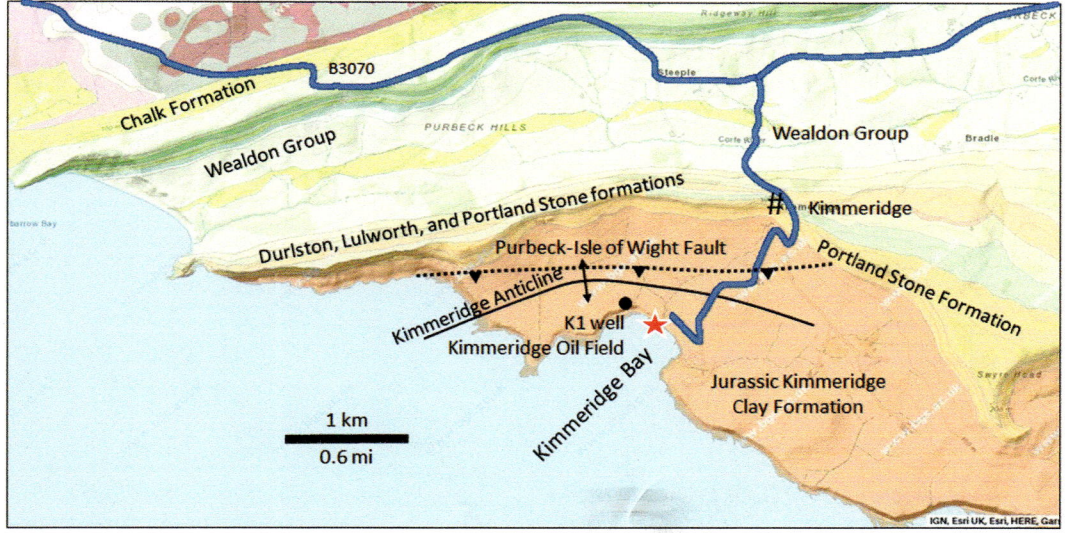

Geologic map of the Kimmeridge Bay area. (Modified after BGS, Geology of Britain Viewer. Contains British Geological Survey materials © UKRI 2022. Base mapping provided by ESRI.)

The horizons around the bay form part of an anticline bounded on the north by an east–west fault and dipping gently to the east, west, and south. The anticline is south of the regional Purbeck Monocline. Drilling in 1959 encountered a small oil reservoir in the Cornbrash Formation. Despite drilling six more wells in Kimmeridge Bay since the 1960s, there is only one producing well (a "nodding donkey" sucker rod pump) at the top of the cliff on the northwest side of the bay. British Petroleum's well initially produced 42 m³/d (350 barrels a day) from the Kimmeridge Oil Field, but current production is no more than 7.8 m³/d (65 barrels a day) from the Cornbrash Limestone at a depth of 320 m (1,050 ft). The Cornbrash Limestone is 27 m (89 ft) thick here and has low matrix porosity (averages 1%) and virtually no permeability; production is from fractures. In 1998 the reservoir was estimated to contain ultimate recoverable reserves of 556,000 m³ (3.5 million barrels; Colley, Kimmeridge Bay; West, 2018).

*Kimmeridge Bay to Winspit Quarry: Return north 3.5 km (2.2 mi) to an unnamed road at the T intersection and turn right (east) toward Church/Knowle/Corfe Castle; continue straight on Tyneham Road; turn right (south) onto A351/East Street; turn right (south) onto B3069; in Kingston bear left (southeast) onto West Street toward Matravers and Swanage; turn right (south) toward Worth Matravers and Swanworth Quarries; in Worth Matravers, turn left (south) onto Winspit Road; continue onto Winspit Bottom. You may have to park and walk the final 300 m (1,000 ft) south to the quarry on the coast. This is **Stop 17, Winspit Quarry** (50.583660, −2.034630), for a total of 16.7 km (10.4 mi; 30 min). The best parking is in Worth Matravers, but this requires you to walk 30–45 minutes down a road marked "Private," although it has public access. Continue along a wide path to the quarry at the coast. There are no facilities.*

STOP 17 WINSPIT QUARRY

Winspit is an abandoned quarry on the cliffs south of Worth Matravers that provided building stone for London. The quarry closed around 1940. During World War II, it was used for naval and air defense. After the war, the caves were opened to the public, but recently many of Winspit's galleries and caves have been closed for public safety (Wikipedia, Winspit).

Portland Limestone Formation, Winspit Quarry.

Falling from the limestone cliffs here or at the coast is a real risk. Rock falls in the cliffs and old quarries are another risk. Climbing should only be undertaken by experienced climbers with proper equipment. Adders have been reported from above the cliffs, but are rarely seen (West, 2013).

Stratigraphy of the Winspit Quarry area. (Information drawn from West, 2013.)

The layers here are nearly horizontal to gently dipping east–southeast. There are extensive galleries east and west of the valley. The succession extends from the Jurassic Kimmeridge Clay at the base through the Portland Beds to the Late Jurassic-Early Cretaceous Purbeck Beds at the top. In fact, this is the last of the Jurassic we will see since the rocks get younger as we go east and we will be in the Cretaceous from here to Dover. The Under Freestone, a layer within the Portland Freestone Series, was the object of the quarrying. The Pond Freestone ("Upper Freestone" or "Top Freestone"), a 2.1 m (7 ft) thick grainy limestone, was considered the best building stone. The House Cap, 2.4 m (8 ft) below the Pond Freestone, is a 1.5–1.8 m (5–6 ft) thick coarse shelly limestone with molds of the giant *Titanites* ammonites. This stone was used for breakwaters.

   In the western part of the quarries, the galleries are in the Under Freestone of the Portland Stone and in the Under Picking Cap. The House Cap forms the roof of these galleries (West, 2013).

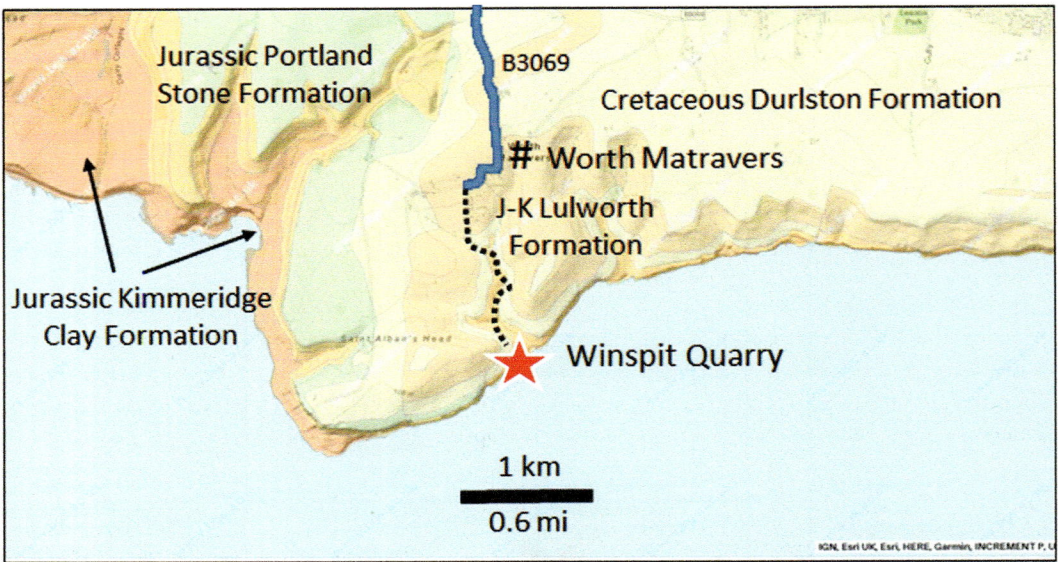

Geologic map of the Winspit Quarry area. (Modified after BGS, Geology of Britain Viewer. Contains British Geological Survey materials © UKRI 2022. Base mapping provided by ESRI.)

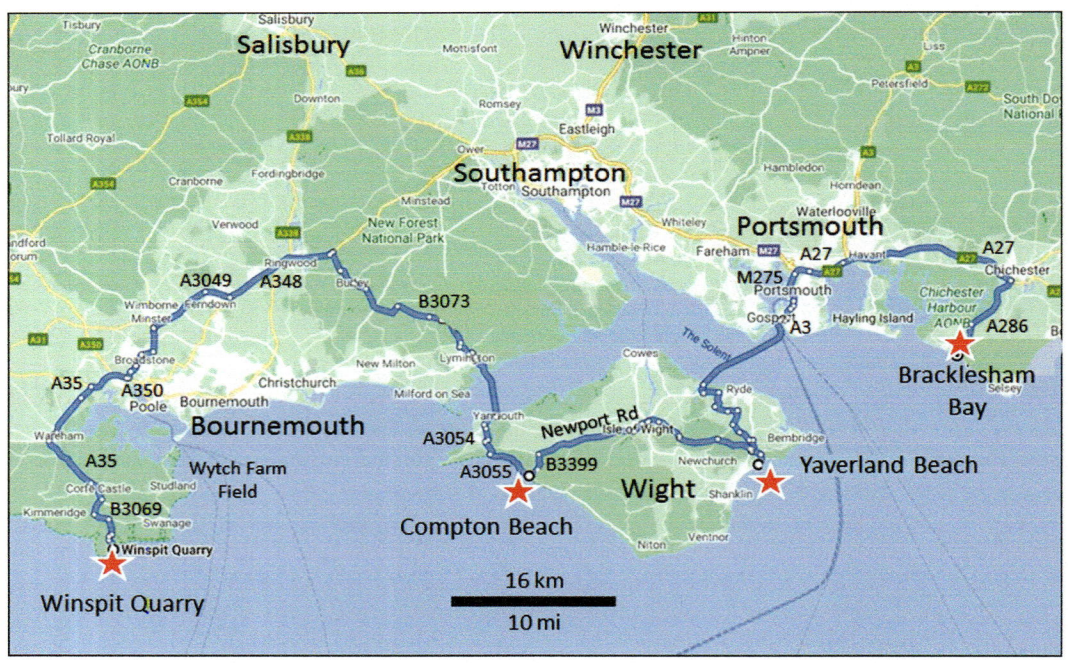

STOP 18 ISLE OF WIGHT

The Isle of Wight, in the English Channel off the south coast is, at 384 km² (148 mi²), the largest island in England. The outstanding exposures of Cretaceous and Paleogene rocks have attracted geologists since the early 1800s. Known locally as "Dinosaur Island," it has not only dinosaur remains but also 15 SSSIs, mainly a result of their geologic importance. The island has many "type sections" (locations where a formation is defined) along the coast exposures, including the sections at Alum Bay on the west and Whitecliff Bay on the east side of the island. Late Mesozoic and early Cenozoic strata are exposed in impressive sea cliffs that extend for a total of 98 km (59 mi) and are a must-see for both amateur and professional geologists.

The island is divided into two parts by an east–west ridge of Cretaceous Chalk Formation that runs from The Needles in the far west to Culver Cliff in the east. The south half of the island is mostly Cretaceous strata, whereas north of the Chalk Formation ridge are mostly Tertiary sandstone and shale with minor limestone. The north part of the island has the most complete Paleogene (Paleocene-Eocene) section in northwest Europe (Hopson, 2011).

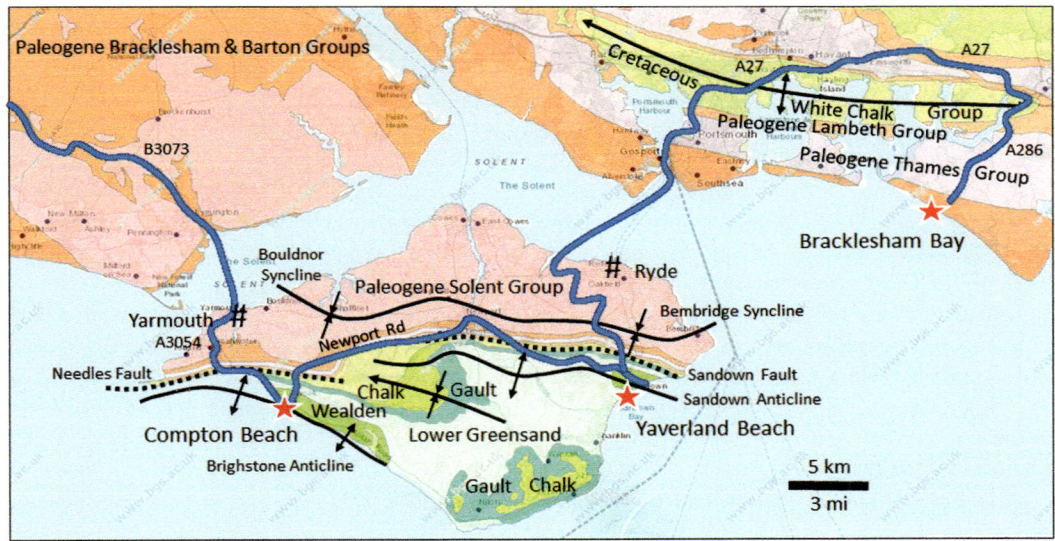

Geologic map of the Isle of Wight to Bracklesham Bay. (Modified after BGS, Geology of Britain Viewer. Contains British Geological Survey materials © UKRI 2022. Base mapping provided by ESRI.)

Fossil-bearing units are plentiful. The Cretaceous Wealden Beds, a fluvial sequence of mudstones and sandstones, are famous for dinosaur fossils and tracks, as seen at Compton Beach. The Lower Greensand above has a good assemblage of marine fossils. The Gault and Upper Greensand formations are fossiliferous and outcrop along the coast at Compton Bay. The Cretaceous Chalk Formation is fossiliferous and well exposed at The Needles in the west and at Culver Cliff north of Yaverland Beach on the east coast. Whereas the Middle Eocene Bracklesham Group is most fossiliferous at Whitecliff Bay (north of Yaverland Beach), the overlying Barton Clay is better exposed and even more fossiliferous at Alum Bay just north of The Needles (West, 2015).

Structurally, the Isle of Wight is in the Wessex Basin, which extends over much of southern England south of the London Syncline and Mendip Hills. This sedimentary basin preserves a thick succession of Permian to Cretaceous rocks above the Paleozoic (Variscan) basement. The basement contains deep-seated structures that formed when the continents of Gondwana and Laurentia (Laurussia) collided, during the Variscan Orogeny, to create the supercontinent Pangea. This period of mountain-building peaked at the end of the Carboniferous. As a consequence of this collision, the Variscan basement rocks were weakly metamorphosed and were cut by several major, northward-directed thrust faults and northwest-oriented right-lateral strike-slip faults (Hopson, 2011).

The Wight-Bray Fault, almost vertical, runs northwest-southeast from near Culver Cliff (Sandown) to Dieppe on the French coast. The Wight-Bray Fault is one of a series of dextral strike-slip faults along the English Channel. These faults defined the margins of basins and sub-basins when compression associated with continental collision was replaced by extensional rifting in Triassic–Jurassic time (Cotterill, Wight Rocks).

During rifting a series of subsiding basins developed, collectively called the Wessex-Channel Basin, extending southwards from near Marlborough/North Wessex Downs to the central English Channel. These include the Pewsey and Weald Basins in the north, and the Portland-Wight Basin (sometimes called the Vectian, Wessex, or Channel Basin) that includes the south half of the Isle of Wight. These basins were separated by the Hampshire-Dieppe High, an east–west uplift that covers the north half of the island. The basins accumulated sediment from the Permian through the Early Cretaceous time (Hopson, 2011; Cotterill, Wight Rocks).

Pre-Cretaceous rocks are all in the subsurface on the Isle of Wight. Deposition of the Triassic, Jurassic, and lowermost Cretaceous strata on the Isle of Wight was controlled by two large east–west-trending and south-dipping normal faults, The Needles Fault in the west and the Sandown Fault to the east. These faults formed during the Triassic–Jurassic rifting of Pangea (early opening of the Atlantic) and may be spatially and genetically related to underlying Variscan thrust faults (Wikipedia, Geology of the Isle of Wight).

Intense cooling of the crust caused rapid subsidence of the fault-bounded basins in the Early Cretaceous. The Early Cretaceous Wealden Beds, the oldest unit at the surface, contain the Wessex and overlying Vectis formations. The Wessex Formation is mainly red mudstones deposited on river floodplains. It also contains river channel sandstones and conglomerates deposited in broad meandering rivers. This is the main dinosaur formation. The Vectis Formation is mainly dark-gray siltstone and mudstone.

Above the Wealden, but still in the Early Cretaceous, the shallow marine Lower Greensand Group is a sequence of claystone, sandstone, and pebbly conglomerate that includes the Sandrock and Carstone/Monk's Bay Sandstone formations. The Sandrock and Carstone are both upward-coarsening cyclic successions.

The dark blue–gray Gault Clay indicates low-energy marine conditions, perhaps 100–200 m deep. The Upper Greensand is a shallow marine, nearshore glauconitic (a green mica mineral) sandstone.

Sea-level rise and continued subsidence of the Portland-Wight Basin produced local deep-water conditions in the Upper Cretaceous. The Lower Chalk Formation begins with a muddy Glauconitic Marl at the base, followed by the Chalk Marl, and then Grey Chalk, with a final thin Plenus Marls at the top. The Lower Chalk has abundant bivalves, sponges, serpulids (tubeworms), and ammonites. Above these are nodular chalks of the Middle Chalk, and finally the brilliant white Upper Chalk. Above the Upper Chalk (White Chalk subgroup) an erosional unconformity spans a time gap of about 15 million years. The erosion indicates an uplift at this time.

Age	Group	Lithology	Formation/member
Oligocene	Solent Group		Cranmore member
			Bouldnor Formation
			Hamstead member
			Bembridge Marls member
			Bembridge Limestone Fm
Eocene			Headon Hill Formation
	Barton Group		Becton Sand Formation
			Chama Sand Formation
			Barton Clay Formation
			Boscombe Sand Formation
	Bracklesham Group		Selsey Sand/Branksome Sand Formations
			Marsh Farm Formation
			Earnley Sand/Upper Poole Formation
			Upper Wittering Formation
			Lower Poole Formation
			Lower Wittering Formation
	Thames Group		Whitecliff Sand member
			Portsmouth Sand member
			London Clay Formation
Pal	Lambeth Group		Reading Formation

100 m
328 ft

Tertiary stratigraphy of the Isle of Wight. (Information drawn from Hopson, 2011.)

The earliest Tertiary sediments on the island are the bright red and mottled muds of the Reading Formation of upper Paleocene age. This was followed by a major marine transgression at the start of the Eocene. The marine London Clay was deposited as tidal flat silty sands and muds and cross-bedded tidal channel sands. The Bracklesham Group contains rhythmic sequences of coarsening-upward mudstone to sandstone: it contains warm water mollusks and the famous " coloured sands" of Alum Bay (Cotterill, Wight Rocks). It is speculated that early uplift along the northern margin of the anticline at Alum Bay resulted in oxidation of sand minerals that tints the sands in shades of red and yellow (West, 2015 Isle of Wight).

Next is the Middle to Late Eocene Barton Group marine claystone and sandstone. Above it, the Headon Hill Formation consists of fresh and brackish water sediments, probably deposited in coastal lagoons and lakes. Several freshwater limestones occur. The Barton Group and Headon Hill Formation indicate the gradual filling of the marine basin.

The Oligocene Bouldnor Formation contains the Bembridge Oyster Beds with fossil mollusks in sandstone. Rocks from the Miocene and younger are absent from the island due to the uplift and erosion related to the Alpine Orogeny (Cotterill, Wight Rocks).

Cretaceous crustal extension was replaced in the Paleogene by crustal compression caused by the collision of the northward-moving African plate with Europe, the Alpine Orogeny. Reactivation of pre-existing normal faults during the Miocene (c. 23–14 Ma) produced inversion structures that define the current outcrop patterns. The downthrown side of the faults became the upthrown side. These former normal faults underwent compressional shortening and are now part of the Portland-Wight Thrust Fault system/Purbeck–Wight Monocline (Hopson, 2011). The Brighstone and Sandown anticlines are both north-leaning asymmetric folds developed above The Needles and Sandown faults, respectively. Both anticlines have gentle southern and near-vertical northern limbs. The northern limbs form monoclines (sharp bends in the rocks). North of the monoclines the Paleogene sequence is deformed into the gentle Bouldnor Syncline (Wikipedia, Geology of the Isle of Wight).

The structure of the island is spectacularly expressed by the vertical Chalk Formation at Culver Cliffs just north of Yaverland Beach and at The Needles/Alum Bay north of Compton Beach. These represent an eastward continuation of the Purbeck–Wight structure that on the island is known as the Isle of Wight Monocline. This structure consists of two *en-echelon* (overlapping, stepped) monoclines associated with the Brighstone and Sandown anticlines separated by a classic fault ramp (a gently-inclined zone between two folds or faults; Hopson, 2011).

The youngest rocks on the island are peats and gravels that cap the cliffs on the south half of the island, are exposed near Newtown Creek on the north coast, and outcrop at Bembridge Forelands, where peats and the famous Bembridge Raised Beach are exposed in the cliffs. They formed between several hundred thousand years ago to just in the last few thousand years (DinosaurIsle. com, Geological History, Isle of Wight).

Mammoths and other large mammals lived on the Ice Age tundra of the Isle of Wight, which wasn't an island at the time because sea level was much lower. Their remains have been found in Pleistocene gravel and loam on both sides of Freshwater Bay, in brickearth (glacial loam or silt) in the Medina Valley near Newport, and in river gravels at Grange Chine near Brighstone, and at Brook Chine (West, 2015).

Winspit Quarry to Compton Beach*: Return north on Winspit Bottom/Winspit Road through Worth Matravers to B3069/West Street; turn left (west) on B3069 and drive to Kingston; turn right (north) onto B3069/Kingston Hill; bear left (northeast) onto A351/East Street; continue north on A351 to the A35; at the roundabout, take the 3rd exit (east) and stay on A35 east; continue straight onto the A350; at the Holes Bay North Roundabout, take the 3rd exit (east onto A3049; at the Tower Park Roundabout, take the 2nd exit (east) onto B3061/Old Wareham Road; at the roundabout, take the 1st exit (northeast) onto B3068/Ringwood Road; at the roundabout, take the 2nd exit (southeast) onto Herbert Ave; turn right (south) onto A3040/Alder Road; at Pottery Junction, take the 2nd exit (east) onto A35/Poole Road; at County Gates Gyratory take the 1st exit (east) onto A338/Wessex Way; take the exit toward Royal Bournemouth/Castlepoint/Christchurch/-Kinson; at the Cooper Dean Roundabout, take the 2nd exit onto A3060/Castle Lane East; at the roundabout in Iford, take the 1st exit (east) onto A35/Christchurch Road; continue on A35 past Hinton Admiral and turn right (southeast) onto B3055; turn right (south) onto Vaggs Lane; at the roundabout in Hordle, take the 1st exit (east) onto Silver Street; continue straight onto Sway Road; in Buckland turn left (north) onto A337/Southampton Road, then at the roundabout, take the 2nd exit onto B3054/Marsh Lane; at the roundabout, take the 1st exit (east) onto B3054/Bridge Road; turn right (south) onto Undershore Road; board the ferry at the Lymington Ferry Terminal; offload* *in Yarmouth and turn right (west) onto A3054; turn left (south) onto Pixley Hill; continue straight on Copse Lane; turn right (west) onto Hooke Hill; at the roundabout, take the 1st exit (south) onto A3055/Stroud Road; at Freshwater Bay turn left (east) to continue on A3055 and drive to Brook Chine National Trust Car Park.* **This is Stop 18.1, Compton Beach** *(50.650888, −1.456029), for a total of 85.3 km (54.7 mi; 2 h 33 min, depending on ferry schedule). Walk west ~670 m (2,200 ft) to dinosaur tracks.* **Note: book a ferry crossing in advance to ensure a spot.**

STOP 18.1 COMPTON BEACH

Compton Beach is particularly famous for dinosaur footprints preserved at the shoreline, and dinosaur bones eroding out of the sea cliffs. For this reason, it is an SSSI. So far over 20 species have been found, including Polacanthus, Brachiosaurus, Iguanodon, and Hypsilophodon. Fossilized dinosaur bones are most commonly black and shiny. Large black dinosaur teeth are sometimes found here too (NationalTrust.org). The best place to look for fossils is in loose gravel and stones winnowed by a receding tide. Do not hammer fossils out of the sea cliffs.

The dinosaur remains are preserved in the Lower Cretaceous Wealden Group on the north flank of the Brighstone Anticline. The mainly fluvial Wealden consists of mottled marls, yellowish sandstones, some shales, and thin limestones (Hopson, 2011; West, 2015). Large three-toed tracks of Iguanodon can be seen at the base of the cliffs just east of Compton Bay National Trust car park at Hanover Point. The theropod tracks are between 30 and 60 cm (1–2 ft) across (NationalTrust.org). There is also a dinosaur trackway in a red clay bed, 150 m (490 ft) out from the cliff at Hanover Point in a south-easterly direction, but it is only accessible at low tide (Chapman, 2017).

Above and north of the Wealden are the Lower and Upper Greensand, then the Chalk Formation at The Needles and Alum Bay. The Lower Greensand is fossiliferous at Compton Bay, with remains of lobsters and mollusks. The Upper Greensand forms prominent cliffs and is chock full of marine fossils. The white Chalk Formation cliffs in north Compton Bay and at The Needles promontory, where they are nearly vertical, also contain marine fossils (West, 2015).

Compton Beach looking north toward The Needles. (Photo courtesy of Mypix, https://commons.wikimedia.org/wiki/File:Compton_Bay,_Isle_of_Wight,_UK.jpg.)

Dinosaur tracks, Compton Beach. Coin for scale. (Photo courtesy of Jim Champion, https://commons.wikimedia.org/wiki/File:Fossilised_dinosaur_footprint_at_Compton_Bay.jpg.)

*Compton Beach to Yaverland Beach: Drive 200 m (650 ft) southeast on A3055 and turn left (northeast) onto Brook Village Road; bear left (north) onto B3399 and drive to the T intersection at The Middle Road; turn right (east) onto B3401/The Middle Road/Newport Road; continue east on B3401 to Shide; turn right (south) on A3020/Blackwater Road; continue straight on A3056; in Downend turn right (east) onto The Downs Road; turn right (south, then east) onto Brading Down Road; continue straight onto Bullys Hill; in Morton turn right (south) onto The Mall; continue straight on Yarbridge; continue straight on Marshcombe Shute; at the roundabout, take the 2nd exit (south) onto B3395/Yaverland Road and drive to the Yaverland Beach Car Park on the left. This is **Stop 18.2, Yaverland Beach** (50.66159, −1.13627), for a total of 30.4 km (18.9 mi; 34 min).*

STOP 18.2 YAVERLAND BEACH

This famous location is well known for dinosaur bones and reptile and fish remains, with the best collecting being after winter and spring high tides that have winnowed the shore. You might want a hammer to split the hard blocks to find bones within. Access to Yaverland Beach is easy from a large car park near the shore. The tides here can be dangerous. You MUST check out tide times before you visit and return to the shore before the tide turns. It is easy to become cut off by the rising tide (Cruickshanks, 2017).

Yaverland Beach to Culver Cliff. Street View northeast.

This site is an SSSI, meaning you can collect loose fossils and rocks, but hammering outcrops is not permitted (Cruickshanks, 2017). The beach fossil history goes back to the time of William Buckland who, just 5 years after identifying Megalosaurus, the World's first dinosaur, in 1829 described an Iguanodon bone. Since then, many vertebrate remains have been found in the Cretaceous Wealden Group at Yaverland. It is also the type locality for the pterosaur (flying reptile) *Caulkicephalus trimicrodon*, discovered by a 7-year-old school boy while on holiday. Under the right conditions, you can see dinosaur tracks in river channel deposits in the middle of the Wealden (Sweetman and Martill, 2019).

Beach outcrops extend from the Wealden Group in the south to the Grey Chalk at Culver Cliff in the north. At Yaverland, the cliff starts with the Wessex and Vectis formations of the Wealden Group. The Wessex Formation is marked by variable-colored beds of sandstones and clays. Vertebrate remains, including the bones and teeth of dinosaurs, crocodiles, turtles, and fish, are found in the Wessex Formation. Silicified remains of the tree-fern Tempskya are widespread in the Wealden Group (West, 2015). A lignite bed shows up as a dark layer with fossil plant material (stems, branches, and cones; Cruickshanks, 2017). The Vectis Formation limestone layers are full of bivalves.

Moving north, you enter the Lower Greensand Group with the Atherfield Clay Formation, which also contains a hard limestone layer; the Perna Bed has abundant bivalves and corals. At the north end of the bay, the Ferruginous Sands are marked by an orange-red colored cliff, followed by Sandrock and Monk's Bay Sandstone at the top of the Lower Greensand Group. Above (north of) these units you encounter the Gault Clay and Upper Greensand. Finally, the headland consists of the Lower Chalk.

Most of the Lower Greensand beds are unfossiliferous at Yaverland. The Gault Clay has an occasional ammonite and some bivalves; at Culver Cliff, the Lower Chalk has the nautilus *Cymatoceras radiatum* and many species of ammonites, bivalves, and sponges. The Upper Greensand Formation has abundant serpulid worm tubes and the large bivalve *Inoceramus lissa*. Shark teeth are occasionally found in the Upper Greensand and Chalk formations. Further up-section, the echinoid *Micraster cortestudinarium* becomes quite common. Most of the fossils can be found on the foreshore during scouring conditions (Cruickshanks, 2017; Sweetman and Martill, 2019).

Beyond this, the Chalk Group exposures along the headland are increasingly difficult and dangerous to access.

*Yaverland Beach to Bracklesham Bay: Return north on B3395; at the roundabout, take the 1st exit (west) onto Marshcombe Shute; in Morton turn right (north) onto A3055/New Road; at the roundabout in north Brading, take the 1st exit (west) onto Coach Lane; continue straight onto West Lane; continue straight onto Green Lane; turn right (north) onto Ashey Road; at the roundabout take the 1st exit (west) onto Carters Road; continue straight onto Stroud Wood Road; turn right (northeast) onto Newnham Road; at the roundabout, take the 1st exit (west) onto A3054/Quarr Hill; turn right (north) onto B3731/Fishbourne Lane; board ferry at the Fishbourne Car & Passenger Terminal; offload at Wightlink Gunwharf Terminal; turn left (north) onto Gunwharf Road; continue straight onto B2154/St George's Road; at Cambridge Junction, take the 1st exit (north) onto A3/Cambridge Road; use the right 2 lanes to bear right onto M275 toward the M27; use the right 2 lanes to merge onto M27 to Brighton and Chichester; continue straight onto A27/Havant Bypass; at the Fishbourne Roundabout, take the 3rd exit onto A27/Chichester Bypass; at the Stockbridge Roundabout, take the 3rd exit (southwest) onto A286/Stockbridge Road; at the roundabout south of Birdham, take the 1st exit (south) onto B2198/Bell Lane; continue straight on Bracklesham Lane; turn left (southeast) onto East Bracklesham Drive and drive to Bracklesham Lane Pay & Display Car Park on the right. This is **Stop 19, Bracklesham Bay** (50.761135, −0.860142), for a total of 65.4 km (40.7 mi; 1 h 56 min, depending on the ferry schedule). Parking is £2.30/2 h; there is a large parking area near toilets and a café. **Note: book a ferry crossing in advance to ensure a spot.***

STOP 19 BRACKLESHAM BAY

Walk down to the shore from the car park. Bracklesham Bay is the best location in the United Kingdom for children to collect fossils, as there are no cliffs and there is a sandy beach, shallow water, easy access, and fossils can be found lying on the sand. This is an SSSI, meaning you can collect loose fossils, but hammering bedrock is not permitted.

At Bracklesham Bay, the Eocene Bracklesham Group Beds are exposed below high tide level. The Bracklesham Group is divided into four formations, all of which are present here. The formations get younger from northwest to southeast: they are the Wittering Formation, the Earnley Sand Formation, the Marsh Farm Formation, and the Selsey Sand Formation (Cruickshanks, 2017).

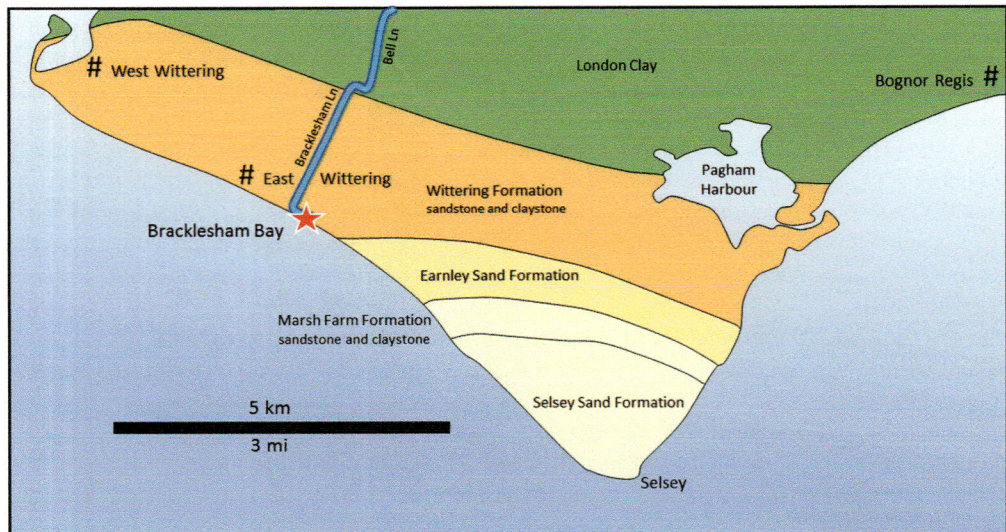

Geologic map of Eocene formations in the Bracklesham Bay area. (Modified after Shepherd and Randell, 2021f.)

The Bracklesham Group of sediments was deposited in the Hampshire Basin around 46 million years ago by rivers that supplied sediment to a large estuary on the North Sea coast. At this time Britain was farther south, in a warm, subtropical environment rich in vegetation and animal life. Southeast from the car park about 1 km (0.6 mi) is the Earnley Sand Formation, a gray, clay-rich unit that contributes the majority of fossils scattered across the sand at low tide. At low tide large pedestals of Earnley Formation clay extend up from the beach, allowing you to see the fossils *in situ*, that is, still in the bedrock (Shepherd and Randell, 2021f).

Among the fossils, you can find are ray and sharks' teeth, bivalve shells (especially *Venericor*), gastropods (*Turritella*), occasional corals, abundant fossilized wood, and lots of large foraminifera (*Nummulites laevigatus*). There are also turtle carapaces, crocodile teeth and bones, and bones of sea snakes, as well as rare bones of birds and mammals. All these fossils tend to appear dark, almost black, against the beach pebbles. They were stained black during the millions of years of burial (Cruickshanks, 2017).

Fossil clam (bivalve), Bracklesham Beach.

The best time to collect fossils is during a falling tide, particularly during the winter months when the sea transports large volumes of sand away from the beach, exposing the underlying fossil-rich clay (Shepherd and Randell, 2021f).

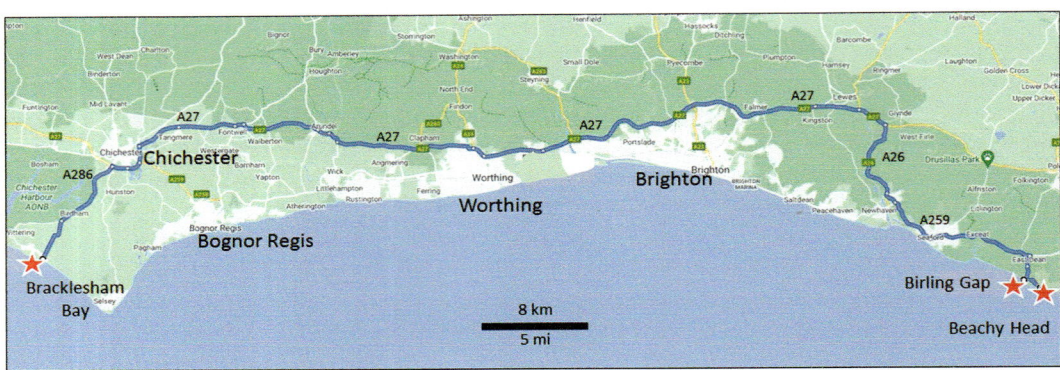

STOP 20 CHALK CLIFFS OF EAST SUSSEX AND SOUTH DOWNS NATIONAL PARK

The chalk cliffs of East Sussex are every bit as stunning as their better-known cousin at Dover. The sea cliffs are mainly limestones of the Chalk Group. The group is divided into the lower Grey Chalk Subgroup and the upper White Chalk Subgroup. The Chalk Group formed in a warm, clear sea that extended across northwest Europe during the Late Cretaceous.

Carboniferous, Permian, and Triassic sediments that form the basement in this area were uplifted and faulted during the Variscan Orogeny, with the land now occupied by East Sussex being a fold belt far from the main deformation located under the present English Channel. In this area, there was little or no metamorphism. The Variscan mountain belt collapsed soon after the orogeny: the former north-directed thrusts were reactivated as normal faults, leading to subsidence of the Weald Basin, part of the regional Wessex Basin. The Weald Basin gently subsided throughout the Jurassic, Cretaceous, and Early Paleogene, leading to a thick succession of sedimentary rocks (Wikipedia, Geology of East Sussex).

The Jurassic-Cretaceous basin was in turn uplifted and inverted during Alpine compression, with the thickest section of sediments now in the core of the Weald Anticline. The South Downs of Sussex, just west of Eastbourne, are on the south flank of the Weald Anticline, a far-field structure of the Miocene Alpine Orogeny.

SOUTH DOWNS NATIONAL PARK

South Downs National Park, England's newest national park, came into being in 2010. The park covers 1,627 km^2 (628 mi^2) between Winchester in the west and Eastbourne in the east. The park was created by joining the East Hampshire Area of Outstanding Natural Beauty with the Sussex Downs Area of Outstanding Natural Beauty, along with surrounding areas. The park is by far the most visited of English National Parks, with about 39 million visits per year.

The coast of East Sussex is where Julius Caesar's armies first landed on English soil and is close to the site of the Norman invasion at Hastings. Many old fortifications still exist, such as the Iron Age hill fort at Cissbury Ring north of Worthing.

A curiosity, the hillside figure of the Long Man of Wilmington, is 72 m (235 ft) tall and appears carved into the underlying chalk. In fact, the figure is formed from white-painted cement blocks and lime mortar. For many years it was thought the figure originated in the Iron Age or even the Neolithic period, but an archaeological investigation in 2003 showed that the figure may have been cut in the 16th or 17th century AD. The modern figure, holding two staves, is not the original figure. It, shown in a drawing made by William Burrell in 1766, shows a man holding a rake and a scythe. Another drawing, by the surveyor John Rowley in 1710, is the oldest date documenting the figure (Wikipedia, Long Man of Wilmington). The Long Man is a mere 5 km (3 mi) northeast of the Litlington White Horse, a chalk hill figure depicting a horse, cut into the chalk of Hindover Hill, that stands 20 m (65 ft) high and 28 m (93 ft) long. This figure was cut in 1924 at the site of an earlier figure carved in either 1838 or 1860. The reason for these carvings is hotly debated but remains unclear.

The national park covers the chalk hills of the South Downs. The park includes the western Weald, a heavily wooded area of sandstone and clay hills and valleys. The area was originally heavily forested but was cleared for the grazing of sheep, giving rise to the typical grass-covered downs seen today.

Geologically, most of the national park lies in the chalk downland. The chalk is a Late Cretaceous limestone, deposited in a shallow sea between 100 million and 66 million years ago. During the mainly Miocene Alpine Orogeny the chalk was uplifted to form the Weald Anticline, part of the regional Weald-Artois Anticline. The chalk forms the dramatic white cliffs at Beachy Head and the Seven Sisters/Birling Gap. These cliffs are an erosional feature formed after the last Ice Age, as sea levels rose and waves began cutting into the uplifted chalk. The chalk itself is a brilliant white limestone consisting of coccoliths, small calcium carbonate plates that make up the spherical shells of coccolithophores, free-floating algal plankton.

Bracklesham Bay to Birling Gap: *Return north on B2198/Bracklesham Lane to Somerly; at the roundabout, take the 3rd exit (northeast) to A286/Main Road; at Stockbridge Roundabout take the 3rd exit (east) onto A27/Chichester Bypass; continue driving east on the A27 to Beddingham; at the Beddingham Roundabout, take the 2nd exit (south) onto A26; in South Heighton take a slight left onto B2109; at the Denton Roundabout, take the 1st exit (southeast) onto A259/Seaford Road; stay on A259 to East dean; turn right (south) onto Gilberts Drive; continue straight on Birling Gap Road to the National Trust car park at Birling Gap. This is* **Stop 20.1, Birling Gap** *(50.742905, 0.201439), for a total of 98.5 km (61.2 mi; 1 h 24 min). There is ample pay-and-display parking for £1.50/h. Facilities are available.*

STOP 20.1 BIRLING GAP

There are steps that provide easy access to the beach. The site has toilets and a pub and I. From the steps, you can walk either east or west, but the best fossil area is to the west (Shimmin, 2012).

In 1966, Birling Gap was established as part of the Sussex Downs Area of Outstanding Natural Beauty and was designated part of the Sussex Heritage Coast in 1973. The site lies in the Seaford to Beachy Head SSSI and contains two Geological Conservation Review sites of national scientific importance.

Birling Gap is situated between the Seven Sisters to the west and Beachy Head to the east. The "gap" is a dry paleo-valley that runs to the coast between high chalk cliffs that form part of the "Seven Sisters," a spectacular coastal landform. The gap provides easy access to the coast in an area otherwise dominated by sea cliffs, and for that reason was popular with smugglers until a coast guard station was built in the early 1800s. The sea cliffs along this stretch of coastline are being undercut by wave action and are retreating at an average annual rate of between 0.5 and 1.5 m (1.5 and 4.5 ft) per year (Moore et al., 2001).

The valley at Birling Gap was probably eroded and filled during the last Ice Age, prior to about 12,000 years ago. It is underlain by "coombe rock," Holocene valley fill deposits composed of fine-grained weathered chalky silt and clay with fragments of chalk, flint, and pebbles. The coombe rock is softer than the surrounding chalk, so it erodes faster (National Trust, Birling Gap).

Birling Gap is near the axis of a west-northwest-trending syncline, with the brilliant white Seaford Chalk Formation exposed in the core (Duperret et al., 2012). The formations between here and Beachy Head dip gently to the west.

This area is not only a stunning example of chalk sea cliffs, but fossils are also common. This is the most accessible chalk location in Sussex. Fossils are mostly found on the foreshore, on the wave-cut platform, or in fallen rocks. Echinoids, sponges, mollusks, and fish remains have all been found.

The Seven Sisters Flint Band is seen on the wave-cut platform at the beach access steps. The Cuckmere Sponge Bed is just above the Flint Band. Specimens of sponges can be seen, and opalized plant remains have sometimes been found in the Sponge Bed. Echinoids, especially micrasters, are common here (Shimmin, 2012).

Seven Sisters looking northwest from Birling Gap.

Fossils found on the beach, Birling Gap.

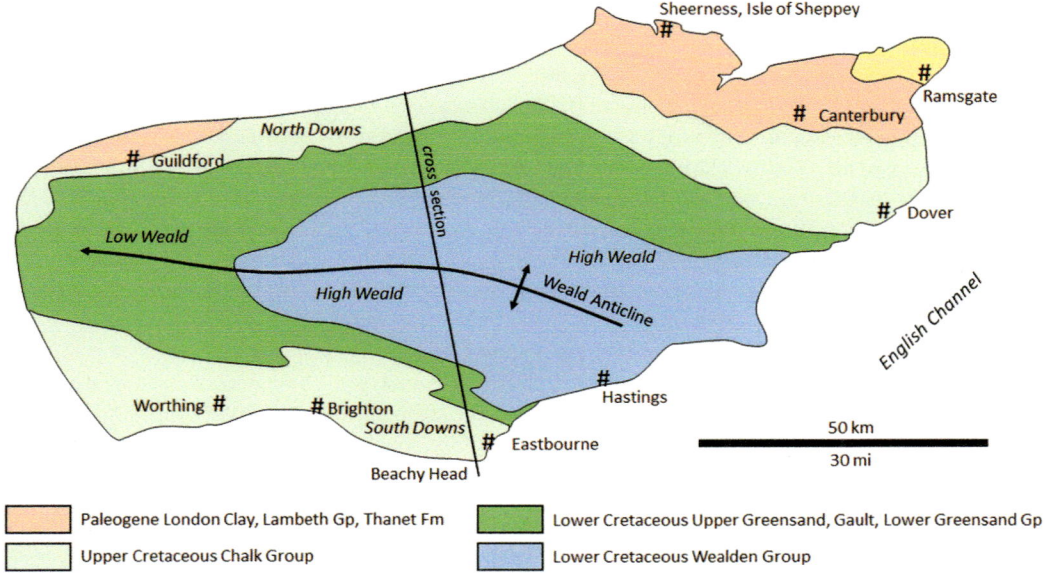

Geologic map of the Weald Basin. (Modified after BGS, Earthwise, http://earthwise.bgs.ac.uk/index.php/File:P902277.jpg.)

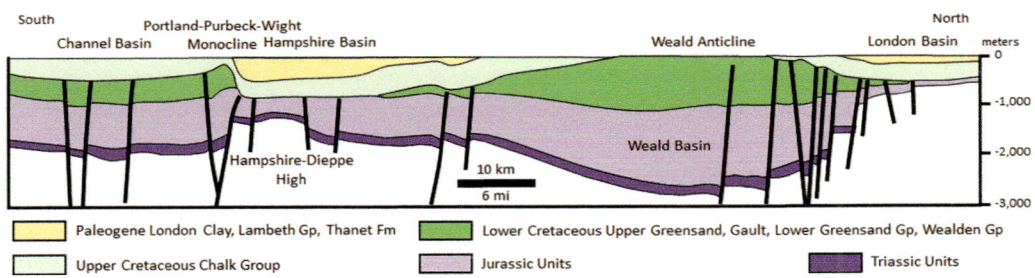

Cross section through the Weald Basin and Channel Basin. (Diagram courtesy of Mikenorton, https://-commons.wikimedia.org/wiki/File:WealdBasinCrossSection.png.)

This site should only be visited on a falling tide, as the sea comes in quickly. You can collect loose rocks and fossils, but hammering on bedrock is not permitted.

*Birling Gap to Beachy Head: Turn right (east) onto Beachy Head Road, and drive 4.4 km (2.7 mi; 7 min) to **Stop 20.2, Beachy Head** (50.741173, 0.252996). There is ample pay-and-park at the Beachy Head car park on the left, as well as toilet facilities.*

STOP 20.2 BEACHY HEAD

Beachy Head is a prominent chalk headland west of Eastbourne and immediately east of the Seven Sisters. The chalk cliffs at Beachy Head reach 162 m (531 ft) at the lighthouse (Duperret et al., 2012; Wikipedia, Beachy Head). The clifftop area is part of the Downland Country Park managed by Eastbourne District Council (Pennington, 2007).

In addition to stunning scenery, it is a fossil hunting locality and an area of active landslides and rockfalls.

From the Beachy Head car park, take the footpath to Cow Gap until you reach the steps that go to the shore. Note that there is no way off of the beach between Cow Gap and Birling Gap. The Cow Gap to Beachy Head section must be visited on a falling tide, as parts of the beach get cut off up to 2.5 h before high tide. Walk west to the lighthouse first, and hunt fossils on your way back to avoid getting caught by the high tide. Fossils are not that abundant in the stretch between Beachy Head and Birling Gap, and high tide reaches all the way to the base of the cliffs (Shimmin, 2017).

The name Beachy Head appeared as "Beauchef" in 1274: it is a corruption of the French words meaning "beautiful headland." It was consistently called Beachy Head by 1724 (Wikipedia, Beachy Head).

Beachy Head provides progressively younger Late Cretaceous (100–66 Ma) chalk exposures as you go from east to west along the base of the cliffs. This is due to the gentle westward dip of the formations on the southwest flank of the Kingston–Beddingham Anticline (Nowell, 2007; Duperret et al., 2012). The earliest deposits in the Beachy Head area belong to the Upper Gault, a dark-gray clay, approximately 103 Ma, exposed around Eastbourne Beach. Above the Gault Clay but below the overlying Chalk is the Upper Greensand, a course sandstone c. 100 Ma.

Overlying the Upper Greensand is the White Chalk Subgroup, represented by the West Melbury Marly Chalk, exposed at the cliff base from Cow Gap to Falling Sands. This unit is overlain by the Zig Zag Chalk Formation followed by the Holywell Nodular Chalk Formation, then the New Pit Chalk Formation, and eventually, toward Belle Tout Lighthouse, the Lewes Nodular Chalk Formation. At Beachy Head lighthouse, the Seaford Chalk Formation lies at the top of the cliff (Shimmin, 2017). The Seaford Chalk Formation constitutes the entire cliff between Beachy Head lighthouse and Birling Gap to the west (Pennington, 2007).

Common fossils in the Gault Clay and overlying Upper Greensand include echinoids, brachiopods, gastropods, and bryozoans (moss animals); less common fossils include ammonites, shark teeth, nautilus, and crustacean burrows lined with fish scales; on rare occasions, belemnites, lobster and fish skeletons, and plant remains can be found (Shepherd and Randell, 2021a).

The chalk at the base of the cliffs has some of the best-exposed reefs in the United Kingdom and Europe (The Geological Society, Seven Sisters & Beachy Head). Fossils in the Chalk occur in the rockfalls behind the lighthouse and between the lighthouse and Cow Gap. Fossils are found in fallen chalk boulders, in chert (or flint) nodules (especially the echinoid *Micraster*, related to sea urchins and sand dollars), and in the shingle. Echinoids, sponges, brachiopods, bivalves, corals, worm tubes, ammonites, and belemnites have all been found (Shimmin, 2017).

The Plenus Marls of the Holywell Chalk are at beach level a short distance east of the lighthouse. At low tide, the Plenus Marls expose rare belemnite guards belonging to *Actinoclamax* (Shepherd and Randell, 2021a).

Interestingly, the silica that formed chert nodules is derived from the remains of sea sponges and siliceous plankton (diatoms and radiolarians). The chert forms as nodules or concretions that grew in the sediment by the migration and precipitation of silica (Shepherd and Randell, 2021a).

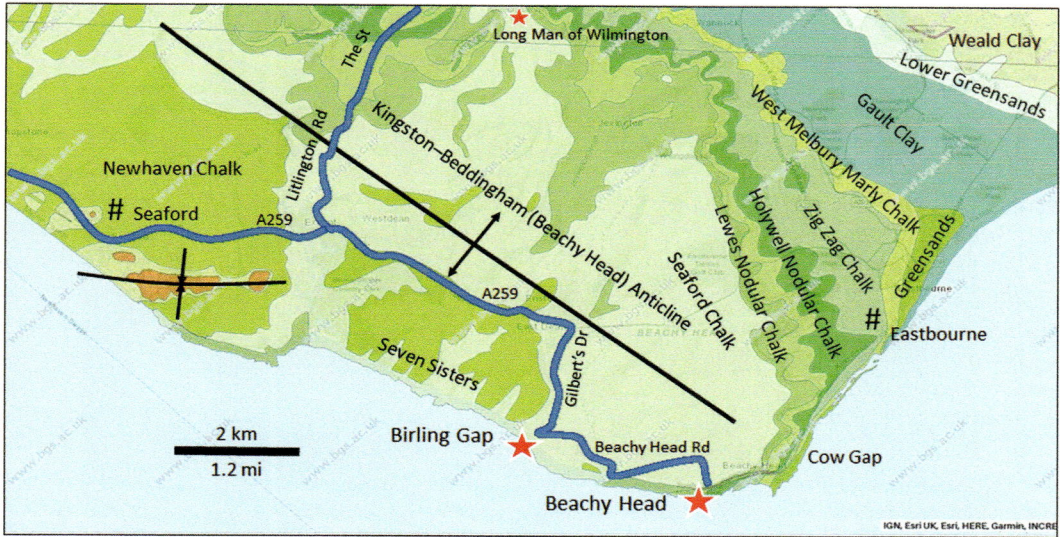

Geologic map of the Seven Sisters-Birling Gap-Beachy Head area. (Modified after BGS, Geology of Britain Viewer. Contains British Geological Survey materials © UKRI 2022. Base mapping provided by ESRI.)

The main source for the chalks was marine plankton. The remains of coccolithophores, planktonic algae, accumulated as calcareous mud on the sea floor. The upward transition from sandstones to claystones and finally limestones in the Beachy Head area record the Late Cretaceous change from a nearshore environment to deep marine conditions far enough offshore to have little or no input of continental sediments (Shepherd and Randell, 2021a).

Today the chalk is at and above sea level because of widespread uplift during the Alpine mountain-building event that peaked around 20–25 Ma. At the end of the last Ice Age, the rise in sea level exposed the chalk to erosion and undercutting by waves, carving the dramatic vertical cliffs above the wave-cut shore platform (Shepherd and Randell, 2021a).

The wave-cut platform at the shore is littered with rockfalls and slides. The lack of confining pressure and the pull of gravity causes deep tension cracks to penetrate the Chalk just behind the cliff face. Given time and undercutting by waves, this leads to spalling of large slabs from the cliff face. This mechanism is assisted by prominent vertical and sub-vertical tectonic joints already in these formations (Pennington, 2007).

This is a Site of Special Scientific Interest. You can collect loose rocks and fossils, but hammering the bedrock is not permitted.

View west from Beachy Head during high tide.

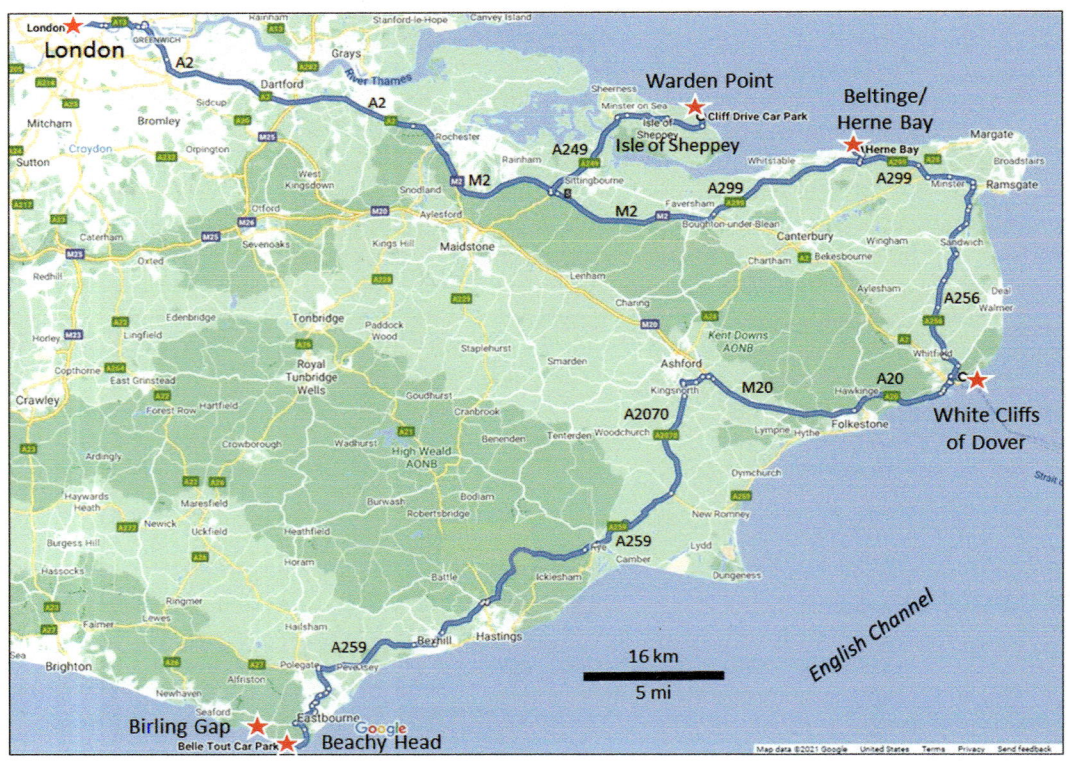

*Beachy Head to White Cliffs of Dover: Continue northeast on Beachy Head Road into Meads/-Eastbourne; continue straight onto Meads Road; turn right (east) onto St John's Road; take a slight left onto South Cliff; turn left (northeast) onto B2103/King Edward's Parade; continue straight on B2106/Grand Parade; at Sovereign Roundabout, take the 3rd exit (north) onto Princes Road; at Langney Roundabout, take the 2nd exit (north) onto B2104/Langney Rise; at the roundabout, take the 2nd exit (northeast) onto B2191/Hide Hollow; continue on B2191 to Pevensey; continue straight on A259/Bexhill Road; continue on A259 to Bexhill and turn left (north) on A269/Combe Valley Way/London Road; continue on A269 to Baldslow; at the Baldslow Roundabout, take the 2nd exit (east) onto A2100/The Ridge W; turn left (north) onto Maplehurst Road; turn right (northeast) onto A28/Westfield Lane; continue on A28 to Broad Oak; turn right (east) onto B2089/Udimore Road; continue on B2089 to Rye where B2089 turns right and becomes Station Approach; Station Approach turns left and becomes Cinque Ports Street; continue on A268 to Skinners Roundabout; at Skinners Roundabout, take the 1st exit (east) onto A259/New Road; continue on A259 to Brenzett; at the Brenzett Roundabout, take the 2nd exit (north) onto A2070; continue north on A2070 to the M20 at Sevington; at the roundabout, take the 4th exit (southeast) onto the M20; in Folkstone, continue straight on the A20 to Dover turn left (north) on A256/Woolcomber Street; turn right (east) onto A258/Castle Hill Road; turn right (east) onto Upper Road and follow the signs to the National Trust car park at **Stop 21, White Cliffs of Dover** (51.131608, 1.338397), for a total of 118 km (73.2 mi; 2 h). There is abundant National Trust parking and facilities available for £5/all day. There are toilets and a café.*

A number of paths go east and west along the cliff top starting at the National Trust car park. The best route for cliff views is to the east by the lowest footpath along the cliff-edge. About 1.3 km (4,280 ft) east a zigzag path descends the cliff face.

STOP 21 WHITE CLIFFS OF DOVER

According to the National Trust, "The White Cliffs of Dover are perhaps most famous as an iconic landmark, the white chalk face a symbol of home and war time defense, but they have so much more to offer: stunning views, a serene walk, a wealth of wildlife, an abundance of history and, most importantly, two tea rooms offering delicious treats."

The cliffs are 110 m (350 ft) high and brilliant white. They are composed of chalk accented by streaks of black flint, deposited during the Late Cretaceous. The cliffs, on both sides of the town of Dover, stretch for 13 km (8 mi). The chalk cliffs are part of the North Downs. A section of coastline including the cliffs was acquired by the National Trust in 2016. The stunning cliffs are part of the Dover to Kingsdown Cliffs SSSI , so you cannot hammer on or otherwise disturb the cliffs.

During the Cretaceous, 100–67 million years ago, the area between Britain in the west and Sweden/Poland in the east was covered by deep tropical seas. Calcareous mud formed on the ocean bottom from the constant rain of dead coccoliths. Up to 500 m (1,640 ft) of Chalk Group sediments were deposited in some areas (Gibson, 2014; Wikipedia, White Cliffs of Dover). The lower half of the cliffs are the Lewes Nodular Chalk Formation; the upper half is the Seaford Chalk Formation.

The white chalk cliffs of Dover as seen from the sea. (Photo courtesy of Natalia Semenova, https://-commons.wikimedia.org/wiki/File:English_coast._The_Cretaceous_rocks_-_%D0%90%D0%BD%D0%B3%D0%BB%D0%B8%D0%B9%D1%81%D0%BA%D0%B8%D0%B9_%D0%B1%D0%B5%D1%80%D0%B5%D0%B3._%D0%9C%D0%B5%D0%BB%D0%BE%D0%B2%D1%8B%D0%B5_%D1%81%D0%BA%D0%B0%D0%BB%D1%8B._-_panoramio.jpg.)

White Cliffs looking north at high tide.

The purity of the chalk indicates it accumulated far from land, mostly free of land-derived sand and silt. Evidence indicates that the nearest land was where Wales is today (Shepherd and Randell, 2021c).

Horizontal bands of dark-colored chert/flint nodules streak the cliffs. The chert is derived from the silica-rich remains of sea sponges and siliceous plankton. Brachiopods, bivalves, and sponges are found in the chalk deposits (Gibson, 2014; Wikipedia, White Cliffs of Dover). Shark's teeth and crinoid stems are harder to find. The most productive and safest place to search for fossils is on the foreshore at low tide.

In some areas, layers of soft, gray chalk known as a "hardground" can be seen. Hardground is thought to indicate periods when sedimentation stopped for a while.

During the Alpine Orogeny, these sea-floor deposits were raised above sea level. Since then, the English Channel has been actively eroding the chalk. From the end of the last Ice Age, the cliffs have been eroding back at a rate of 0.23–0.30 m (8.7 in to 1 ft) per year (Wikipedia, White Cliffs of Dover).

White Cliffs showing nodule bands (thin dark streaks) in the chalk.

In the North Sea, the chalk is an important oil and gas reservoir (think giant fields like Ekofisk), but it can also act as a seal, trapping hydrocarbons in underlying formations. Unless the chalk is fractured or has good porosity due to later dissolution, the coccoliths that make up the chalk have virtually no permeability or porosity. Permeability is the interconnectedness of the pores or open spaces in a rock, and is necessary for hydrocarbons to migrate from their source to a reservoir rock (Gibson, 2014).

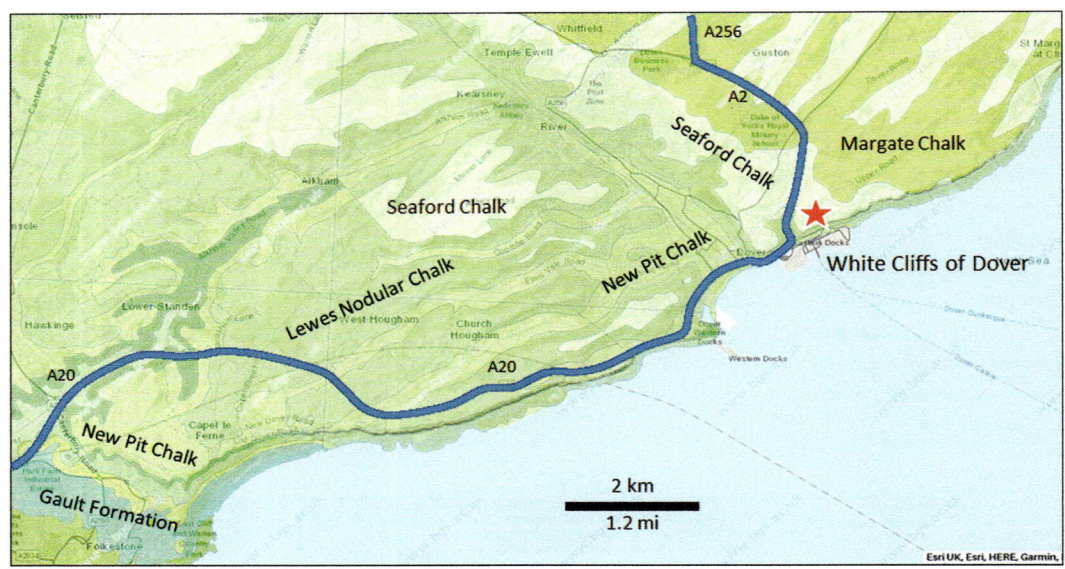

Geologic map of the Dover area. (Modified after BGS, Geology of Britain Viewer. Contains British Geological Survey materials © UKRI 2022. Base mapping provided by ESRI.)

White Cliffs to Herne Bay: *Return west on Upper Road to A258; turn right (northeast) on A258 and drive to the A2; at the Duke of York's Roundabout, take the 1st exit (northwest) onto the A2 and drive to the Whitfield Interchange S; take the 3rd exit (north) onto A256 and drive to Sevenscore/Cliffsend; at the Sevenscore Roundabout, take the 2nd exit (north) onto the A299; follow the A299 to Hawthorne; turn left (south) toward Heat In Hand Road and then right (north) onto Heat In Hand Road; continue straight on Sweechbridge Road; turn left (west) on Reculver Road; turn right (northeast) on Reculver Drive and drive to the Ocean View car park at the beach. This is* **Stop 22, Beltinge/Herne Bay** *(51.373763, 1.167035), for a total of 45.8 km (28.5 mi; 42 min). A footpath takes you directly onto the shore. There is ample pay-by-phone parking, but no facilities at this stop.*

STOP 22 BELTINGE/HERNE BAY

Beltinge (Herne Bay) is renowned as one of the best locations to collect sharks' teeth in the United Kingdom. This location is best visited during extremely low tides (Langham and Cruickshanks, 2017).

The strata are inclined about 3° to the west, bringing progressively younger deposits to beach level as you walk west. Walking east, toward Reculver, the Thanet Formation is sometimes exposed if the beach sand has been washed away. The Thanet Formation contains a number of bivalve and gastropod shells. The Thanet deposits were laid down during the latest Paleocene in a warm shallow sea far enough from shore, perhaps 30–50 km (20–30 mi), that only mud was deposited (Shepherd and Randell, 2021d).

Age		Formation	Description
Eocene	Thames Group	London Clay Fm	Gray marine clay
		Harwich Fm / Blackheath Beds	Fine-grained marginal marine sandstone, rounded black flint pebbles
Paleocene	Lambeth	Upnor Fm	Shallow marine sandstone-siltstone
		Thanet Sand Fm	Shallow marine sandstone

Stratigraphy in the cliffs at Herne Bay. (Information drawn from Clement, 2012; Shepherd and Randell, 2021d.)

At Beltinge, the Beltinge Fish Bed of the Paleocene Upnor Formation is at beach level. The shark, fish, and reptile fossils come from this gray–green sandy clay. Only a small area is exposed. The Early Eocene London Clay is exposed at the extreme west end of the bay. The exposures are not very good, being largely covered by sand.

Most fossils are found immediately shoreward and on either side of the beach access point. Fossils are usually found washed up on the foreshore or within the shingle. Time your visit for at least 2 h before the tide turns. Go as far out as the tide will let you. The collecting area is in the section of beach between the groynes on either side of the concrete steps. When the tide is low enough, a small "island" consisting mainly of pebbles appears. Collectors rarely wade out to this site, so teeth accumulate among the pebbles (Langham and Cruickshanks, 2017; Shepherd and Randell, 2021d).

There are teeth from about 24 species of shark, ray, and other fish, as well as crocodile and turtle remains. Vertebrae of sharks and fish can also be found. The most common species of shark fossil is the sand shark, *Stratiolamia macrota*. The next most abundant is the sand tiger shark, *Carcharias hopei*. The largest teeth found at Beltinge are those of a large extinct mackerel shark, *Otodus obliquus*. This shark was approximately 8–10 m (26–33 ft) in length, and some think it may be the ancestor of the modern great white shark. Some of the shark teeth found at Beltinge are very small and are only found by sieving the sediments from the pebble bed (Langham and Cruickshanks, 2017).

London Clay
Harwich
Upnor
Thanet

Herne Bay looking east. Thanet Sand is from beach level to mid-cliff; the Upnor and Harwich formations make up most of the upper half of the slope.

This site is a site of special scienic interest. Hammering bedrock is not permitted.

Food, drink, and toilets are available in the Beltinge town center less than 2 km (1.2 mi) away.

Herne Bay to Warden Point: Turn left (south) onto Oceanview; continue straight on Glenbervie Drive; turn left (southeast) on Reculver Drive; turn right (south) on Sweechbridge Drive; cross the A299 and turn left (east) onto the A299/Thanet Way access ramp; drive west on A299; keep right to continue straight on the M2 toward London; at Junction 5 take the A249 exit to Sittingbourne/Maidstone; at the roundabout, take the 3rd exit (north) onto A249/Maidstone Road; on the Isle of Sheppey roundabout, take the 2nd exit (north, then east) onto the A2500/Lower Road; at the Eastchurch Roundabout, take the 2nd exit (south and east) onto B2231/Rowetts Way; turn left (north) onto Warden Bay Road; continue straight onto Jetty Road and drive to the Cliff Drive car park (free; no facilities). This is **Stop 23, Warden Point** *(51.409129, 0.908102), for a total of 66.9 km (41.5 mi; 49 min). Walk northwest along the shore up to 6.5 km (4 mi) for fossil collecting.*

STOP 23 WARDEN POINT, SHEPPEY SHORE, AND THAMES ESTUARY

Fossil collectors have been sifting the London Clay of Sheppey for over 300 years (Ward, A Field Guide to the London Clay of Sheppey). Still today Warden Point is internationally renowned as the best and most popular site for collecting London Clay fossils. It has easy access, and fresh fossils are constantly being washed out by the waves and tides. The site is known for its wide variety of fossils, including remains of turtles, lobsters, crabs, sharks' teeth, snakes, crocodiles, mollusks, and plants (Cruickshanks, 2017).

The Isle of Sheppey is underlain by Tertiary deposits lying unconformably above the northeast-dipping Upper Cretaceous chalk that forms the North Downs. The Lower Eocene London Clay underlies the alluvium of the Medway Estuary and the Swale marshes and is exposed in the core of the London Basin, including here along the north shore of the Isle of Sheppey.

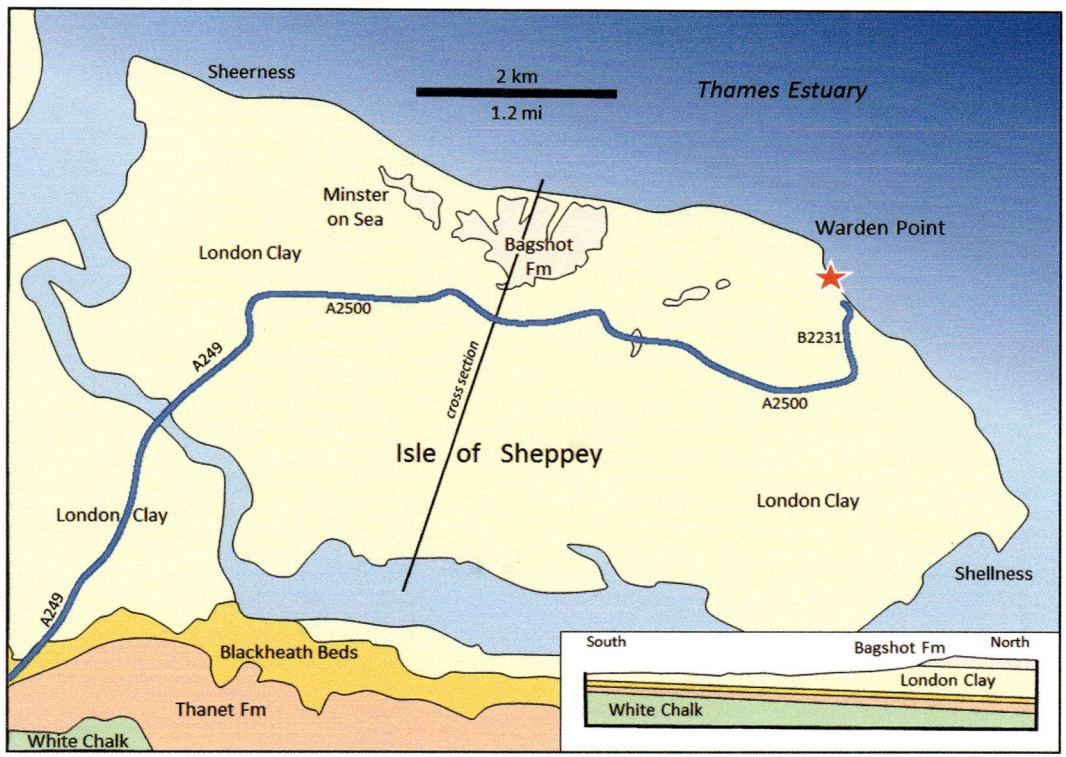

Geology of the Isle of Sheppey. (Map units derived from Young, 2007; BGS Geology of Britain Viewer. Contains British Geological Survey materials © UKRI 2022.)

The thickest exposure of London Clay on the island occurs at Warden Point, where it forms cliffs 42–46 m (138–151 ft) high (Young, 2007). The cliffs expose over 50 m (164 ft) of London Clay Formation capped by Middle Eocene Virginia Water Formation of the Bracklesham Group (Lower Bagshot Beds), and thin Pleistocene gravels. The London Clay itself is a gray silty clay widely distributed across the London and Hampshire basins. It overlies the Paleocene Woolwich or Oldhaven Formations (Ward, A Field Guide to the London Clay of Sheppey).

The gently northeast-inclined London Clay that is exposed between Warden Point and Minster to the northwest represents about a million years of sedimentation at c. 52–51 Ma. At that time the Isle of Sheppey lay beneath a warm, low-energy shallow sea, and the nearest land was perhaps 50 km (30 mi) distant (Shepherd and Randell, 2021e). Fringing the coast was a swamp, as seen by abundant Nipa palm fruits and species related to mangroves (Ward, A Field Guide to the London Clay of Sheppey).

Life in this Eocene sea flourished, while the relatively near landmass was covered by lush tropical vegetation, providing a habitat for mammals, birds, and insects. Evidence of marine life found in the clay included bivalves, crustaceans, nautiluses, gastropods, and shark teeth and vertebrae. Terrestrial animals include snake skeletons and bird skulls, and a variety of land-derived twigs, fruits, and seeds that were probably carried out to sea by tidal currents.

Cliffs of London Clay exposed at Warden Point. View west.

The London Clay exposed on the foreshore and in the cliffs is subject to intense washing by the sea, such that the fossil material on the beach is replenished throughout the year. The best fossil hunting is along a 3 km (2 mi) stretch of coast from Warden Point northwest to Minster (Shepherd and Randell, 2021e). Fossils can be found year-round, and finding good specimens is not dependent on beach or tidal conditions. Some claim that the best time to collect is during the summer when the clay is harder and accumulates more shingle. Others say to look during the winter and spring or after storms when new material is exposed (Cruickshanks, 2017; Shepherd and Randell, 2021e).

Many fossils are found in nodules and in the areas of shingle, exposed along the foreshore. It is recommended that you get down on hands and knees and scan the shingle. Look for erosion-resistant pyrite and phosphate nodules. Unfortunately, pyrite is highly reactive to changing humidity and, unless preserved, will begin to decompose within a few weeks (Cruickshanks, 2017; Shepherd and Randell, 2021e).

Larger fossils are generally found along the upper beach after the sea has scoured the cliffs. Fossils here are mostly found in phosphate nodules (Cruickshanks, 2017). As well, collectors looking for small invertebrates, sharks' teeth, mollusks, and seeds should scan fresh accumulations of pyrite nodules high up the beach. On the other hand, nodules with crab, lobster, or large vertebrate remains are usually found lower on the beach, on the wave-cut platform, where it is muddier and stickier… don't lose your boots! (Ward, A Field Guide to the London Clay of Sheppey).

A selection of shark and ray teeth, fish vertebrae, and others found on the shore near Warden Point. (From Shepherd and Randell, 2021e.)

Among the objects of interest are cream-colored phosphate nodules that stick out from the clay. Fossils occur in many of these nodules, and a small percent have complete organisms, usually crabs and lobsters (Shepherd and Randell, 2021e). Bring a hammer to crack open the nodules.

Warden Point is an SSSI . Collecting loose fossils is permitted; digging them out of the cliffs is not.

THAMES ESTUARY

From here back to London we are driving along the Thames Estuary. The name "Thames" is ultimately derived from the Sanskrit word *Tamasa*, meaning "dark water." The River Thames rises in the Costwold Hills near Trewsbury Mead, 108.5 m (356 ft) above sea level, and flows east through Oxford, Reading, and London to join the North Sea in the Thames Estuary. Whereas the origins of the river are lost in the mists of time, it must have initiated as this area was rising above sea level during the Alpine Orogeny (between 50 and 30 Ma, and perhaps as young as 14 Ma). During the last Ice Age, the river was a tributary to the Rhine, which it joined just east of the present-day estuary in an area now beneath the North Sea. A combination of glaciers to the north, tilting of southeast England, and sinking of the area around London has gradually displaced the river southward to its present course. As the glaciers melted and sea level rose, Britain was cut off from the continent and the link to the Rhine was severed, probably around 8,500 years ago (Hart, 2009; Jamison, 2008).

The Thames is tidal up to Teddington Lock. It is essentially fresh water as far downstream as Battersea.

From the source to sea, the river is 344 km (215 mi), as calculated using the pre-1964 seaward limit of the Port of London Authority (Hart, 2009).

Through much of London, and again east and south of Canvey Island, the Thames and its estuary are developed on the Eocene (56–48 Ma) London Clay. The Cretaceous White Chalk Group is at or just below the surface between London City Airport and Canvey Island. The river flows along the axis of the London Syncline.

Satellite image of the Thames Estuary showing sandbanks (brown) and sediment-laden waters (brown and green swirling patterns). (NASA Operational Land Imager, 28 April 2013, https://commons.wikimedia.org/-wiki/File:Thames_Estuary_and_Wind_Farms_from_Space_NASA.jpg.)

It is estimated that the Thames carries 300,000 tonnes (331,000 tons) of sediment, mostly silt and clay, each year. Most of this sediment is deposited near the river mouth as mud flats and sandbanks (Hart, 2009). The river delta is classified as a combination estuary and tide-dominated delta (Wikipedia, River Delta). The estuary is classified (NOAA, Classifying Estuaries) as a drowned-river valley similar to the Hudson River in New York or the Seine in France. The delta is characterized by submerged sandbanks shaped by fluvial and tidal scouring and deposition. The estuary has a tidal range of 4 m (13 ft) just east of the Tideway. The greater Thames Estuary has extensive mudflats, low-lying beaches, and salt marshes such as the North Kent Marshes and the Essex Marshes (Wikipedia, Thames Estuary).

Sunset from the Isle of Sheppey. (Photo courtesy of Helen Brown / Sunset over Isle of Sheppey / CC BY-SA 2.0, https://commons.wikimedia.org/wiki/File:Sunset_over_Isle_of_Sheppey_-_geograph.org.uk_-_1040382.jpg.)

Warden Point to London: *Return via A249 west and south to the M2; continue straight on the A2 west toward London for about 90 km (56 mi; 1 h 25 min).*

We are once again on the outskirts of London. We have gone from the Eocene to the Precambrian and back, traversing the breadth of England from east to west, all while exploring canyons and quarries, sea cliffs and Cheddar caves, mines and monuments, the birthplaces of stratigraphy and paleontology, and dinosaur stomping grounds. We visited the brilliant coastal chalk cliffs of Southdown National Park, passed through 14 Sites of Special Scientific Interest, and examined the geology of five UNESCO World Heritage Sites including the Jurassic Coast, the Roman Baths, and Stonehenge. There's a lot packed into this small area, and it's all well worth seeing.

REFERENCES

Anderson, M.W., and A. Morris. 2004. The puzzle of axis-normal magnetic lineations in folded low-grade sediments (Bude Formation, SW England). In Martín-Hernández, F., Lüneburg, C.M., Aubourg, C., and Jackson, M. (eds), *Magnetic Fabric: Methods and Applications*. Geological Society, London, Special Publications, v. 238, pp. 175–190.

Baker, L., D. Clements, V. Marks, and P. Rainey. 2016. *Green Chain Walk Geotrail*. Geodiversity Partnership, London, 23 p.

Barron, A.J.M., T.H. Sheppard, R.W. Gallois, P.R.M. Hobbs, and N.J.P. Smith. 2015. Geology of the Bath Area: Applied Geology: Building Stone and Bulk Minerals. Accessed 20 October 2021, http://earthwise.bgs.ac.uk/index.php/Geology_of_the_Bath_area:_Applied_geology:_building_stone_and_bulk_minerals

Bevins, R. 2012. Stonehenge. The Geological Society. Accessed 20 October 2021, https://www.geolsoc.org.uk/GeositesStonehenge

BGS, Cheddar Gorge. Accessed 21 October 2022, https://www2.bgs.ac.uk/mendips/localities/cheddar.html

BGS, Earthwise. Geological Sketch Map of the Wealden District. Accessed 20 October 2021, http://earthwise. bgs.ac.uk/index.php/File:P902277.jpg

BGS, Geology of Britain Viewer. Accessed 20 October 2021, https://mapapps.bgs.ac.uk/geologyofbritain/ home.html

BGS Lexicon. Accessed 20 October 2021, https://www.bgs.ac.uk/technologies/the-bgs-lexicon-of-named-rock-units/

Blue Suede. 2010. Delabole Slate. Accessed 20 October 2021, https://www.geocaching.com/geocache/ GC2GDZJ_delabole-slate

Bristow, C.M. 2014. The geology of the building and decorative stones of Cornwall, UK. In Cassar, J., M.G. Winter, B.R. Marker, N.R.G. Walton, D.C. Entwisle, E.N. Bromhead, and J.W.N. Smith (eds), *Stone in Historic Buildings: Characterization and Performance.* Geological Society, London, Special Publication, 391, pp. 93–120.

Bromley, A.V. 1991. Field excursion to the Lizard Complex, 5th January 1991. *Proceedings of the Ussher Society,* v. 7, p. 428.

Brook, D., D. Clements, G. Lucy, L. Baker, N. Stevenson, P. Collins, P. Rainey, and V. Marks. 2014. Guide to London's Geological Sites. 104 p. Accessed 20 October 2021, http://londongeopartnership.org.uk/wp/-wp-content/uploads/2018/08/Guide-to-Londons-Geological-Sites-Final-Sept-2014-reduced-size.pdf

Butler, R. Kilve. The Geological Society. Accessed 20 October 2021, https://www.geolsoc.org.uk/GeositesKilve

Chaffee, J. 2012. The Geology of Dorset: The Jurassic Rocks. Accessed 20 October 2021, https://www. dorsetlife.co.uk/2012/03/the-geology-of-dorset-the-jurassic-rocks/

Chapman, C. 2017. UK Fossil Collecting, Compton Bay. Accessed 23 October 2021, https://ukfossils.co.uk/ 2016/06/17/compton-bay/

Chippindale, C. 2004. Stonehenge Complete, 3rd edition. Thames & Hudson, London and New York. 312 p.

Clements, D. 2012. The Geology of London. *GeoExPro,* v. 9, no. 3, 10 p.

Clements, D. 2014. The Geology of London, Introduction. London, Geologists' Association Guide No. 68, pp. 1–17.

Coastal Landscape. 2019. Land's End. Accessed 20 October 2021, https://landsend-landmark.co.uk/nature/ coastal-landscape/

Colley, H. Kimmeridge Bay. The Geological Society. Accessed 20 October 2021, https://www.geolsoc.org. uk/GeositesKimmeridge

Colley, H. Lulworth. The Geological Society. Accessed 20 October 2021, https://www.geolsoc.org.uk/ GeositesLulworth

Cook, T. 2018. Travels in Geology: Mesozoic Masterpiece: England's Jurassic Coast. Earth Magazine. Accessed 20 October 2021, https://www.earthmagazine.org/article/travels-geology-mesozoic-masterpiece-englands-jurassic-coast/

Cornwall Beaches. 2021. Porthmeor Cove. Accessed 20 October 2021, https://www.cornwall-beaches.co.uk/ west-cornwall/porthmeor-cove.htm

Cornwall Council. 2018. Building Stone - Minerals Safeguarding DPD Evidence Report. Truro, 13 p.

Cotterill, M. Wight Rocks – An Introduction. Accessed 20 October 2021, http://iwnhas.org/groups/geology/ wight-rock-an-introduction/

Cruickshanks, A. 2017a. Warden Point (Isle of Sheppey). UK Fossil Collecting. Accessed 20 October 2021, https://ukfossils.co.uk/2012/01/24/warden-point/

Cruickshanks, C. 2017b. UK Fossil Collecting, Watchet. Accessed 23 October 2021, https://ukfossils. co.uk/2004/08/25/watchet/

Cruickshanks, A. 2018. UK Fossil Collecting, Charmouth. Accessed 23 October 2021, https://ukfossils.co.uk/ 2008/08/07/charmouth-dorset/

Delabole Slate. 2013. History. Accessed 20 October 2021, http://www.delaboleslate.co.uk/history.asp

Dinosaurisle.com, Geological History. Accessed 21 October 2021, http://www.dinosaurisle.com/geological_ history.aspx

Discovering Black Down, Geology. Accessed 21 October 2022, http://www.discoveringblackdown.org. uk/geology

Duperret, A., S. Vandycke, R.N. Mortimore, and A. Genter. 2012. How plate tectonics is recorded in chalk deposits along the eastern English Channel in Normandy (France) and Sussex (UK). *Tectonophysics,* v. 581, pp. 163–181.

Ellison, R.A., M.A. Woods, D.J. Allen, A. Forster, T.C. Pharoah, and C. King. 2004. *Geology of London.* British Geological Survey, Keyworth, Nottingham, 114 p.

Evans, J. 2014. Mineral Collecting Holiday in Cornwall 9–16 May 2014. Accessed 21 October 2021, https://www.mindat.org/article.php/1952/Mineral+collecting+holiday+in+Cornwall+9-16+May+2014

Friend, P. 2008. *Southern England: The Geology and Scenery of Lowland England*. Collins, London, 320 p.

Fuller, J.G.C.M. 2006. Smith Field Trip 1- John Strachey, William Smith, and the Strata of England, 1719–1801, Notes for Field Excursions to the Birthplace of Stratigraphy. The Geological Society. Accessed 21 October 2021, https://www.geolsoc.org.uk/Geoscientist/Archive/July-2007/Smiths-other-debt/Smith-field-trip-1

Gallois, R.W. 2004. The Kimmeridge Clay: The most intensively studied formation in Britain. *Open University Geological Journal*, v. 25, pt. 2, 14 p.

Gallois, R.W. 2006. The geology of the hot springs at Bath Spa, Somerset. *Geoscience in South-West England*, The Ussher Society, v. 11, pp. 168–173.

Gallois, R.W. 2014. The position of the Permo-Triassic boundary in Devon, UK. *Geoscience in South-West England*, v. 19, 25 p.

Gallois, R.W. 2019. The stratigraphy of the Permo-Triassic rocks of the Dorset and East Devon Coast World Heritage Site, UK. *Proceedings of the Geologists Association*, v. 130, pp. 274–293.

Gallois, R.W., and G. Davis. 2001. Saving Lyme Regis from the sea: Recent geological investigations at Lyme Regis, Dorset. *Geoscience in South-West England*, v. 10, pp. 183–190.

Geevor Tin Mine Home Page. Accessed 21 October 2021, https://geevor.com/

Geology Page. 2018. Millook Haven Beach. Accessed 21 October 2021, http://www.geologypage.com/2018/12/millook-haven-beach-england.html

Gibson, R. 2014. Chalk Cliffs of Dover. Accessed 21 October 2021, http://historyoftheearthcalendar.blogspot.com/2014/11/november-3-chalk-cliffs-of-dover.html

Google. Fact File: Lulworth Cove. Accessed 21 October 2021, https://sites.google.com/site/southcoastgeomorphology/home/case-study-lulworth-cove

Green, C.P. 1997. Stonehenge: Geology and prehistory. *Proceedings of the Geologists' Association*, v. 108, no. 1, pp. 1–10.

Greengrove, C., and S. Davies-Vollum. 2006a. Bath. The Real Jurassic Park: Geological Explorations in Southwest England. University of Washington, Tacoma. TESC 417. Accessed 21 October 2021, http://courses.washington.edu/uwtgeo06/locations/bath/bath.html

Greengrove, C., and S. Davies-Vollum. 2006b. Exmouth. The Real Jurassic Park: Geological Explorations in Southwest England. University of Washington, Tacoma. TESC 417. Accessed 21 October 2021, http://courses.washington.edu/uwtgeo06/locations/exmouth/exmouth.html

Greengrove, C., and S. Davies-Vollum. 2006c. Kimmeridge Bay. The Real Jurassic Park: Geological Explorations in Southwest England. University of Washington, Tacoma. TESC 417. Accessed 21 October 2021, http://courses.washington.edu/uwtgeo06/locations/kimmerdige_bay/kimmerdige_bay.html

Greengrove, C., and S. Davies-Vollum. 2006d. Lyme Regis. The Real Jurassic Park: Geological Explorations in Southwest England. University of Washington, Tacoma. TESC 417. Accessed 21 October 2021, http://courses.washington.edu/uwtgeo06/locations/lyme_regis/lyme_regis.html

Hand Luggage Only. 2020. How to go Fossil Hunting on The Jurassic Coast, England. Accessed 21 October 2021, https://handluggageonly.co.uk/2018/05/08/how-to-go-fossil-hunting-on-the-jurassic-coast-england/

Hart, D. 2009. The River Thames – Its Geology, Geography and Vital Statistics from Source to Sea. Accessed 21 October 2021, http://www.the-river-thames.co.uk/thames.htm

Harvey, T., and J. Gray. 2013. The Hydrocarbon Prospectivity of Britain's Onshore Basins. Department of Energy and Climate Change, Promote UK 2014, London, 86 p.

Havaej, M., J. Coggan, D. Stead, and D. Elmo. 2015. A Combined Remote Sensing–Numerical Modelling Approach to the Stability Analysis of Delabole Slate Quarry, Cornwall, UK. Rock Mechanics and Rock Engineering August, Springer, Vienna, 19 p.

Hecht, C.A. 1992. The variscan evolution of the Culm Basin, south-west England. *Proceedings of the Ussher Society*, v. 8, pp. 33–38.

Heritage Services – Bath & NE Somerset Council. 2018. Conservation of the King's Bath. https://www.heritagedaily.com/2018/10/conservation-of-the-kings-bath/122030

Higgs, R. 2015. Turbidite Reservoir Analog Field Trips: Bude Formation, Cornwall, UK. GeoClastica Ltd. Accessed 21 October 2021, http://www.geoclastica.com/ReservoirAnalogFieldTrips.htm

Higgs, R. 2019. Bude's Geology. Bude Tourist Board. Accessed 20 October 2021, https://www.visitbude.info/blog/budes-geology/

History Channel, London. Accessed 17 October 2022, https://www.history.com/topics/british-history/london-england

Hollick, L.M., R.K. Shail, and B.E. Leveridge. 2006. Devonian rift-related sedimentation and Variscan tectonics – New data on the Looe and Gramscatho basins from the resurvey of the Newquay District. *Geoscience in South-West England*, v. 11, pp. 191–198.

Hopson, P. 2011. The geological history of the Isle of Wight: An overview of the 'diamond in Britain's geological crown.' *Proceedings of the Geologists' Association*, v. 122, no. 5, pp. 745–763.

Hunt, B. 2015. Great British Stone: Bath Stone. Accessed 21 October 2021, https://www.stonespecialist.com/news/heritage/great-british-stone-bath-stone

iWalk Cornwall. Circular Walks Visiting Porthmeor Cove. Accessed 23 October 2021, https://www.iwalkcornwall.co.uk/walks/beach/porthmeor_cove

Ixer, R., R. Bevins, and D. Pirrie. 2020. Provenancing the stones. *Current Archaeology*, v. 366, pp. 34–41. Accessed 21 October 2021, https://pure.southwales.ac.uk/en/publications/provenancing-the-stones(2baf334b-9380-46aa-8a41-1f1d27f42f2b).html

Jackson, A.A. 2015. Bedrock Geology UK South: Carboniferous. BGS, Earthwise. Accessed 21 October 2021, http://earthwise.bgs.ac.uk/index.php?title=Bedrock_Geology_UK_South:_Carboniferous&oldid=20172.

Jamison, D.J. 2008. Geology of the Thames. Accessed 21 October 2021, http://www.thamesdiscovery.org/riverpedia/geology-of-the-thames

Johns, B.S., and L.E. Jackson. 2012. Stonehenge's Mysterious Stones. Earth Magazine. Accessed 21 October 2021, https://www.earthmagazine.org/article/stonehenges-mysterious-stones/

Jurassic Coast Trust. Fossil Collecting along the Jurassic Coast. Accessed 21 October 2021, https://jurassiccoast.org/visit/fossil-collecting/

Kim, A. 2015. The Earth Story. Accessed 21 October 2021, https://the-earth-story.com/post/132338272895/millook-haven-spectacular-folding-situated-in#:~:text=This%20dramatic%20folding%20happened%20at, sediments%20within%20were%20intensely%20folded

Kovacs, D. 2019. Discovering Mary Anning's Jurassic Coast. *Reservoir of the Canadian Society of Petroleum Geologists*, v. 46, no. 6, p. 21.

Kratinová, Z., J. Ježek, K. Schulmann, F. Hrouda, and R.K. Shail. 2003. The role of regional tectonics and magma flow coupling versus magmatic processes in generating contrasting magmatic fabrics within the Land's End Granite, Cornwall. *Geoscience in South-West England*, v. 10, pp. 442–448.

Kratinová, Z., J. Ježek, K. Schulmann, F. Hrouda, R.K. Shail, and O. Lexa. 2010. Noncoaxial K-feldspar and AMS subfabrics in the Land's End granite, Cornwall: Evidence of magmatic fabric decoupling during late deformation and matrix crystallization. *Journal of Geophysical Research*, v. 115, 21 p.

Langham, L., and A. Cruickshanks. 2017. Beltinge (Herne Bay), UK Fossil Collecting. Accessed 22 October 2021, https://ukfossils.co.uk/2012/01/24/herne-bay/

Leveridge, B.E., and R.K. Shail. 2011. The Gramscatho Basin, south Cornwall, UK: Devonian active margin successions. *Proceedings of the Geologists' Association*, v. 122, pp. 568–615.

LymeRegis.com. The Jurassic Coast of Lyme Regis. Accessed 22 October 2021, http://www.lymeregis.com/history/geolog_fossil.htm

Manz, L.A. 2015. William "Strata" Smith. Geo News. Accessed 22 October 2021, https://www.dmr.nd.gov/ndgs/documents/newsletter/Summer%202015/William%20Strata%20Smith.pdf

Mikenorton. Weald Basin Cross Section. Accessed 22 October 2021, https://commons.wikimedia.org/wiki/File:WealdBasinCrossSection.png

Moore, J. McM., and N. Jackson. 1977. Structure and mineralization in the Cligga granite stock, Cornwall. *Journal of the Geological Society*, v. 133, pp. 467–480.

Moore, R., T. Collins, and A. King. 2001. The Geology and Geomorphology at Birling Gap, East Sussex: The Scientific Case for Geo-Conservation. *36th DEFRA Conference of River and Coastal Engineers*, Keele University, Keele, 20–22 June 2001, pp. 02.3.1–02.3.12.

Mortimore, R. 2018. The Geology Durdle Door, Dorset - Chalk Stratigraphy, Sedimentology and Tectonic Structure. Open University Geological Society Wessex Group, Field Excursion Sunday 13th May 2018. Accessed 22 October 2021, https://ougs.org/files/wsx/reports/Durdle_Door_13_5_18.pdf

Mortimore, R. 2019. Late Cretaceous stratigraphy, sediments and structure: Gems of the Dorset and East Devon Coast World Heritage Site (Jurassic Coast), England. *Proceedings of the Geologists' Association*, v. 130, pp. 406–450.

Muller, A., R. Seltmann, C. Halls, W. Siebel, P. Dulski, T. Jeffries, J. Spratt, and A. Kronz. 2006. The magmatic evolution of the Land's End pluton, Cornwall, and associated pre-enrichment of metals. *Ore Geology Reviews*, v. 28, pp. 329–367.

Nance, R.D., G. Gutiérrez-Alonso, J.D. Keppie, U. Linnemann, J.B. Murphy, C. Quesada, R.A. Strachan, and N.H. Woodcock. 2010. Evolution of the Rheic Ocean. *Gondwana Research*, v. 17, pp. 194–222.

Nationaltrust.org. Dynamic Shorelines and Shifting Shores at Birling Gap. Accessed 22 October 2021, https://www.nationaltrust.org.uk/birling-gap-and-the-seven-sisters/features/managing-change-at-birling-gap

Nationaltrust.org. Searching for Fossils around Compton Beach. Accessed 22 October 2021, https://www.nationaltrust.org.uk/compton-bay-and-downs/features/searching-for-fossils-around-compton-bay

NOAA. Classifying Estuaries: By Geology. Accessed 22 October 2021, https://oceanservice.noaa.gov/education/tutorial_estuaries/est04_geology.html#:~:text=The%20four%20major%20types%20of,built%2C%20tectonic%2C%20and%20fjords.&text=Fjords%20are%20steep%2Dwalled%20river,seawater%20as%20the%20glaciers%20retreated

Nottle, P. 2006. Cligga Head, 23rd July 2006. Open University Geological Society. Accessed 22 October 2021, https://ougs.org/southwest/event-reports/174/cligga-head/

Nowell, D. 2007. Chalk and landscape of the South Downs, England. Blackwell Publishing Ltd., *Geology Today*, v. 23, no. 4, July–August, pp. 147–152.

Nowell, D.A.G. 1998. The Geology of Lulworth Cove, Dorset. Blackwell Publishing Ltd., *Geology Today*, v. 14, no 2, March–April, pp. 71–74.

Pearson, M.P., R. Bevins, R. Ixer, J. Pollard, C. Richards, K. Welham, B. Chan, K. Edinborough, D. Hamilton, R. Macphail, D. Schlee, J-L Schwenninger, E. Simmons, and M. Smith. 2015. Craig Rhos-y-felin: a Welsh bluestone megalith quarry for Stonehenge. *Antiquity*, v. 89, pp. 1331–1352.

Pennington, C. 2007. Landslides at Beachy Head, Sussex. Accessed 22 October 2021, http://nora.nerc.ac.uk/id/eprint/514096/1/Landslides_BeachyHead.pdf

Pownall, J.M., D.J. Waters, M.P. Searle, R.K. Shail, and L.J. Robb. 2012. Shallow laccolithic emplacement of the Land's End and Tregonning granites, Cornwall, UK: Evidence from aureole field relations and P-T modeling of cordierite-anthophyllite Hornfels. *Geosphere*, December 2012, v. 8, no. 6, pp. 1467–1504.

Prudden, H. 2004. Somerset Geology – A Good Rock Guide. Accessed 22 October 2021, https://people.bath.ac.uk/exxbgs/Somerset_Good_Rock_Guide.pdf

Pythian, H. 2006. The Building Stones of Bath. Open University Geological Society. Accessed 22 October 2021, https://ougs.org/southwest/event-reports/134/the-building-stones-of-bath/

Rock Doc Travel. In the Footsteps of William Smith, Father of Modern Geology. Accessed 22 October 2021, https://www.rocdoctravel.com/england-birth-of-modern-geology

Ross, D. Lizard Heritage Coast. Accessed 23 October 2021, https://www.britainexpress.com/countryside/coast/lizard.htm

Ross, Penwith Heritage Coast. Accessed 23 October 2021, https://www.britainexpress.com/countryside/coast/penwith.htm

Scotese, C.R. 2013. Map Folio of the Toarcian (179.3 Ma), Paleomap Project. Accessed 23 October 2021, https://www.researchgate.net/publication/272944258_Map_Folio_39_Early_Jurassic_Toarcian_1793_Ma

Scott, M. 2008. William Smith. NASA Earth Observatory. Accessed 23 October 2021, https://earthobservatory.nasa.gov/features/WilliamSmith

Shepherd, R., and R. Randell. 2021a. Beachy Head. Accessed 21 October 2021, http://www.discoveringfossils.co.uk/beachy-head-east-sussex/beachy-head-main/

Shepherd, R., and R. Randell. 2021b. Charmouth. Accessed 21 October 2021, http://www.discoveringfossils.co.uk/charmouth-dorset/

Shepherd, R., and R. Randell. 2021c. Dover. Accessed 21 October 2021, http://www.discoveringfossils.co.uk/dover-kent/

Shepherd, R., and R. Randell. 2021d. Herne Bay. Accessed 21 October 2021, http://www.discoveringfossils.co.uk/herne-bay-kent/

Shepherd, R., and R. Randell. 2021e. Warden Point. Accessed 21 October 2021, http://www.discoveringfossils.co.uk/warden-point-isle-of-sheppey/

Shepherd, R., and R. Randell. 2021f. Bracklesham Bay (West Sussex). Accessed 21 October 2021, http://www.discoveringfossils.co.uk/bracklesham-bay/

Shimmin, J. 2012. Birling Gap. UK Fossil Collecting. Accessed 21 October 2021, https://ukfossils.co.uk/2012/01/24/birling-gap/

Shimmin, J. 2017. Beachy Head. UK Fossil Collecting. Accessed 21 October 2021, https://ukfossils.co.uk/2012/01/24/beachy-head/

Simmons, M. 2016. Wessex Basin: Petroleum Geology 101 in the Field. Neftex Exploration Insights. Accessed 21 October 2021, file:///C:/WORD/Publishing/GeoTours/5_References/3_Europe/2-3-4_UK/4_UK_Southern/3_Lizard-Wight/Simmons_2016_WessexBasinResources.pdf

Somersetshire Coal Canal Society. Somersetshire Coal Canal – An Educational Guide. Accessed 23 October 2021, https://www.coalcanal.org/sccs/educ.htm

Stevenson, N., and H. Colley. 2012. Permian Red Sandstone Cliffs, Dawlish. The Geological Society. Accessed 23 October 2021, https://www.geolsoc.org.uk/GeositesDawlish

Sweetman, S.C., and D.M. Martill. 2019. The Cretaceous Succession between Yaverland and Culver Cliff. Accessed 23 October 2021, https://www.researchgate.net/publication/335635759_THE_CRETACEOUS_SUCCESSION_BETWEEN_YAVERLAND_AND_CULVER_CLIFF

The Geological Society. Geevor Tin Mine. Accessed 23 October 2021, https://www.geolsoc.org.uk/Geosites Geevor

The Geological Society. Lulworth and Wessex Basin. Accessed 23 October 2021, https://www.geolsoc.org.uk/Policy-and-Media/Outreach/Plate-Tectonic-Stories/Lulworth-Cove

The Geological Society. Millook Haven. Accessed 23 October 2021, https://www.geolsoc.org.uk/GeositesMillook

The Geological Society. Seven Sisters and Beachy Head. Accessed 23 October 2021, https://www.geolsoc.org.uk/GeositesSevenSisters

The Geological Society. Slate, Cornwall. Accessed 23 October 2021, https://www.geolsoc.org.uk/ks3/gsl/education/resources/rockcycle/page3692.html

Themes, S. 2020. Cligga Head Mine. Explore Cornwall. Accessed 23 October 2021, https://explorecornwall.org/cligga-head-mine/

Thomas, G. 2019. Guide to the Lizard Peninsula. Accessed 23 October 2021, https://www.classic.co.uk/nas/places-to-go/guide-to-the-lizard-peninsula-2207.html

Trenchard, T. 2020. Why the Jurassic Coast Is One of the Best Fossil-Collecting Sites on Earth. Smithsonian Magazine. Accessed 23 October 2021, https://www.smithsonianmag.com/travel/why-jurassic-coast-is-one-best-fossil-collecting-sites-on-earth-180975003/

Truro School. 2020. Geology Trip from Cligga Head to Perranporth. Accessed 23 October 2021, https://www.truroschool.com/latest-news/geology-trip-from-cligga-head-to-perranporth/

UCMP Berkeley. Mary Anning (1799–1847). Accessed 23 October 2021, https://ucmp.berkeley.edu/history/anning.html

UKAF. East Beach (Stonebarrow), Charmouth, Dorset. Accessed 23 October 2021, https://ukafh.com/wp-content/uploads/2018/08/Charmouth.pdf

Underhill, J.R., and R. Stoneley. 1998. Introduction to the development, evolution and petroleum geology of the Wessex Basin. In Underhill, J.R. (ed.), *Development, Evolution and Petroleum Geology of the Wessex Basin*. Geological Society, London, Special Publications, v. 133, pp. 1–18.

Variscan Coast. 2021. Cligga Head. Accessed 23 October 2021, https://variscancoast.co.uk/cligga

Walsh, J. 2011. British Slate Forum, Cornish Slate. Accessed 23 October 2021, http://www.britishslateforum.co.uk/2011/03/

Ward, D.J. A Field Guide to the London Clay of Sheppey. Accessed 23 October 2021, http://www.sheppeyfossils.com/pages/pdf/Field_Guide_Sheppey.pdf

Webster, S. 2017. West Somerset Coast Field Trip - Tectonic Inversion, 1st October 2017. Open University Geological Society. Accessed 23 October 2021, https://ougs.org/southwest/event-reports/380/west-somerset-coast-field-trip-tectonic-inversion/

West, I.M. 2013. Winspit and Seacombe, Isle of Purbeck, Dorset, Geological Field Guide. Accessed 24 October 2021, https://wessexcoastgeology.soton.ac.uk/Winspit-Seacombe.htm

West, I.M. 2014. Durdle Door, Dorset; Geology of the Wessex Coast, Geological Field Guide. Accessed 24 October 2021, https://wessexcoastgeology.soton.ac.uk/durdle.htm

West, I.M. 2015. Introduction to the Geology of the Isle of Wight, Geological Field Guide. Accessed 24 October 2021, https://wessexcoastgeology.soton.ac.uk/wight.htm

West, I.M. 2016a. Teignmouth to Dawlish, Devon; Geology of the Wessex Coast, Geological Field Guide. Accessed 24 October 2021, https://wessexcoastgeology.soton.ac.uk/Teignmouth-Dawlish.htm

West, I.M. 2016b. The Fossil Forest Exposure, Part 1- Introduction; Geology of the Wessex Coast, Geological Field Guide. Accessed 24 October 2021, https://wessexcoastgeology.soton.ac.uk/Fossil-Forest.htm

West, I.M. 2017. Stair Hole Near Lulworth Cove; Geology of the Wessex Coast, Geological Field Guide. Accessed 24 October 2021, https://wessexcoastgeology.soton.ac.uk/Stair-Hole-Lulworth.htm

West, I.M. 2018. Kimmeridge - Kimmeridge Clay and Kimmeridge Bay; Geology of the Wessex Coast. Geological Field Guide. Accessed 24 October 2021, https://wessexcoastgeology.soton.ac.uk/Kimmeridge-Bay.htm

West, I.M. 2019. Cheddar Gorge - Geological Field Guide, Supplement to Geology of the Wessex Coast. Accessed 24 October 2021, https://wessexcoastgeology.soton.ac.uk/Cheddar-Gorge.htm

West, I.M. 2020. Lulworth Cove, Dorset - Introduction - First and Main Part; Geology of the Wessex Coast of Southern England. Accessed 24 October 2021, https://wessexcoastgeology.soton.ac.uk/Lulworth-Cove-Introduction.htm

West, I.M., and R. Gallois. 2019. Lyme Regis, Westward – Monmouth Beach, Pinhay Bay & Landslide. Accessed 21 October 2021, https://wessexcoastgeology.soton.ac.uk/Lyme-Regis-Westward.htm

Whaley, J. 2010. Cornwall's geological treasures. *GEO ExPRO*, v. 7, no. 2, 12 p.

Wikipedia, Bath Stone. Accessed 24 October 2021, https://en.wikipedia.org/wiki/Bath_stone

Wikipedia, Beachy Head. Accessed 24 October 2021, https://en.wikipedia.org/wiki/Beachy_Head

Wikipedia, Britannia. Accessed 24 October 2021, https://en.wikipedia.org/wiki/Britannia

Wikipedia, Bude. Accessed 24 October 2021, https://en.wikipedia.org/wiki/Bude

Wikipedia, Cheddar Cheese. Accessed 24 October 2021, https://en.wikipedia.org/wiki/Cheddar_cheese

Wikipedia, Cornish Diaspora. Accessed 24 October 2021, https://en.wikipedia.org/wiki/Cornish_diaspora

Wikipedia, Delabole. Accessed 24 October 2021, https://en.wikipedia.org/wiki/Delabole

Wikipedia, Durdle Door. Accessed 24 October 2021, https://en.wikipedia.org/wiki/Durdle_Door

Wikipedia, Geevor Tin Mine. Accessed 24 October 2021, https://en.wikipedia.org/wiki/Geevor_Tin_Mine

Wikipedia, Geology of Cornwall. Accessed 24 October 2021, https://en.wikipedia.org/wiki/Geology_of_Cornwall

Wikipedia, Geology of East Sussex. Accessed 24 October 2021, https://en.wikipedia.org/wiki/Geology_of_East_Sussex

Wikipedia, Geology of the Isle of Wight. Accessed 24 October 2021, https://en.wikipedia.org/wiki/Geology_of_the_Isle_of_Wight

Wikipedia, Kilve. Accessed 24 October 2021, https://en.wikipedia.org/wiki/Kilve

Wikipedia, London. Accessed 17 October 2022, https://en.wikipedia.org/wiki/History_of_London

Wikipedia, Long Man of Wilmington. Accessed 24 October 2021, https://en.wikipedia.org/wiki/Long_Man_of_Wilmington

Wikipedia, Lyme Regis. Accessed 24 October 2021, https://en.wikipedia.org/wiki/Lyme_Regis

Wikipedia, Mary Anning. Accessed 24 October 2021, https://en.wikipedia.org/wiki/Mary_Anning

Wikipedia, Roman Baths (Bath). Accessed 24 October 2021, https://en.wikipedia.org/wiki/Roman_Baths_(Bath)

Wikipedia, Somerset Coal Canal. Accessed 24 October 2021, https://en.wikipedia.org/wiki/Somerset_Coal_Canal

Wikipedia, Somerset Coalfield. Accessed 24 October 2021, https://en.wikipedia.org/wiki/Somerset_Coalfield

Wikipedia, Stonehenge. Accessed 24 October 2021, https://en.wikipedia.org/wiki/Stonehenge

Wikipedia, Thames Estuary. Accessed 24 October 2021, https://en.wikipedia.org/wiki/Thames_Estuary

Wikipedia, Wessex Basin. Accessed 24 October 2021, https://en.wikipedia.org/wiki/Wessex_Basin

Wikipedia, White Cliffs of Dover. Accessed 24 October 2021, https://en.wikipedia.org/wiki/White_Cliffs_of_Dover

Wikipedia, Winspit. Accessed 24 October 2021, https://en.wikipedia.org/wiki/White_Cliffs_of_Dover

Wilson, G. 2014. Tin-Tungsten Ore. Turnstone Geological. Accessed 24 October 2021, http://www.turnstone.ca/rom152co.htm

Winchester, S. 2001. *The Map That Changed the World – William Smith and the Birth of Modern Geology.* HarperCollins Publishers, New York, 329 p.

Young, C. 2007. The geology and landscape of the Isle of Sheppey. *Transactions of the Kent Field Club*, v. 18, Doddington, Sittingbourne, Kent, pp. 47–66.

Index